Joint AIRAPT-25th & EHPRG-53rd International Conference on High Pressure Science and Technology 2015

Journal of Physics: Conference Series Volume 950

Madrid, Spain
30 August – 4 September 2015

ISBN: 978-1-5108-5612-7
ISSN: 1742-6588

Printed from e-media with permission by:

Curran Associates, Inc.
57 Morehouse Lane
Red Hook, NY 12571

Some format issues inherent in the e-media version may also appear in this print version.

Printed by Curran Associates, Inc. (2018)

For permission requests, please contact the Institute of Physics
at the address below.

Institute of Physics
Dirac House, Temple Back
Bristol BS1 6BE UK

Phone: 44 1 17 929 7481
Fax: 44 1 17 920 0979

techtracking@iop.org

Additional copies of this publication are available from:

Curran Associates, Inc.
57 Morehouse Lane
Red Hook, NY 12571 USA
Phone: 845-758-0400
Fax: 845-758-2633
Email: curran@proceedings.com
Web: www.proceedings.com

Table of contents

Volume 950

Joint AIRAPT-25th & EHPRG-53rd International Conference on High Pressure Science and Technology

30 August to 4 September 2015, Madrid, Spain

Accepted papers received: 26 September 2017
Published online: 7 November 2017

Preface

Plenary

Keynote

Papers

Bio and Life Sciences, Soft Matter

Food Science and Technology

Theory

Spectroscopic and Structural Studies

Condensed Matter I

Condensed Matter II

Dynamical Pressure

Geo- and Planetary Sciences

Chemistry and Molecular Systems

Preface

The "Joint AIRAPT-25 & EHPRG-53" International Conference on High Pressure Science and Technology was held in Madrid from August 30th to September 4th, 2015. The scientific program consisted of 483 scheduled presentations organized into 12 topical areas and included two plenary, three award and one AIRAPT 50th-anniversary lectures, along with 30 keynote and 147 talks, and 300 poster contributions. The scientific focus of the Joint Conference was on fundamental and applied research topics related to High Pressure Science and Technology.

442 scientists and engineers from 31 countries registered at the conference. The attendee countries represented included: Argentina (1), Austria (2), Belgium (2), Brazil (4), Canada (7), China (2), Czech Republic (6), Denmark (2), Estonia (1), France (41), Germany (47), Greece (2), Hungary (5), India (5), Israel (9), Italy (13), Japan (71), Lithuania (1), Netherlands (1), Poland (13), Portugal (7), Russia (18), Slovakia (2), South Korea (1), Spain (72), Sweden (5), Switzerland (3), Taiwan (1), Turkey (3), United Kingdom (39) and United States of America (38).

The Joint Conference was organized by the "Matter at High Pressure" Team (MALTA-Consolider). The Local Executive Committee was:

V. G. Baonza, Universidad Complutense de Madrid, Chairman
M. Taravillo, Universidad Complutense de Madrid, Secretary
J. Sánchez-Benítez, Universidad Complutense de Madrid, Vice-Secretary
B. Guignon, Universidad Complutense de Madrid, Vice-Secretary
O. R. Montoro, Universidad Complutense de Madrid, Vice-Secretary

The Organizing Committee was:

F. Rodríguez, Universidad de Cantabria, Santander, Spain
J. González, Universidad de Cantabria, Santander, Spain
A. Segura, Universidad de Valencia, Burjassot, Spain
D. Errandonea, Universidad de Valencia, Burjassot, Spain
F. J. Manjón, Universidad Politécnica de Valencia, Valencia, Spain
J. M. Recio, Universidad de Oviedo, Oviedo, Spain
V. Luaña, Universidad de Oviedo, Oviedo, Spain
B. Guamis, Universidad Autónoma de Barcelona, Bellaterra, Spain
J. Saldo, Universidad Autónoma de Barcelona, Bellaterra, Spain
A. Muñoz, Universidad de La Laguna, Tenerife, Spain
V. Lavín, Universidad de La Laguna, Tenerife, Spain
J. Andrés, Universidad Jaime I, Castellón, Spain
A. Beltrán, Universidad Jaime I, Castellón, Spain
P. D. Sanz, Instituto de Ciencia y Tecnología de Alimentos y Nutrición-CSIC, Madrid, Spain
O. Prieto-Ballesteros, Centro de Astrobiología, CSIC-INTA, Madrid, Spain

The International Advisory Committee was formed by the AIRAPT Executive Committee and EHPRG Committee, along with the following scientists:

G. G. N. Angilella, Università di Catania, Catania, Italy
J. Paul Attfield, University of Edinburgh, Edinburgh, Scotland
A. Delgado, Friedrich-Alexander-Universität, Erlangen-Nürnberg, Germany
L. S. Dubrovinsky, Universität Bayreuth, Bayreuth, Germany
D. J. Dunstan, University of London, London, England
G. I. Kanel, Institute for High Temperatures, Russian Academy of Sciences, Moscow, Russia
H.-K. Mao, Geophysical Laboratory, Carnegie Institution of Washington, Washington, USA
M. I. McMahon, University of Edinburgh, Edinburgh, Scotland
I. F. Silvera, Harvard University, Cambridge, USA
B. Sundqvist, Umeå University, Umeå, Sweden
J. S. Tse, University of Saskatchewan, Saskatoon, Canada

IOP Conf. Series: Journal of Physics: Conf. Series **950** (2017) 011001 doi:10.1088/1742-6596/950/1/011001

We would like to express our appreciation to the many people who contributed to the success of the conference: the plenary and keynote speakers, the authors, the participants, the session chairs, and the members of the Committees who nominated plenary and keynote speakers. The editors are especially grateful to those who reviewed the manuscripts included in this special issue.

A very special thank you is due to José Manuel Recio and David Abassi for arranging the hugely successful Summer Under Pressure School, which was attended by 70 students, and to the speakers (Valentín G. Baonza, Miriam Marqués, Julia Contreras García, Konstantin V. Kamenev, Stefan Klotz, Julio Pellicer, Fernando Rodríguez, Alfonso San Miguel, Raymond Jeanloz, Margherita Citroni, Maurizio Mattesini, Roland Winter, David Flores, and John S. Tse) for their generous participation.

The organizers would like to acknowledge conference sponsorship from Sanchez Technologies, Almax-Easylab, Micro Support, Betsa, Technodiamant, Taylor & Francis, American Elements, Ypsicon S.L., Microbeam S.A., Jasco Spain, Syntek, Hiperbaric, Fisher Scientific, Nova Swiss, and IOP Conference Series. The local organizes express their gratitude to the host institutions (MALTA-Consolider Team, Universidad Complutense de Madrid, Facultad de Ciencias Químicas, Instituto de Geociencias UCM-CSIC, Real Sociedad Española de Química, Universidad Rey Juan Carlos and Ayuntamiento de Madrid) for their invaluable support.

The organizers also thank the volunteers (the orange T-shirt team) who willingly gave of their time, support, and encouragement throughout this conference. A very special thank you goes to the people of Viajes El Corte Inglés (Sandra Fernández, Marta Tuquet, Marta Pamplona, and Carlos Romeo) for their invaluable help with registration and conference attendee services.

Alfonso Muñoz
Bérengère Guignon
Fernando Rodriguez
Francisco Javier Manjón
MercedesTaravillo
Valentín G Baonza

IOP Conf. Series: Journal of Physics: Conf. Series **950** (2017) 011002 doi:10.1088/1742-6596/950/1/011002

Foreword

It was under a severe thunderstorm that we opened the "Joint AIRAPT-25 & EHPRG-53" International Conference on High Pressure Science and Technology held in Madrid from August 30th to September 4th, 2015. Probably, that was a sky signal to announce that the large shock-wave community was there representing the outstanding role they play in our field. And indeed they do, as it was clearly demonstrated during the whole meeting with a balanced participation of dynamic and static compression contributions within the wide range of topics covered in this successful conference. The strong efforts we had invested to have with us all the potential areas touching the high-pressure field were amply compensated by the presence and high quality of scientists spanning a wide range of topics: Bio- and Life Sciences, Food Science and Technology, Metrology and Industrial Applications, Theory and Computational Modeling, Spectroscopic and Structural Studies, Condensed Matter, Geo- and Planetary Sciences, Nanoscience and Nanotechnology, Materials Science and Energetic, and Materials Chemistry and Molecular Systems. By no doubt, it was demonstrated that High-Pressure is a unique "hot" field of research displaying both multi- and inter-disciplinary features, which makes cooperation not only a natural but also a necessary ingredient for a prosperous investigation. The "Joint AIRAPT-25 & EHPRG-25" conference was a faithful reflection of this scenario.

Our plenary speakers convey the spirit of the conference in a brilliant way from different perspectives. Raymond Jeanloz and Roland Winter were in charge to cope the challenge of presenting panoramic views and latest results of two extreme realms, hard matter and planetary interiors the former, soft matter and biomolecular systems the later. Paul Loubeyre, Ross T. Howie and Oleksandr O. Kurakevych (Bridgman, Jamieson and EHPRG awardees, respectively) were provided with the same plenary time for their talks. The first two focused their participation on one of the most candent and long-living cases of study in high-pressure science: dense hydrogen. Kurakeviych illustrated the technological benefits of incorporating high-pressure techniques in materials science by disclosing how silicon-based allotropes can be useful for industrial applications. These are just representative examples of the rich variety of themes that under the umbrella of thirty careful selected keynote speakers were developed in a number of parallel sessions covering all the above mentioned topics. Last but not least, Karl Syassen showed us the beautiful journey of the history of AIRAPT over the last 50 years to celebrate the 25th AIRAPT conference.

And the merit for the success of the conference was, of course, on the very high and active participation of attendees coming from all over the world. More than one hundred and fifty oral contributions and near three hundred posters clearly illustrate the effervescence of high-pressure science and technology nowadays. These proceedings are a good sample of this propitious atmosphere!

We are aware of the many endeavors still pending to spread high-pressure in academia and laboratories. And we know that young researchers need to be instructed to incorporate high-pressure science and technology in their investigations. With these goals in mind, a "Summer Under Pressure School (SUPS)" was scheduled just two days before the opening of the conference. Following the scheme of our previous Spanish High-Pressure Schools, SUPS was split into Theory, Experimental Techniques and Applications blocks with fourteen one-hour lectures delivered by experts in all the different topics. They included plenary speakers of the "Joint AIRAPT-25 & EHPRG-25" conference, as well as former and current presidents of the EHPRG. It was a successful school with full occupancy (more than seventy students) revealing that the good health of High-Pressure will continue at least during some next generations.

We would like to finish this Foreword by thanking all the speakers and participants, along with exhibitors and sponsors for making it possible to fulfill the leit-motiv of the conference organizers, Madrid: a place for pressure, a place for pleasure!

The MALTA-Consolider Team

IOP Conf. Series: Journal of Physics: Conf. Series **950** (2017) 011003 doi:10.1088/1742-6596/950/1/011003

Peer review statement

All papers published in this volume of *Journal of Physics: Conference Series* have been peer reviewed through processes administered by the proceedings Editors. Reviews were conducted by expert referees to the professional and scientific standards expected of a proceedings journal published by IOP Publishing.

IOP Conf. Series: Journal of Physics: Conf. Series **950** (2017) 022001 doi:10.1088/1742-6596/950/2/022001

From Earth to Stars: Toward Atomic Pressures and Kilovolt Chemistry

Raymond Jeanloz [a]*

[a] *University of California, Berkeley, CA, USA*

Combinations of static and dynamic experiments are providing high-quality data well into the TPa regime, comparable to the quantum force stabilizing the electron within its atomic orbital (2.3 TPa). These are the pressures required for understanding the constitution and dynamics of planetary interiors, from Earth (0.36 TPa) to Jupiter (7 TPa) and beyond.

Many of the laboratory results are corroborated by geophysical observations, but key discoveries are only possible because of experiment and complementary theory. Specifically, high-pressure experiments indicate that 1) Earth's center is as hot as the Sun's surface; 2) the primary materials making up our planet are now largely documented; 3) iron undergoes an electronic collapse deep within Earth's rocky mantle; 4) the boundary between the mantle and the iron-alloy liquid core is likely the most chemically active region of our planet; and 5) the "solid" inner core is convectively unstable (tectonically active). New studies support the use of glasses for understanding the effects of pressure on melt structure, of importance because the chemical separation (and associated heat transfer) of planetary interiors is largely determined by the movement of fluids, yet their pressure-dependent structures can determine the buoyancy of melts.

For giant planets, experiments offer crucial evidence for the theoretically predicted separation of helium as well as of carbon from the hydrogen-rich planetary fluid, allowing settling (differentiation) that determines the planet's internal structure and acts as a source of internal heat. Metallization has been documented for a wide variety of elements and compounds at high pressures (0.1-1 TPa) and temperatures (10^3-10^4 K), from hydrogen, helium and noble gases to salts and oxides, with implications for understanding the interiors of giant and supergiant planets, as well as the violent conditions – characterized by late-stage giant impacts – under which planets are formed. Contrary to early theoretical expectations, high temperature can help induce metallization.

Dynamic experiments are required for the highest pressures now reached, and there is great interest in using ramp (shock-free) compression and pre-compression of samples for tuning the final pressure–temperature states in order to achieve the highest densities (lowest temperatures) possible. Experiments are for the first time documenting material compression in the regime characterized by Thomas-Fermi-Dirac theory, for example.

Current work is extending laboratory capability toward Gbar pressures (0.1 PPa), exceeding the atomic unit of pressure (29.4 TPa) at which core electrons start playing an important role in chemical bonding: the regime of kilovolt chemistry, for which the energy of compression is $-P\Delta V \sim$ keV. These conditions are relevant to the interiors of brown dwarfs, objects transitional between supergiant planets and stars, and there is the potential of characterizing material properties and chemical bonding in dense plasmas (Coulomb jelly) at the onset of thermonuclear fusion.

*jeanloz@berkeley.edu
Keywords: planetary interiors, quantum pressure, kilovolt chemistry

IOP Conf. Series: Journal of Physics: Conf. Series **950** (2017) 022002 doi:10.1088/1742-6596/950/2/022002

Exploring the conformational space and dynamics of biomolecular systems using pressure perturbation

*Roland Winter**

TU Dortmund University, Physical Chemistry I - Biophysical Chemistry, Otto-Hahn Straße 6, D-44227 Dortmund, Germany

Decades of studies have contributed invaluable insights into the forces controlling the structure and dynamics of biomolecules and their response to changes in temperature and solution conditions. Understanding of the effects of high hydrostatic pressure (HHP) in contrast, has lagged behind because high-pressure conditions have been more difficult to access in nature and in the laboratory. This is a pity because pressure, as an important physical-chemical parameter next to temperature and composition, can be used as a gateway to gain new and fundamental insights into the structure, dynamics and function of biomolecular systems. Pressure perturbation serves as an important tool to mechanistically explore the conformational and free-energy landscape of biomolecular systems and to reveal the effects of pressure on biosystems exposed to environmental high-pressure stress.[1,2] Apart from this, HHP has an additional important biological component - life under extreme conditions: The greatest portion of our biosphere on Earth is in the realm of environmental extremes, including the cold and high pressure habitats of marine depths (70 % of the surface of the Earth is covered by oceans, and the average pressure on the ocean floor is 40 MPa), and pressures up to 100 MPa and more are reached in the deep sea and sub-seafloor crust. Hence, there are also extraordinary prospects for using the pressure variable to deepen our understanding of life, touching areas such as deep sea biology, prebiotic chemistry, and astrobiology. Moreover, the consequential insights obtained in the course of fundamental pressure studies may be essential for a broad range of other disciplines and applications, including food science (*e.g.*, high-pressure processing) and biotechnology (*e.g.*, the exploitation of enzymes and metabolism of deep sea bacteria for novel biotechnological processes).

Here, we review the underlying principal action of pressure on biomolecules with emphasis on lipid membranes, proteins, DNA, RNA, and their assemblies along with some highlighted experiments.[1-6] In particular, we discuss how pressure affects the structure, phase behavior and dynamics of lipid membranes (including prebiotic lipid assemblies) and proteins of varying complexity. By combining X-ray and neutron scattering techniques with liquid-state theoretical approaches, the effects of molecular crowding, cosolvents and pressure on the intermolecular interaction potential, solvational properties and dynamics of proteins will also be discussed.[7-11] Applying the pressure-jump relaxation technique in combination with time-resolved synchrotron X-ray diffraction and spectroscopies, the kinetics of lipid phase transformations and protein folding and polymerization reactions was also investigated. We then review the promising role of the pressure perturbation approach to study high energy conformational and functional substates in proteins, and we elaborate on the use of pressure modulation to understand protein aggregation (*e.g.* insulin) and polymerization (*e.g.* actin) reactions.[12-15] This approach reveals new insights into the pre-aggregated regime as well as mechanistic details about concurrent polymerization pathways and the differential stability of the protein filaments formed. Finally, we present some recent results using pressure perturbation to explore membrane-associated biomolecular assemblies and uncover membrane-mediated conformational substates of proteins, furnishing unprecedented information on proteo-lipid interactions (*e.g.*, of signalling proteins like Ras, or membrane-associated enzymes like phospholipase A2).[16,17]

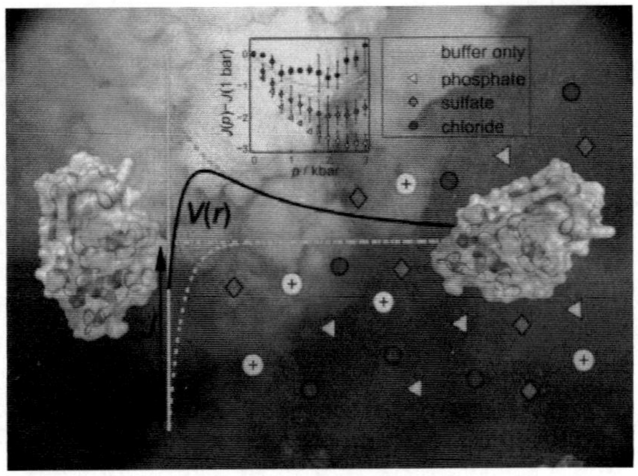

References

[1] I. Daniel, P. Oger, R. Winter, *Chem. Soc. Rev.* **35,** 858 (2006).
[2] R. Mishra, R. Winter, *Angew. Chem. Int. Ed.* **47,** 6518 (2008).
[3] C. Conn, R. Winter, J. M. Seddon, R. H. Templer, *Phys. Rev. Lett.* **96,** 108102 (2006).
[4] S. Kapoor, ..., R. Winter, *Angew. Chem. Intern. Ed.* **53,** 8397 (2014).
[5] C. Schuabb, M. Berghaus, C. Rosin, R. Winter, *ChemPhysChem* **16,** 138 (2015).
[6] J. Seeliger, N. Erwin, C. Rosin, M. Kahse, K. Weise, R. Winter, *Phys. Chem. Chem. Phys.* **17,** 7507 (2015).
[7] M. Erlkamp, S. Grobelny, R. Winter, *Phys. Chem. Chem. Phys.* **16,** 5965 (2014).
[8] M. A. Schroer, ..., R. Winter, *Phys. Rev. Lett.* **106,** 178102 (2011).
[9] M. A. Schroer, ..., R. Winter, *Angew. Chem. Int. Ed.* **50,** 11413 (2011).
[10] J. Möller, ..., R. Winter, *Phys. Rev. Lett.* **112,** 02810 (2014).
[11] M. Erlkamp, J. Marion, N. Martinez, C. Czeslik, J. Peters, R. Winter, *J. Phys. Chem. B* **119,** 4842 (2015).
[12] N. Javid, K. Vogtt, C. Krywka, M. Tolan, R. Winter, *Phys. Rev. Lett.* **99,** 028101 (2007).
[13] S. Grudzielanek, V. Smirnovas, R. Winter, *J. Mol. Biol.* **356,** 497 (2006).
[14] C. Rosin, M. Erlkamp, J. von der Ecken, S. Raunser, R. Winter, *Biophys. J.* **107,** 2982 (2014).
[15] C. Rosin, K. Estel, J. Hälker, R. Winter, *ChemPhysChem* **16,** 1379 (2015).
[16] S. Kapoor, ..., R. Winter, *Proc. Natl. Acad. Sci. U.S.A.* **109,** 460 (2012).
[17] S. Kapoor, A. Werkmüller, R. S. Goody, H. Waldmann, R. Winter, *J. Am. Chem. Soc.* **135,** 6149 (2013).

*roland.winter@tu-dortmund.de
Keywords. Pressure-jump relaxation, proteins, lipid membranes, aggregation

Loading Hydrogen and Helium in the Diamond Anvil Cell: the First 30 Years

Paul Loubeyre [a]

[a] *Commissariat à l'Energie Atomique, Centre DAM/DIF, Arpajon, 91297, France*

The combination of three major innovations have helped the advent of current high pressure physics, which aims at revealing the properties of matter in the multi-100 GPa range as finely as if under ambient pressure. Namely: the invention of the diamond anvil cell in 1958,[1] the loading of helium in 1979,[2] and the use of third generation synchrotrons, in the mid 80's. The results on dense hydrogen and helium certainly exemplify what hence has been achieved.

I was very fortunate to do my research training under the supervision of J. M. Besson after he had just achieved the loading of helium in the DAC. At the end of the 80's, I was initiated to the use of synchrotron by H. K. Mao to measure the structural properties of helium under pressure, using the NSLS facility. Later we have been among the first users of the 3rd generation synchrotron source, ESRF, to measure the equation of state of dense hydrogen.[3]

In my presentation, I will review what have been the important questions motivating high pressure studies on hydrogen & helium during 30 years, what kind of technical developments were required, which results were disclosed by us and others and what are the challenges for the next ten years.

References

[1] W. A. Bassett Diamond anvil cell 50th birthday. *High Press. Res.* **29**, 163 (2009).
[2] J. M. Besson, J. P. Pinceaux, Melting of helium a room temperature and high pressure. *Science* **206**, 1073 (1979).
[3] P. Loubeyre, R. LeToullec, D. Hausermann, M. Hanfland, R. J. Hemley, H. K. Mao, L. W. Finger, X-ray diffraction and equation of state of hydrogen at Megabar pressures. *Nature* **383**, 702 (1996).

*paul.loubeyre@cea.fr
Keywords: Hydrogen, helium, DAC

IOP Conf. Series: Journal of Physics: Conf. Series **950** (2017) 022004 doi:10.1088/1742-6596/950/2/022004

The AIRAPT arose as a scientific forum for discussions aiming to promote contacts and cooperation between scientists and organizations of different countries, to collect and disseminate of information, having the advancement of science and technology in the field of high pressure as main objective.

Nowadays the field has experienced an astonishing growth in an open multidisciplinary environment that rather contributed to the creation of different interdisciplinary teams of excellence, being able to face the important challenges posed by high-pressure research. Great achievements have been possible towards higher pressures, nowadays making possible experiments in the TPa range in our laboratories, the improvement and adaptation of different characterization techniques of matter under extreme conditions of pressure and temperature or the development of efficient *ab initio* methods with capabilities to explain and eventually predict new physical phenomena and materials design. All decisively contributed to the advance of science and understanding of nature, with high pressure as common *leitmotiv*.

On the 50th Anniversary of the AIRAPT, this talk will is intended to commemorate the enthusiastic and encouraging work of high-pressure pioneers and to motivate young scientists to join us and continue the challenge of exploring compressed matter as a way to unveil new phenomena and materials and a better understanding of our world.

IOP Conf. Series: Journal of Physics: Conf. Series **950** (2017) 022005 doi:10.1088/1742-6596/950/2/022005

Experiments on Dense Hydrogen

Ross T. Howie [a,b]*, *Philip Dalladay-Simpson* [a] *and Eugene Gregoryanz* [a]

[a] *CSEC, The University of Edinburgh, Edinburgh, UK*
[b] *HPSTAR, Shanghai, China*

Hydrogen is the lightest, simplest and most abundant element in the universe, and as such the behavior of hydrogen at the most extreme conditions is of vital importance to many fields of science. There has been remarkable progress in high pressure experimental studies of hydrogen over the past 5 years, from the discovery of phase IV[1,2] and its characterization[3,4] to the possibility of Anderson localization in dense isotopic mixtures.[5] In this talk I will present work that will further elucidate properties of the high pressure, high temperature phase diagram of hydrogen.

References

[1] M.I. Eremets, I.A. Troyan, *Nat. Mat.* **10**, 927, (2011).
[2] R.T. Howie, C.L. Guillaume, T. Scheler, A.F. Goncharov, E. Gregoryanz, *Phys. Rev. Lett.*, **108**, 125501 (2012).
[3] R.T. Howie, T. Scheler, C. L. Guillaume, and E. Gregoryanz, *Phys. Rev. B* **86**, 214104 (2012).
[4] C.S. Zha, Z. Liu, M. Ahart, R. Boehler, R.J. Hemley, *Phys Rev Lett* **110** 217402 (2013).
[5] R.T. Howie, I.B. Magdau, G.J. Ackland, A.F. Goncharov, E. Gregoryanz, *Phys. Rev. Lett.*, 113, 175501 (2014).

* ross.howie@ed.ac.uk, ross.howie@hpstar.ac.cn
Keywords: Hydrogen

Exploring silicon allotropy by high pressure – high temperature conditions

O. O. Kurakevych [a]*, *Y. Le Godec* [a], *T. A. Strobel* [b], *D. Y. Kim* [b], *W. A. Crichton* [c], *J. Guignard* [c]

[a] *IMPMC, UPMC Sorbonne Universités, UMR CNRS 7590, Muséum National d'Histoire Naturelle, IRD UMR 206, 75005 Paris, France*
[b] *Geophysical Laboratory, Carnegie Institution of Washington, Washington, DC 20015, United States*
[c] *European Synchrotron Radiation Facility, 71, Avenue des Martyrs, Grenoble 38043, France*

Silicon is the second abundant element, after oxygen, in the earth crust. It is essential for today's electronics because of its ability to show various electronic behaviors that allow covering the numerous fields of cutting-edge applications. Moreover, silicon is not a pollutant and, therefore, an ideal candidate to replace the actual materials in photovoltaics, like compounds based on the arsenic and heavy metals. It has not replaced them so far because Si is an indirect gap semiconductor and cannot absorb directly the solar photons without thermal agitations of crystal lattice (phonons). This puts it apart from the next-generation applications (light diode, high-performance transistor). That is why we attempted to create a silicon with direct gap that can absorb and emit light. Our recent high-pressure studies of the chemical interaction and phase transformations in the Na-Si system,[1] revealed a number of interesting routes to new and known silicon compounds[1] and allotropes[2] (Figure). The pressure-temperature range of their formation is suitable for large-volume synthesis and future industrial scaling. The variety of properties observed (e.g. quasi-direct bandgap of open-framework allotrope Si_{24})[2] allow us to suggest future industrial applications.

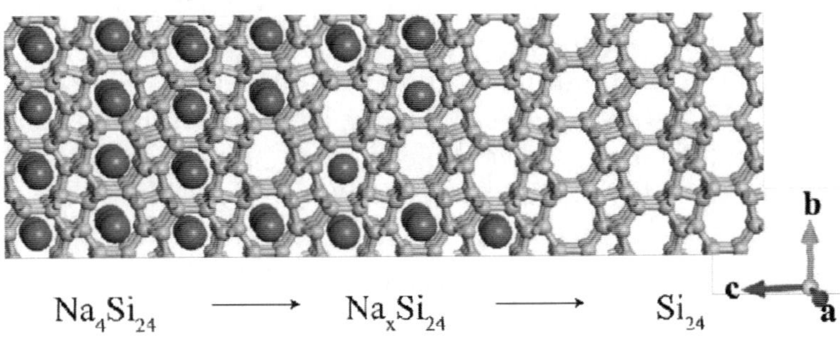

Figure. Crystal structures of Na_4Si_{24}, Si_{24} and intermediate compounds. @ Nature Materials

References
[1] O.O. Kurakevych, T.A. Strobel, D.Y. Kim, T. Muramatsu, V.V. Struzhkin, *Cryst. Growth Des.* **13**, 303 (2013).
[2] D.Y. Kim, S. Stefanoski, O.O. Kurakevych, T.A. Strobel, *Nat. Mater.* **14**, 169 (2015).

* oleksandr.kurakevych@impmc.jussieu.fr
Keywords: Silicon, allotropy, bandgap, optoelectronics

IOP Conf. Series: Journal of Physics: Conf. Series **950** (2017) 032001 doi:10.1088/1742-6596/950/3/032001

Phase Stability Behaviour of Rare Earth Orthovanadates under High Pressure

*Alka B. Garg**

High Pressure and Synchrotron Radiation Physics Division, Bhabha Atomic Research Centre, Mumbai, India 400085

Rare earth orthovanadates, RVO_4 ($R^{3+}V^{5+}O_4^{2-}$; R= rare earth element including Y and Sc) belongs to the family of ABO_4 type compounds. At ambient pressure and temperature conditions, all the RVO_4 compounds crystallize in zircon structure (except $LaVO_4$ which can also be stabilized in monoclinic monazite structure). The basic building blocks of tetragonal zircon structure (space group: $I4_1/amd$; Z=4) are VO_4 tetrahedra and RO_8 polyhedra which extend parallel to c axis and VO_4 chains are joined laterally by edge sharing RO_8. The V–O bond distance remains nearly the same for the entire lanthanide series, and the R–VO_4 interaction is predominantly ionic. The RVO_4 series represents an ideal system for studying the interaction between the sublattice of magnetic ions and the ligands of the host lattice. From technological point of view also these zircon-type orthovanadates have important applications as cathodoluminescence, thermophosphors, scintillators, and laser-host materials. They can also be useful for the development of green technologies through applications like photocatalytic hydrogen production. As is well known that pressure, an important thermodynamic variable can change the inter-atomic distances in the solids by an order of magnitude which dramatically alter the electronic properties, break existing bonds, or forming new chemical bonds which intern leads to variety of pressure-induced phenomena such as metallisation, amorphization, superconductivity and polymerisation. Hence, compression provides a unique possibility to control the structure and properties of materials. Usually the zircon-structured materials are known to transform to a denser (~10%) low symmetry tetragonal scheelite structure (space group $I4_1/a$) under pressure. On further compression this scheelite phase becomes unstable and system transforms to low symmetry monoclinic structure. In this talk evolution of equation of states and other structural details for various phases of RVO_4 compounds studied using synchrotron based X-ray diffraction and Raman spectroscopic measurements will be presented.

* *alkagarg@barc.gov.in*

Multiple orders in high pressure perovskites

*Angel M. Arévalo López**

Centre for Science at Extreme Conditions & School of Chemistry,
University of Edinburgh, King's Buildings, Edinburgh, EH9 3FD, United Kingdom

The discovery of new materials with appealing functional properties is challenging and such properties are often observed in complex oxides containing transition metals in intermediate or unusual oxidation states which must be stabilised under special synthetic conditions.

High-pressure synthesis conditions ("hard") is a great tool to lock instabilities, such as unusual oxidation states or coordination environments, into denser phases that are metastable when recovered to ambient conditions, thus leading to unusual properties.

Manipulating and tuning the structures and physical behaviour of complex oxides by modifying their anion lattices has gain interest in recent years. This has been promoted by the use of binary metal hydrides as low temperature reducing agents ("soft" chemistry), thus allowing the access to unprecedented structures and unusual oxidation states, for instance as in $SrFeO_2$.[1] Hence, it is attractive to explore combined "hard–soft" routes to novel materials by partially relieving the instability of a high-pressure precursor through post-synthesis modification.[2]

We have demonstrated this "hard-soft" approach in the high pressure perovskites $SrCrO_3$ and $CaCrO_3$. 15R-$SrCrO_{2.8}$ forms a long-period Cr^{3+}/Cr^{4+} charge-density wave which gives rise to a spin-density wave-type modulation of the magnetic moments with a long, doubled c-axis, periodicity ($2c \approx 69$ Å).[2] $CaCrO_3$ reduction promotes oxygen vacancy ordering along the $[001]_c$ direction and generates two new $CaCrO_{3-\delta}$ phases (δ=0.3333 and 0.5). Both phases show alternating tetrahedral (T) and octahedral (O) layers in an OOT and OTOT sequence.[3]

The success of this "hard-soft" chemistry approach in chromium perovskites along with some ternary Mn oxides ($MnBO_3$) with interesting properties obtained from high pressure will be presented and discussed.

References

[1] Y. Tsujimoto *et al. Nature.* **450**, 1062 (2007).
[2] A. M. Arévalo-López *et al. Angew. Chem. Int. Ed.* **51**, 10791 (2012).
[3] A. M. Arévalo-López *et al. J. Mat. Chem. C.* **2**, 9364 (2014).

* aalopez@staffmail.ed.ac.uk
Keywords: Perovskites, high pressure synthesis, neutron diffraction, magnetic properties

IOP Conf. Series: Journal of Physics: Conf. Series **950** (2017) 032003 doi:10.1088/1742-6596/950/3/032003

Pressure induced phase transition in correlated oxides and simple metals: Mott and charge-transfer insulators

*Angel Rubio**

Max Planck Institute for the Structure and Dynamics of Matter, Hamburg, Germany
NanoBio Spectroscopy Group and ETSF Scientific Development Centre, Universidad del País Vasco, UPV/EHU San Sebastián, Spain
FHI Max-Planck-Gesellschaft, Berlin, Germany

High pressure can radically change the behavior of simple substances. Consider sodium. At ambient pressures it has a symmetric crystal structure. Under large enough pressure, it changes from a shiny metal to a transparent insulator with several low-symmetry crystal structures. At very high pressures, the core electrons overlap and, because of the Pauli exclusion principle, give rise to less symmetric charge distributions. Coulomb repulsion between core and valence electrons localize charge in interstitial positions instead of at the nuclei, eventually leading to an insulating state. In this talk we will review the recent advances within density-functional and many-body based schemes to describe spectroscopic properties of complex systems with special emphasis to modeling the pressure (or light) induced phase transitions in correlated oxides and simple metals. Our first-principles calculations show that at pressures above the metal-insulator transition, sodium should be transparent in one polarization direction but reflective, like a normal metal, in the other. Photoluminescence experiments along the two axes of polarization should elucidate the nature of the bound exciton as well as the crystal structure at high pressures. We extended those studies to adder the pressure induce Mott insulator transition in MnO and well as the insulator-to-metal phase transition of VO_2. For the later we show that the band gap in monoclinic phase is extremely sensitive to small changes in the occupation of the localized d bands of V atoms. In particular, the photo-induced hole doping in VO_2 can strongly alter the dynamical screening, which then leads to a collapse of the band gap. Our results support the experimental findings and point to the electronic origin of the insulator-to-metal phase transition of monoclinic VO_2 in the time resolve photoelectron experiments.

Acknowledgements: This work was supported the European Research Council Advanced Grant DYNamo (ERC-2010-AdG-267374), MINECO (FIS2013-46159-C3-1-P), Grupos Consolidados UPV/EHU del Gobierno Vasco (IT578-13) and European Community FP7 project CRONOS (Grant number 280879-2) and COST Actions CM1204 (XLIC) and MP1306 (EUSpec).

Work done in Collaboration with: L. Xian, M. Gatti, P. Cudazzo, lya Tokatly, D. Wegkamp, M. Herzog, M. Wolf, J. Stähler

*angel.rubio@mpsd.mpg.de

IOP Conf. Series: Journal of Physics: Conf. Series **950** (2017) 032004 doi:10.1088/1742-6596/950/3/032004

High pressure experiments on icy moon materials and planetary geophysics

Carmen Sanchez-Valle [a,*], *Chloe Yao* [b], *Frederic Deschamps* [c], *Davide Mantegazzi* [b]

[a] *Institute for Mineralogy, Westfälische Wilhelms-Universität Münster, Münster, Germany*
[b] *Institute for Mineralogy and Petrology, ETH Zurich, Zurich, Switzerland.*
[c] *Institute of Earth Science, Academia Sinica, Taipei, Taiwan*

The icy moons of Jovian planets are remarkable in their diversity and distinct evolutionary paths, and represent a challenge for our understanding of planetary formation and evolution.[1] Although some of these planetary bodies have been subject to close-up geophysical exploration, their formation conditions, structure, geophysical evolution and current state remain. Current and past spacecraft missions (e.g., Galileo and Cassini-Huygens) have provided unique information about the bulk and surface properties of icy moons. The expansion on the geophysical datasets has encompassed progresses on geodynamical models able to compute the heat fluxes through the satellite and to reconstruct their thermal history.[2] However, the ability of these models to interpret unambiguous the information encoded in the spacecraft observations is limited by the paucity of laboratory data on the properties of candidate icy moon materials at relevant pressure-temperature conditions.

In this contribution we will present geodynamical models supported by new laboratory experiments to constrain the effect of volatile compounds on the crystallization of primordial oceans and on the thermal evolution of large icy moons. Specifically, we will describe new experiments designed to determine the melting temperature of ice phases (Ih and VI) and the fluid density in the binary methanol-water system (0-20 wt% methanol) using Brillouin scattering spectroscopy.[3,4] Methanol has been recently detected in several comets and at the surface of Europa, and may be an important chemical component of subsurface oceans. The explored pressure-temperature range, 230 - 300 K and 10^{-4} - 1.2 GPa, spans the conditions that prevail in the icy crust and primordial ocean of Titan. The experiments were conducted in a membrane-type diamond anvil cell outfitted for low temperature studies employing an in-house designed Peltier cooling system. Melting and crystallization in the system were monitored by changes in the Brillouin spectral features and in the pressure dependence of the measured sound velocities. The density of fluids $\rho(P, T, x)$ in the binary methanol-water system was determined from the inversion of sound velocities measured in the fluids as a function of pressure along isotherms from 250 to 300 K. The results are combined with advanced numerical modeling to examine the effect of methanol on the thermo-chemical evolution of the subsurface ocean. The implications of these results for the thermal and structural evolution of icy moons, with particular applications to Titan, will be further discussed. To conclude, we will survey the gaps in knowledge on the thermos-physical properties of candidate icy moon materials. This review will set the bases to outline comprehensive laboratory research programs on material properties to support geophysical models and to define new exploration strategies for ongoing and future missions (*e.g.*, JUICE).

References

[1] H. Hussmann, C. Sotin, J. Lunine, Interiors and Evolution of Icy Satellites (Chapter 15, p. 509-539). In: *Treatise on Geophysics Vol. 10: Planets and Moons.* G. Schubert (Eds.), Elseviers, UK (2007).
[2] C. Tobie, C., D. Gautier, F. Hersant, *Astrophys. J.* **752**, 125 (2012).
[3] D. Mantegazzi, C. Sanchez-Valle, E. Reusser, T. Driesner, *J. Chem. Phys.* **137**, 224501 (2012).
[4] C. Sanchez-Valle, C. Structure and Thermodynamics of subduction zone fluids from spectroscopic studies (Chapter 8, p. 265-309). In: *Reviews in Mineralogy and Geochemistry Vol. 76: Thermodynamics of geothermal fluids.* A. Stefansson, T. Driesner, P. Bennezeth (Eds.), MSA, USA (2013).

*sanchezm@uni-muenster.de
Keywords: icy moons, methanol-water, low temperature, Brillouin spectroscopy

IOP Conf. Series: Journal of Physics: Conf. Series **950** (2017) 032005 doi:10.1088/1742-6596/950/3/032005

Magnetism at high pressure probed with x-ray resonant absorption and scattering techniques

Daniel Haskel *

Advanced Photon Source, Argonne National Laboratory. Argonne IL 60439, USA

The occurrence of spontaneous ordering of electronic magnetic moments in a solid is intimately connected to the overlap of their electronic wave functions. Altering this overlap by the application of pressure has profound effects on both direct and indirect exchange (magnetic) interactions responsible for magnetic ordering phenomena. Coupled with the high-brilliance, penetration power, and polarization/energy tunability of synchrotron radiation, x-ray studies of magnetism at high pressure in the diamond anvil cell provide a unique opportunity to tune, hence help decipher, complex exchange interactions in magnetic materials of current technological and fundamental interest.

We demonstrate the uniqueness of this approach by reviewing methodologies and instrumentation that enable polarization-dependent x-ray resonant absorption and scattering studies of magnetic materials under simultaneous external stimuli of high pressure (1 Mbar), magnetic field (6.5 Tesla) and low temperature (1.4 K). Recent advances derived from this approach in our understanding of magnetic interactions in a number of complex magnetic materials of current interest, are highlighted.

* haskel@aps.anl.gov
Keywords: Magnetism, x-ray resonant techniques

High pressure Raman studies of single- and bi-layer graphene on copper

K. Filintoglou[a], *D. Kokkinos*[a], *D. Christofilos*[b]*, *J. Arvanitidis*[a], *O. Frank*[c], *M. Kalbac*[c], *K. Papagelis*[d], *S. Ves*[a], *G. A. Kourouklis*[b]

[a] *Physics Department, Aristotle University of Thessaloniki, 54124 Thessaloniki, Greece*
[b] *Chemical Engineering, Aristotle University of Thessaloniki, 54124 Thessaloniki, Greece*
[c] *J. Heyrovsky Institute of Physical Chemistry, CZ-18223 Prague 8, Czech Republic*
[d] *Materials Science Department, University of Patras, 26504 Rio Patras, Greece*

The applicability of graphene towards the revolutionization of electronics passes through efficient, reproducible and large area-high quality production of at least single- and bi-layer graphene (SLG and BLG). The use of catalytic copper substrates have laid the foundations for this kind of production. Although transfer of graphene onto silicon substrate is the next step to applications, the study of the quality, adhesion, charge transfer and other properties of the as-prepared graphene on copper is of fundamental interest. In this direction, the use of isotopically marked BLG is critical in order to discriminate the properties of the two layers. High hydrostatic pressure experiments of graphene can be viewed as peculiar friction studies. Graphene adheres on the substrate and responds to its surface contraction due to friction and the increasing normal force. Such conditions are practically probing friction/adherence and charge transfer effects depending on the graphene-substrate spacing and interaction. Raman spectroscopy has been established as the tool of choice in the research of graphene and industrial characterization. Strong and easily exploitable characteristic bands are the D, G and 2D. The usual minimum requirement of good quality SLG is the intensity ratio of the I_{2D}/I_G being above 2. However, based on our ambient conditions measurements we believe that the ω_{2D}-ω_G correlation plot[1] is potentially a better indicator of SLG or BLG. BLG lies in the "forbidden" up-left region of the plot and the top layer appears less stressed.

The samples under investigation were prepared by appropriate sequential introduction of $^{12}CH_4$ and $^{13}CH_4$ in the CVD chamber operating at 1000 °C resulting mainly in BLG with bottom and top layers consisting of ^{13}C and ^{12}C carbon isotope atoms, respectively.[2,3] More specifically, four characteristic sample areas are identifiable by mapping and careful analysis of the frequencies and line shapes of the Raman bands: SLG and BLG solely consisting of ^{13}C as well as $^{12}C/^{13}C$ BLG layers with turbostratic or AB (bernal) stacking.

Here, high pressure measurements concern SLG and turbostratic BLG on areas of the same sample using the polar glycerol and the non-polar 1:1 FC70-FC77 Fluorinert™ pressure transmitting media (PTM). The former solidifies at ~5.5 GPa and the latter at less than 1 GPa. The corresponding ω_{2D}-ω_G plot with pressure as a parameter is indicative of negligible or non-existent charge transfer in both the polar and non-polar medium as well as in the liquid or solidified PTM state as we previously observed for SLG.[4] As long as the PTM remains liquid, the pressure response of the frequencies is reversible suggesting that graphene layers still adhere on the substrate. On the other hand, solidification of the PTM and the concomitant stress gradients result, sooner or later, in the detachment and quality deterioration of the layers. The ω_G pressure slopes are in the range 9-12 cm^{-1}/GPa. In BLG, the top layer exhibits consistently smaller or at best equal values to those of the bottom one, in accordance with substrate stress being transmitted through the interaction between graphene layers. Still, the ω_G pressure slope always remains below ~15 cm^{-1}/GPa that would be expected from the surface contraction, suggesting non-perfect graphene adherence.

References

[1] J. E. Lee, G. Ahn, J. Shim, Y. S. Lee, and S. Ryu, *Nat. Commun.* **3**, 1024 (2012).
[2] P. T. Araujo, O. Frank, D. L. Mafra, W. Fang, J. Kong et al., *Sci. Rep.* **3**, 2061 (2013).
[3] J. Ek-Weis, S. Costa, O. Frank, M. Kalbac, *J. Phys. Chem. Lett.* **5**, 549 (2014).
[4] K. Filintoglou, N. Papadopoulos, J. Arvanitidis, D. Christofilos et al., *Phys. Rev. B* **88**, 045418 (2013).

*christof@eng.auth.gr
Keywords: graphene, Raman, high pressure

IOP Conf. Series: Journal of Physics: Conf. Series **950** (2017) 032007 doi:10.1088/1742-6596/950/3/032007

Current trends in research and applications of dynamic high-pressure

Dominique Chevalier-Lucia *, *Laetitia Picart-Palmade, Eliane Dumay*

University of Montpellier, Joint Research Unit Agropolymer Engineering and Emerging Technologies
CC023 Place E. Bataillon, 34095 Montpellier cedex 5, France

The interest in dynamic high-pressure, also called ultra high-pressure homogenization, in food, cosmetic and pharmaceutical areas significantly grew during the last decade as a result of the latest developments in high pressure technology and promising results on different applications. A particular attention was obviously drawn on the effect of dynamic high pressure on microorganism and enzyme inactivation in various liquid foods. The focus was also placed on the specific innovative functionalities developed from the structural modifications that dynamic high pressure could induce, and particularly to the processing of submicron emulsions and protein engineering by dynamic high pressure.

The presentation highlights the current trends in dynamic high pressure research, provides a summary of the available findings on the technological aspects of dynamic high pressure and gives a closer look to the opportunities to produce efficient delivery vehicles for bioactive molecules with this technology.

*Dominique.chevalier-lucia@umontpellier.fr

Keywords: high pressure homogenization, dynamic high pressure, continuous processing, delivery vehicles.

A watchmaker's precision at high-pressure reveals $CeCu_2Si_2$ and $CeAu_2Si_2$ *p-T* phase diagrams

Gaétan Giriat [a]*, *Gernot Werner Scheerer* [a], *Zhi Ren* [a], *Pablo Pedrazzini* [b], *Gerard Lapertot* [c], *Didier Jaccard* [a]

[a] *DPMC - University of Geneva, 24 Quai Ernest-Ansermet, 1211 Geneva 4, Switzerland*
[b] *Lab. Bajas Temperaturas and Instituto Balseiro - CNEA, 8400 San Carlos de Bariloche, Argentina*
[c] *SPSMS, UMR-E 9001, CEA-INAC/UJF-Grenoble 1, 38054 Grenoble, France*

Nearly forty years after the discovery of the first heavy fermion (HF) superconductor, $CeCu_2Si_2$,[1] high-pressure (HP) investigations continue to provide information on the above and also enable the discovery of new Ce-based HF superconductors. Taking advantage of multiprobe HP setups we examined the resistivity, ac-magnetic susceptibility, ac-calorimetry and thermopower of $CeCu_2Si_2$ and $CeAu_2Si_2$ single crystals up to 27 GPa. We confirmed the existing *p-T* phase diagram of the former[2,3] based on bulk properties and found pressure-induced HF superconductivity in the latter.
The ac-susceptibility study of $CeCu_2Si_2$ up to 5 GPa indicates that the superconducting (SC) transition signature observed from the magnetic field expulsion coincides with the specific heat jump and the zero resistivity. Below T_c, the diamagnetic signal is huge with a signal-to-noise ratio of 10^3 and corresponds to that expected for the sample volume. Thus the pressure dependence of T_c ($T_c^{max} \approx 2$ K at 4 GPa), follows the bulk transition contrary to what is suggested by a previous report.[4] Furthermore, the broadening of the resistive SC transition in the intermediate pressure range $2 \leq p \leq 4$ GPa results essentially from phenomena linked to the surface quality of the sample as shown by a later investigation.
In the antiferromagnet $CeAu_2Si_2$ ($T_N \approx 10$K), HP resistivity, ac-calorimetry and thermopower measurements revealed a highly unusual interplay between superconductivity and magnetism. We recently reported on pressure-induced superconductivity with $T_c^{max} \approx 2.5$ K near $p_c \approx 22.5$ GPa occurring deep inside the antiferromagnetic phase and over a broad pressure range from 11.8 to 26.5 GPa leading to a giant overlap of the magnetic and pressure-induced SC phases.[5] In addition, both the bulk T_c and T_N are strongly enhanced (by a factor 2) with increasing pressure from 16.7 to to 20.2 GPa. Subsequently we were also able to reveal a magnetic structure rearrangement below T_N which follows T_c and is strongly enhanced at p_c when T_N vanishes.

References

[1] F. Steglich, J. Aarts, C.D. Bredl, W. Lieke, D. Meschede, W. Franz, H. Schäfer, Phys. Rev. Lett. **43**, 1892 (1979).
[2] G. Seyfarth, A.-S. Rüetschi, K. Sengupta, A. Georges, D. Jaccard, S. Watanabe, K. Miyake, Phys. Rev. B **85**, 205105 (2012).
[3] G. Giriat, Z. Ren, P. Pedrazzini, D. Jaccard, Solid State Commun. **209–210**, 55 (2015).
[4] F. Thomas, C. Ayache, I.A. Fomine, J. Thomasson, C. Geibel, J. Phys.: Condens. Matter **8**, L51 (1996).
[5] Z. Ren, L. V. Pourovskii, G. Giriat, G. Lapertot. A. Georges, and D. Jaccard, Phys. Rev. X **4**, 031055 (2014).

* gaetan.giriat@unige.ch
Keywords: Heavy fermion, High pressure, Superconductivity, Magnetic susceptibility

Putting pressure on resolving phase relationships between polymorphs of pharmaceuticals

Ivo B. Rietveld [a,*], *Maria Barrio* [b], *René Céolin* [a], *Josep-Lluìs Tamarit* [b]

[a] *Faculté de Pharmacie, Université Paris Descartes, Paris, France*
[b] *Departament de Fisica I Enginyeria Nuclear, Universitat Politècnica de Catalunya, Barcelona, Spain*

Although Bridgman studied crystalline polymorphism as a function of pressure for many small organic molecules, in the pharmaceutical industry, pressure as a variable of the Gibbs energy to determine polymorph stability is rarely used. For a long time, even crystalline polymorphism has been neglected by the pharmaceutical industry, even though they were not unaware of its existence (see chloramphenicol palmitate in 1967).[1] Ever since the notorious case of ritonavir and following the adage "better safe than sorry", the EMEA and the FDA require crystalline polymorph screening and stability testing for new drug molecules. Whereas polymorph screening can be a difficult task, as it depends in part on trial and error, the stability of observed polymorphs is in principle strictly described by thermodynamics. The minimum of the Gibbs energy determines the most stable state and the two characteristic variables of the Gibbs energy are temperature and pressure. Whereas temperature is obviously important for a drug formulation, pressure is considered negligible even if the French name for a tablet is "comprimé" (= compressed). Thus pressure may actually be important in industrial processes such as grinding and compression, but it can also be used as a tool to obtain a complete description of the polymorph behavior, a so-called pressure - temperature phase diagram. Our group uses two approaches to obtain the behavior of a system as a function of pressure. The first method is by direct measurement in a high-pressure differential thermal analyzer and by X-ray under pressure (synchrotron). The second approach is by using standard laboratory measurements such as differential scanning calorimetry and X-ray diffraction as a function of temperature, which in combination with the Clapeyron equation (eq. 1) can lead to pressure-temperature phase diagrams of the polymorphism of drug molecules.

$$\frac{dP}{dT}=\frac{\Delta S}{\Delta V}=\frac{\Delta H}{T\Delta V} \tag{1}$$

With dP/dT the slope of a phase equilibrium in a pressure - temperature phase diagram, ΔS the entropy associated to the phase change, ΔV the volume change, ΔH the enthalpy change, and T the temperature at which the previous quantities have been obtained.

As an example case, the phase diagram of the two polymorphs of ritonavir will be constructed, which surprisingly has never been published before.[2] Other examples to demonstrate the topological method (using the Clapeyron equation) and the construction of phase diagrams by direct measurements will be provided in the presentation. It will be shown that although direct measurement may be better, making use of the topological method will lead to the same overall conclusions in the phase behavior of crystalline polymorphs and may be a good alternative, when high-pressure measurements are inaccessible or not feasible.

References

[1] A.J. Aguiar, J. Jr. Krc, A.W. Kinkel, J.C. Samyn, *J. Pharm. Sci.* **56**, 847 (1967).
[2] R. Céolin, I.B. Rietveld, *Ann. Pharm. Fr.* **73**, 22 (2015)

*ivo.rietveld@parisdescartes.fr
Keywords: ritonavir, topological method, phase diagram, physical stability

IOP Conf. Series: Journal of Physics: Conf. Series **950** (2017) 032010 doi:10.1088/1742-6596/950/3/032010

New Gas Hydrates at High Pressure

D.M. Amos [a], M.E. Donnelly [a], C.L. Bull [b], A. Falenty [c], A. Hermann [a], W.F. Kuhs [c] and J.S. Loveday [a]

[a] SUPA, School of Physics and Astronomy and Centre for Science at Extreme Conditions, The University of Edinburgh, JCMB, Peter Guthrie Tait Road, Edinburgh EH9 3FD, UK
[b] ISIS Facility Rutherford Appleton Laboratory, Chilton OX11 0QX, UK
[c] GZG, Abteilung Kristallographie, Universität Göttingen, Göttingen, Germany

Simple gas hydrates are known to adopt seven basic structures depending on the conditions of pressure, temperature and guest species. When taken at face value this may seem to be a rich range of structures, but when compared with other types of network structure it is remarkably limited. For example, there are over two hundred known zeolite structures. In my talk I will describe recent work by the Edinburgh hydrate group which has identified three more hydrate structures. One of these is a chiral channel structure which is quite unlike any other hydrate structure, and which is not found silica or zeolite systems but is instead related to a Zintl-phase structure.

* j.loveday@ed.ac.uk
Keywords: water, liquids, molecular physics, phase diagram, carbon-dioxide, structure

Dynamic Compression Experiments Using Lasers

Jon Eggert[a]*

[a]*Lawrence Livermore National Laboratory, Livermore, USA*

Dynamic compression using lasers has become an important technique in high-pressure science. Lasers offer many advantages, including access to the most extreme pressures available in the laboratory, excellent control of the time-dependent drive, variable drive area, small sample size, and rapid shot rate. Large lasers capable of reaching pressures in the TPa regime have been built around the world, and several FEL and synchrotron x-ray light sources have or are planning laser drivers.[1] These lasers are becoming true user-access facilities aiming to be accessible to a broad user community. I will describe several common experiments and diagnostics used in laser-compression experiments, highlight some of our group's recent results, and offer my vision for the future of the field.

The primary attribute of the laser-driven compression platform lies in the extreme flexibility of stress-loading paths and compression geometries available. Single shock, multiple shock, pure ramp, various shock-ramp combinations, unsupported shocks, and purposefully-unsteady shocks are all finding applications in laser-compression experiments. Monochromatic or broad-band x-rays, generated with exquisite timing precision by the laser itself, are employed for a variety of diffraction, radiography, absorption, and phase-contrast imaging experiments. Using standard diagnostics, laser platforms are used to measure the, stress-strain EOS, melting, phase-transitions, Gruneissen-parameter, sound-speed, crystal-structure, temperature, and non-equilibrium dynamics of compressed materials.

Near-term progress for the laser-compression platform involve the inclusion of impressive x-ray probe sources. Several X-ray FELs and 3rd and 4th generation synchrotrons are installing ~100 J shaped-pulse laser drivers on specialized beam lines.[1] Perhaps the most exciting advance in laser-drive technology is the reduction of shot repetition rate for TPa-class experiments from 1 every 0.5 to 12 hours to shots every few minutes of even shots operating at 10 Hz using high-energy class diode-pumped solid-state lasers (HEC-DPSSL). These advances require a paradigm shift from very-expensive, drive-limited target designs, to rapid-fire, low-cost, easily-manipulated targets. This emerging paradigm will be coupled to a shift from exclusive-access facilities, to a broad-based, user-centric future for laser-driven compression experiments. The future for laser-driven compression using economical drivers, targets, and probes accessible to a broad range of users is very bright.

This work was performed under the auspices of the U.S. Department of Energy by Lawrence Livermore National Laboratory under Contract DE-AC52-07NA27344.

References

[1] http://portal.slac.stanford.edu/sites/lcls_public/instruments/mec, http://www.dcs-aps.wsu.edu, http://www.xfel.eu/research/instruments/hed.

*eggert1@llnl.gov
Keywords: laser, compression, shock, ramp

IOP Conf. Series: Journal of Physics: Conf. Series **950** (2017) 032012 doi:10.1088/1742-6596/950/3/032012

Microbial cannibalism and recovery of bacteria injured by high hydrostatic pressure

Kazutaka Yamamoto [a*], *Kazuya Morimatsu* [b], *Yoshiko Nakaura* [a], *Takashi Inaoka* [a], *Keitarou Kimura* [a]

[a] *National Food Research Institute, NARO, Tsukuba, Ibaraki 305-8642, Japan*
[b] *Ehime University, Matsuyama, Ehime 790-8566, Japan*

High hydrostatic pressure (HHP) treatment can inactivate bacteria. Lethal inactivation of bacteria can be referred to as pasteurization or sterilization. On the other hand, sublethal inactivation is considered to be injury. Up until now, HHP treatment has been applied to food processing for pasteurization of pathogenic and/or spoilage bacteria. Thus, HHP-treated foods aiming at microbial control are commercialized in the world as juice, dip paste, meat products, and so forth. However, the injury of bacteria has not been studied sufficiently, and it is necessary to reveal the bacterial injury for better food safety in HHP inactivation of food microbes.
It is reported that *Escherichia coli* ATCC25922 cells suspended in phosphate-buffered saline (PBS) at around 10^8 CFU/ml were inactivated sublethally by HHPs at 500, 550, and 600 MPa and that the injured bacteria recovered close to 10^8 CFU/ml when the cell suspensions were stored at 4 °C for 5 days and thereafter moved to an environment at 25 °C. No recovery was observed when the cell suspensions were stored at 37 °C after a 5-day storage at 4 °C. It was indicated that HHP-injured *E. coli* cells did not prefer to live at 37 °C although the temperature is optimum for the native cells. In addition, the recovery was not complete, and the number of cells after recovery reached about a half of the initial viable cell counts (10^8 CFU/ml). Since PBS does not contain any primary nutrients such as saccharides, proteins, and lipids, the cells cannot propagate in PBS but may recover without using nutrients. Furthermore, it was assumed that the recovered cells might eat dead cells for propagation. However, even in the experimental system simply comprising *E. coli* cells, phosphates, and other ions, it was difficult to estimate the numbers of both recovered cells and dead cells.
In this study, "microbial cannibalism" during the storage of HHP-treated *E. coli* cells was studied to give further insight into the bacterial injury and recovery after HHP-treatment of foods. Live cells (10^2 CFU/ml) suspended in dead cell suspension (10^9 CFU/ml) did not grow at 4 °C for 10 days. When stored at 10 °C, the cells decreased at 4 days and reached undetected after 10 days. When stored at 15 °C, the cells did not increase until 0.5 day but increased after 1 day. Thereafter, it increased gradually until 6 days and reached a plateau at about 10^7 CFU/ml. At 25 °C, the cells increased rapidly until 1 day and thereafter remained almost constant at 10^7 - 10^8 CFU/ml. It was indicated that the microbial cannibalism was largely dependent on the temperature and that 10^2 live cells may propagate to 10^7 - 10^8 by ingesting dead cells. The decrease in the number of live cells at 10 °C may endorse that the live cells cannot ingest dead cells at 10 °C or lower temperatures.
HHP-treated cell samples (10^9 CFU/ml) were as well stored at 5, 10, 15, and 25 °C. Immediately after the treatment, no colony was observed for each sample. When stored at 5 °C, the number of cells increased gradually after 6 h and reached a plateau at about 10^2 CFU/ml after 3 days. When stored at 10 °C, the increment was a bit faster than that at 5 °C and reached a plateau at about 10^3 CFU/ml after 1 day. It was indicated that the injured cells recovered at 5 °C and that the recovery ratio (= recovered cells / inoculated cells) was only 10^{-7} (= $10^2/10^9$).

References

[1] S. Koseki, K. Yamamoto, *Int. J. Food Microbiol.* **110**, 108-111 (2006).

* kazutaka@affrc.go.jp
Keywords: pressure, microbial cannibalism, injury

IOP Conf. Series: Journal of Physics: Conf. Series **950** (2017) 032013 doi:10.1088/1742-6596/950/3/032013

Structural stability of proteins under pressure

*László Smeller**

Dept. Biophysics and Radiation Biology Semmelweis University, Budapest, Hungary

Pressure and temperature stability of proteins are governed by several factors. The phenomenological thermodynamic theory predicts broad pressure and thermodynamic stability regions when the changes of the compressibility and of heat capacity due to unfolding are small.[1] From structural point of view, proteins are stabilized by several factors, like disulphide bridges, prosthetic groups (e.g. porphyrine for myoglobin and horseradish peroxidase)[2] stabilizing ions, (like Ca^{2+} in case of parvalbumin).[3] We investigated several proteins with high pressure FTIR spectroscopy, utilizing the diamond anvil cell. This report will focus on the relation of the pressure stability and the secondary structure content of the proteins. A few of highly pressure stable proteins were already found, like the horseradish peroxidase[2], the MjHSP 16.5 heat shock protein,[4] the green fluorescence protein (GFP).[5] We investigated the orange variant of the GFP, which is further modified to have a large stokes shift.[6] To our knowledge this is the most pressure stable protein studied so far. It did not unfold at 20 kbar pressure at 40 ºC. Besides its high pressure stability its heat unfolding temperature is also above 100 ºC. Although the secondary structure disrupts only under very extreme conditions, distortions in the secondary and tertiary structure are clearly visible in the elastic range. The high unfolding pressure makes the elastic range exceptionally wide in this case, which allows the characterization of the small distortions of the structure as reported by the changes in the fluorescence spectrum. The reason for the high pressure stability will be discussed for these proteins, taking into account different structural stabilizing effects.

References

[1] L. Smeller, *Biochim. Biophys. Acta - Protein Struct. Molec. Enzymol.* **1595**, 11 (2002).
[2] L. Smeller, and J. Fidy, *Biophysical Journal* **82,** 426 (2002).
[3] J. Somkuti, M. Bublin, H. Breiteneder, L. Smeller, *Biochemistry* **51,** 5903 (2012).
[4] F. Tölgyesi, C.S. Böde, L. Smeller, D.R. Kim, K.K. Kim, K. Heremans, J. Fildy, *Cellular and Molecular Biology* **50**, 361 (2004).
[5] C. H. Scheyhing, F, Meersman, M. A. Ehrmann, K. Heremans, and R. F. Vogel, *Biopolymers*, **65**, 244 (2002).
[6] D. M. Shcherbakova, M. A. Hink, L. Joosen, T. W. J. Gadella and V. V. Verkhusha, *J Am Chem Soc.* **134**, 7913 (2012).

* laszlo.smeller@eok.sote.hu
Keywords: Proteins, pressure stability, FTIR spectroscopy, fluorescence spectroscopy

Graphene oxide and simple molecules at high pressure: new perspectives for 2D nanoconfined chemistry of carbon based materials

Matteo Ceppatelli [a,b]*, *Demetrio Scelta* [a,b], *Giulia Tuci* [a], *Giuliano Giambastiani* [a,c], *Michael Hanfland* [d], *Roberto Bini* [a,b,e]

[a]*ICCOM-CNR, Institute of Chemistry of OrganoMetallic Compounds, National Research Council of Italy, Sesto Fiorentino (FI), Italy*
[b]*LENS, European Laboratory for Non-Linear Spectroscopy, Sesto Fiorentino (FI), Italy*
[c]*KFU, Kazan Federal University, Kazan, Russia*
[d]*ESRF, European Synchrotron Radiation Facility, Grenoble, France*
[e]*Dipartimento di Chimica "Ugo Schiff", Università degli Studi di Firenze, Sesto Fiorentino (FI), Italy*

Graphene oxide (GO) has recently emerged as a versatile platform material for the large-scale synthesis of graphene and for the chemical functionalization of related 2D materials. In particular N-doping of these systems is a highly relevant topic for potential applications to energetic and environmental issues.[1] From the structural point of view, the intriguing properties of GO are related to its layered structure featuring sufficiently large interlayer spacing for molecular insertion.[2] Whereas a large variety of methods have been reported to explore the reactive behavior of GO at ambient pressure, mainly at high temperature in gas phase or in solution, with the only exception of a few structural studies, the high pressure chemistry of this material remains substantially unexplored.

The idea of this study is to use pressure for inserting molecule between the GO layers and realize high density conditions, where bond breaking and reconstruction can occur, leading to the incorporation of heteroatoms or to the functionalization of the carbon framework by molecular fragments. For this purpose the room temperature high pressure behavior of pure GO and of GO in presence of atomic and molecular systems of increasing reactive character, such as Ar, N_2 and NH_3, were studied using a membrane Diamond Anvil Cell (DAC) by means of X-ray diffraction, FTIR and Raman spectroscopy. Electronic photoexcitation was also employed to generate highly reactive species in high density conditions for activating chemical reactivity.

The presented results indicate the substantial stability of the underlying layered structure of GO at the investigated pressure and provide indication for the high pressure incorporation of nitrogen functionalities within the carbon framework, thus suggesting an innovative approach for the chemical functionalization of nanostructured graphene related systems and opening new perspectives for the synthesis of 2D advanced functional materials.[4]

References

[1] H. Wang, T. Maiyalagan, X. Wang *ACS Catal.* **2**, 781 (2012).
[2] D. R. Dreyer, A. D. Todd, C. W. Bielawski Chem. Soc. Rev. **43**, 5288 (2014).
[3] D. Talyzin, V. L. Solozhenko, O. O. Kurakevych, T. Szabó, I. Dékáni, A. Kurnosov, and V. Dmitriev *Angew. Chem. Int. Ed.* **47**, 8268 (2008).
[4] M. Ceppatelli, D. Scelta, G. Tuci, G. Giambastiani, M. Hanfland, R. Bini, *Carbon* **93**, 484 (2015).

*ceppa@lens.unifi.it and matteo.ceppatelli@iccom.cnr.it
Keywords: graphene oxide, argon, nitrogen, ammonia

IOP Conf. Series: Journal of Physics: Conf. Series **950** (2017) 032015 doi:10.1088/1742-6596/950/3/032015

Spectroscopic and structural studies: effect of high pressure on selected polymers, glasses and clays

Naira Maria Balzaretti [a]*, *Maria Alexandra Puerto Medina* [a], *Jackeline Barbosa Brito* [a], *Silvio Buchner* [b], *Claudio Antonio Perottoni* [c], *Tania Maria Haas Costa* [a], *João Alziro Herz da Jornada* [d]

[a] *Universidade Federal do Rio Grande do Sul, Porto Alegre, RS, Brazil*
[b] *Universidade de Ciências Básicas da Saúde de Porto Alegre, Porto Alegre, RS, Brazil*
[c] *Universidade de Caxias do Sul, Caxias do Sul, RS, Brazil*
[d] *INMETRO, Rio de Janeiro, Brazil*

The significant changes induced by high pressure in the interatomic distances of condensed matter can generate very interesting phenomena in the fields of physical chemistry, geology and biology. These changes would depend on the structural constraints and they can be monitored by Raman and infrared spectroscopy as well as by X-ray diffraction. This work presents a compilation of results related to the effect of high pressure at room temperature or at high temperatures on selected polymers, glasses and clays investigated using a diamond anvil cell and a toroidal high pressure apparatus. In the field of polymers, it was observed a fully reversible conformational disorder induced by high pressure on paraffin[1] and 5-azacytosine[2]. For polyacetylene (PA), an irreversible structural modification for *cis*-PA was observed above 5 GPa while the modifications induced on *trans*-PA were reversible.[3] The pyrolysis of methyl groups at the surface of aerosil cold sintered under high pressure induced the formation of luminescent carbon nanodots.[4] In the field of glasses, the structural, mechanical and optical properties of lithium dissilicate were significantly altered by densification at high pressure. It was observed, for instance, the formation of a distinct amorphous phase after densification at 7.7 GPa at room temperature.[5,6] In the field of clays, it was observed that the structure of montmorillonite under high pressure at room temperature underwent reversible changes and the intrinsic water content was preserved up to 13 GPa.[7]

References

[1] M.A.P. Medina, N.M. Balzaretti. *Vib. Spectr.* **75**, 93 (2014).
[2] M.A. P. Medina, F. Miotto, J. Catafesta, C.A. Perottoni, N.M. Balzaretti, *Vib. Spectr.* doi:10.1016/jvibspec.2015.04.001
[3] N.M. Balzaretti, C.A. Perottoni, J.A.H. da Jornada, *J. Raman Spectr.* **34**, 259 (2003).
[4] J.B. Brito, T.M.H. Costa, F.S. Rodembusch, A.S. Konowalow, R. Reis, N.M. Balzaretti, J.A.H. da Jornada, *Carbon* (accepted for publication).
[5] S. Buchner, A.S. Pereira, J.C. Lima, N.M. Balzaretti, *J. Non-Crystall. Solids* **387**, 112 (2014).
[6] S. Buchner, M.B. Pereira, N.M. Balzaretti, *Opt. Mater.*, **34**, 826 (2011).
[7] F.G. Alabarse, R.V. Conceição, N.M. Balzaretti, F. Schenato, A.M. Xavier, *Appl. Clay Sci.* **51**, 202 (2011).

* naira@if.ufrgs.br
Keywords: high pressure, Raman, infrared spectroscopy, x-ray diffraction

Coping with the pressure: Microbial adaptation to high hydrostatic pressure in deep sea hydrothermal vents

Phil Oger *

Laboratoire de Géologie de Lyon, Ecole Normale Supérieure de Lyon, Lyon, France

Most of Earth's biotopes are located in the deep-sea and the deep subsurface under elevated hydrostatic pressures (HHP) inhibitory to surface organisms. HHP has numerous effects on organisms and cellular components. In many ways, these resemble that of an increase or a decrease in temperature, such as protein denaturation, membrane destabilization, alteration of transcription and translation. HHP is however required for optimal activity of deep-environment adapted microbes (piezophiles). Several HHP-adapted microbes, or piezophiles, have been isolated, but our understanding of their adaptation mechanisms to HHP remains limited. Piezophiles can be classified in two main groups, depending on their optimal growth temperature: psychrophiles, which optimal growth temperature is below 15 °C (most of these are bacteria and have been isolated from the deep-sea or deep-sea sediments), and thermophiles, which optimal growth temperature is higher than 60 °C (most of these are archaea and have been isolated from hydrothermal settings (vents or seeps)).

Experimental evidence on macromolecules shows that HHP has a different impact depending of the macromolecule. For example, DNA and lipids are stabilized under physiological pressures, while multimeric proteins tend to be destabilized. However, for these three types of macromolecules, HHP has a similar negative impact on the cellular function. In piezosensitve organisms such as the bacterium *E. coli*, HHP inhibits the cell division at ca. 500 bars, replication and translation at ca. 700 bars, while some function may still be active at higher pressures.

To date, three adaptive strategies to HHP have been proposed in piezophilic bacteria.[1] First, piezophilic bacteria express a specific set of genes under HHP. This is the case for example in *S. violacea* which expresses a different respiratory chain at low or high pressure conditions. Second, piezophiles regulate the expression of their transcription to compensate for the loss of stability/activity of their proteome. A strong induction of membrane associated proteins has been shown in the moderate piezophile *P. profundum* for example. Third, in the two above mentioned piezophiles, there is evidence of structural adaptation of part of the proteome, which leads to an increased stability of the proteins under HHP.

The adaptive strategy in thermophilic piezophiles is not yet fully understood. However, several lines of evidence show that it involves at least the regulation of the transcription of the genome as well as the expression of specific genes under HHP.[2] To date, there is no direct evidence for the structural adaptation in proteins of hyperthermophilic piezophiles, although several indirect evidences clearly demonstrate it.[3,4] In addition, the adaptation to HHP in hyperthermophilic piezophiles may also involve structural modifications in the membrane structure as well as the accumulation of osmolytes to maintain protein folding and activity.

References

[1] Oger, P.M. and Jebbar, M. *Res Microbiol* **161**, 799 (2010).
[2] Vannier, P. et al. *Res Microbiol*, in press (2015).
[3] Peters, J. et al. *Zeit. Phys. Chem.* **228**, 1121 (2014).
[4] Cario, A. *et al. Biochimie*, in press (2015).

* poger@ens-lyon.fr
Keywords: High Hydrostatic Pressure, Adaptation, Piezophiles, Hydrothermal Vents

IOP Conf. Series: Journal of Physics: Conf. Series **950** (2017) 032017 doi:10.1088/1742-6596/950/3/032017

Inelastic X ray scattering under pressure to probe the quantum phase transition in the transition metal dichalcogenides

*Maxime Leroux[a], Ion Errea[b,c], Mathieu Le Tacon[d], Sophia-Michaela Souliou [d], Gaston Garbarino [e], Laurent Cario [f], Alexey Bosak [e], Francesco Mauri [c], Matteo Calandra [c], Pierre Rodière [a]**

[a] *Institut Néel, CNRS, Grenoble, France*
[b] *IMPMC, Paris, France*
[c] *UPV/EHU and DIPC San Sebastian, Spain*
[d] *Max-Planck-Institut , Stuttgart, Germany*
[e] *ESRF, Grenoble, France*
[f] *IMN, Nantes, France*

Transition metal dichalcogenides are systems in which the interplay of strong electron – electron and electron-phonon interactions give rise to a wide variety of electronic states at low temperature ranging from metal, Mott-insulator or an electronic spatial modulation : charge density wave. The temperature below which the charge density wave appears (T_{cdw}) can be tuned continuously down to 0K by applying an external parameter such as pressure. In the vicinity of this point a superconducting state appears.
One of the key ingredient of both the superconducting and charge density wave states is the phonon spectrum. Indeed, a charge density wave is associated to a periodic lattice distortion stabilised by a strong electron-phonon coupling, and, in most cases, phonons are responsible for the formation of Cooper pairs.
Recently, we performed inelastic X-ray scattering up to 16 GPa and down to very low temperature. Temperature and pressure dependences of the soft mode associated to the charge density wave have been measured in several transition metal dichalcogenides among them $NbSe_2$ and NbS_2. A strong anisotropy of the phonon dispersions along different crystallographic directions has been observed. In NbS_2 at ambient pressure, we observed a huge temperature dependence of the phonon softening. By comparing it with the temperature dependence of the phonon spectra of $NbSe_2$ under pressure, we show that NbS_2 is at the vicinity of a charge density wave instability. By explicitly taking into account anharmonic effects, we obtain an accurate, quantitative, description of the (P,T) dependence of the phonon spectrum, and accounts for the rapid destruction of the CDW under pressure by zero mode vibrations - or quantum fluctuations - of the lattice. At least, this pressure and temperature dependence could be general to other dichalcogenides for which electron-electron correlations are stronger.

References

[1] M. Leroux & al. *Phys. Rev. B* **86** 155125 (2012).
[2] M. Leroux & al; to be published

* pierre.rodiere@neel.cnrs.fr

Pressure on Correlated Materials: Transport in iron and implications for the geodynamo, and electronic transitions in iron compounds

R. E. Cohen [a,b]*, *Peng Zhang* [a], *Kristjan Haule* [c]

[a] *Geophysical Laboratory, Carnegie Institution for Science, Washington, D.C. USA*
[b] *University College London, Gower Street, WC1E 6BT, London, United Kingdom*
[c] *Rutgers University, New Jersey, USA*

Under pressures in the Earth, iron in minerals behaves very differently from simple elements, such as magnesium, even though Fe^{2+} and Mg^{2+} easily substitute for each other at low pressures. I will discuss the behaviour of FeO and Fe under high pressures and temperatures using modern electronic structure methods. FeO is an insulator at ambient conditions, but at high pressures and temperatures we found it becomes a metal.[1] We are investigating these properties using a method beyond the standard model, density functional theory (DFT), for Electronic structure. We are using self-consistent DFT/Dynamical Mean Field Theory (DMFT),[2] which includes quantum and thermal fluctuations, multiplets, and all-orders of local electronic correlations. We find that electron correlations are important even in the generation Of Earth's magnetic field. Earth's magnetic field was believed to arise from thermal convection of molten iron alloy in Earth's outer core, but density functional theory (DFT) calculations suggested that the conductivity of iron is too high to support thermal convection,[3-6] so that new geodynamo models were being developed.[7,8] The DFT computations for resistivity were based on the scattering of electrons off of atomic vibrations, or electron-phonon (e-p) scattering. We applied self-consistent Density functional Theory plus Dynamical mean-field Theory (DFT+DMFT)[2] to Iron and found that at high temperatures electron-electron (e-e) scattering is comparable to the e-p scattering, bringing theory into agreement with experiments and solving the transport problem in Earth's core, consistent with the conventional thermal geodynamo.[9] How electron correlations change with pressure, and how this affects material properties, will be discussed. This work is supported by the US National Science Foundation and the ERC Advanced grant ToMCaT.

References

[1] K. Ohta, R. E. Cohen, K. Hirose, K. Haule, K. Shimizu, and Y. Ohishi, Phys. Rev. Lett. **108** (2012).
[2] K. Haule, C. H. Yee, and K. Kim, Phys. Rev. B **81** (2010).
[3] X. Sha and R. E. Cohen, J. Phys. Cond. Matt. **23**, 075401 (2011).
[4] N. de Koker, G. Steinle-Neumann, and V. Vlcek, Proc Natl Acad Sci U S A **109**, 4070 (2012).
[5] H. Gomi, K. Ohta, K. Hirose, S. Labrosse, R. Caracas, M. J. Verstraete, and J. W. Hernlund, Phys. Earth Planet. Int. **224**, 88 (2013).
[6] M. Pozzo, C. Davies, D. Gubbins, and D. Alfe, Nature **485**, 355 (2012).
[7] B. Buffett, Nature **485**, 319 (2012).
[8] P. Olson, Science **342**, 431 (2013).
[9] P. Zhang, R. E. Cohen, and K. Haule, Nature **517**, 605 (2015).

*rcohen@carnegiescience.edu
Keywords: Electronic structure, geodynamo, electrical conductivity, metal/insulator transition

Rigid liquid –to– dense gas transition at supercritical pressures

V.V. Brazhkin [a]*, *Yu.D. Fomin* [a], *V.N. Ryzhov* [a], *E.N. Tsiok* [a], *Kostya Trachenko* [b]

[a] *Institute for High Pressure Physics, Russian Academy of Sciences, Troitsk, Moscow, Russia*
[b] *South East Physics Network and School of Physics, Queen Mary University of London, Mile End Road, London, E1 4NS, UK*

A liquid near the melting curve has much more in common with a solid than with a gas – this interesting result has been increasingly appreciated recently in the area of liquids, although not widely by physics community in general. For example, a liquid supports transverse collective modes at high frequency that endow the liquid with shear rigidity at that frequency. Many physical properties, including heat capacity, heat conductivity, electronic conductivity and so on, change only weakly as a result of melting despite the loss of the long-range order. Close to the melting curve, this behavior is observed at very high pressures and temperatures well above the critical point. To denote this state of the liquid, the term "rigid liquid" was proposed.[1] In a rigid liquid, particle motion consists of fairly rare jumps over the activation barrier and oscillatory motion in between the jumps. This is reflected in relaxation time τ, the time between two consecutive particle jumps, being larger than the shortest vibration period τ_0 ($\tau_0=2\pi/\omega_0$, where ω_0 is the maximal frequency of transverse-like modes). Recently, we have shown that the condition $\tau \approx \tau_0$ defines a line, the "Frenkel line" on the pressure-temperature (or density-temperature) phase diagram that separates the state of the rigid-liquid from the dense gas.[1-5] Crossing the Frenkel line on temperature increase results in the disappearance of shear rigidity at all frequencies, specific heat reaching $2k_B$, and, importantly, the qualitative change of temperature dependence of key system properties such as structure factor features, heat capacity, speed of sound, diffusion coefficient, viscosity and thermal conductivity.

Here we present new results of liquid-to-dense gas transition at very high pressures for real fluids such as iron, water, CO_2, CH_4. Both experimental and computer simulation data will be presented. The interconnections of the Frenkel line with the metal-semiconductor transition and different version of percolation lines will be also considered. The data of the location of the lines of liquid-to-dense gas transition at supercritical pressures are very important for supercritical technologies.

References

[1] Brazhkin, V.V, Fomin, Yu.D., Lyapin, A.G. et al., Phys.Rev. E 2012, **85,** 031203.
[2] Brazhkin, V.V., Fomin, Yu.D., Lyapin, A.G. et al, JETP Lett. 2012, **95**, 164.
[3] Brazhkin, V.V., Lyapin, A.G., Ryzhov, V.N. et al. Physics-Uspekhi, 2012, **55**(11), 1061.
[4] Brazhkin, V.V. & Trachenko, K. Phys. Today, 2012, **65**(11), doi: 10.1063/PT.3.1796.
[5] V.V. Brazhkin, Yu.D. Fomin, A.G. Lyapin, V.N. Ryzhov, E.N. Tsiok, K. Trachenko, Phys. Rev. Lett., 2013, **111**, 145901.

* brazhkin@hppi.troitsk.ru
Keywords: supercritical fluids, dynamical crossover, phase diagram

IOP Conf. Series: Journal of Physics: Conf. Series **950** (2017) 032020 doi:10.1088/1742-6596/950/3/032020

Unusual Magnetic Fields of Uranus and Neptune: Metallic Fluid Hydrogen

W. J. Nellis

Department of Physics, Harvard University. Cambridge, MA, USA

The Voyager 2 spacecraft discovered the unusual non-dipolar and non-axisymmetric magnetic fields of the Ice Giants Uranus and Neptune (U/N) in the 1980's. Those field shapes are unique in the Solar System and their cause has been a major scientific question since their discovery. Their likely explanation lies in electrical conductivities of fluids that generate those magnetic fields. External magnetic fields of U/N are made by dynamo action of convecting, electrically conducting fluids relatively close to their surfaces at pressures (P) and temperatures (T) up to a few 100 GPa and several 1000 K, respectively. Relevant P/Ts are achieved in a laboratory by adiabatic shock compression and quasi-isentropic multiple-shock compression. Over the past ~30 years a substantial database of measured electrical conductivities of representative planetary fluids at representative planetary P/Ts has been accumulated. Although neither exact compositions nor P/Ts in U/N are known, this database, gravitational moments measured by Voyager 2, estimated chemical abundances of the Solar System[1] and the recent discovery that rotational motion of Earth's strong, rock mantle is strongly coupled to its fluid outer core suggest: (i) There is little nebular ``Ice" in the Ice Giants. (ii) Magnetic fields of U/N are made by convecting metallic fluid H relatively close to the outer surfaces of U/N. (iii) Thus, it is reasonable to observe non-dipolar non-axisymmetric magnetic fields. (iv) Those field shapes are probably caused by decoupling of rotational motion of U/N from convective motions in their dynamos, unlike Earth with strong coupling between its mantle rotation and convective fluid Fe outer core and a dipolar field.[2] Basically, the magnetic field of Earth reverses from time to time because of fluctuations in the dynamo. The strong coupling either causes the dipolar magnetic field to remain virtually parallel to the axis of rotation or, if a fluctuation is so great that the magnetic field essentially reverses, the strong coupling causes the reversing dipolar magnetic field to align virtually anti-parallel to the axis of rotation. In U/N the weak fluid H/He envelops are weakly coupled to convection in the fluid core and so the magnetic fields probably wander over the age of the Solar System with magnetic field shapes and orientations dictated by fluctuations in their convecting fluid interiors. The full paper on this work is published.[3]

References

[1] D. Arnett, *Supernovae and Nucleosynthesis*, (Princeton, Princeton University Press, 1996).
[2] R. Hide et al, in *Relating Geophysical Structures and Processes: The Jeffreys Volume*, Geophysical Monograph 76, eds. K. Aki and R. Dmowska, IUGG Vol. **16** (2013).
[3] W. J. Nellis, *Modern Physics Letters B* **29**, 1430018 (29 pages) (2015).

*nellis@physics.harvard.edu

Beyond GGA: benefits from hybrid HSE06 functional for correct description of crystal and electronic structure and lattice dynamics of solids at ambient and elevated pressure

Wojciech Grochala [a]*

[a]*Center of New Technologies, Univ. of Warsaw, Zwirki i Wigury 93, 02089 Warsaw Poland*

Selected cases will be presented of the superior performance of the HSE06 / PBEsol density functional calculations in reproduction of crystal and electronic structure as well as lattice dynamics of extended solids. Particular attention will be devoted to the cases when the "industrial standard" GGA-PBE calculations fail badly and provide misleading *i.e.* erroneous (qualitatively or quantitatively distorted) answers.[1] The fascinating world of high pressure-driven phenomena[2] (Figure), including the important issue of metallization of the chemical elements and compounds, will be addressed.[1]

Application of the HSE06/PBEsol tandem will then be extended to the seemingly well-established cases of polymorphism of the chemical elements, notably that of carbon.[3] If time allows, selected aspects of the graphite-diamond transformation would be discussed.[4]

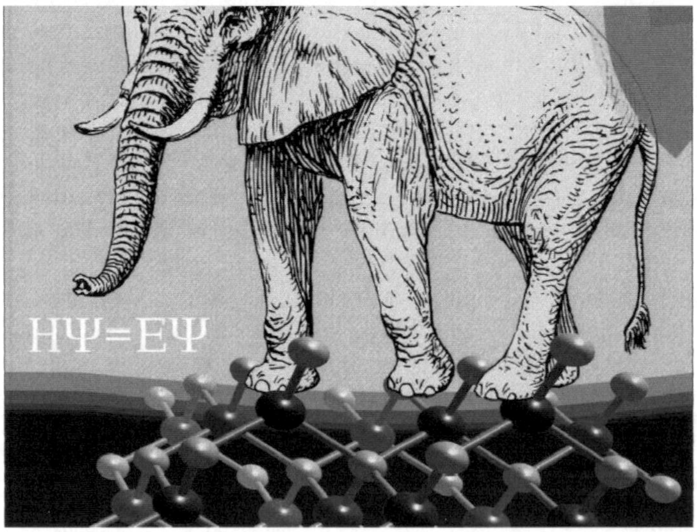

References

[1] E. Zurek, W. Grochala, *Phys. Chem. Chem. Phys.* **17**, 2887 (2015).
[2] W. Grochala, J. Feng, R. Hoffmann, N. W. Ashcroft, *Angew. Chem. Int. Ed. Engl.* **46**, 3620 (2007).
[3] W. Grochala, *Angew. Chem. Int. Ed. Engl.* **53**, 3680 (2014).
[4] W. Grochala, M. Derzsi, submitted (2015).

*w.grochala@cent.uw.edu.pl
Keywords: density functional theory, hybrid functionals, lattice distortions, electronic structure

New ordered perovskite-structure oxides synthesized by high-pressure technique

Yuichi Shimakawa*

Institute for Chemical Research, Kyoto University
and Japan Science and Technology Agency, CREST
Uji, Kyoto 611-0011, Japan

Cation ordering in transition-metal oxides often drastically modifies their physical and chemical properties. We here focus on some ordered perovskite-structure oxides, in which transition-metal ions are included at the A and B sites in perovskite ABO_3 in ordered manners like $AA'_3B_4O_{12}$ and $AA'_3B_2B'_2O_{12}$. In such compounds A-A, A-B, and B-B interactions compete and/or cooperate with each other and play important roles in giving rise to wide variety of functional properties. Note that many of this structure type compounds are stabilized under high pressure conditions.

We can see unusual A-site magnetism in $AA'_3B_4O_{12}$ with magnetic ions at the A' site and nonmagnetic ions at the B site. We also recently found a new A-and-B-site-ordered quadruple perovskite $CaCu_3Fe_2Sb_2O_{12}$ with magnetic Fe^{3+} at the B site and nonmagnetic Sb^{5+} at the B' site. The B-site Fe^{3+} spin sublattice adapts a tetrahedral framework and the Fe^{3+}-Fe^{3+} antiferromagnetic interaction causes geometrical spin frustration as seen in the double perovskite Ca_2FeSbO_6. With the introduction of Cu^{2+} into the A' site, the frustration is relieved by strong antiferromagnetic A'(Cu^{2+})-B(Fe^{3+}) interaction, leading to a ferrimagnetic ordering below 160 K. When B'-site Sb^{5+} was replaced with Re^{5+}, another A-and-B-site-ordered quadruple perovskite $CaCu_3Fe_2Re_2O_{12}$ was obtained. Strong antiferromagnetic A'(Cu^{2+})-B'(Re^{5+}) interaction overcomes the A'(Cu^{2+})-B(Fe^{3+}) interaction, leading to a ferrimagnetism with the ferromagnetic A'(Cu^{2+})-B(Fe^{3+}) spin arrangement below 560 K. More importantly, the electronic structure of $CaCu_3Fe_2Re_2O_{12}$ is half metallic and the compound shows large magnetoresistance by the spin-dependent transport.

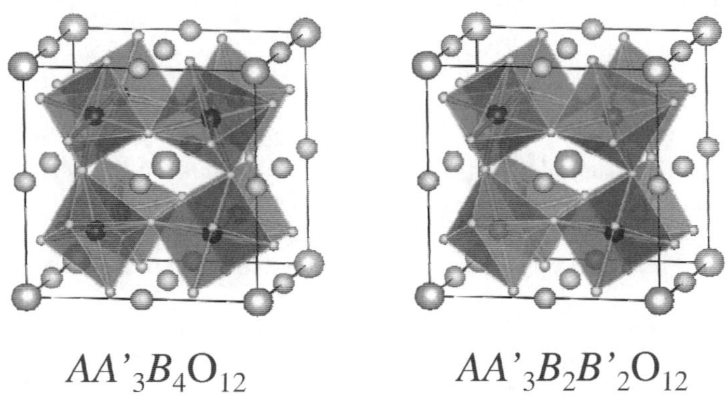

Figure Crystal structures of ordered perovskites.

References

[1] Y. Shimakawa (Review), *Inorg. Chem.* **47**, 8562 (2008).
[2] Y. Shimakawa and T. Saito (Feature Article), *Phys. Status Solidi* **B**, 249, 423 (2012).
[3] Y. Shimakawa and M. Takano (Review), *Z. Anorg. Allg. Chem.*, **635**, 1882 (2009).
[4] Y. Shimakawa and M. Mizumaki (Topical Review), *J. Phys. C: Cond. Matter.* **26**, 473203 (2014).

*shimak@scl.kyoto-u.ac.jp
Keywords: Transition-metal oxides, high-pressure synthesis, functional properties

IOP Conf. Series: Journal of Physics: Conf. Series **950** (2017) 042001 doi:10.1088/1742-6596/950/4/042001

High pressure spectrofluorimetry – a tool to determine protein-ligand binding volume

G Skvarnavičius[1], Z Toleikis[1], M Grigaliūnas[1], J Smirnovienė[1], P Norvaišas[1], P Cimmperman[1,2], D Matulis[1], and V Petrauskas[1]

[1]Department of Biothermodynamics and Drug Design, Institute of Biotechnology, Vilnius University, Vilnius, Lithuania

[2]Baltic Institute of Advanced Technology, Vilnius, Lithuania

E-mail: vytautas.petrauskas@bti.vu.lt

Abstract. The change in protein volume observed upon protein-ligand interaction (termed as the binding volume) is an important but largely neglected thermodynamic parameter from the perspective of both fundamental science and potential applications in the development of specific protein ligands. The binding volume is the pressure derivative of the Gibbs energy, thus elevated pressure is required to determine the volumetric properties of proteins. Here we describe the use of high-pressure spectrofluorimetry to determine both unfolding and ligand binding-induced volume changes of a protein. The degree of protein unfolding at elevated pressures was monitored by an intrinsic tryptophan fluorescence. Different approaches of experimental fluorescence spectra analysis are described and the impact on the quality of thermodynamic parameters is discussed.

1. Introduction

Despite the laborious nature of pressure techniques, numerous studies were devoted to determine the thermodynamic properties of pressure-induced protein unfolding [1–13]. Relatively high pressures are required to determine the volumetric properties of proteins, and that is probably the most serious obstacle for the pressure to become a standard descriptor of the thermodynamic state of a protein.

Of particular interest is the change in volume related to protein-ligand interaction, hereafter termed as protein-ligand binding (or reaction) volume. A majority of high pressure studies were devoted to reveal the thermodynamics of protein unfolding/refolding reaction and the dissociation of multimeric proteins under pressure. However, relatively little attention has been paid to the volume changes resulting from the interaction between a protein and small molecule [4, 12, 14–19].

High pressure spectrofluorimetry has been extensively used during the past several decades [20] and helped to reveal various aspects of protein folding and stability. Here we describe how high pressure fluorescence could be used to determine protein-ligand binding volume. We continue the development and validation of the method on several isoforms of human carbonic anhydrase (CA) – a protein involved in cancer progression and therapy. We describe two different approaches of experimental fluorescence spectra analysis and discuss their impact on the reliability of calculated thermodynamic parameters.

2. Materials and Methods

Three proteins, namely CA I, CA II, and CA XIII, which are isoforms of human carbonic anhydrase (CA), were used to illustrate the determination of protein-ligand binding volume by fluorescent pressure shift assay (FPSA) technique. The FPSA was previously described using other protein-ligand systems [12, 19, 21]. The protein unfolding experiments were performed using an ISS PC1 photon counting spectrofluorimeter equipped with a high-pressure cell connected to a hydrostatic pump. The pressure of 380 MPa was reached by gradual increments of 20 MPa while maintaining the constant temperature of 25 °C by circulating water bath. The system was allowed to equilibrate for approximately two minutes after each pressure increment prior to recording the fluorescence spectrum. The intrinsic fluorescence of tryptophan residues at atmospheric and elevated pressures was observed by exciting the tryptophan at 295 nm and recording the emission spectra from 320 to 400 nm. A quartz cuvette was filled with the aqueous protein solution containing 3 µM CA, 0–1.5 M guanidine hydrochloride (GndHCl), 1% dimethyl sulfoxide (DMSO), 10 mM Bis-Tris buffer, *p*H 7.0 and covered with sealing film to separate the protein solution from water in the pressure cell. Various concentrations of acetazolamide (AZM) ligand was used to determine the stability of CA against pressure-induced denaturation.

3. Model

The pressure-induced unfolding profiles of a protein, $f(p)$, can be described by an equation

$$f(p) = f_N + \frac{f_U - f_N}{1 + \exp(\Delta G(p)/RT)}, \tag{1}$$

where R is the universal molar gas constant and T – absolute temperature. Here we use an assumption that there are only two protein states at equilibrium – native (N) and unfolded (U) – and the transition between states is stepwise (i.e., without any intermediate states). If protein unfolding is monitored by fluorescence intensity at a fixed wavelength, λ, the parameters f_N and f_U denote the experimental fluorescence yields for the native and unfolded protein states, respectively (solid square curves in Figure 1 (c) and (d)). Equation (1) is also valid, if protein unfolding profile is obtained from the center of spectral mass (CSM) [4], which is defined as

$$\lambda_{CSM} = \frac{\sum_i f_i \lambda_i}{\sum_i f_i}, \tag{2}$$

where f_i is the fluorescence intensity at a wavelength λ_i and the summation is over entire spectrum range. The Gibbs energy of a protein, ΔG, as a function of pressure, p, at a constant temperature is [2, 10]

$$\Delta G = \Delta G_0 + \Delta V_0(p - p_0) + \frac{\Delta\beta}{2}(p - p_0)^2 . \tag{3}$$

ΔG_0, ΔV_0 and $\Delta\beta$ denote changes of the standard state thermodynamic parameters of protein unfolding – the Gibbs energy, volume and compressibility factor, respectively. The pressure value at the midpoint of protein unfolding curve (i.e., when both native and unfolded protein concentrations are equal) is referred to as the melting pressure, p_m. The addition of ligand to the protein solution usually changes the value of its melting pressure. The relationship between the concentration of added ligand, L_t, total protein concentration, M_t, and the midpoint of unfolding transition is [21]

$$L_t = (\exp(-\Delta G_U/RT) - 1)\left(\frac{M_t}{2\exp(-\Delta G_U/RT)} + \frac{1}{\exp(-\Delta G_b/RT)}\right), \tag{4}$$

where indexes U and b stand for the changes in Gibbs energy related to protein unfolding and protein-ligand binding, respectively:

$$\Delta G_x = \Delta G_{0\_x} + \Delta V_{0\_x}(p_m - p_0) + \frac{\Delta\beta_x}{2}(p_m - p_0)^2; \quad x = U, b. \tag{5}$$

The relationship (4) is often referred to as a dosing curve. The main parameter of interest – protein-ligand binding volume, ΔV_{0_b} – can be obtained by fitting experimental dosing curve with Equation (4).

4. Results and Discussion

To determine the protein-ligand binding volume a series of pressure-induced protein unfolding experiments at various concentrations of added ligand have to be performed. Typical fluorescence spectra of CA II and CA I proteins at various pressures are shown in the Figure 1 (a) and (b), respectively. Native state of the protein dominates at low pressure, which is manifested by the peak position of tryptophan fluorescence at emission wavelength approximately equal to 332 nm [22]. The increase of pressure promotes the unfolded states of a protein and shifts the maximum of emission spectra, λ_{max}, towards 350–353 nm. This wavelength range corresponds to the fluorescence peak of tryptophan residues in a highly polar environment.

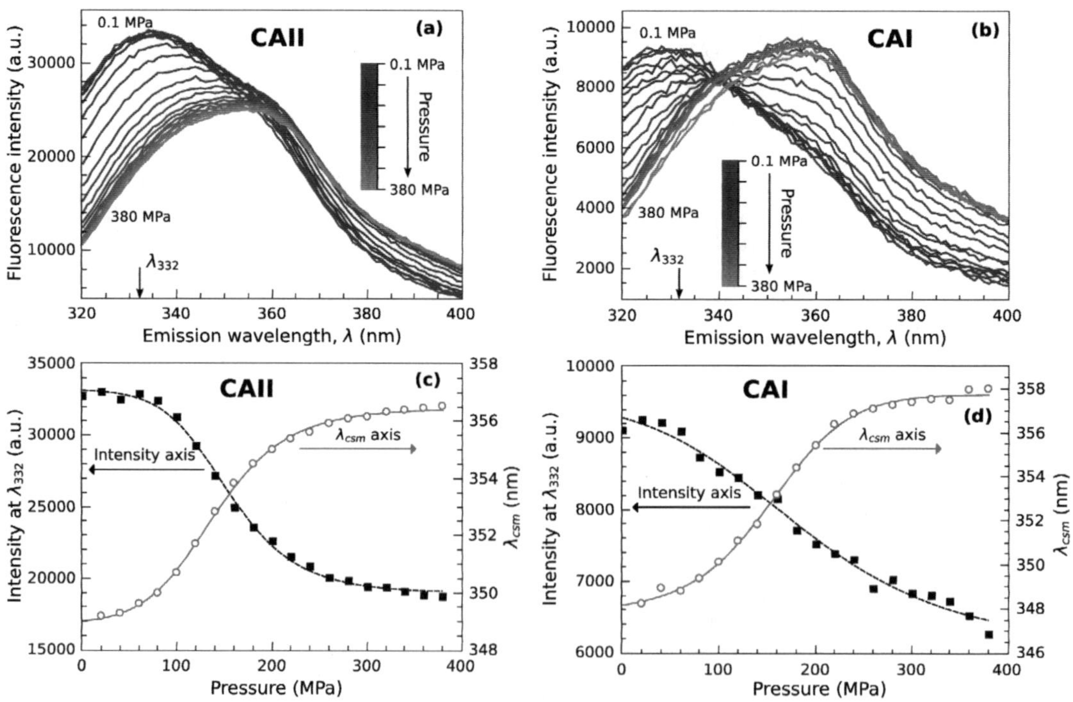

Figure 1. Intrinsic tryptophan fluorescence spectra at various pressures and pressure-induced unfolding profiles of CA II and CA I proteins. Lines in (c) and (d) are fits to Equation (1), λ_{CSM} data points are calculated from the corresponding spectra using Equation (2).

Several types of unfolding profiles could be obtained from the spectral information shown in Figure 1 (a) and (b), including the pressure-induced shift in either λ_{CSM} (Equation (2)) or λ_{max}, or by following the change in fluorescence intensity at a particular wavelength, I_λ.

Both CSM and λ_{max} unfolding profiles require an entire tryptophan fluorescence spectrum and thus are more time consuming than I_λ, which could be obtained from the fluorescence yield measurements at a single wavelength. I_λ unfolding profile is shown by black solid squares, while calculated values of λ_{CSM} versus pressure are represented by open circles in Figure 1 (c) and (d). Our results show that for some proteins (e.g., CA II) both methods provide reliable unfolding profiles while in some cases (e.g., CA I) these profiles could not be used interchangeably to precisely determine the thermodynamic parameters of unfolding (compare fluorescence intensity curves of CA II and CA I). It was possible to fit the unfolding transition of CA I using I_λ profile, but the thermodynamic parameters were obtained with lower accuracy than in the case of CA II. For example, the relative errors of fitted values of protein unfolding volume ΔV_{0_U} were 27% (I_λ profile) and 22% (λ_{CSM} profile) for CA II protein, while that values for CA I protein were 91% and 36%, respectively. In case of pressure-induced unfolding of CA XIII protein, the transition to unfolded state was almost hindered in the I_λ profiles (not shown here). In contrary to I_λ profile, the shift in λ_{CSM} of CA XIII denaturation demonstrated unambiguous unfolding transition.

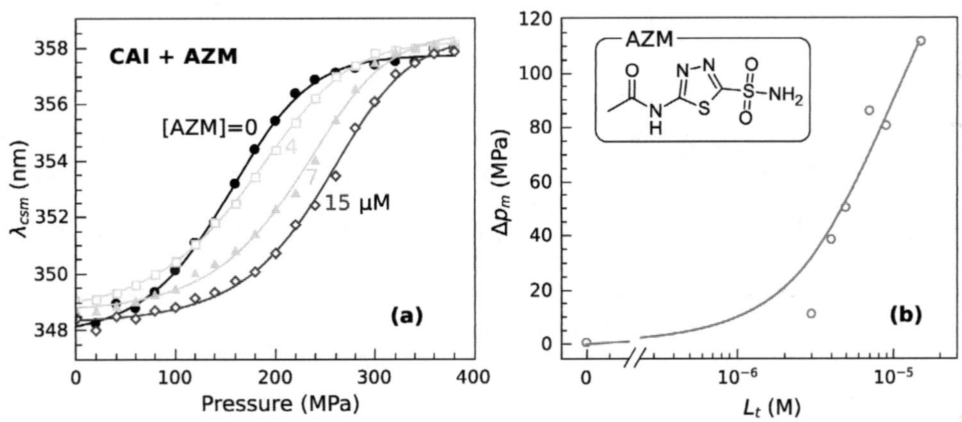

Figure 2. (a) Pressure-induced unfolding profiles of CA I at various concentrations of added ligand AZM (lines are fits to Equation (1)). All protein solutions contained 1.2 M of GndHCl. (b) The shift of protein melting pressure, Δp_m, versus total concentration of added ligand (line is the fit to Equation (4)).

The next step in the determination of protein ligand binding volume is the analysis of protein-ligand dosing curve, which is obtained by plotting the shift of p_m values versus total concentration of added ligand. Higher hydrostatic pressure is necessary to reach the protein melting point upon addition of stabilizing ligand. λ_{CSM} unfolding profiles of CA I titration by acetazolamide (AZM) are shown in Figure 2 (a). The titration of CA I protein solution with AZM yields the dosing curve shown in Figure 2 (b). The binding volume can be determined by fitting data with model Equation (4). In the shown example of CA I – AZM system (Figure 2 (b)) the fitted value of ΔV_{0_b} is equal to (-32 ± 4) ml/mol.

Many proteins can withstand relatively high pressures without being unfolded. An addition of stabilizing ligands requires even higher hydrostatic pressures to unfold a protein. The unfolding pressure values are often higher than 400 MPa, which is the limit of our experimental equipment. Partial protein unfolding or unfolding profiles without a post-transition baseline result in unreliable fitting parameters. Thus, chaotropic agent guanidine hydrochloride was used to assist the pressure-induced protein unfolding. Many previous reports on chemical denaturation

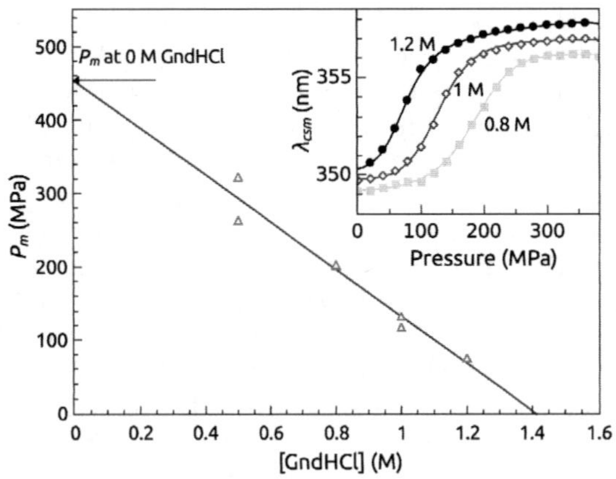

Figure 3. Dependence of CA XIII melting pressure on the concentration of GndHCl. Inset: pressure-induced unfolding profiles of CA XIII at 0.8 M, 1.0 M and 1.2 M of added GndHCl.

of proteins (see, e.g., [23, 24]) show that dependence of the Gibbs energy of unfolding on the concentration of denaturant appears to be linear. The same law is applicable for the melting pressure:

$$p_m([\text{GndHCl}]) = p_m(0) + m \times [\text{GndHCl}], \quad (6)$$

where m is the proportionality constant and $p_m(0)$ is the melting pressure at zero GndHCl concentration, which could be obtained by extrapolation of the linear fit as it is shown in Figure 3.

In conclusion, the protein-ligand binding volume can be determined using fluorescent pressure shift assay technique. The unfolding profiles obtained from the shifts of CSM ensure the quality of these profiles and subsequently the accuracy of fitted thermodynamic parameters of protein unfolding. For some proteins I_λ unfolding profiles with an accuracy similar to that of CSM could be used; in such cases considerably less time is required to obtain the protein unfolding curve.

Acknowledgments

This research was funded by a grant (No. MIP-004/2014) from the Research Council of Lithuania.

References

[1] Brandts J F, Oliveira R J and Westort C 1970 *Biochemistry* **9** 1038–1047
[2] Hawley S A 1971 *Biochemistry* **10** 2436–2442
[3] Zipp A and Kauzmann W 1973 *Biochemistry* **12** 4217–4228
[4] Royer C A, Weber G, Daly T J and Matthews K S 1986 *Biochemistry* **25** 8308–8315
[5] Silva J L and Weber G 1993 *Annual Review of Physical Chemistry* **44** 89–113
[6] Chalikian T V and Breslauer K J 1996 *Biopolymers* **39** 619–626
[7] Frye K J and Royer C A 1998 *Protein Science : A Publication of the Protein Society* **7** 2217–2222
[8] Lin L N, Brandts J F, Brandts J M and Plotnikov V 2002 *Analytical Biochemistry* **302** 144–160
[9] Balny C, Masson P and Heremans K 2002 *Biochimica et Biophysica Acta (BBA) - Protein Structure and Molecular Enzymology* **1595** 3–10
[10] Meersman F, Dobson C M and Heremans K 2006 *Chemical Society Reviews* **35** 908–917
[11] Schweiker K L, Fitz V W and Makhatadze G I 2009 *Biochemistry* **48** 10846–10851
[12] Toleikis Z, Cimmperman P, Petrauskas V and Matulis D 2011 *Analytical Biochemistry* **413** 171–178
[13] de Oliveira G A P and Silva J L 2015 *Proceedings of the National Academy of Sciences* **112** E2775–E2784
[14] Li T M, Hook III J W, Drickamer H G and Weber G 1976 *Biochemistry* **15** 5571–5580
[15] Li T M, Hook J W, Drickamer H G and Weber G 1976 *Biochemistry* **15** 3205–3211
[16] Heremans K 1982 *Annual Review of Biophysics and Bioengineering* **11** 1–21
[17] Torgerson P M, Drickame H G and Weber G 1979 *Biochemistry* **18** 3079–3083
[18] Torgerson P M, Drickamer H G and Weber G 1980 *Biochemistry* **19** 3957–3960

[19] Petrauskas V, Gylytė J, Toleikis Z, Cimmperman P and Matulis D 2013 *European Biophysics Journal* **42** 355–362
[20] Royer C A 2006 *Chemical Reviews* **106** 1769–1784
[21] Toleikis Z, Cimmperman P, Petrauskas V and Matulis D 2012 *The Journal of Chemical Thermodynamics* **52** 24–29
[22] Reshetnyak Y K and Burstein E A 2001 *Biophysical Journal* **81** 1710–1734
[23] Greene R F and Pace C N 1974 *Journal of Biological Chemistry* **249** 5388–5393
[24] Myers J K, Nick Pace C and Martin Scholtz J 1995 *Protein Science* **4** 2138–2148

IOP Conf. Series: Journal of Physics: Conf. Series **950** (2017) 042002 doi:10.1088/1742-6596/950/4/042002

Tolerance of *Artemia* to static and shock pressure loading

B C Fitzmaurice[1], G J Appleby-Thomas[1], J D Painter[1], F Ono[2], P F McMillan[2], R Hazael[2] and F Meersman[2]

[1]Cranfield Defence and Security, Cranfield University, Shrivenham, Swindon, SN6 8LA, United Kingdom
[2]Department of Chemistry, University College London, 20 Gordon Street, London WC1H 0AJ, United Kingdom

E-mail: b.fitzmaurice@cranfield.ac.uk

Abstract. Hydrostatic and hydrodynamic pressure loading has been applied to unicellular organisms for a number of years due to interest from food technology and extremophile communities. There is also an emerging interest in the response of multicellular organisms to high pressure conditions. *Artemia salina* is one such organism. Previous experiments have shown a marked difference in the hatching rate of these organisms after exposure to different magnitudes of pressure, with hydrostatic tests showing hatching rates at pressures up to several GPa, compared to dynamic loading that resulted in comparatively low survival rates at lower pressure magnitudes. In order to begin to investigate the origin of this difference, the work presented here has focussed on the response of *Artemia salina* to (quasi) one-dimensional shock loading. Such experiments were carried out using the plate-impact technique in order to create a planar shock front. *Artemia* cysts were investigated in this manner along with freshly hatched larvae (nauplii). The nauplii and cysts were observed post-shock using optical microscopy to detect motility or hatching, respectively. Hatching rates of 18% were recorded at pressures reaching 1.5 GPa, as determined with the aid of numerical models. Subjecting *Artemia* to quasi-one-dimensional shock loading offers a way to more thoroughly explore the shock pressure ranges these organisms can survive.

1. Introduction

Artemia salina, commonly known as brine shrimp, is a parthenogenetic branchiopod crustacean that has long been used in studies of a number of aquatic organisms as a food source [1, 2] as well as for bioassays to test for toxicity in various systems [3, 4]. A hatching rate of 24-48 hr in optimum conditions for *Artemia* cysts make them ideal for testing in a laboratory environment. *Artemia* are typically ovoviviparous; they are born as free-swimming nauplii. In less favorable environments, embryos are developed oviparously; in cysts that wait to hatch until conditions are stable. This dormant state in which encysted embryos reside is known as diapause where they experience reduced metabolic activity [1]. Once they are exposed to water they typically exhibit a hatching rate of approximately 90%, with a minimum of ~75% [5], under normal conditions They are capable of surviving a number of stressors, including salinity and temperature that can vary substantially, with an optimum salt concentration in most *Artemia* species of 60 gL^{-1} and an optimum temperature of 25 °C in laboratory conditions [1]. The temperatures for nauplii viability, however, have a rather considerable range of 5-40 °C [1]. In addition, cysts have been found to remain viable at even more extreme temperatures [6] along with certain evident enzyme activities that are maintained. For example, protease activity in the cyst shells of *Artemia franciscana* has been detected, although at a reduced rate, following a 15 min exposure to 100 °C [7].

Artemia exposure to high pressures extending into the multi GPa range has also been investigated more recently. Quasi-hydrostatic tests carried out on *Artemia salina* cysts in fluorinert medium by Ono *et al.* [8] found hatching rates of 80–90% after exposure of several dozen examples to a pressure of 7.5 GPa for up to 48 hr. In contrast, Udagawa and Suzuki [9] showed that shock waves with a pressures in the range of 25-100 MPa produced by underwater detonations resulted in cyst hatching rates of < 2.5% after 48 hr observation [9]. The marked difference between these findings could be related to the nature of the pressure loading, along with the different timescales over which the pressures were applied when compared with the biochemical or physiological changes that determine hatching.

The study presented here focuses on the response of *Artemia salina* nauplii and cysts to quasi-one-dimensional shock pressures. This is in keeping with similar work carried out on *Escherichia coli* [10] in order to analyse the effects of shock pressure on biological systems without the effects of a multi-dimensional wave front. Both hydrostatic and hydrodynamic pressure investigations have been carried out over a number of years to ascertain the survivability of single- to multicellular organisms following pressurisation to several hundred MPa or into the multi-GPa range. The results are of interest to fields ranging from food and agricultural products sterilisation and preservation, deep subsurface biology and exobiology through to panspermia and the origins of life [11, 12].

2. Experimental method

The response of *Artemia salina* was tested at three different shock pressures; 0.78, 0.96 and 1.5 GPa. Dried *Artemia salina* cysts were obtained from Sciento® and used for both cyst and nauplii shock loading experiments. Hatched nauplii were attained by immersing several cysts in a 3% saline solution at 25 °C for 48 hr in a water bath. Sample sizes of 100 cysts were chosen for shock loading at each pressure, while 100 hatched nauplii were examined at the lowest pressure only. These sample sizes were divided into subsets of 20 for the purpose of encapsulating them during the shock. Multiple experiments with reduced sample sizes prevented overcrowding within the sealed capsule used during shock pressure loading. Comparing the results from different runs also provided a measure of statistical significance.

The shock loading experimental set-up, outlined in figure 1, included the plate-impact technique with aluminium flyer plates carried out on a 50 mm bore single stage gas gun for quasi-one-dimensionally loading the *Artemia* samples. The sample capsule assembly was described previously [13]. A capsule containing 20 cysts or nauplii was filled to overflowing with 3% saline solution to avoid any cavitation in the sample during the shock. The cavity in the larger Al capsule was then filled with 20% ballistic gelatin to attenuate the shock and minimise rarefaction waves that may move back through the sample.

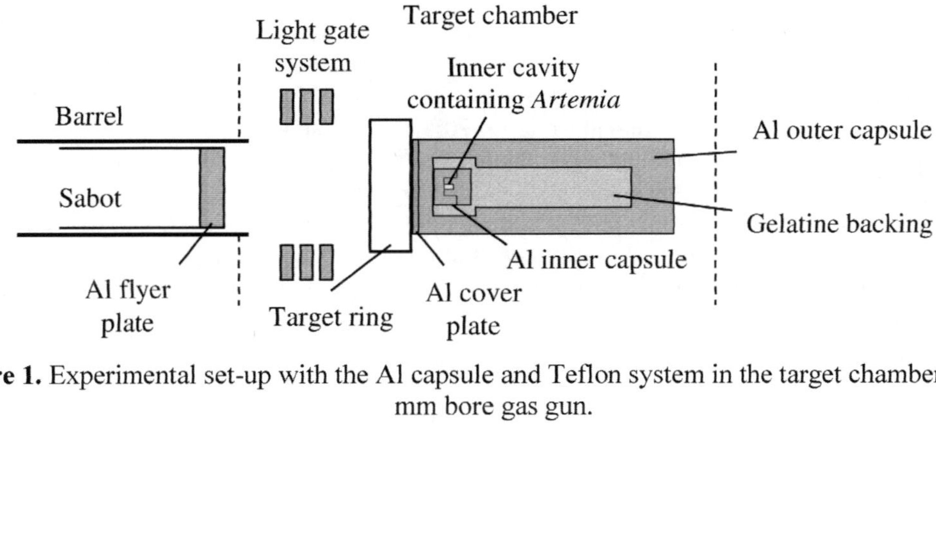

Figure 1. Experimental set-up with the Al capsule and Teflon system in the target chamber of the 50 mm bore gas gun.

In place of pressure gauges, peak shock pressures attained for each sample of cysts and nauplii were evaluated using previously validated numerical models [13]. Our estimates were based on a Lagrangian model of the capsule system implemented within ANSYS® Autodyn. The materials properties used were as listed by Leighs *et al.* [13] and followed the same set-up as used during the shock loading experiments. In order to validate the models the initial shock pressure for one experiment at both the lowest and highest impact velocities were measured using manganin pressure gauges. Subsequent comparison indicated that the pressures derived from the numerical simulations were reliable (figure 2). Fluctuations appearing in the stress-time trace for the experimental data were attributed to wave reflections occurring within the lid of the outer capsule.

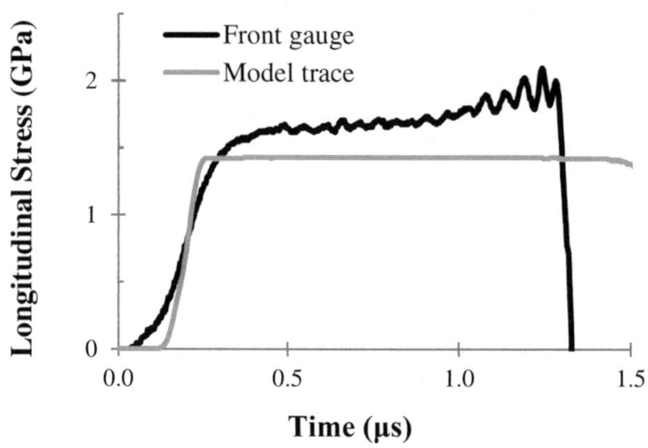

Figure 2. Shock traces from experimental data and numerical model for highest pressure with 230 ms^{-1} impact velocity.

In order to maintain *Artemia* under anaerobic conditions, it was ensured that the nauplii and cysts remained sealed inside their capsules for < 2 hr for each experiment. Control samples were encapsulated for the same period of time. The samples were immediately examined post-shock using light microscopy to search for any visible external damage to the cysts and nauplii. They were then studied after 24 and 48 hr to determine hatching rates and observe motility of the nauplii.

3. Results

Following incubation of the shocked cysts, the hatching rates (i.e., emergence of the embryo from the hatching membrane) were determined after 24 hr and 48 hr in each case (table 1, figure 2). The 'breaking' stage of the cysts was also recorded to observe the emergence of the embryo from the shell (figures 3 and 4). These rates were systematically higher than the hatching rates in each case. At the lowest pressure reached, 0.78 GPa, a breaking rate of 75 % was attained after 48 hr, while applying a pressure of 0.96 GPa resulted in a breaking rate of 70 %. The maximum pressure achieved during shock loading was 1.5 GPa and this led to the lowest breaking rate of 43 %. The hatching rates also showed similar decrease (26, 23 and 18%) with increasing peak pressures. One sample from each different shock pressure was also analysed after 14 days of incubation to search for any further hatched nauplii. However, no additional hatching was observed. Peak temperatures attained during the shock runs were determined through the numerical models (table 1). These ranged from 41°C at 0.78 GPa to 50°C at 1.5 GPa.

Table 1. Breaking and hatching rates of *Artemia salina* cysts observed for each shock pressure after both 24 and 48 hr

Impact velocity (ms^{-1})	Pressure (GPa)	Cyst breaking frequency (%)		Hatching frequency (%)		Shock peak temperature (°C)
		24 hr	48 hr	24 hr	48 hr	
135	0.78	60	75 (± 3)	16	26 (± 3)	41
153	0.96	59	70 (± 4)	12	23 (± 3)	42
230	1.5	30	43 (± 3)	4	18 (± 2)	50

In addition, hatched nauplii were studied following shock loading at 0.78 GPa but only at the lowest pressure due to the temperature increase of the samples during the shock. The motility of their antennae was analysed after shock and observed after 24 and 48 hr for 1 min each. 20% demonstrated motility of their antennae throughout this time, but with a noticeable lack of overall motion compared to nauplii hatched from shock loaded cysts. Many did not appear to have any significant structural damage post shock, as illustrated by figure 5. Nauplii that were found to be unmoving but appeared to be structurally intact were left for up to 14 days in order to check for any subsequent movement of their antennae; however, none was observed in any case.

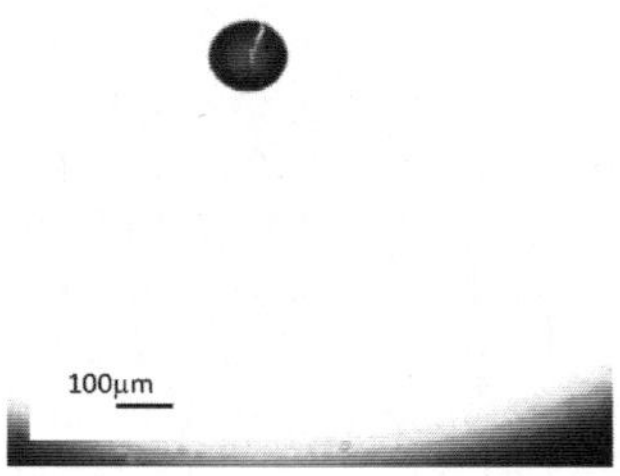

Figure 3. Breaking cyst after shock loading at 0.96 GPa.

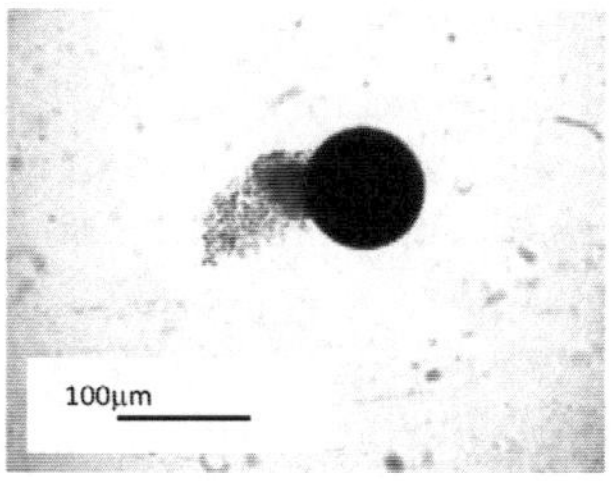

Figure 4. Emerging embryo from cyst after shock loading at 0.96 GPa.

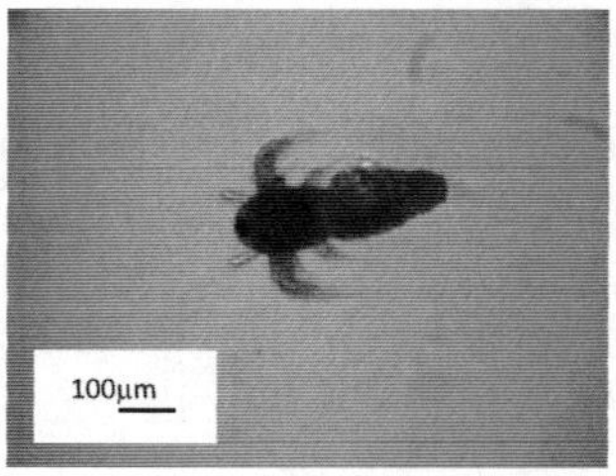

Figure 5. Hatched nauplius after shock loading at 0.78 GPa.

4. Discussion

Exposure of *Artemia* cysts to shock pressures of 0.78 and 0.96 GPa caused a decrease in both cyst breaking and hatching rates, whereas a much larger effect was observed following 1.5 GPa shock. While the nauplii that hatched successfully following 1.5 GPa exposure appeared to be largely undamaged, a small number demonstrated impaired motility. Our results confirm the observation that *Artemia* cysts appear to be significantly more sensitive to quasi-one-dimensional shock compression than static pressurisation into the multi-GPa range. Our experiments resulted in considerably greater hatching rates than had been observed in previous shock experiments, carried out at substantially lower peak shock pressures. However, the previous work used non-planar wave fronts that could have played a role in the reduced hatching success. More studies will be required to elucidate the relative roles played by biochemical changes and the mechanical response of the cyst coating to shock *vs* static pressurisation in determining the rate of successful hatching, and subsequent properties of the hatched nauplii. In order to avoid complications due to heating of the samples during shock compression, we only carried out a single series of experiments subjecting hatched nauplii to 0.78 GPa pressure. Studies of antenna motion indicated that motility was considerably reduced.

In the case of the cyst shells, visible light microscopy showed that in some instances, the breaking stage was initiated, but was not completed (figure 3). Little damage appeared to have occurred to the external structure of the cysts, even for those that did not hatch after 48 hr. This could indicate that delayed hatching or possible death of the embryos contained inside the cysts might be due to some internal biochemical mechanism controlling their shock response, such as particular genes being activated in the encysted embryos but not in the hatched nauplii. Some such genes involved in the production of proteins for embryo protection have already been identified, including p26, but it is not yet known how these genes respond to pressure loading. It is clear that future work lies with studying the internal mechanisms that govern *Artemia salina* response to shock as well as static pressure.

5. Conclusions

By applying quasi-one-dimensional shock wave pressures in the range of 0.78-1.5 GPa, breaking and hatching rates of *Artemia salina* cysts were found to decrease with increasing pressure, unlike static compression results that maintained 80-90% hatching rates after exposure to 7.5 GPa for up to 48 hr. The enhanced effect of shock *vs* static pressurization in reducing *Artemia* cyst hatching is in general agreement with previous studies, although the hatching rates found here were significantly greater than those seen following shock from underwater detonation waves. This implied that the nature of the wave front must play an important role in survival and hatching probability of the cysts, certainly by affecting the mechanical stress fields applied to the cyst envelope. It is apparent that shock and static pressurisation also certainly affect the biochemical and biophysical state of the encysted embryo and these effects could be studied by future genomic and proteomics investigations.

Acknowledgements

The authors would like to thank Mr Andrew Roberts and Dr David Wood for their invaluable assistance in carrying out these experiments. Contributions of PFM and RH were supported by grant RPG-350 from the Leverhulme Trust.

References

[1] Gajardo G M and Beardmore J A 2012 *Front. Physiol.* **3** 1
[2] Grabner M, Wieser, W and Lackner R 1981 *Aquaculture* **26** (1–2) 85
[3] Milhem M M, Al-Hiyasat A S and Darmani H 2008 *J. Appl. Oral Sci.* **16** (4) 297
[4] Caldwell G S, Bentley M G and Olive P J W 2003 *Toxicon.* **42** (3) 301
[5] Liu Y-L, Zhao Y, Dai Z-M, Chen H-M and Yang W-J 2009 *J. Biol. Chem.* **284** (25) 16931

[6] Miller D and McLennan A G 1988 *J. Therm. Biol.* **13** (3) 119
[7] Stabili L, Miglietta A M and Belmonte G 1999 *J. Exp. Mar. Bio. Ecol.* **237** (2) 291
[8] Ono F, Minami K, Saigusa M, Matsushima Y, Mori Y, Takarabe K, Saini N L and Yamashita M 2010 *J. Chem. Phys. Solids* **71** (8) 1127
[9] Udagawa Y and Suzuki M 2013 *Trans. Japan Soc. Mech. Eng. Part B* **79** (801) 804
[10] Leighs J A, Appleby-Thomas G J, Wood D C, Goff M J, Hameed A and Hazell P J 2014 *J. Phys. Conf. Ser.* **500** (18) 182026
[11] Dasgupta R 2013 *Rev. Mineral. Geochem.*, **75** (1) 183
[12] Rios A C 2015 *PNAS* **112** (3) 643
[13] Leighs J A, Appleby-Thomas G J, Stennett C, Hameed A, Wilgeroth J M and Hazell P J 2012 *Rev. Sci. Instrum.* **83** (11) 115113

IOP Conf. Series: Journal of Physics: Conf. Series **950** (2017) 042003 doi:10.1088/1742-6596/950/4/042003

Effects of pressure, temperature, and treatment time on oscillatory rheological properties and non-isothermal gelatinization kinetics of chickpea (Cicer arietinum L.) flour slurry

W Canet[1], F J Cuesta[2], R Fuentes[1] and M D Alvarez[1]

[1]Department of Characterization, Quality, and Safety, Institute of Food Science, Technology and Nutrition (ICTAN-CSIC), José Antonio Novais 10, 28040 Madrid, Spain
[2]Department of Products, Institute of Food Science, Technology and Nutrition (ICTAN-CSIC), José Antonio Novais 10, 28040 Madrid, Spain

E-mail: wenceslao@ictan.csic.es

Abstract. The effect of high hydrostatic pressure (HHP), temperature at pressurization, and treatment time at three levels (200, 400, 600 MPa; 10, 25, 50 °C; 5, 15, 25 min) on the gelatinization of chickpea flour (CF) slurry during non-isothermal heating from 25 to 95 °C and gel rheological properties after subsequent cooling to 25 °C was investigated. CF gelatinization kinetics from the cross-over of elastic modulus (G') and viscous modulus (G'') to 95 °C were considered for rate estimation. Zero-order reaction kinetics adequately described the CF gelatinization process. Structure development rate (dG'/dt) is described by two exponential functions with activation energies ranging from 51.2 to 576.6 and –128.9 to 538.9 kJ mol^{-1} for downward and upward gelatinization curves, respectively. Changes in dG'/dt vs. temperature seem to be closely related to the degree of gelatinization induced by HHP pre-treatment.

1. Introduction

All the factors such as pressure, temperature at pressurization, treatment time, starch type, and other major components (protein and fiber) interact and contribute to the effect of pressure-induced gelatinization of starch. There are also significant differences in the structural and rheological properties between heated and pressurized starches [1]. In addition to starch, CF slurries contain a relatively large amount of protein (20.64 g 100 g^{-1} of flour), which could give a second endothermic peak on heating in water [2]. Differential scanning calorimetry (DSC) measurement could not detect gelatinization temperature for all studied specimen and the endotherm depends upon starch to water ratio and some other factors [3]. According to the author just cited, rheometric measurement (small amplitude oscillation shear (SAOS) measurement) has advantages over DSC and can successfully detect the gelatinization temperature in addition to the gel point during non-isothermal heating and provide reliable information on gelatinization reaction kinetics. A suitable selection of HHP treatment at appropriate levels can produce chickpea pastes that possess the desirable handling properties in terms of sheeting, flattening, rolling, and taking form leading to preparation of products. Differences between pressure- and temperature-induced gelatinization offer benefits with respect to new product development [1].

The objective of this work was to evaluate the combined effects of treatment pressure (0.1, 200, 400, and 600 MPa), temperature at pressurization (10, 25, and 50 °C), and treatment time (5, 15, and 25 min) on the subsequent temperature-induced gelatinization of CF slurry. Unpressurized and HHP-treated CF slurries were then pasted under non-isothermal (from 25 up to 95 °C) heating conditions and the kinetics of CF gelatinization were established by an oscillatory rheological approach. After heating, CF slurries were cooled to 25 °C and then evaluated for changes in their viscoelastic measurements.

2. Materials and Methods

2.1. Slurry used in the experiments

Spanish chickpea (*C. arietinum* 'Castellano') flour was a commercially available product donated by the Los Pisones flour milling company (Zamora, Spain). CF slurries were prepared at a concentration of 1:5 flour-to-water ratio.

2.2. High hydrostatic pressure treatment

CF slurries (200 ml) were vacuum packaged in a very low gas permeability bag type, Doypack® (Polyskin XL, Flexibles Hispania, S.L.). HHP treatment was performed using a Stansted Fluid Power Iso-lab 900 High Pressure Food Processor (Model: FPG7100:9/2C, Stansted Fluid Power Ltd., Harlow, Essex, UK). Four packed samples were inserted into the pressure unit filled with water, and treated at pressures of 200, 400, or 600 MPa. Pressure was increased at 500 MPa min^{-1} and maintained at the desired level for holding times of 5, 15, and 25 min; the decompression time was less than 4 s. The temperature of the pressure unit vessel was thermostatically controlled at 10, 25, and 50 °C throughout the different treatment combinations. Representations of the variation of pressure and temperature vs. time during HHP treatment carried out at 600 MPa and 25 °C for 15 min is shown in figure 1. Increases of up to a maximum of 8 °C ± 1 °C, 13 °C ± 2 °C, or 17 °C ± 2 °C at 200, 400, or 600 MPa, respectively, due to compressive heating, were observed in the temperature of the pressuring fluid, but they were transient and equilibrated at 10, 25, and 50 ± 2.5 °C during the holding period at those pressure levels. After HHP treatment, samples were immediately stored in a refrigerator at 4 °C for 24 h before further use.

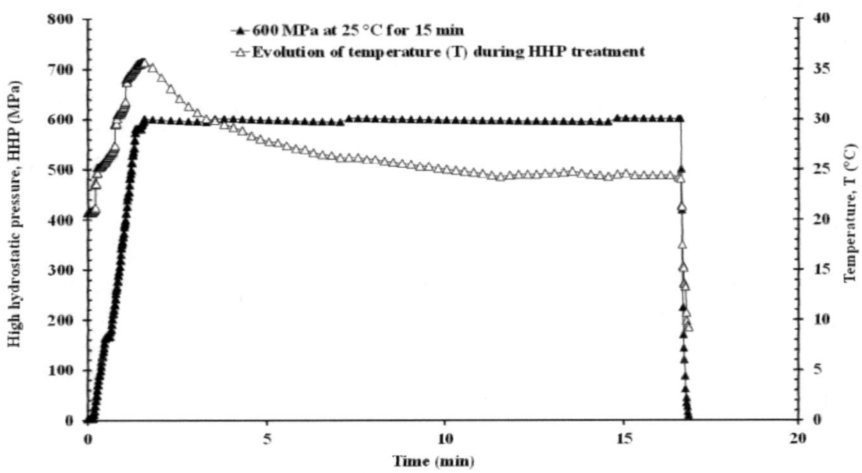

Figure 1. Representation of the temperature and pressure vs. time variation during HHP treatment (600 MPa at 25 °C for 15 min).

2.3. Rheological measurements

A Kinexus pro rotational rheometer (Malvern Instruments Ltd, Worcestershire, UK) was used to conduct SAOS measurements using concentric cylinder geometry. Temperature sweeps were recorded from 25 to 95 °C at a heating rate of 2 °C min^{-1} and at a frequency (*f*) of 1 Hz, with controlled shear stress at 1 Pa. Values of elastic modulus (*G*', Pa) and viscous modulus (*G*", Pa) were recorded. Onset elastic modulus at 25 °C ($G'_{initial}$, Pa), starting gel point temperature (T_{gel}, °C), considered as the temperature where there was cross-over of *G*' and *G*", temperature where *G*' achieved its maximum value ($T_{G'max}$, °C), and G'_{max} (Pa) value itself were also obtained. After heating, the sample was immediately cooled down to 25 °C. Mechanical spectra at 25 °C in the linear region at variable angular frequency (ω) from 1 to 100 rad s^{-1} (~0.16 to 16 Hz) were also recorded.

3. Results and Discussion

3.1. Kinetic modeling of rheological data during non-isothermal heating

The non-isothermal kinetic is based on a combination of the Arrhenius equation and the time-temperature relationship [4]. The kinetic equation is converted in terms of rheological parameters (*G*' and *dG*'), as described by other authors [5-7]. However, in this study, it was observed that the structure development rate was better described as the sum of two reaction processes for the case of $n = 0$. The constants *a*, *b*, *c*, and *d* from this sum of two reactions were determined in each process by nonlinear regressions. Therefore, the structure development rate (*dG'/dt*) is described by two exponential functions, and zero-order reaction kinetics adequately described the CF gelatinization process. A coming complete paper will illustrate the detailed description of the kinetic modeling developed to obtain the kinetic parameters. By representing both *G*' and *dG'/dt* vs. temperature, it was possible to appreciate a significant change in the profiles of the CF slurries as the pressure applied increased. Figure 2 shows the changes in both *G*' and *dG'/dt* vs. temperature (°C) for four different CF slurries: untreated (figs. 2a, 2e), pre-treated with 200 MPa at 50 °C for 5 min (figs. 2b, 2f), pre-treated with 400 MPa at 25 °C for 5 min (figs. 2c, 2g), and pre-treated with 600 MPa and 50 °C for 25 min (figs. 2d, 2h). For the samples pre-treated with 0.1, 200 and 400 MPa, a continuous increase in *G*' was observed with increasing temperature from the T_{gel} up to 95 °C. On the other hand, for the unpressurized CF slurry, *dG'/dt* decreased with temperature, and exhibited a minimum value (T_0) that was followed by a continuous increase upon further heating. CF slurries pre-treated with 200 MPa at 50 °C for 5 min and with 400 MPa at 25 °C for 5 min followed the same pattern, although the increase after T_0 was smaller with increasing pressure, probably in proportion to the extent of HHP-induced starch gelatinization. T_0 was considered as the end point for the first downward curve and the beginning of the second curve, being the inflection point of the experimental curve *dG'/dt* vs. temperature. In addition, generally T_0 was higher for pressurized CF slurries as compared with unpressurized sample (84.3 °C, fig. 2e). Except for HHP treatments at 600 MPa and 50 °C for 15 and 25 min, at constant temperature and treatment time, T_0 values tended to increase with increasing pressure. Indeed, as stated by Ahmed et al. [8] for the case of the change of *G*' vs. temperature during non-isothermal heating of blended dough, the curve above the T_0 resembles the gelatinization similar to DSC curve.

On the other hand, the T_0 values increased with increasing pressure up to 400 MPa. As a result, when heating pre-treated CF slurries subjected to 200 and 400 MPa, at the various temperature and time combinations, the second phase of the gelatinization process took place in a narrower temperature range than when heating at ambient pressure. As pre-treatment with HHP changed the granule structure and crystallinity in varying degrees, it changed the behavior of CF slurries upon subsequent pasting. In contrast, it was found that after pressurization at 600 MPa and 50 °C for 25 min, *G*' increased markedly up to 82 °C, followed by a drop in magnitude. As a consequence, for this CF slurry, the *dG'/dt* was positive up to $T_{G'max}$ (82 °C), where it became negative. At first, this change in the behavior of CF slurry pre-treated with 600 MPa and 50 °C for 25 min may possibly indicate that complete starch gelatinization was induced by this severe HHP treatment.

IOP Conf. Series: Journal of Physics: Conf. Series **950** (2017) 042003 doi:10.1088/1742-6596/950/4/042003

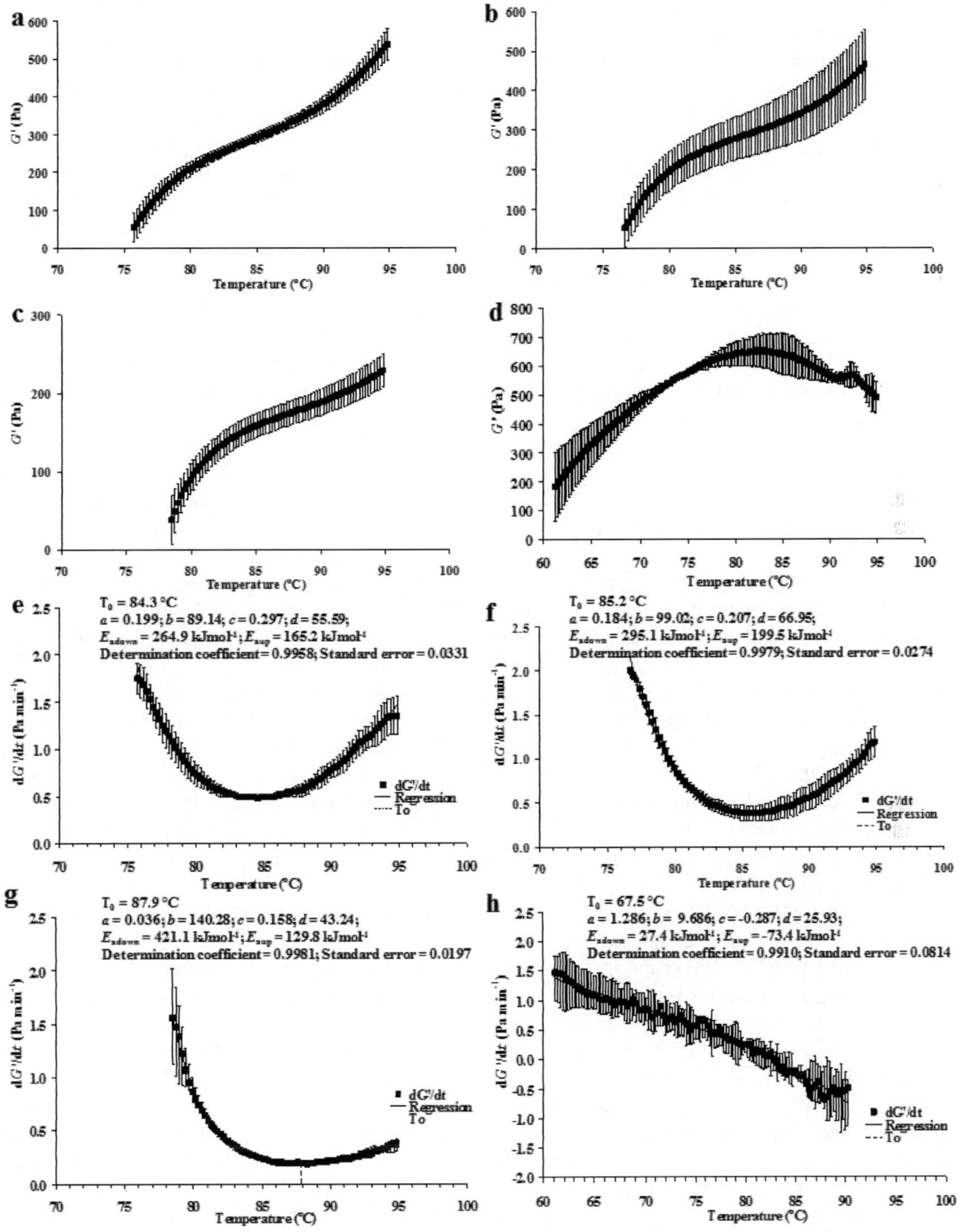

Figure 2. Variation of both *G'* and *dG'/dt* vs. temperature (from starting gel point temperature to the temperature where *G'* achieved its maximum value) for: (a, e) unpressurized CF slurry; (b, f) CF slurry pre-treated at 200 MPa and 50 °C for 5 min; (c, g) CF slurry pre-treated at 400 MPa and 25 °C for 5 min; (d, h) CF slurry pre-treated at 600 MPa and 50 °C for 25 min. Mean values of six measurements ± error bars.

A previous study also showed that complete gelatinization (100%) was associated with CF slurries at the same concentration after HHP with 600 MPa at 25 °C for 15 min [9]. Nevertheless, likely protein denaturation, aggregation or gelation as well affected the behavior of this sample. From the sum of the two reaction processes, the corresponding activation energies for the downward (E_{adown}) and upward (E_{aup}) dG'/dt vs. temperature curves were obtained for all the cases. The goodness of fit for the two exponential functions was very high, with the coefficients of determination (R^2) and standard errors (SE) ranging between 0.985 and 1.000, and between 0.017 and 0.126, respectively. As could be expected from the shape of the curves, the worst fits of experimental data corresponded to the CF samples HHP-treated at 600 MPa and 50 °C for 25 min. The E_{adown} and E_{aup} values for the untreated case were 247.8 kJ mol^{-1} and 173.5 kJ mol^{-1}, respectively, evidencing that for single temperature-induced gelatinization of unpressurized CF slurry a higher activation energy is required for granule swelling than for leaching of solubilized amylose/amylopectin. Remarkably, the lowest E_{adown} and E_{aup} values were obtained for CF slurries treated with 600 MPa at 50 °C for 25 min. Likewise, a significant increase in E_{adown} was observed with increasing pressure for CF slurries pre-treated at 10 °C for 5 min and at 25 °C for 5, 15, and 25 min. It seems logical that higher minimum energy is required to start the swelling (pre-gelatinization) of an increasingly smaller amount of remaining ungelatinized starch granules with increasing pressure.

3.2. Effect of pressure, temperature, and treatment time on kinetic and gelatinization parameters of CF slurry under non-isothermal heating and on rheological properties of heat-induced CF gel after cooling to 25 °C

G'_{max} values decreased sharply between 200 and 400 MPa and increased between 400 and 600 MPa, indicating that there might be a critical pressure level between 400 and 600 MPa for the HHP-treated CF slurries. In CF slurry, the formation of protein networks induced by HHP might in part explain the increase of the solid character observed in the samples pressurized at the highest pressure level. On the other hand, after cooling down to 25 °C, pressure combined with heating resulted in reduction in CF paste rigidity. Increasing temperature during pressurization also increased the degree of gelatinization of slurries pressurized at 400 and 600 MPa, and consequently the CF gels induced subsequently had inferior mechanical strength. However, the temperature effect was more significant at constant pressure of 600 MPa for 15 min. The elasticity of CF slurry and gel treated with 200 and 400 MPa at 10, 25, and 50 °C decreased with increasing holding time during pressurization in proportion to the extent of HHP-induced gelatinization. This result corroborates that HHP-induced gelatinization is also a time-dependent process [10, 11].

In contrast, rigidity also increased at higher values of pressure, temperature and holding time. This result might reflect that a breakdown of granules caused an increase in elasticity and viscosity of samples HHP-treated at 600 MPa and 50 °C for 25 min with significant leaching of amylose, and therefore a higher extent of recrystallization occurring outside the starch granules. The formation of protein networks induced by HHP might in part explain the increase of the solid character observed in this sample. Disulfide cross-linking of proteins within rice granules has also been linked with altered pasting properties and waxy corn starches with elevated internal protein content [12].

Between kinetic and gelatinization parameters and rheological properties, the highest positive correlation was found between G'_{max} and G'' (r = 0.809), evidencing that the higher the maximum value of elasticity reached during non-isothermal heating, the higher the viscosity of the gels was after cooling. However, correlations between T_{gel} and $T_{G'max}$ and G', G'', and η^* were all significant but negative. The highest correlation was between T_{gel} and G'' (r = –0.567), reflecting that the lower the onset temperature of gelatinization was, the higher the viscoelastic properties of the CF gels were.

4. Conclusion

A two-step mechanism for heat-induced gelatinization process of HHP-treated CF slurry was observed from the dG'/dt vs. temperature. Activation energy values (E_{adown} and E_{aup}) derived for the two phases seem to be related to the degree of swelling (E_{adown}) and amylose leaching together with amylopectin breakdown (E_{aup}), both induced by the HHP treatment. Although the effect of pressure was dependent on temperature and time, the impact of pressure on CF gelatinization was much more significant. The HHP-treated CF slurry exhibited solid-like behavior upon heating and cooling. Increasing temperature increased the degree of gelatinization of slurries pressurized at 400 and 600 MPa, and accordingly the CF gels induced subsequently had inferior mechanical strength. Elasticity of CF slurries treated at 200 and 400 MPa and 10, 25, and 50 °C decreased with increasing holding time in proportion to the extent of HHP-induced gelatinization. Pre-treatment with 200 and 400 MPa, at constant temperature and treatment time, caused restricted swelling (as indicated by an increase in E_{adown} and decrease in G'_{max} values) and restricted granule disintegration (as indicated by a higher T_{gel}). An increased rapid retrogradation of leached amylose (as indicated by a higher increase in G' values on cooling) due to disintegration might explain the significant increase in either G'_{max} or G' upon cooling for CF slurries pre-treated at 600 MPa. Results obtained for CF slurries pre-treated with 600 MPa and 50 °C for 25 min could be explained by various mechanisms and probably one mechanism is not the sole origin for the altered rheological properties. Further research is necessary between 400 and 600 MPa, and between 25 and 50 °C. Overall, CF starch granules with different degree of gelatinization can be obtained by HHP pre-treatment with 200 and 400 MPa at 10 to 50 °C and 5 to 25 min, to obtain CF-based products with easier handling properties.

Acknowledgments
Authors wish to thank the Spanish Ministry of Economy and Competitiveness for financial support (AGL2011-28569).

References

[1] Vallons K J R, Ryan L A M and Arendt E K 2014 *Crit. Rev. Food Sci. Nutr.* **54** 399-409
[2] Meares C A, Bogracheva T Y, Hill S E and Hedley C L 2004 *Starch/Stärke* **56** 215-224
[3] Ahmed J 2012 *J. Food Eng.* **109** 321-328
[4] Rhim J W, Nunes R V, Jones V A and Swartzel K R 1989. *J. Food Sci.* **54** 446-450
[5] Ahmed J, Ramaswamy H S and Alli I 2008 *Food Hydrocoll.* **22** 278-287
[6] Yoon W B, Gunasekaran S, Park J W 2004 *J. Food Sci.* **69** E338-E343
[7] Alvarez M D, Fuentes R, Olivares M D, Cuesta F J, Canet W 2014 *J. Food Eng.* **136** 9-18
[8] Ahmed J, Almusallam A S, Al-Salman F, AbdulRahman M H, Al-Salem E 2013 *LWT-Food Sci.Technol.* **51** 409-416
[9] Alvarez M D, Fuentes R, Olivares M D, Canet W 2014. *Innov. Food Sci. Emerg. Technol.* **21** 12-23
[10] Bauer B A, Knorr D 2005 *J. Food Eng.* **68** 329-334
[11] Stolt M, Oinonen S, Autio K 2001 *Innov. Food Sci. Emerg. Technol.* **1** 167-175
[12] Han X Z, Campanella O H, Guan H, Keeling P L, Hamaker B 2002 *Carbohydr. Polym.* **49** 323-330

IOP Conf. Series: Journal of Physics: Conf. Series **950** (2017) 042004 doi:10.1088/1742-6596/950/4/042004

Effect of solvent type and high pressure treatment on the extraction of *Gomphrena globosa* L. bioactive compounds

L Fernandes[1,2,3], E Ramalhosa[1], J A Pereira[1], S Casal[2] and J A Saraiva[3]

[1] Mountain Research Centre (CIMO) – School of Agriculture, Polytechnic Institute of Bragança, Campus de Stª Apolónia, Apartado 1172, 5301-855 Bragança
[2] LAQV@REQUIMTE/Laboratory of Bromatology and Hydrology, Faculty of Pharmacy, Porto University, R. D. Manuel II, Apartado 55142, 4051-401 Porto
[3] QOPNA - Organic Chemistry, Natural and AgriFood Products, University of Aveiro, Campus Universitário de Santiago, 3810-193 Aveiro.

E-mail: jorgesaraiva@ua.pt

Abstract. The present study aimed to compare the influence of different extraction solvents (water, methanol, water:acetone (6:4, v/v)), methods (heating (37 °C, 30 min) or high pressure (HP) (300 or 500 MPa) and extraction time (7.5 or 15 min)) on flavonoids, hydrolysable tannins and antioxidant activity (Total Reducing Capacity (TRC), DPPH Free Radical Scavenging Activity and Reducing Power) of *Gomphrena globosa* L. flower extracts. The water:acetone extracts obtained by heating had the highest values of flavonoids, hydrolysable tannins and antioxidant activity. When applying HP, variable results were obtained. Still, the application of HP to water allowed to extract more hydrolysable tannins, as well as to obtain extracts with higher antioxidant activity than with heating, but no significant alterations were observed with methanol. In conclusion, both solvent and extraction method influence the content of bioactive compounds, being HP treatment a promising method to obtain enriched aqueous extracts in line with the principles of green-chemistry.

1. Introduction

Gomphrena spp. are edible, ornamental and medicinal plants commonly known as Globe Amaranth or Bachelor Button. The flowers of *Gomphrena globosa* L. have medicinal potential because they are rich in bioactive compounds such as betacyanins, betalaines and flavonoids. This flower has been studied by some researchers, who had identified phenolic compounds such as hydroxycinnamic acids and flavonoids [1], being kaempferol-3-*O*-(6-rhamnosyl) hexoside and kaempferol-3-*O*-hexoside the main compounds determined in decoction extracts [2].

In recent years the search for cheap and abundant sources of natural antioxidants has been increasing. There are many reports about different extraction methods for edible flowers such as Soxhlet [3-4], ultrasonics [4], supercritical fluid extraction [5] and solid phase microextraction [6]. These methods are based on an appropriate selection of solvent and energy input to increase chemical solubility and mass transfer rate [7]; however, high energy consumption, long extraction times and relatively low extraction yields can be observed [8]. High pressures (HP) have been recently applied to extract bioactive ingredients from plant materials [8, 9], taking advantage of time saving, higher extraction yields, fewer impurities in the extraction solution, minimal heat and thermal degradation of the activity and structure of bioactive components.

Few studies have been performed on the antioxidant activity and bioactive compounds of *Gomphrena globosa* L. specie [1, 10, 11]; however none of these studies had applied HP.

In order to increase the knowledge on this subject, the present study aimed to perform a physicochemical characterization of dry *Gomphrena globosa* L. flowers obtained in a Portuguese herbal shop, followed by a comparison on the content of flavonoids, hydrolysable tannins, and antioxidant activity (Total Reducing Capacity, DPPH free radical scavenging activity and Reducing Power) of flower extracts obtained with different extraction solvents (water, methanol, water:acetone (6:4; v/v)) and by different extractive methods (heating under agitation at 37 °C for 30 min or high pressures 300 or 500 MPa at different times 7.5 or 15 min).

2. Materials and methods

2.1. Plant material

Dry Globe Amaranth flowers (*Gomphrena globosa* L.) were purchased from a local herbalist in Bragança city (Portugal) in bags of 40 grams (Figure 1).

Figure 1. *Gomphrena globosa* L. flowers.

2.2. Physicochemical characterization

The following parameters were evaluated in the flowers: weight, width, length and protein content. The width and length of ten flowers were measured with a digital caliper (Powerfix, Leeds, UK) and the weight in a digital balance (Kern, Balingen, Germany).

Protein content was determined by the Kjeldahl method using the AOAC procedure [12] using one gram of sample. The content of nitrogen was multiplied by the coefficient 6.25 and expressed as crude protein in %.

2.3. Flowers extraction

2.3.1. Heating under agitation. Extraction was based on the method described by Li et al. [13] with slight modifications. Approximately, 5 grams of flowers were crushed and mixed with 100 mL of solvent (water, methanol and water:acetone (6:4, v/v)), at 37 °C, for 30 min under agitation (IKA, RCT Model B, Staufen, Germany) at 900 rpm. Subsequently, the supernatant was collected and the process was repeated twice. Then, the combined extracts were filtered, frozen and placed in the lyophilizer (Scanvac, Coolsafe, Lynge, Denmark) for 2 days. In the case of the methanol extracts, these were placed on a rotary evaporator at 40 °C. The extracts obtained were weighed and redissolved with the different solvents to a concentration of 50 mg extract/mL, covered with aluminium foil and stored under freezing until further analysis.

2.3.2. High pressure (HP). Five-gram of flowers powder was mixed with 100 ml of each solvent, as reported above, in polyethylene bags. The HP treatments were carried in an Hiperbaric (Burgos, Spain) equipment with 55 L of vessel volume. After the selected high hydrostatic pressure treatments (300 and 500 MPa and holding times of 7.5 and 15 min) at room temperature, the mixtures were filtered through filter paper. Assays were done in duplicate. The water and water:acetone extracts were preserved as indicated above.

2.4. Flavonoids and hydrolysable tannins
Total flavonoid and hydrolysable tannins contents were determined by the method described by Viuda-Martos et al. [14] and Elfalleh et al. [15] respectively.

2.5. Antioxidant activity
Antioxidant activity of the flower extracts was determined by the following methods: total reducing capacity (TRC), DPPH free radical scavenging activity and Reducing Power, by the procedure described by Delgado et al. [16] with some modifications.

2.6. Statistical analysis
The Statistic SPSS software, version 18.0 (SPSS Inc., Chicago, USA), was used for the statistical treatment of the data.

3. Results and discussion
3.1. Physicochemical characterization
The dry *Gomphrena globosa* flowers had a mean width of 1.47 cm and length of 1.55 cm (Table 1), with a low average weight of 0.26 g each. The flower heads were colourful with purple shades (Figure 1).

Table 1. Physicochemical characterization of *Gomphrena globosa* flowers, n=10

Parameter	Values	Min	Max
Weight (g)	0.26±0.07	0.16	0.35
Width (cm)	1.47±0.36	1.40	1.70
Length (cm)	1.55±0.14	0.90	1.90
Crude protein (%, dry weight)	9.41±0.12	---	----

Gomphrena globosa flowers had a crude protein content of 94.1 g/kg (dry weight) that was within the range of the three edible flowers studied by Navarro-González et al. [17] (79-186 g/kg dry weight for *Tagetes erecta* and *Tropaeolum majus*, respectively).

3.2. Flavonoids and hydrolysable tannins
The extracted flavonoids and hydrolysable tannins showed significant differences among the five extraction conditions, as well as between solvents (Figure 2).

The water:acetone extracts obtained by heating had the highest amounts of flavonoids and hydrolysable tannins (Figure 2A, B). So, the highest molecular weight compounds such as tannins and flavonoids are better extracted by using aqueous organic solvent solutions.

Among HP treatments, the highest value of flavonoids was also obtained with water:acetone (P300/15 min with a value of 7.4 mg of QE/g dry flower). Concerning methanol, the extracts obtained after classical heating were not significant different to those obtained with HP. For water the highest amounts were extracted with HP treatments, except for the binomial P500/15 min that gave the lowest flavonoid content; however, the HP results were not statistically different to that obtained with the heating treatment. Thus, the solvent effect is more relevant than the treatment applied, showing that different solvents extract distinct compounds. With the exception of water:acetone solvent, similar concentrations of flavonoids were obtained between HP treatment and the traditional method with water (green solvent) or methanol, suggesting similar extraction efficiencies. On contrary, when using water:acetone solvent the heating treatment gave the best results.

For hydrolysable tannins, when performing methanolic extractions, the highest value was obtained with the heating treatment. Within the HP treatments, the highest values were observed for longer times, with the binominals P300/15 and P500/15. For water, the highest value was obtained with the P300/15 HP treatment. Nevertheless, the extracts obtained with the other HP treatments were not significantly different to that prepared by heating under agitation.

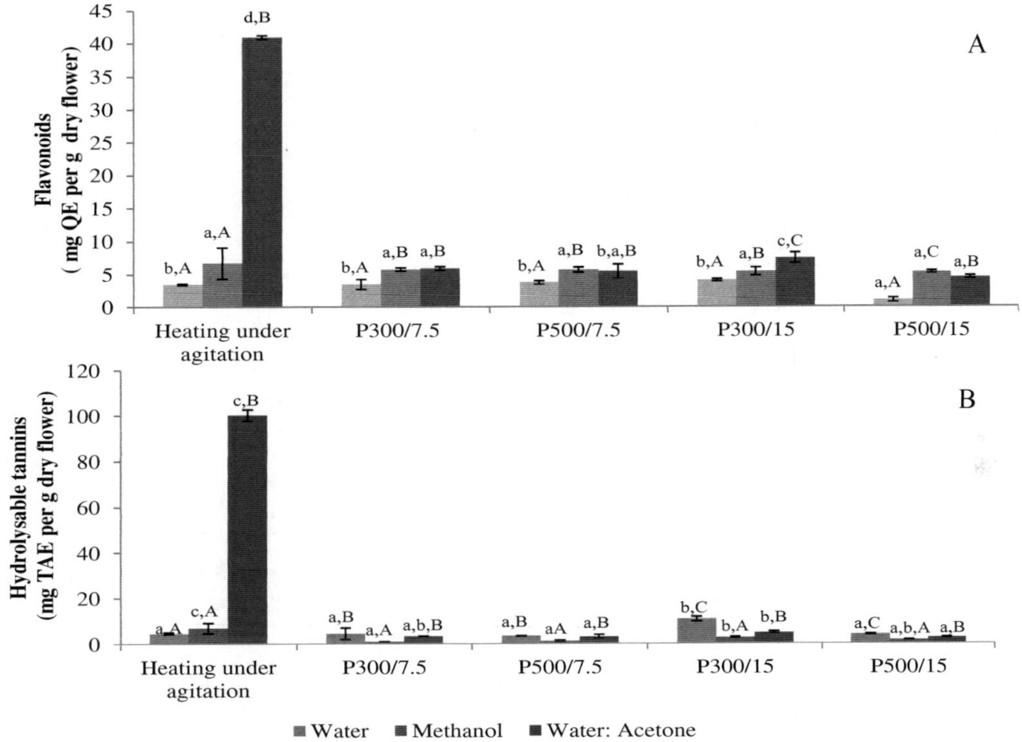

Figure 2. Flavonoids and hydrolysable tannins contents of flower extracts obtained after application of different extraction methods and solvents. Values with the same lowercase letter are not statistically different (p>0.05) in what concerns treatment into the same solvent. Values with same uppercase letter are not statistically different (p>0.05) in what concerns solvent into the same treatment.

3.3. Antioxidant activity

3.3.1. Total Reducing Capacity (TRC). The TRC of *Gomphrena globosa* L. flowers extracts varied between 1.50 to 48.60 mg GAE/g dry flower, obtained after HP treatment (P500/15) and heating treatment with water and water:acetone (6:4, v/v), respectively (Table 2). Among the solvents used, water:acetone (6:4, v/v) extracts showed the highest value of TRC, followed by methanol by heating. Within the HP treatments, the extracts with the highest TRC were again obtained with water:acetone at P300/15. These results suggested that a mixture with water and an organic solvent such as water:acetone, can be the best way to extract compounds with Reducing Capacity in this flower specie, such as phenols. This result is in agreement with Kuźma et al. [18], who reported that higher phenolic contents and antioxidant activity were obtained with aqueous organic solvents than with the respective absolute organic solvents, as methanol, ethanol, acetone and ethyl acetate. By adding water the polarity increases and so the aqueous organic solvents are able to extract both high and low polarity compounds [19]. For methanol the highest value was observed with the heating treatment. Regarding water, the HP and heating treatments were not statistically different.

Table 2. Total Reducing Capacity, EC_{50} DPPH and EC_{50} Reducing Power values of flower extracts obtained with different solvents and extraction methods.*

Assay	Methods	Water	Methanol	Water:acetone
TRC (mg GAE per g dry flower)	Heating	$1.73\pm0.07^{a,A}$	$22.6\pm0.79^{c,B}$	$48.60\pm1.87^{c,C}$
	P300/7.5	$1.69\pm0.26^{a,A}$	$2.55\pm0.25^{b,B}$	$7.58\pm0.31^{a,C}$
	P500/7.5	$1.57\pm0.12^{a,A}$	$2.24\pm0.22^{a,b,B}$	$7.25\pm0.94^{a,C}$
	P300/15	$1.67\pm0.11^{a,A}$	$1.90\pm0.20^{a,A}$	$9.99\pm0.70^{b,B}$
	P500/15	$1.50\pm0.05^{a,A}$	$2.03\pm0.26^{a,b,B}$	$6.31\pm0.47^{a,C}$
EC_{50} DPPH (mg extract per mL)	Heating	$3.74\pm0.20^{c,C}$	$1.79\pm0.18^{a,B}$	$0.95\pm0.04^{a,A}$
	P300/7.5	$2.44\pm0.49^{b,B}$	$2.21\pm0.48^{a,B}$	$1.10\pm0.02^{b,A}$
	P500/7.5	$1.74\pm0.28^{a,B}$	$2.90\pm0.31^{b,C}$	$0.97\pm0.12^{a,b,A}$
	P300/15	$2.09\pm0.24^{a,b,B}$	$2.71\pm0.27^{b,C}$	$0.97\pm0.05^{a,b,A}$
	P500/15	$1.58\pm0.03^{a,B}$	$1.77\pm0.32^{a,C}$	$0.87\pm0.08^{a,A}$
EC_{50} Reducing Power (mg extract per mL)	Heating	$3.01\pm0.01^{b,C}$	$1.12\pm0.01^{a,B}$	$0.08\pm0.01^{a,A}$
	P300/7.5	$2.34\pm0.08^{a,C}$	$1.72\pm0.06^{b,A}$	$1.90\pm0.11^{c,B}$
	P500/7.5	$2.50\pm0.13^{a,C}$	$1.84\pm0.05^{c,A}$	$2.26\pm0.04^{d,B}$
	P300/15	$2.35\pm0.13^{a,C}$	$1.80\pm0.06^{b,c,B}$	$1.61\pm0.05^{b,A}$
	P500/15	$2.45\pm0.09^{a,C}$	$1.88\pm0.06^{c,A}$	$2.29\pm0.03^{d,B}$

*Values are expressed as: Mean±Standard deviation. Values with the same lowercase letter are not statistically different ($p>0.05$) in what concerns treatment into the same solvent. Values with the same uppercase letter are not statistically different ($p>0.05$) in what concerns solvent into the same treatment.

3.3.2. DPPH (2,2-diphenyl-1-picrylhydrazyl) radical scavenging activity. The water:acetone extracts were again those that presented the lowest EC_{50} values of DPPH free radical scavenging activity, suggesting higher antioxidant activity than with other solvents (Table 2). For water, the lowest values of EC_{50} were obtained with all conditions of HP treatment, while for methanol the lowest values were obtained with the heating treatment and the two HP treatments of P300/7.5 and P500/15. Our results of EC_{50} for the methanolic extracts were lower than the values obtained by Roriz et al. [20] (4.87 mg/mL) after using methanol at 25 °C for 1 hour, but higher than the value reported by Hamiduzzaman and Azam [11] of 0.020 mg/mL for the whole plant after an extraction at room temperature but during 15 days, what might explain the difference obtained.

Regarding HP treatments, again different results were obtained with different time and pressure combinations, as well as solvents. Nevertheless, the lowest EC_{50} value of DPPH was obtained at P500/15 for all solvents. So, this binomial time/pressure showed the highest antioxidant potential.

When comparing the two extraction methods (heating *versus* HP) applied to water (green solvent), the HP treatments always originated lower EC_{50} values than the conventional one. Moreover, when the pressure increased from 300 to 500 MPa in water, the EC_{50} also decreased. This can be explained because when pressure is applied, the permeability of the cells increases, enabling increased solvent permeation in the cells and therefore the extraction will be more efficient [9]. Thus, HP can have higher extraction efficiency and can greatly shorten the extraction time. Other advantage of HP treatment is to operate at room temperature without any heating process, so there is a lower energy expense, while the bioactivity of the extracted compounds is preserved.

3.3.3. Reducing Power. The EC_{50} values of reducing power showed again that antioxidant activity of the extracts is strongly dependent on the extraction solvent. The water:acetone extract obtained by heating had the lowest EC_{50} value, followed by the HP treatment (P300/15). With methanol, the lowest value of EC_{50} was obtained with the heating treatment, while for water the lowest values were obtained after any of the HP treatments.

4. Conclusion

In summary, the results showed that *Gomphrena globosa* L. flowers are a promising source of natural antioxidants, with the solvent and conditions used in the extraction having direct influence on the content of bioactive compounds. Among the solvents tested, water:acetone was found to be the most efficient solvent to extract bioactive compounds, originating extracts with higher antioxidant activity, followed by methanol and water, for both extractive methods. Our results also showed that the application of HP can be a promising method to extract more natural antioxidants with green solvents, like water, from natural sources such as *Gomphrena globosa* L.. However, in the future more studies must be performed to identify the individual compounds extracted after application of different time/pressure binomials to better understand their extractability.

Acknowledgements

The authors acknowledge the Portuguese Foundation for Science and Technology (FCT, Portugal) for the financial support provided by the research grant SFRH/BD/95853/2013 and FCT/MEC for the financial support to QOPNA research Unit (FCT UID/QUI/00062/2013) and REQUIMTE (UID/QUI/50006/2013) through national funds and when applicable co-financed by the FEDER, within the PT2020 Partnership Agreement.

References

[1] Ferreres F, Gil-Izquierdo A, Valentão P and Andrade P B 2011 *Rapid Commun Mass Spectrom.* **25** 3441–3446
[2] Silva L R, Valentão P, Faria J, Ferreres F, Sousa C, Gil-Izquierdo A, Pinho B R and Andrade P B. 2012 *Food Chem.* **135** 756-763
[3] Hirulkar N B and Agrawal M 2010 *Int. J. Pharm. Biol. Arch.* **1** 478-484
[4] Otterbach A and Wenclawiak B W 1999 *Fresenius J. Anal. Chem.* **365** 472-474
[5] Hamburger M, Baumann D and Adler S 2004 *Phytochem. Anal.* **15** 46-54
[6] Dong L, Wang J, Deng C and Shen X 2007 *J. Sep. Sci.* **30** 86-89.
[7] Zhang S, Junjie Z and Changzhen W 2004 *Int. J. Pharm.* **278** 471-474.
[8] Jun X, Deji S, Shou Z, Bingbing L, Ye L and Rui Z 2009 *Int. J. Pharm.* **382** 139-143
[9] Jun X 2009 *J. Food Eng.* **94** 105-109
[10] Saikia L R and Upadhyaya S 2011 *Int. J. Pharm. Bio Sci.* **2** 383-388
[11] Hamiduzzaman Md and Azam A T M Z 2012 *Bangladesh Pharm. J.* **15** 183-185
[12] AOAC. 1995. Official methods of analysis 16 Ed.. Arlington V A, (USA:Association of Official Analytical Chemists).
[13] Li A-N, Li S, Li H-B, Xu D-P, Xu X-R and Chen F 2014 *J. Funct. Foods.* **6** 319-330
[14] Viuda-Martos M, Ruiz-Navajas Y, Fernández-López J, Sendra E, Sayas-Barberá E and Pérez-Álvarez J A 2011 *Food Res. Int.* **44** 1217-1223
[15] Elfalleh W, Hannachi H, Tlili N, Yahia Y, Nasri N and Ferchichi A 2012 *J. Med. Plants Res.* **6** 4724-4730
[16] Delgado T, Malheiro R, Pereira J A and Ramalhosa E 2010 *Ind. Crop Prod.* **32** 621-626
[17] Navarro-González I, González-Barrio R, García-Valverde V, Bautista-Ortín A B and Periago M J 2015 *Int. J. Mol. Sci.* **16** 805-822
[18] Kuźma P, Drużyńska B and Obiedziński M 2014 *Acta Sci. Pol. Technol. Aliment.* **13** 145-154
[19] Sultana B, Anwar F and Ashraf M 2009 *Molecules* **14** 2167-2180
[20] Roriz C L, Barros L, Carvalho A M, Santos-Buelga C and Ferreira I C F R 2014 *Food Res. Int.* **62** 684-693

IOP Conf. Series: Journal of Physics: Conf. Series **950** (2017) 042005 doi:10.1088/1742-6596/950/4/042005

Effect of added sugar and ascorbic acid on the anthocyanin content of high pressure processed strawberry juices during storage

B Salamon, V Farkas, Gy Kenesei and I Dalmadi

Department of Refrigeration and Livestock Products Technology, Corvinus University of Budapest, 43-45 Ménesi Street, Budapest, 1118, Hungary

E-mail: istvan.dalmadi@uni-corvinus.hu

Abstract. Berries have high nutritional value and can be processed in many kinds of ways. Their pigments (anthocyanins, flavonoids, carotenoids) have antioxidant properties, effectively neutralize the health-damaging free radicals. High hydrostatic pressure (HHP) technology is a minimal processing technique which is a promising alternative solution instead of traditional preservation technologies. Low molecular weight materials such as colour pigments are well preserved by application of HHP. However, the effect can be influenced by the composition of the treated food matrix. The available scientific information related to the impact of sugar and ascorbic acid content on the preservation of anthocyanins in the samples is controversial. Thus, the aim of our study was to determine the effect of HHP treatment parameters (pressure, treatment time) on the preservation of the anthocyanin content of strawberry juice supplemented by different amounts of sugar and ascorbic acid. 2^n type factorial experimental design was used to evaluate the effect of four factors (refraction index, ascorbic acid, pressure, treatment time) on the residual content of total anthocyanins immediately after HHP treatment and after 21 days storage at room temperature.

1. Introduction

The berries, such as strawberries, contain prominent amounts of natural and beneficial antioxidant compounds. These help the immune system in proper functioning and are designed to protect the health of the body from dangerous free radicals as results of different stress effects [1]. Unfortunately, these compounds are very sensitive and easily lose their stability during processing and storage as an effect of various environmental factors (e.g. high temperature, pH, UV radiation). In addition, the composition of the berry itself and the composition of the product made of it also affect the stability of anthocyanins. Other research results regarding the effects of ascorbic acid and sugar content on the anthocyanin content are controversial. Number of researchers reported the negative impact of ascorbic acid on anthocyanins [2-6], while others showed its protective effect [7] and there are also studies where researchers were unable to demonstrate any significant effect of ascorbic acid on anthocyanins [8].

Similar discrepancies are observed in case of saccharides. While some studies have demonstrated that the presence of saccharides increased the pigment loss [9-11], other sources identified anthocyanin-stabilizing effect of the higher sugar content [12-15]. However, according to Malien-Aubert the anthocyanin content is not influenced by the addition of saccharide [16]. In our previous study we examined the effect of addition of five different kinds of sugar to strawberry purées and it was found that in terms of coloration, glucose was the most protective after treatments and also during storage [17].

IOP Conf. Series: Journal of Physics: Conf. Series **950** (2017) 042005 doi:10.1088/1742-6596/950/4/042005

Consequently, the conventional processing of strawberries may cause degradation of the anthocyanins, since the industry applies large quantities of sugar and a small amount of ascorbic acid for preservation in numerous cases. Therefore, in this area it would be worth to use new types of mild food processing techniques. High hydrostatic pressure (HHP) treatment is a preservative method where liquid or solid foods are exposed to hydrostatic pressure from 100-800 MPa. The advantage of HHP treatment is that, according to the Pascal's principle, the hydrostatic pressure is transmitted instantaneously and uniformly (isostatically) throughout the whole product, because the food, that is packed in flexible packaging material, preferably free of air, is immersed into a pressure transmitting fluid. Thus the effect of pressure treatment does not depend on the size or shape of the food in contrast to other preservation methods [18]. Phenols, like anthocyanins, prove to be quite resistant to HHP treatments [19]. Salamon found that the 400 MPa and 600 MPa treated strawberry purées had well preserved colour stability for 3-4 weeks, even at 20 °C storage as well, and it did not influence the consumer's perception [20]. During this period they could have been safely presented in the commercial market. Based on these our research goal was to determine the effect of different pressure levels and different treatment times on anthocyanin content of the sugar and/or the ascorbic acid supplemented strawberry juices.

2. Material and methods

2.1. Sample preparation, treatment and storage

Frozen strawberries were used in our experiments. Strawberries were thawed at room temperature then purée was prepared by an automatic sieve. Then the purée was centrifuged twice for 20 minutes at 1000 rpm and 4 °C and the supernatant was used for the treatments because in this case the residual oxygen in the fibres cannot affect the anthocyanin content. The obtained fiber-free juice had pH value of 3.6; and refraction index was 8.5 °Brix. Subsequently, each sample was supplemented with glucose up to 10, 15 and 20 °Brix, and with ascorbic acid to 0.015 and 0.03% on relative weight as shown in Table 1. 20 mL of samples were placed into plastic bags and vacuum sealed. The foil pouches filled with strawberry juices were treated by RESATO FPU 100-2000 high hydrostatic pressure unit at 400, 500 and 600 MPa pressure values for 5, 12.5 and 20 minutes. The treatments were made at room temperature.

Table 1. The examined factor levels in the experimental design

Factor	Minimum (-1)	Center (0)	Maximum (+1)
(A) Ascorbic acid addition (%)	0	0.015	0.03
(B) °Brix	10	15	20
(C) Pressure (MPa)	400	500	600
(D) Treatment time (min)	5	12.5	20

After treatment strawberry juices were stored in water bath controlled by thermostat (Contraves C40 P) for 21 days at 20 °C. During this period and at this storage temperature, changes were visually perceptible but the aspect was still acceptable from the commercial viability point of view.

2.2. Determination of anthocyanin content

To determine the anthocyanin content of the differently treated, supplemented and stored samples, 1 g of each sample was measured into 50 mL volumetric flask. This flask was filled with hydrochloric-ethanol without loss and well shaken. Thereafter they were kept in a dark place for half an hour then filtered through glass wool, and the absorbance was measured by a spectrophotometer (Hitachi U-2900) at 530 nm wherein hydrochloride alcohol was the blank sample.

Hydrochloride alcohol preparation: 20 mL of conc. HCl was measured into 1000 mL volumetric flask and filled with 96% ethanol to the mark.

$$Anthocyanin\,content\ (mg/kg) = \frac{15 \cdot Absorbance \cdot dilution}{weight\ of\ the\ sample}$$

2.3. Statistical analyses

The measurements were compiled on the basis of type 2^n full factorial experimental design, and the measurements were evaluated by Unscrambler 9.0 software. The implementation of the 2^n full factorial experimental design determines a response function which allows interpolating all internal points of the factor space with the expected amount of the preserved anthocyanin content. We gave a central point of the experiment plan for two reasons: to check the stability of the process and measure the inherent variability and besides these to monitor the curvature (different than linear function). The factors have effects independently from each other in the rarest cases. The change in the settings and effects of one factor can entirely change the effects of other factors. It is therefore necessary to change all the factors together in the same time, and not one by one to search optimum.

The response functions of the four factors:

$$R(A,B,C,D) = M + \beta_A*A + \beta_B*B + \beta_C*C + \beta_D*D + \beta_{AB}*A*B + \beta_{AC}*A*C + \beta_{AD}*A*D + \beta_{BC}*B*C + \beta_{BD}*B*D + \beta_{CD}*C*D$$

where:

- M: average effect (Intercept)
- β_A, β_B, β_C, β_D: β-coefficients describing the main effects of the factors
- A,B,C,D: the factors
- β_{AB}, β_{AC}, β_{AD}, β_{BC}, β_{BD}, β_{CD}: β-coefficients of pairwise interactions of factors

3. Results and discussion

As the addition of sugar and ascorbic acid changes the absolute amount of anthocyanin content in the strawberry juice the change of the relative anthocyanin content was monitored in our assessment. On 0 day the post-HHP treatment status was compared to pre-treatment condition (always the cohesive sugar and ascorbic acid combinations), and in the case of the stored samples the 21 day results were compared to the results of the 0 day treated sample.

It could be observed on the residual anthocyanin contents that the direct effect of high hydrostatic pressure treatment was much smaller compared to the changes due to 21 days storage at 20 °C. While on 0 day maximum 20% decrease was observed in the amount of total anthocyanin content, the reduction was approx. 30-55% during storage at room temperature. It draws attention to the high rate of changes that occur during storage and that how important it is to store the products under appropriate conditions after the processing.

If we evaluate the factorial experimental design we can quantify the effect of each factor. We obtained the results shown in Table 2. The table includes the statistical results of both examined periods. The p-values refer to the significance level of the impact of each factor while the β-coefficients indicate the direction and magnitude of the effects. The established model and the factors are significant if the associated p-value is less than 0.05 (i.e., smaller than the type I error). The factors where significant effects were found are highlighted in the table by bold characters.

If we examine the fitting of the post-treatment model (shown in Figure 1 marked with rhombus) we can see that the correlation between the measured and calculated data was only $r^2 = 0.76$. Besides, slope (that is ideally one, but here is 0.6), offset (which is 0.39 instead of the ideal zero) and the value of calibration error (which is also non-zero) are also indicated in Figure 1 to show the goodness of fitting. It can be seen that in some cases these values vary widely from the expected ideal values. The fitted model was not significant as it can be seen in Table 2., since the p-value of the post-treatment model was much higher than 0.05 ($p = 0.4037$). When we assess the effect of the factors (Table 2) we can observe that only the ascorbic acid addition had significant effect ($p = 0.033$). In case of a significant factor the β-coefficient value is also important because it suggests what kind of changes (positive or negative) can be expected if we change that given factor. In this case the β-coefficient is positive thus we can say that ascorbic acid supplementation had a positive impact on the residual anthocyanin content of strawberry juice samples. Consequently increased ascorbic acid content preserved greater amounts of anthocyanins. Changes were not significantly affected by the other factors and their interactions.

Table 2. Statistical results of the 2^n full factorial experimental design

Main effects and	After HHP treatment		After 21 day of storage	
interactions	p-value	β -coefficient	p-value	β -coefficient
Model	0.4037	-	**0.009**	-
Intercept	**0**	0.971	**0**	0.553
Ascorbic acid addition (A)	**0.033**	0.482	**0.0087**	-0.303
°Brix (B)	0.7828	-0.0016	**0.0035**	0.0108
Pressure (C)	0.8537	-0.0000535	0.0728	0.000272
Treatment time (D)	0.6112	-0.00198	0.1844	0.00255
AB	0.2703	-0.0296	0.0556	-0.0262
AC	0.2877	0.0284	0.0807	0.0234
AD	0.5541	-0.0154	**0.0145**	0.0364
BC	0.9727	0.000882	0.4703	0.00889
BD	0.2035	-0.0346	**0.0406**	0.0286
CD	0.8669	-0.00432	0.8608	0.00212

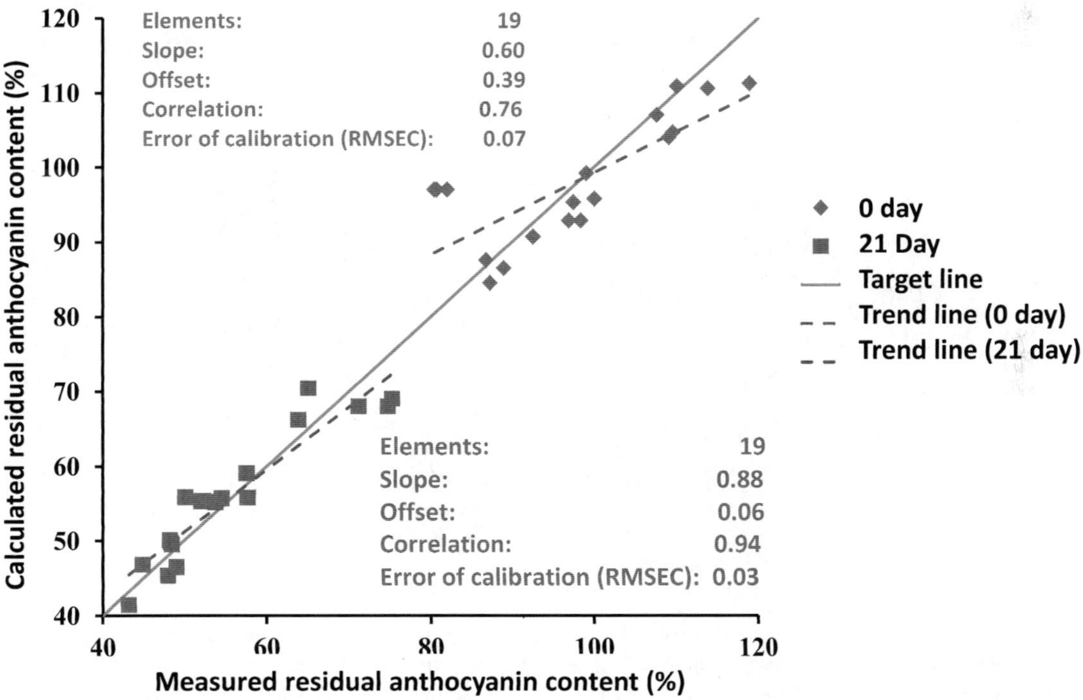

Figure 1. Fitting of the calculated residual anthocyanin contents to the original measurement data, assessed by the established models of the 2^n full factorial experimental design

The results after the 21 days of storage show (Figure 1. marked with square) that the fitted model is considered to be significant ($p = 0.009$). The correlation coefficient calculated between the measured and calculated data highly increased ($r^2 = 0.939$), and the slope was also closer to the ideal value of one and the offset to zero. The calibration error was also reduced to 0.03%. In view of the main effects, the effect of added ascorbic acid was the most interesting one. The effect was significant ($p = 0.0087$),

IOP Conf. Series: Journal of Physics: Conf. Series **950** (2017) 042005 doi:10.1088/1742-6596/950/4/042005

however the β-coefficient was negative (-0.303). So, contrary to the direct effects of the pressure treatments, the storage period had a negative effect on the preservation of the anthocyanin content. This dominates if we examine the combined effect of the sugar content and ascorbic acid (AB) which is very close to be significant ($p = 0.06$). Cao and his co-workers also determined the same negative effect of added ascorbic acid during storage [21]. They found correlation $r = 0.923$ and $r = 0.954$ between the ascorbic acid and anthocyanin content decrease in fibrous strawberry juice, and $r = 0.882$ and $r = 0.967$, respectively, for strawberry juice without fiber in case of storage at 4 and 25 °C. Torres and his co-workers found correlation between ascorbic acid and cyanide-3-glucoside in blood orange juice which were $r = 0.79$ and $r = 0.9$ at 4 and 20 °C storage temperature indicating a strong interaction during the refrigerated storage [22]. As for the main effects, only the ascorbic acid supplementation and the sugar supplementation had significant effect with positive β-coefficient ($p = 0.0035$, β-coeff. = 0.0108) in our experiment. However it should be noted that the effect of pressure treatment remained only marginal ($p = 0.07$) from the statistically proved level. If we evaluate the interaction effects, the interaction of ascorbic acid supplementation and the treatment time and also the interaction of sugar supplementation and treatment time had significant effects. Thus to estimate the anthocyanin content occurring during 21 days we can define the following equation:

$$R(A,B,C,D) = 0.553 + (-0.303*A) + 0.0108*B + 0.0364*A*D + 0.0268*B*D$$

The graphical representation of the model created by the prescribed correlation is shown in Figure 2. The parallel contour lines illustrate that the effect of the pressure values during the treatments are negligible (Figure 2a) and the presence of interactions can be assumed from the twisting of the response surfaces (Figure 2b).

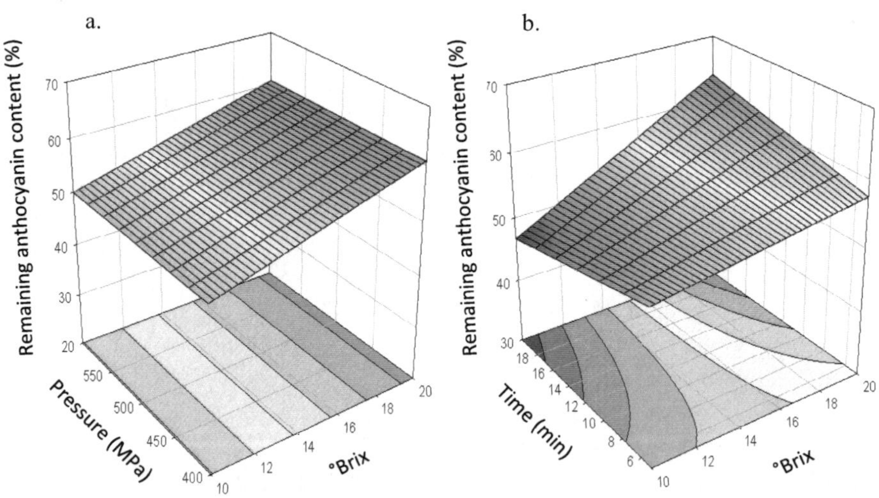

Figure 2. The pairwise comparisons of changes in the middle point of the examined factor space, assessed by the prescribed model based on the significant effects of the factorial experimental design (ascorbic acid supplementation: 0.015%, °Brix: 15%, pressure: 500 MPa, treatment time: 12.5 min)

4. Conclusions

We can conclude that after the high pressure treatments the various pressure levels, treatment times and the sugar supplementations did not have significant effects on the decrease of the anthocyanin content in strawberry juices. But the addition of ascorbic acid has influenced the anthocyanin preservation. It was observed that samples with higher ascorbic acid content had also higher anthocyanin content after the treatments. The statistical analysis also showed it was significant. Thus it can be established, that in case of strawberry juices the ascorbic acid has a protective role in the preservation of the initial anthocyanin content during the high pressure treatments.

After 21 days of storage at 20 °C, the anthocyanin content of the samples considerably decreased, often nearly halved compared to the initial anthocyanin content. The different pressure levels and treatment times had no significant effect on these changes based on statistical analysis, in contrast of added ascorbic acid and sugar supplementation. Interestingly, during storage the added ascorbic acid significantly reduced the anthocyanin content that remained after pressure treatment of strawberry juices. Consequently the ascorbic acid content has an antagonist effect on anthocyanin content in strawberry juices, after high pressure treatment helps to preserve the initial anthocyanin content, but during storage reduces that.

References

[1] Farkas V, Dalmadi I. and Balla Cs. 2013 *Élelmiszer Tudomány Technológia* **LXVII(4)** 9–14
[2] Markakis P 1982 Stability of anthocyanins in foods *Anthocyanins as food colors* ed P. Markakis (New York Academic Press) chapter 6 pp 163–80
[3] García-Viguera C and Bridle P 1999 *Food Chemistry* **64(1)** 21–26
[4] Iversen C K 1999 *Journal of Food Science* **64(1)** 37–41
[5] De Rosso V V and Mercadante A Z 2007 *Food Chemistry* **103(3)** 935–43
[6] Tiwari B K, O'Donnell C P, Patras A, Brunton N and Cullen P J 2009 *European Food Research and Technology* **228(5)** 717–24
[7] Kaack K and Austed T 1998 *Plant Foods for Human Nutrition* **52(3)** 187–98
[8] Rababah T M, Ereifej K I and Howard L 2005 *Journal of Agricultural and Food Chemistry* **53(11)** 4444–47
[9] Markakis P, Livingston G E and Fellers C R 1957 *Journal of Food Science* **22(2)** 117–30
[10] Daravingas G and Cain R F 1968 *Journal of Food Science* **33(2)** 138–42
[11] Dyrby M, Westergaard N and Stapelfeldt H 2001 *Food Chemistry* **72(4)** 431–37
[12] Erlandson J A and Wrolstad R E 1972 *Journal of Food Science* **37(4)** 592–95
[13] Wrolstad R E, Skrede G, Lea P and Enersen G 1990 *Journal of Food Science* **55(4)** 1064–65
[14] Watanabe Y, Fang X, Adachi S, Fukami H and Matsuno R 2004 *LWT-Food Science and Technology* **37(4)** 395–400
[15] Tsai P-J, Hsieh Y-Y and Huang T-C 2004 *Journal of Agricultural and Food Chemistry* **52(10)** 3097–99
[16] Malien-Aubert C, Dangles O and Amiot M J 2001 *Journal of Agricultural and Food Chemistry* **49(1)** 170–76
[17] Elek M and Salamon B 2010 Nagy hidrosztatikus nyomással kezelt szamócapürék színstabilitásának vizsgálata *MÉTE XVIII. OTDK* (Szeged) 117-21
[18] Farkas V and Dalmadi I 2013 *Élelmiszer Tudomány Technológia* **LXVII(3)** 10–13
[19] Patras A, Brunton N P, Da Pieve S and Butler F 2009 *Innovative Food Science & Emerging Technologies* **10(3)** 308–13
[20] Salamon B. 2011 *Élelmiszer Tudomány Technológia* **LXV. (1)** 14–17
[21] Cao X, Bi X, Huang W, Wu J, Hu X and Liao X 2012 *Innovative Food Science and Emerging Technologies* **16** 181–90
[22] Torres B, Tiwari B K, Patras A, Cullen P J, Brunton N and O'Donnell C P 2011 *Innovative Food Science and Emerging Technologies* **12(2)** 93–97

IOP Conf. Series: Journal of Physics: Conf. Series **950** (2017) 042006 doi:10.1088/1742-6596/950/4/042006

Combined effects of high hydrostatic pressure and sodium nitrite on color, water holding capacity and texture of frankfurter

G Jonas, B Csehi, P Palotas, A Toth, Gy Kenesei, K Pasztor-Huszar and L Friedrich

Department of Refrigeration and Livestock Products Technology, Faculty of Food Science, Szent István University, 1118 Budapest, Ménesi str. 43-45, Hungary

E-mail: Jonas.Gabor@etk.szie.hu

Abstract. The aim of this study was to investigate the effect of sodium nitrite and high hydrostatic pressure on the color, water holding capacity (WHC) and texture characteristics of frankfurter. Three hundred, 450 and 600 MPa (5 minutes; 20 °C) and 50, 75, 100 and 125 ppm (calculated on weight of meat) sodium nitrite were applied. Parameters were measured right after the pressure treatment. Data were evaluated with two-way analysis of variance (p 0.05) with pressure levels and sodium nitrite amounts as factors. Nitrite reduction significantly increased lightness (L*) and resulted in decreased redness (a*) value. The pressure treatments decreased the lightness at all nitrite concentrations and did not significantly affect the red color of frankfurters. Fifty and 75 ppm nitrite and pressurization at 300 or 450 MPa improved the water holding property of frankfurter. The pressure treatment did not significantly affect the WHC but changing the nitrite amount had significant effect on it. Interactive effect occurred between pressure levels and nitrite concentrations for hardness. The pressure treatment increased and the nitrite reduction decreased hardness. Significant changes were found in cohesiveness at 450 and 600 MPa in frankfurters containing 50 and 75 ppm nitrite: pressure treatment at higher levels and nitrite reduction decreased the value of cohesiveness.

1. Introduction

Frankfurters belong to the group of classical raw-cooked meat products, originated more than 100 years ago and now enjoy worldwide popularity [1]. Sodium nitrite ($NaNO_2$) mixed in salt (NaCl) are important additives of meat products (e.g. frankfurter) to provide color, flavor, antioxidant and antimicrobial activity and texture properties [2]. However, nitrite in meat products have been identified as a potential health risk that can generate methemoglobin, which is a recognizable sign of nitrite toxicity in humans and may react with certain amines in foods to produce carcinogenic N-nitrosocompounds such as nitrosamines [3,4]. In view of the potential health risks and the increasing demand for healthier products, the use of nitrite in meat products faces a challenge to reduce the content of nitrite [5].

High hydrostatic pressure (HHP) treatment is a non-thermal technology to preserve foods and food products. The high hydrostatic pressure treatment is also known as 'minimal processing technology' because it retains more original properties of foods than the conventionally used technologies (e.g. heat treatment) [6]. Based on earlier findings the industrial application of HHP began in the 1990's

[7]. At present more than 300 HHP equipment are in use worldwide [8]. In the last two decades, a lot of studies have shown the efficacy of this technology in inactivation of pathogenic microorganisms and increasing the shelf-life of meat products [9]. High pressure has an effect on proteins in meat products and can modify their functional properties by protein denaturation, aggregation or gelation [10]. Due to changes in protein structure HHP suggests a process to tenderize meat or produce innovative and additive-free meat products [11, 12].

HHP has proved an alternative to reduce the salt level in frankfurter and sausage [13, 14]. Macfarlane *et al.* [15] and Mandava *et al.* [16] found that high pressure below 200 MPa increases the water holding and binding properties of meat batters at reduced salt concentration. High pressure treatment results in changes in fresh meat color. The lightness (L* value) increases right after the pressure treatment (above 250 MPa), whereas the redness (a* value) decreased at 400 to 500 MPa resulting in gray-brown color [17]. In case of meat products the color changes are acceptable and depending on the water content and a_w-value [18]. However, these changes are not too relevant if the products are further processed for example cooked before eating. Sikes *et al.* [14] investigated the effects of HHP on texture of low-salt beef sausage batters and found that at all salt concentrations (0% to 2%), the hardness and gumminess of pressure-treated (up to 400 MPa, 10 °C, 2 min) samples were higher compared to the non-pressurized ones. There was greater acceptability in terms of both appearance and texture of pressure treated low-salt sausages in comparison with the non-pressurized samples.

In this sense the high pressure treatment could be used to improve the functional properties and safety of meat products with reduced additive (e.g. nitrite) content. Impact of sodium chloride and phosphates under high pressure was investigated by Crehan *et al.* [13], Tintchev *et al.* [19] and Villamonte *et al.* [20]. However, there is no clear relationship between the applied sodium nitrite amount and high hydrostatic pressure treatment. Therefore the aim of this study was to investigate the effect of sodium nitrite and high hydrostatic pressure on color, WHC and texture characteristics of frankfurter.

2. Materials and methods

2.1. Preparation of frankfurter

Frankfurter samples were prepared from pork muscle (pork shoulder) and back-fat that were obtained from a local producer. The muscle was ground (diameter 4 mm) and mixed with salt (sodium chloride, NaCl; 2.0% weight of meat), polyphosphate (tetrasodium pyrophosphate, $Na_4P_2O_7$; 0.4% of meat), ascorbate (sodium L ascorbate; 0.5% weight of meat), ice (70% weight of meat) and pork back-fat (40% weight of meat). The amount of added nitrite (sodium nitrite, $NaNO_2$) was 0.005% (50 ppm), 0.0075% (75 ppm), 0.01% (100 ppm), 0.0125% (125 ppm) weight of meat. The raw meat batters were relaxed between 4-6 °C for 30 minutes then filled into polyethylene casing (diameter 100 mm). The frankfurters were hand-linked and cooked in a Kerres CS 305 smoker cabin to core temperature of 72 °C. The cooked frankfurters were showered by cold water then stored overnight at 4-6 °C. Before pressure treatment the frankfurters were vacuum packed without removing the casing. Three frankfurter samples from each nitrite concentration were pressure treated in independent cycles.

2.2. Pressure treatment

After heat treatment the frankfurters were pressurized at 300, 450 and 600 MPa for 5 minutes in RESATO FPU-100-2000 (Resato International B.V, Netherland) high pressure equipment. Pressure gradient was 100 MPa/min. Pressure treatment was carried out at room temperature (approx. 20 °C) and sample temperature variations due to adiabatic heat were considered as intrinsically included inside pressure factor (reproducible sample thermal history linked to pressure level). Zero (0) MPa means unpressurized frankfurters.

2.3. WHC determination
The WHC was characterized as released water from frankfurter samples and was measured according to Grau [21]. An amount of sample between 200 and 300 mg was exactly weighed and put on a known weight 2500 mm^2 (50x50 mm) area filter paper. The filter paper and the sample were placed between two glass plates and pressurized with 0.5 kg weight for 5 min then the filter paper was dried. The appeared spot (moisture from sample) was cut out from the filter paper and the paper without spot was weighted. The released water from sample was calculated as the proportion of area and weight of spot and sample and was expressed in [mm^2/mg] dimension. Three replicates were analyzed from each pressure treatment-nitrite content combinations.

2.4. Color measurement
Objective color was measured using a Minolta ChromaMeter CR-400 (Konica Minolta Inc., Japan) on the cut internal surface of frankfurter. The measured values were expressed in CIE Lab as L* (lightness), a* (redness) and b* (yellowness). Appearance of frankfurter is basically determined by its lightness and redness. Therefore lightness (L*) and redness (a*) parameters were evaluated. Nine replicates were analyzed from each combination.

2.5. Texture profile analysis (TPA)
Texture of pressure treated frankfurters were measured using SMS TA. XT Plus (Stable Micro Systems Ltd., United Kingdom) texture analyzer. Three cylinders (diameter=18 mm; height=20 mm) were cut from each frankfurter and axially compressed to 70% of their original height in two cycles. Load cell was 500 N and crosshead speed was 2 mm/s. Evaluated texture parameters were the hardness (N) and cohesiveness (dimensionless). Three replicates were analyzed from each pressure treatment-nitrite content combinations.

2.6. Statistical analysis
The data were analyzed using IBM SPSS 22.0 software. Statistical analysis was performed by analysis of variance (ANOVA) with pressure levels and nitrite concentrations as factors. Difference was considered to be statistically significant at $p<0.05$.

3. Results and discussion
3.1. Water holding capacity
According to Macfarlane *et al.* [15] pressure treatment at 150 MPa improved the water holding property of heat treated beef patty and decreasing WHC was found between 300 and 600 MPa. These results are in part consistent with ours. Figure 1 shows the released water of heat treated then pressurized frankfurters prepared with varying nitrite concentrations. Changing the nitrite concentration had an effect on amount of released water of pressurized samples. Fifty and 75 ppm nitrite and pressurization at 300 or 450 MPa decreased the amount of released water. However, pressure treatments slightly increased the amount of released water in case of frankfurter prepared with 125 ppm nitrite. Based on statistical results pressure treatment had no significant effect on released water ($p=0.302$). However, changing the nitrite concentration had significant effect on the released water ($p=0.020$) of frankfurters in case of all pressure treatments. Interaction between pressure level and nitrite concentration was not observed ($p=0.208$).

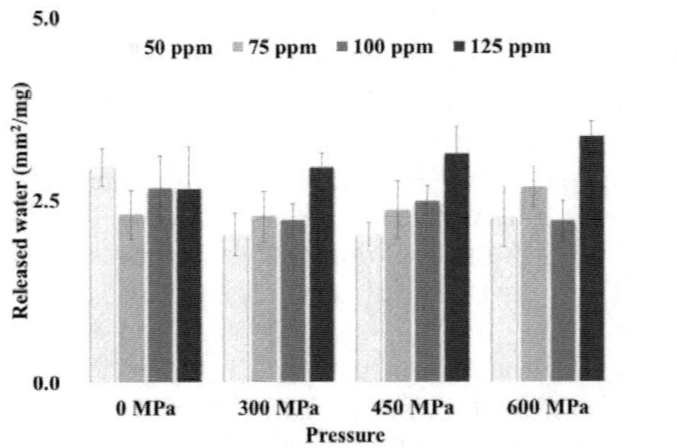

Figure 1. Effect of 50, 75, 100, 125 ppm nitrite and 0, 300, 450 and 600 MPa pressure treatment on released water of frankfurter

3.2. Instrumental color

Appearance of frankfurter is basically determined by its lightness and redness. Therefore the lightness (L*) and redness (a*) parameters were evaluated. Figure 2 and figure 3 show the objective lightness (L*) and redness (a*) of heat-treated and pressurized frankfurters prepared with various nitrite concentrations. Nitrite reduction increased the lightness value (L*) of frankfurter at each pressure level. Application of 50 and 75 ppm nitrite resulted in 'paler' appearance than 100 or 125 ppm. Difference of lightness values between 100 and 125 ppm frankfurters was negligible. The pressure treatments decreased the lightness of frankfurters in case of each nitrite concentration. Based on statistical results pressure treatment at 450 and 600 MPa significantly decreased the lightness of frankfurters ($p<0.05$) and the nitrite reduction significantly increased the lightness ($p<0.05$). Nitrite reduction resulted in decrease of redness (a*) independently of the pressure treatment. This, taking into account the role of nitrite in promoting red color development was in line with our expectations. Nitrite reduction in frankfurter samples resulted in significantly lower red color values ($p<0.05$). High pressure treatment of meat pigment (myoglobin) caused partial denaturation and discolouration [22]. In our study the pressure treatments did not significantly affect the red color of frankfurters ($p=0.365$). It is explicable since the effect of high pressure treatment on myoglobin depends on the temperature at which the pressure treatment occurs [23]. Interactive effect on lightness ($p=0.320$) and red color ($p=0.912$) was not observed between nitrite amount and pressure treatment.

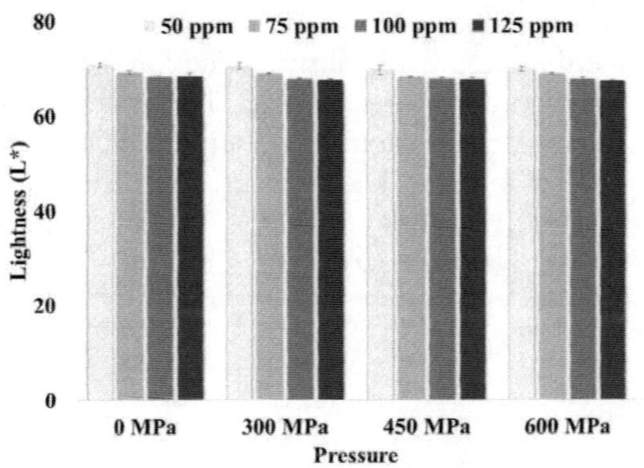

Figure 2. Effect of 50, 75, 100, 125 ppm nitrite and 0, 300, 450 and 600 MPa pressure treatment on lightness (L*) of frankfurter's internal surface

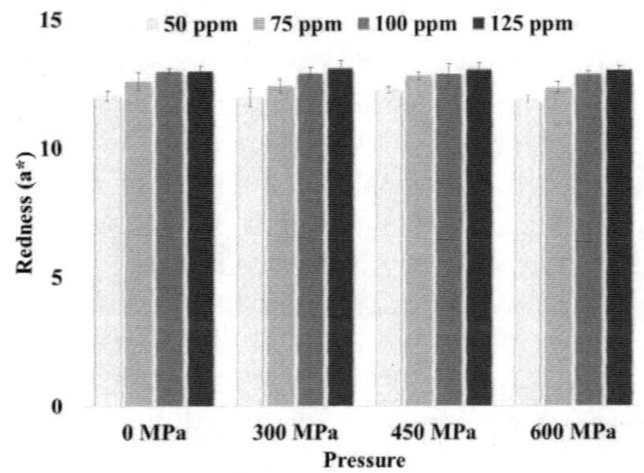

Figure 3. Effect of 50, 75, 100, 125 ppm nitrite and 0, 300, 450 and 600 MPa pressure treatment on redness (a*) of frankfurter's internal surface

3.3. *Texture profile analysis*

Changing the nitrite amount and pressure treatment had effect on hardness of frankfurter (figure 4). The effect was statistically verifiable in case of both parameters. Crehan *et al.* [13] found that pressure treatment at 300 MPa resulted in improved hardness compared to non-pressurized frankfurter. Hardness of frankfurters prepared with 100 and 125 ppm nitrite gave similar results. However, hardness of 50 and 75 ppm nitrite frankfurters did not change. At 450 MPa pressure level a decrease in hardness was seen at 100 and 125 ppm levels while in case of lower nitrite concentrations an increase in hardness was observed. Based on statistical results interactive effect occurred between pressure treatment and nitrite concentrations for hardness ($p=0.009$). Pressure treatment can improve the cohesiveness in meat products [15]. However, our results presented that pressure treatment at higher levels and nitrite reduction decreased the cohesiveness (figure 5). Significant decrease was found in cohesiveness at 450 and 600 MPa in frankfurters containing 50 and 75 ppm nitrite. At these pressure levels a tendency was seen: reducing the amount of nitrite resulted a product with less cohesive texture. Effects of pressure treatment and nitrite concentrations were significant ($p<0.05$) on cohesiveness but between the factors no statistically verified interactions ($p=0.053$) were shown.

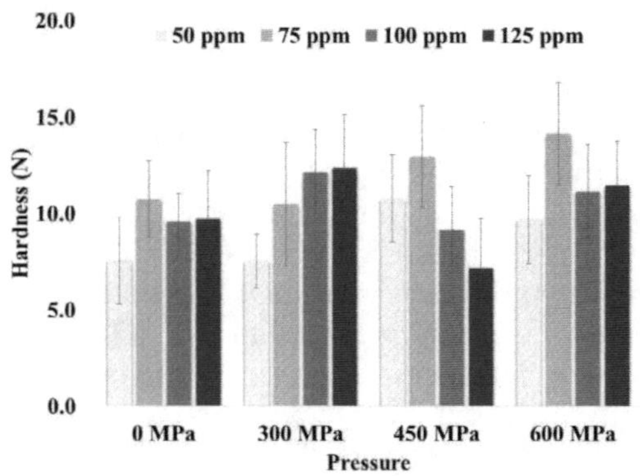

Figure 4. Effect of 50, 75, 100, 125 ppm nitrite and 0, 300, 450 and 600 MPa pressure treatment on hardness of frankfurter

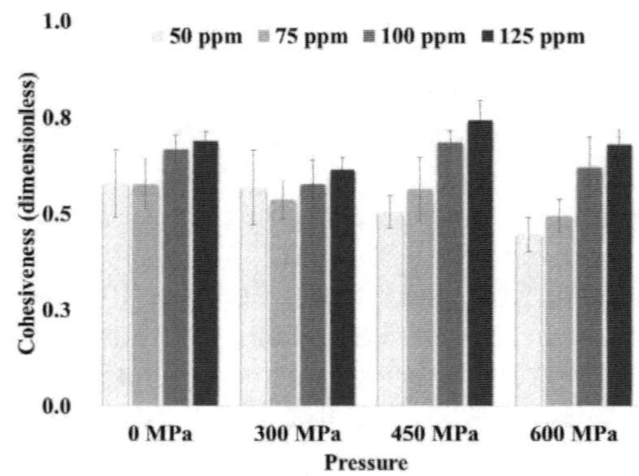

Figure 5. Effect of 50, 75, 100, 125 ppm nitrite and 0, 300, 450 and 600 MPa pressure treatment on cohesiveness of frankfurter

4. Conclusion

Our experiment showed that nitrite reduction combined with high pressure to 450 MPa decreased the amount of released water and improved the hardness of frankfurter. However, pressure treatment at higher level and nitrite reduction resulted in a decrease of cohesiveness. In terms of lightness antagonist effect was seen. Pressure treatment decreased the lightness while nitrite reduction increased it. Changing in redness was caused by the nitrite alone. The high pressure treatment did not affect the red color. Combination of 75 ppm nitrite and 450 MPa resulted in acceptable product in the viewpoint of investigated functional properties right after the pressure treatment. The high pressure treatment showed promising results that could allow the production of frankfurter with less sodium nitrite. However, further experiments need to be carried out to investigate the microbial effects and changes of functional properties and shelf-life.

Acknowledgement

This research was supported by TÁMOP 4.2.1/B-09/1/KMR/-2010-0005.

References

[1] Heinz G and Hautzinger P 2007 *Meat Processing Technology For Small- To Medium-scale Producers* (Bangkok: FAO Regional Office For Asia And The Pacific Publication) pp. 127 – 128

[2] Sindelar J J and Milkowski A L 2011 *Sodium Nitrite In Processed Meat And Poultry Meats: A Review of Curing And Examining the Risk/Benefit if its Use* Number 3 (Champaign, USA: AMSA Educational Foundation Publication)
[3] Chow C K and Hong C B 2002 *Toxicology* **180** 195 – 207
[4] Cassens R G 1997 *Food Chem.* **59** 561 – 566
[5] Zhou C, Wang J, Wang H, Zhang L and Cai K 2012 *Food Sci. Biotechnol.* **21(1)** 267 – 272
[6] Rastogi N K, Raghavarao K S, Balasubramaniam V M, Niranjan, K and Knorr D 2007 *Crit. Rev. Food Sci. Nutr.* **47** 69 – 112
[7] Hite B H 1899 *West Virginia Agricultural Experiment Station Bulletin* (Morgantown, VA, USA) **58** 15 – 35
[8] Flores Corrochano D, Onrubia Monedero M, Wilches Perez D and Tonello Samson C 2015 Trends in high pressure processing equipment and products. *Book of Abstracts. Joint AIRAPT-25 & EHPRG-53 Int. Conf. on High Pres. Sci. Technol. (Madrid, Spain, 30 Aug. - 4 Sept. 2015)* p 160.
[9] Vercammen A, Vanoirbeek K G A, Lurquin I, Steen L, Goemaere O, Szczepaniak S, Paelinck H, Hendrickx M E G and Michiels C W 2011 *Innov. Food Sci. Emerg. Technol.* **12** 407 – 415
[10] Messens W, van Camp J and Huyghebaert A 1997 *Trends Food Sci. Tech.* **8** 107 – 112
[11] Jung S, de Lamballerie-Anton M and Ghoul M 2000 *Lebensm. Wiss. Technol.* **33** 313 – 319
[12] Sun X D and Holley R A 2010 *J. Food Sci.* **75** R17 – R23
[13] Crehan C M, Troy D J and Buckley D J 2000 *Meat Sci.* **55** 123 – 130
[14] Sikes A L, Tobin A B and Tume R K 2009 *Innov. Food Sci. Emerg. Technol.* **10** 405 – 412
[15] Macfarlane J J, McKenzie I J, Turner R H and Jones P N 1984 *Meat Sci.* **10** 307 – 320
[16] Mandava R, Fernández I and Juillerat M A 1994 Effect of high hydrostatic pressure on sausage batters *Proc.of 40th Int. Congr. on Meat Sci. Technol. (The Hague, Netherlands)*
[17] Tintchev F, Wackerbarth H, Kuhlmann U, Toepfl S, Knorr D, Hildebrandt P and Heinz V 2010 *Ann. N.Y. Acad. Sci.* **1189** 34 – 42
[18] Ferrini G, Comaposada J, Arnau J and Gou P 2012 *Innov. Food Sci. Emerg. Technol.* **13** 69 – 74
[19] Tintchev F, Bindrich U, Toepfl S, Strijowski U, Heinz V and Knorr D 2013 *Meat Sci.* **94** 376 – 387
[20] Villamonte G, Simonin H, Duranton F, Chéret R and de Lamballerie M 2013 *Innov. Food Sci. Emerg. Technol.* **18** 15 – 23
[21] Grau R and Hamm R 1953 *Naturwissenschaften* **40** 29 – 30
[22] Defaye A B, Ledward D A, MacDougall D B and Tester R F 1995 *Food Chem.* **52** 19 – 22
[23] Campus M 2010 *Food Eng. Rev.* **2** 256 – 273

IOP Conf. Series: Journal of Physics: Conf. Series **950** (2017) 042007 doi:10.1088/1742-6596/950/4/042007

Thermograms of the combined High Hydrostatic Pressure and Sous-vide treated *Longissimus dorsi* of pork

Gy Kenesei, G Jónás, B Salamon, I Dalmadi

Corvinus University of Budapest, Department of Refrigeration and Livestock Products Technology, Budapest, Hungary

E-mail: kenesei.gyorgy@etk.szie.hu

Abstract. In this work, slices of *Longissimus dorsi* of pork was used as raw material to establish the effects of the sous-vide technology and the high hydrostatic pressure treatments (and their combinations) on meat. The state of the proteins in meat has a very important effect on several quality parameters of the product, such as weight loss, water holding capacity, organoleptic properties. Therefore it is important to follow and analyse the denaturation of the protein content during food processing. The samples were cooked sous-vide (60 °C, 5-480 minutes) or pressurized (100-600 MPa, 5 minutes, room temperature). Also two steps treatments were studied combining both technologies, applying high hydrostatic pressure treatment (300 or 600 MPa, 5 minutes, room temperature) after or previous to sous-vide cooking (60 °C, 30 minutes). The changes in the condition of meat proteins were followed by a differential scanning calorimeter. The DSC curves were analysed using the unit's own software where denaturation heat was determined. Thermograms show through the change of the sample's protein state the dissimilar effect of the treatments. Using the Polar Qualification System -previously proved to be effective with NIR measurements- the spectral information was reduced to a two dimensional polar co-ordinate system where each DSC curve is represented by a "quality point". As a new experiment the applied PQS data reduction method compared to the traditional thermal analysis data processing gave us less information on the differences of our samples although the results are promising as we were able to detect the same trends and characteristics.

1. Introduction

At the development of modern food processes, it is important to ensure the microbiological stability of the product with only the slightest damage in the quality parameters. This can be achieved by applying the principle of hurdle technology [1]. The lethality of low temperature heat treatment could be enhanced by combining it with other physical treatments. Sous-vide cooking as a mild heat treatment and high hydrostatic pressure technology seem to be promising for this purpose. The sous-vide technology is already applied not only in small scale but in the catering industry as well [2]. The optimal time-temperature parameters are highly dependent on the raw material [3]. It results juicy, perfectly cooked products with minimal nutrient loss and high organoleptic value [4]. With keeping the heat treatment at a minimal level, we face several microbiological problems [2]. The application of HHP as a secondary treatment could be a solution only in the case if it is not altering the excellent properties of the original sous-vide product. Only a few studies can be found in literature investigating the effects of the combined treatment of sous-vide and high hydrostatic pressure. Pineapple [5], fish [6,7] and chicken meat patties [8] were investigated previously. In meat, proteins are the second most important constituents after water. The state of the proteins in meat products has a very important role and affects several quality parameters of the product, such as weight loss, water holding capacity, sensory properties. The protein components of the meat react differently in the case of heat induced denaturation as well as at the pressure induced changes [9,10]. The Differential Scanning Calorimeter can detect the kinetic changes of the proteins [11] and the preserved quality of the meat [12,13]. Therefore this work presents a detailed study on the behaviour of the proteins in pork under different one and two step food processing technologies. The aim also was to analyse the power of the PQS data transformation method using the DSC curves as spectral information.

2. Materials and methods

2.1. Sample preparation

The whole pork chops (*Longissimus dorsi*) were purchased at a wholesale market (m= 3.78 – 3.93 kg). After being prepared to be free of surface fat, ligament and connective tissue each meat was cut into slices (2 cm) transversally to the fiber direction. The weight of each sample was 60-70 g. Samples were vacuum sealed in 90 μm PA/PE plastic pouches using a Multivac C100 V.S. machine. All treated samples were kept frozen at -24 °C as the DSC measurements took several days.

2.2. Sous-vide treatment (SV)

Heat treatment was carried out in a water bath (Labor Műszeripari Művek LP507/1). Samples were cooked sous-vide at 60 °C, the treatment time was 5, 15, 30, 60, 120, 240 or 480 minutes. Samples were cooled in iced water immediately after cooking.

2.3. High hydrostatic pressure treatment (HHP)

Meat samples were pressurized at 100, 200, 300, 400, 500, or 600 MPa for 5 minutes at room temperature in a RESATO FPU-100-2000 HHP unit (Resato International B.V., the Netherlands). The pressure transmission fluid was glycol-oil mixture (Resato PG fluid, Roden, Holland). The pressure build-up rate was 100 MPa/minute. Build-up and decompression times were not included in the treatment time. The initial temperature of the pressure transmission fluid was 18.5 °C, the adiabatic temperature change of the system (samples and the pressure transmission fluid) was under 14 °C.

2.4. Combined treatments

Following the hurdle principle, we used both technologies. Heat treatment was carried out at 60 °C for 30 minutes and high hydrostatic pressure was set at 300 and 600 MPa for 5 minutes. The combinations were: heat after pressure treatment and pressure after heat treatment at both pressure levels. Treatment parameters were determined by previous heat transfer calculations and practical experience for the sous-vide cooking and two borderline pressure values were chosen for the high hydrostatic pressure treatment. The 300 MPa is around the entry treatment level where protein denaturation and microbiological inactivation is already observed while the 600 MPa is a generally applied pressure value in the food industry [10]. As the PA/PE pouch packaging of the raw material is suitable for both technologies cross-contamination is avoided during treatments.

2.5. DSC analysis

The changes in the state of meat proteins were followed by a SETARAM micro DSC III differential scanning calorimeter. The scanned temperature range was 20-95 °C at a heating rate of 1 °C/minute. The DSC curves were analysed in the 35-90 °C range using the unit's own Calisto software. The total denaturation enthalpy of each meat sample was determined. The enthalpy is calculated with integration – it is equal with the area in between the DSC curve and the baseline.

2.6. PQS data analysis

The recorded DSC curves were analysed using the PQS32 Evaluation Software (MetriNIR Research, Development and Service Co., Budapest, Hungary). This data reduction method was primarily used with NIR measurements [14-16] and was previously applied with promising results to analyse DSC data of cocoa butter and its constituents [17]. In PQS the spectrum is plotted in a two-dimensional polar co-ordinate system (X-Y). The quality points – the center of gravity of the polar spectrum- is obtained from the DSC thermograms in the range of 35-90 °C. The properties of a sample can be characterized by the location of its quality point in the polar quality plane. The distance between them can describe the similarity, the smaller the value the more the samples are identical and vice-versa.

3. Results and discussion

The shape of the DSC curves characterize well the sample and its composition. The untreated (raw) pork sample has three well distinguished peaks. The myosin protein group can be found at around 50-55 °C, the sarcoplasm proteins and collagen at around 60-65 °C and the actin at 70-75 °C. Just after the 5 minute heat treatment, proteins denatured below the temperature of 55-60 °C, while the other components remain intact. As the treatment time is increased more important protein quantities are damaged or denatured even in the ranges higher than 60 °C. After 480 minutes nearly all protein groups are affected by the heat. (Figure 1.a)

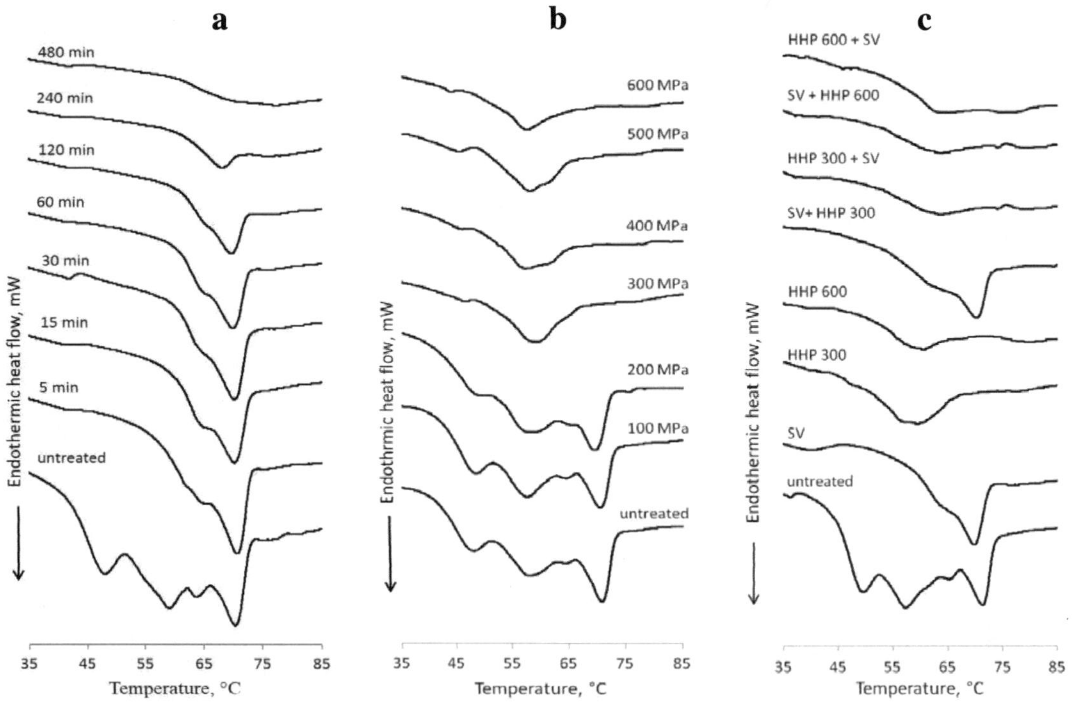

Figure 1. DSC thermograms of the sous-vide cooked at 60 °C (**a**), the high hydrostatic pressure treated for 5 minutes at room temperature (**b**), and the combined treated (**c**) meat samples (SV = sous-vide cooked at 60 °C for 30 minutes, HHP300 = 300 MPa high hydrostatic pressure treated for 5 minutes at room temperature, HHP600 = 600 MPa high hydrostatic pressure treated for 5 minutes at room temperature.

In the case of high hydrostatic treatment the most important observation is that the HHP treatment has effect on different protein groups than the heat treatment induced denaturation. We can also observe that there is a borderline between 200 and 300 MPa. Below this pressure value the treatment has practically no effect on the meat samples. Ongoing from 300 MPa denaturation can be seen at the myosin and actin groups. The collagen and sarcoplasmic proteins are still stable and the pressure has only a slight effect on them. (Figure 1.b)

On the results of the combined treatments we have to remark that the order of the treatments has a very important role concerning the protein state of the samples. The application of the heat treatment in combination with the pressure treatment is still resulting a partly denatured meat sample. This could be observed also at 600 MPa but when the lower pressure level (300 MPa) was used the difference between the order of the treatments is much more important. (Figure 1.c)

The calculated enthalpy values are in accordance with the visual observations as it can be seen on the diagrams. (Figure 2.a,b,c)

The measured enthalpy values were 2.795 J/g at the raw sample, while at the sous-vide treated samples we measured 1.847; 1.243; 1.055; 0.892; 0.459; 0.287; 0.143 J/g at the 5, 15, 30, 60, 120, 240, 480 minute samples respectively that results a decrease in accordance with treatment time. (Figure 2.a) The enthalpy of the pressure treated samples were 2.671; 2.315; 1.068; 0.902; 0.796; 0.489 J/g at 100, 200, 300, 400, 500, and 600 MPa samples respectively. (Figure 2.b)

To compare the two preservation methods (heat and pressure) the 300 MPa (5 minutes) pressure treated sample needed approximately the same denaturation enthalpy as the 30 minute heat treatment at 60 °C while they show very different characteristics. The pressure treated sample in physical appearance is very similar to the raw meat while the sous-vide sample shows the typical signs of cooked meat (not shown). The measurement of the combined treated samples resulted the most important data of this study. When the pressure treatment was applied previous to sous-vide cooking the enthalpy values show no difference between the two pressure levels. In the reverse order (heat treatment followed by the pressure treatment) the pressure value determined the result.

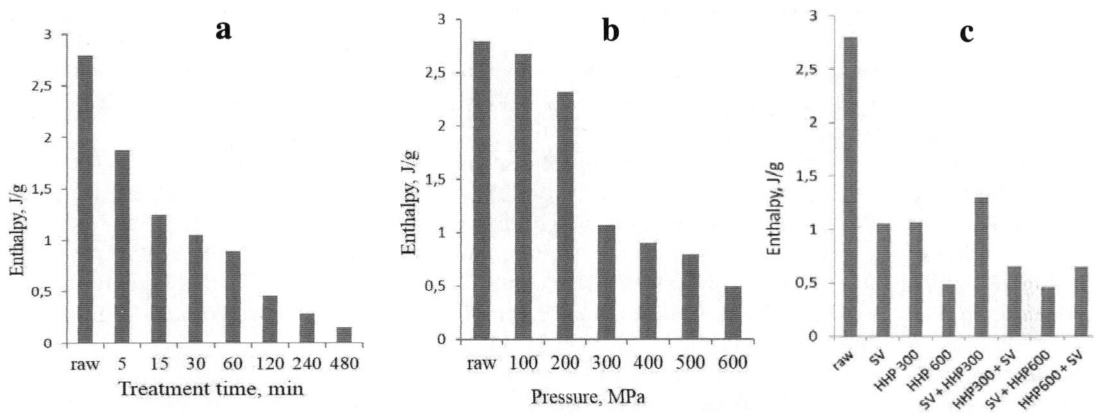

Figure 2. Denaturation enthalpy values of the sous-vide cooked at 60 °C (a), the high hydrostatic pressure treated for 5 minutes at room temperature (b), and the two step combined treated (c) meat samples (SV = sous-vide cooked at 60 °C for 30 minutes, HHP300 = 300 MPa high hydrostatic pressure treated for 5 minutes at room temperature, HHP600 = 600 MPa high hydrostatic pressure treated for 5 minutes at room temperature).

The PQS quality points in the X-Y polar co-ordinate system are able to visualize the major changes in the protein state effectively. (Figure 3.a) The raw sample is well separated from the cooked ones on the X axis, while the cooked samples show a vertical linear pattern as the Y value increase the treatment time was longer so the heat impact was more important. The 5 minute heat treatment is already well separated from the raw sample indicating the fast structural changes in the protein groups. The next four treatment times (15, 30, 60, 120 minutes) form a subgroup showing the similarities already observed on the DSC curves above (Figure 1.a). In between this group the PQS show no difference. This result is derived from the similar forms of the curves with the same well defined peaks. The 240 and 480 minute samples are on the same axis as all the heat treated ones but as they contain mostly denatured proteins they are well separated in the polar quality plane. The apparent contradiction that the 480 minutes cooked sample is next to the raw sample in the X-Y polar space is due to the PQS data transformation as it's curve is more "balanced" with the three main peaks therefore it is similar in this polar space to the 480 minute cooked sample's curve that has no significant peak. That results a nearer position to the raw sample than the ones treated for shorter time and having only one but well defined peak.

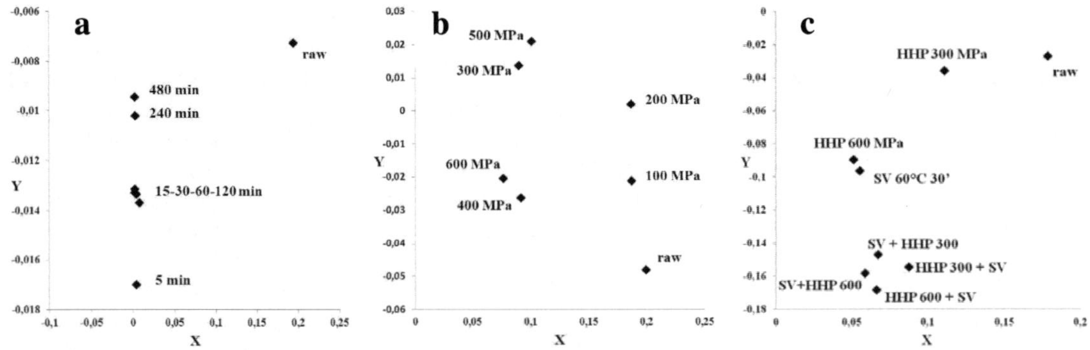

Figure 3. PQS quality points in the polar quality plane of the sous-vide samples cooked at 60 °C (a), the pressure treated samples (b), and the combined treated (c) meat samples.

The co-ordinates of the HHP treatments similarly to the literature data and the above seen spectra show two well defined subgroups separated by the X axis values. (Figure 3.b) The raw, the 100 MPa and the 200 MPa treatments are in the first group with minimal or no changes in the protein state. The more important pressure value treatments (300, 400, 500 and 600 MPa) are forming the second group, where denaturation already altered the protein structure.

The results of the combined treatments show new scientific data. The differences due to the order of the two minimal processing procedures can be observed. By all means, the 300 MPa HHP-treated sample is the nearest to the raw meat as it underwent only a mild treatment and most of their proteins were kept in a native state. The position of the quality points of the simple (one step heat or pressure) treatments reflect their dissimilar character. The combined treated samples form a subgroup well separated from the single treatments but the distance within the group is not so important. (Figure 3.c)

4. Conclusions

Based on the results it can be concluded that both data processing methods were capable of distinguishing the different pork samples treated by a single technology at different levels or the combinations of sous-vide treatments and HHP processes. Using the DSC's own software only one data, the enthalpy value was calculated for each sample. The PQS method used the DSC curves like spectral information and positioned the samples in the X-Y polar space according to the structural changes of the sample's protein content. These changes could be monitored objectively using the PQS data transformation. With this unsupervised data processing method groups could be formed where similar characteristics are observed. The results of this experimental work can help in the process design and also point on the importance of the treatment order of the combined minimal processing technologies.

Acknowledgments

The authors wish to thank for -making the HHP treatments possible- for TÁMOP 4.2.1/B/09/1/KMR/-2010-0005 program of the National Development Agency, Hungary.

References

[1] Leistner L 2000 *Int. J. Food Microbiol.* **55** 181-186

[2] Creed P G and Reeve W 1998 *Sous Vide and Cook-Chill Processing for the Food Industry* ed S Ghazala (Gaithersburg, Maryland: Aspen Publishers Inc.) chapter 1 pp 39-41
[3] Sheard M A and Rodger C 1995 *Food Control* **6** **(1)** 53-56
[4] Baldwin D E 2012 *Int. J. Gastronomy Food Sci.* **1** 15-30
[5] Chakraborty S, Srinivasa Rao P and Mishra H N 2015 *Innov. Food Sci. Emerg. Technol.* **28** 10–21
[6] Espinosa M C, Díaz P, Linares M B, Teruel M R, and Garrido M D 2015 *LWT – Food Sci. Technol.* **64** **(2)** 657–662
[7] Picouet P A, Cofan-Carbo S, Vilaseca H, Carboné Ballbè L and Castells P 2011 *Innov. Food Sci. Emerg. Technol.* **12** 26–31
[8] Zheng H, Xiong G, Han M, Deng S, Xu X and Zhou G 2015 *Innov. Food Sci. Emerg. Technol.* **30** 8-14
[9] Tornberg E 2005 *Meat Sci.* **70** **(3)** 493-508
[10] Cheftel J C and Culioli J 1997 *Meat Sci.* **46** **(3)** 211-236
[11] Skipnes D, Van der Plancken I, Van Loey A and Hendrickx M E 2008 *J. Food Eng.* **85** 51–58
[12] Biliaderis C G 1983 *Food Chem.* **10** 239-265
[13] Farkas J 1994 *Élelmiszervizsgálati Közlemények* **40** **(3)** 180-189
[14] Kaffka K J and Gyarmati L S 1998 *J. Near Infrared Spec.* **6** 191–200
[15] Bázár Gy 2011 *Application possibilities of near infrared spectroscopy in the qualification of pork, meat products and lard* PhD theses Kaposvár University Fac. of Animal Science
[16] Bázár Gy, Szabó A and Romvári R 2010 *Food Control* **21** 992-997
[17] Kerti K 2000 *Thermoanalytical analysis of cocoa butter and its alternatives* PhD thesis Szent István University

IOP Conf. Series: Journal of Physics: Conf. Series **950** (2017) 042008 doi:10.1088/1742-6596/950/4/042008

HHP treatment of liquid egg at 200-350 MPa

A Tóth[1], Cs Németh[2], P Palotás[1], J Surányi[1], I Zeke[1], B Csehi[1], L A Castillo[1], L Friedrich[1] and Cs Balla[1]

[1]Department of Refrigeration and Livestock Products Technology, Szent István University, Budapest, Ménesi út 43-45, 1118 Hungary
[2]Capriovus Ltd., Szigetcsép, Dunasor, 073/72. hrsz., 2317 Hungary

E-mail: csaba.nemeth@capriovus.hu

Abstract. High hydrostatic pressure (HHP) treatment of egg proteins partially limits their sensitivity to pressure. According to the literature, at the 450 MPa level, denaturation of some proteins sets in to the extent that sensory and functional characteristics are impacted. This study involved treating liquid egg (egg white, yolk, and melange) at less than the above-mentioned value, after which the microbiological effect was examined. For the study, pressure pouches were filled with 100ml of raw liquid egg per pouch. Then the samples were treated at 200, 250, 300 and 350 MPa. In each case, the level was reached by increasing pressure at a rate of 100 MPa/min. Measurements were taken at the Corvinus University of Budapest, Faculty of Food Science, Dept. of Refrigeration and Livestock Products Technology RESATO FPU 100-2000 equipment. Denaturation was determined with calorimetric (DSC) tests. From our results, it appears that even at 250 MPa pressure treatment, the viable cell count decreases. Further, it can be said that microbe count went down in the egg white samples at 300-350 MPa, below the impact level. Significant denaturation was not detected during our examinations. In summary, we state that the most HHP-sensitive liquid egg type, egg white, can be pressure treated to reduce microbe count at a level less than that which causes denaturation. Microbe reduction was smaller in yolk and melange, so higher pressure values are appropriate for these products.

1. Introduction

Eggs have been classified as nature's original functional food [1]. Eggs are very important in food for emulsification, leavening, smoothness and flavour, to name a few. Owing to these facts eggs are used in many food products. At present there is a tendency to use more and more eggshell processed products where term egg products refers to eggs that are removed from their shells, like liquid egg products, or boiled eggs [2, 3].

Non-thermal processes might be an excellent alternative to overcome the problems associated to the changes in egg functional properties [4]. Similar to thermal pasteurization, non-thermal pasteurization of eggs is challenging and despite substantial efforts, and none of the non-thermal technologies has been commercialized for liquid egg products, like pulsed electric fields [5], electron beam radiation [6] and gamma radiation, or high hydrostatic pressure [7, 8].

The recent development of alternative processing technologies for food preservation such as high hydrostatic pressure (HHP), offers potential to inactivate microorganisms, reduce loss of essential nutrients, and contribute to the development of novel egg products[8 - 11]. While the mechanism of high pressure-induced protein unfolding is not completely understood, but there is evidence of structural changes in protein molecules that are distinct from those caused by thermal or chemical treatment. The underlying mechanism of pressure-induced protein denaturation involves water penetration into cavities within the molecule resulting in varying population of molecular conformations [12, 13].

The initial microflora of raw liquid egg consist of diverse Gram-negative and Gram-positive bacteria that originate from the shell, or occasional infected egg, or processing equipment (e.g.: egg breakers, pipes, shell filters), as well as from those who handle eggs. In order to produce ultrapasteurized egg products of good bacteriological quality, it is essential both to use eggs within a few hours of their being laid by dedicated flocks of hens and to pay critical attention to the cleanliness and hygiene of the processing equipment [14, 15].

The aim of the present paper is to study the effect of HHP treatment of liquid egg products (liquid whole egg LWE, liquid egg white LEW and liquid egg yolk LEY) between 200 and 350 MPa. Microbiological changes, color and protein denaturation are inspected.

2. Materials and methods

2.1. Samples preparing

Three different liquid egg products were used in our experiment: liquid egg white (LEW), liquid whole egg (LWE) and liquid egg yolk (LEY). All of the samples were taken from production line of Capriovus Ltd (Szigetcsép, Hungary) directly after homogenization.

Samples were stored at refrigerated temperature 2-5°C, a maximum of 24 hours. From every samples 100 mL were packed in polyethylene bags. For every pressure range and for control 3 packages were taken from prepared samples, so altogether 45 bags were prepared for measurements.

2.2. Methods

HHP treatments were carried out at 200, 250, 300 and 350 MPa for 5 min holding time, pressure increases were in every case 100 MPa/min, on a RESATO FPU 100-2000 HHP equipment. Treatment and measurements were at Corvinus University of Budapest (new name: Szent István University), Faculty of Food Science, Department of Refrigeration and Livestock Products Technology.

For color measurement Minolta CR 200 colorimeter was used. From L* (brightness), a* (red-green hue) and b* (yellow-blue hue) values color differences (ΔE_{ab}^*) were counted to untreated control samples.

Thermophysical, calorimetric properties were examined on Micro DSC III (differential scanning calorimeter). In each case approximately 580 mg of samples were taken, reference was distilled water. Speed of heating was 1.5 °C/min, temperature of measuring was 95 °C and speed of cooling was 1.5 °C/min, controlled by SetSoft2000. Callisto 7.6 software was used to evaluate DSC thermograms.

In case of microbiological load of samples mesophilic aerobe cell counts were inspected. As usual, TGA agar and plat clasting were used with an incubation of 30 °C and 48 hours.

Statistical evaluation of results was carried out by paired t-tests.

3. Results and discussion

3.1. Color

Brightness of LWE increased after HHP, higher applied pressures caused higher L* values. By contrast a* and b* decreased effected by increasing pressure. Changes were statistically not significant.

In case of LEW L* increased, therefore samples after processing became brighter, it can be caused of protein's denaturation, or agglomeration. Values of a* increased and b* decreased, but there were no statistically significant.

LEY's color showed a tendency like LWE. Color differences of samples to controls is shown in table 1. Color difference ΔE_{ab}^* defines how visible differences are between color of objects for human eyes. Called our results there were mostly hardly visible, or visible changes in color of treated samples. Based on the results obtained slight color changes are very prospering for future industrial applications, because visible sensorial attributes may not be influenced by HHP treament.

Table 1. The color differences (ΔE_{ab}^*) of samples compared to untreated controls and explanation of classification.

ΔE_{ab}^*	LWE	LEW	LEY
200 MPa	0.38	1.40	1.41
250 MPa	1.04	1.73	0.38
300 MPa	2.44	0.58	0.78
350 MPa	3.51	2.09	0.52

ΔE_{ab}^*	classification
0-0.5	not visible
0.5-1.5	barely visible
1.5-3	visible
3-6	clearly visible
6-	very clearly visible

3.2 Calorimetric parameters

In case of LEW and LWE two denaturation points were observed, that is shown on figure 1 and figure 2 respectively. Enthalpy of denaturation for LEW and temperature of denaturation were decreasing after HHP treatment and increasing of pressure range in both denaturation points. But there were not statistically significant differences, that means, that there is no, with DSC detectable, protein denaturation caused by HHP to 350 MPa.

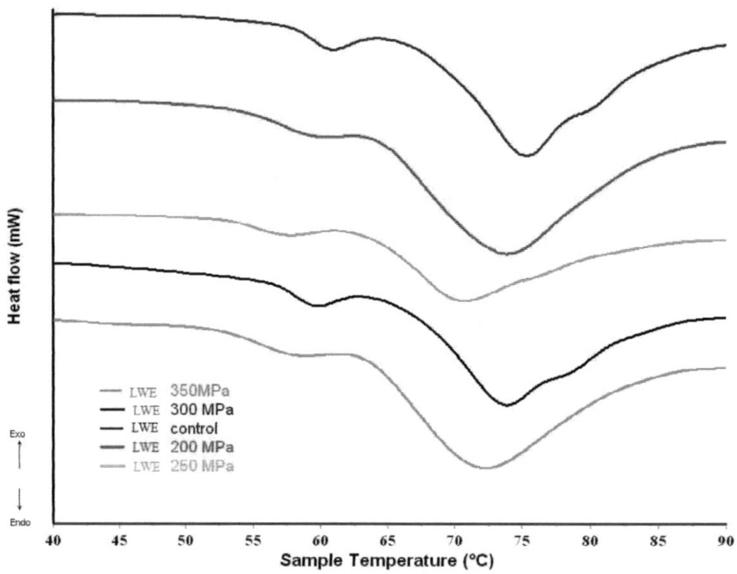

Figure 1. Normalized thermogram of LWE after processings and control

Enthalpy of denaturation was decreasing with increasing pressure of processing's, but the temperature of denaturation showed no interpretable tendency, 3-4°C differences between the samples may be caused by inhomogeneity of LEW . Statistically significant differences were not found.

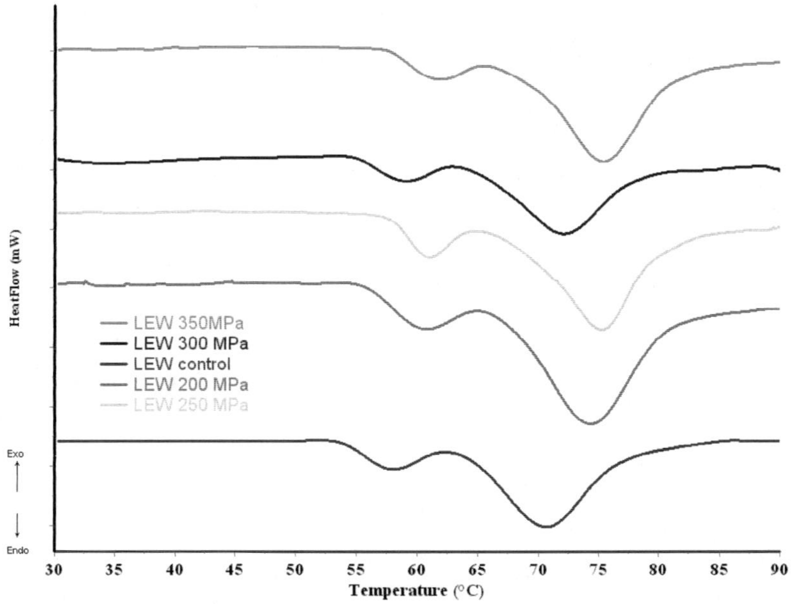

Figure 2. Normalized thermogram of LEW after processing and control

Thermogram of LEY is shown in figure 3, DSC detected only one denaturation point of LEY samples. The enthalpy of denaturation was decreased by increasing pressure, but the main difference was between untreated control and treated samples. Because of decreased enthalpy can be assumed that during HHP treatment proteins are denatured. Temperature changed only 4-5 °C that is insignificant.

Overall we can be seen that our applied HHP treatments do protein denaturation only in the case of LEY, which is detectable by DSC.

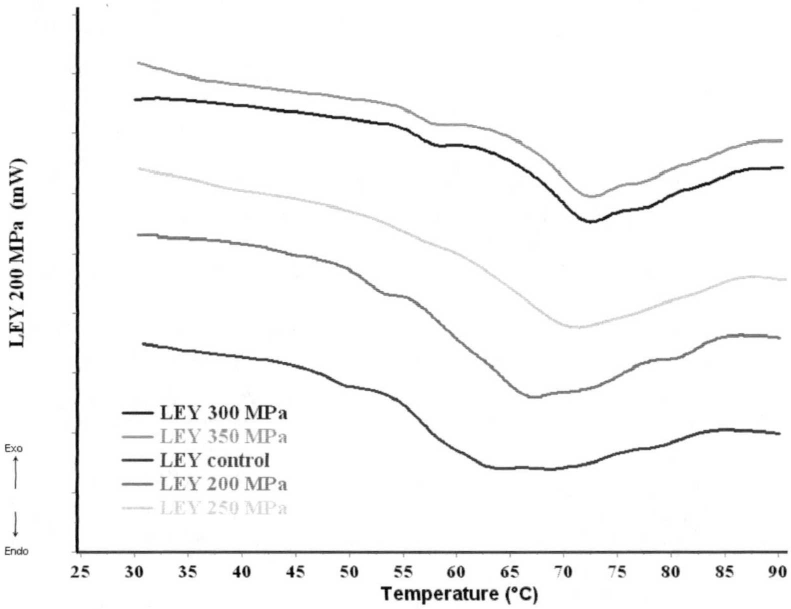

Figure 3. Normalized thermogram of LEY after processing and control

3.3. *Microbiological load*

Mesophyll aerobe cell count of samples is shown in figure 4. Least decreasing of microbiological load was observed in LWE samples, this was only one log magnitude. In case of LEW and LEY the microbiological inactivation effect of HHP was 3 log magnitude.

Our microbiological results (excepted LEY, however microbiological load is still acceptable for commercialization) are promising than increasing pressure ranges effect increasing microbial inactivation, by LEW samples there was no detectable microbiological load at 300 MPa already.

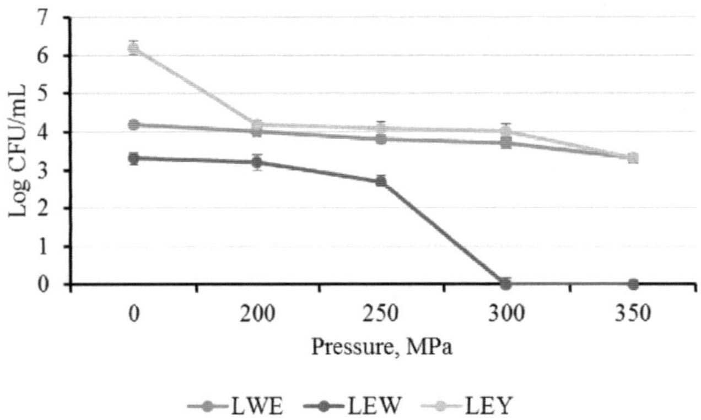

Figure 4. Mesophilic aerobic cell count of LWE, LEW and LEY caused by HHP treatment

4. Conclusion

Our results show that HHP treatment of liquid egg products is a prosperous opportunity for preservation. HHP caused no detectable protein denaturation in LEW and LWE between 200 and 350 MPa, which may offer the same techno-functional properties like raw products (e.g. foaming ability). Color of every samples was not significant changed that is very important because of consumer's choices.

In our study microbiological inactivation effect of HHP is shown in case of egg products. Examined low pressure ranges can effect already 3 log magnitude decreasing in microbiological load of LEW and LEY. Increasing of pressure we can enhance effectivity of HHP treatment and microbiological safe products can be produced.

References

[1] Ponce E, Pla R, Sendra E, Guamis B, and Mor-Mur M. 1999 Destruction of Salmonella enteritidis inoculated in liquid whole egg by high hydrostatic pressure: Comparative study in selective and non-selective media *Food Microbiology* **16** 357–365

[2] Kumbár V, Nedomova Š, Strnková J and Buchar J 2015 Effect of egg storage duration on the rheology of liquid egg products *Journal of Food Engineering* **156** 45–54

[3] Hasler C M 2000 The changing face of functional foods *J. Am. Coll. Nutr.* **19** (5), 499–506

[4] de Souza P A, Müller A, Beniaich A, Mayer-Miebach E, Oehlke K, Stahl M, Greiner R and Fernández A 2015 Functional properties and nutritional composition of liquid egg products treated in a coiled tube UV-C reactor *Innovative Food Science & Emerging Technologies* **32** 156-164

[5] Amiali M, Ngadi M O, Smith J P and Raghavan V G S 2006 *Journal of Food* **71** 88–94

[6] Wong Y C, Herald T J and Hachmeister K A 1996 Comparison between irradiated and thermally pasteurized liquid egg whites on functional, physical and microbiological properties *Poultry Science* **75** 803–808

[7] Andrássy É, Farkas J, Seregély Z S, Dalmadi I, Tuboly E and Lebovics V 2006 *Acta Alimentaria* **35** (3), 305–318

[8] Ponce E, Pla R, Capellas M, Guamis B, and Mor-Mur M 1998 Changes of hen eggs and their components caused by non-thermal pasteurizing treatments II. Some non-microbiological

effects of gamma irradiation or hydrostatic pressure processing on liquid egg white and egg yolk *Food Microbiology* Inactivation of Escherichia coli inoculated in liquid whole egg by high hydrostatic pressure **15** 265–272

[9] Ponce E, Pla R, Mor-Mur M, Gervilla R, and Guamis B 1998 Inactivation of Listeria innocua inoculated in liquid whole egg by high hydrostatic pressure *Journal of Food Protection* **61** 119–122

[10] San Martin M F, Barbosa - Cánovas G V and Swanson B G 2002 Food processing by high hydrostatic pressure *Critical Reviews in Food Science and Nutrition* **42** 627–645

[11] Shebuski J R and Freier T A 2009 Microbiological Spoilage of eggs and egg products Compendium of the Microbiological Spoilage of Foods and Beverages, *Food Microbiology and Safety* ed Sperber W H and Doyle M P (London: Springer Science) pp 121-134

[12] Hoppe A, Jung S, Patnaik A and Zeece M G 2013 Effect of high pressure treatment on egg white protein digestibility and peptide products *Innovative Food Science and Emerging Technologies* **17** 54–62.

[13] Silva J L, Foguel D and Royer C A 2001 Pressure provides new insights into protein folding, dynamics and structure *Trends in Biochemical Sciences* **26** 612–618

[14] Delves-Broughton J 2014 Encyclopedia of food microbiology 1 Microbiology of Egg Products ed Batt C A and Tortorello M L (London: Academic Press) pp 617-621.

[15] Hayashi R, Kawamura Y, Nakasa T and Okinaka O 1989 Application of high pressure to food processing: Pressurization of egg white and yolk, and properties of gels formed *Agricultural and Biological Chemistry* **53** 2935–2939

IOP Conf. Series: Journal of Physics: Conf. Series **950** (2017) 042009 doi:10.1088/1742-6596/950/4/042009

Crystal structure and superconductivity in atomic hydrogen: Deformation between $I4_1/amd$ and $Fddd$

T Ishikawa[1], H Nagara[2], T Oda[3], N Suzuki[4] and K Shimizu[1]

[1] Center for Science and Technology under Extreme Conditions, Graduated School of Engineering Science, Osaka University, 1-3 Machikaneyama, Toyonaka, Osaka 560-8531, Japan

[2] Department of Applied Mathematics and Physics, Tottori University, 4-104 Koyama-cho-minami, Tottori, Tottori 680-8552, Japan

[3] Institute of Science and Engineering, Kanazawa University, Kakuma, Kanazawa, Ishikawa 920-1192, Japan

[4] Department of Pure and Applied Physics, Faculty of Engineering Science, Kansai University, 3-3-35 Yamate, Suita, Osaka 564-8680, Japan

E-mail: ishikawa@stec.es.osaka-u.ac.jp

Abstract. We investigated crystal structures of solid metallic hydrogen using the potential energy surface trekking for structure search. We applied this technique to a tetragonal $I4_1/amd$ structure at pressures of 500 and 600 GPa and obtained the transformation into multiple orthorhombic $Fddd$ structures, which are formed by distortions in the ab plane of $I4_1/amd$. The potential barriers are easily surmounted by few trekking steps, which indicates that in solid metallic hydrogen crystal structure is softened with respect to the distortion and is easily fluctuated among the $I4_1/amd$ and $Fddd$ structures. Calculated superconducting critical temperatures show 269 K for $I4_1/amd$ and 263 K for $Fddd$ at 500 GPa. The structures are softened and the electron-phonon coupling are enhanced with pressurization to 600 GPa. As the results, the superconducting critical temperature is increased to 281 K for $I4_1/amd$, whereas it is decreased to 252 K for $Fddd$ owing to its larger phonon softening than that of $I4_1/amd$.

1. Introduction

Solid hydrogen has theoretically been predicted to be a room-temperature superconductor [1–3], and its metallic phases have been explored for decades. High-pressure experimental studies have revealed at least four solid phases in pressure region to 315 GPa [4]. The H_2 molecules are in a spherical rotational state on the close-packed lattice (phase I) at low pressures [5], and it transforms into an anisotropic one (phase II) at around 110 GPa owing to the increase of the intermolecular interactions [6]. By further compression below 125 K, another state (phase III) emerges at around 155 GPa [7, 8], accompanying large discontinuities in vibron frequencies. Recently, a phase IV has been discovered at 220 GPa and 300 K by Raman and visible transmission spectroscopy, where the structure has been thought to form a mixture of molecular and atomic hydrogens [4]. Furthermore, it has been found that the phase IV takes the transition into a new phase at around 255 GPa and 480 K, which is possibly based on melting [9]. Above 220 GPa, conductive hydrogen has experimentally been observed [10], but complete metalization has not been obtained.

First-principles calculations predict that a hexagonal $P6_3/m$ is the most stable structure up to around 105 GPa of candidate structures and then a monoclinic $C2/c$ of a layered structure

IOP Conf. Series: Journal of Physics: Conf. Series **950** (2017) 042009 doi:10.1088/1742-6596/950/4/042009

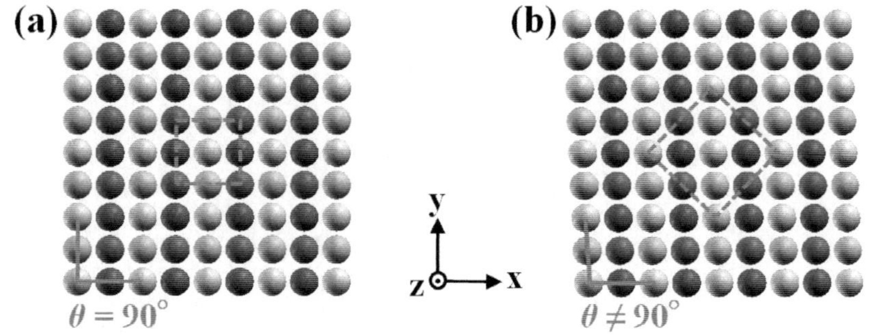

Figure 1. Projection of crystal structures for solid metallic hydrogen onto the xy plane: (a) tetragonal $I4_1/amd$ and (b) orthorhombic $Fddd$, in which the H_2 molecules are completely dissociated. The unit cells are shown by a broken line and the atoms with different z values are represented by different colors. The structures are characterized by distortion of an angle θ: $\theta = 90°$ for $I4_1/amd$ and $\theta \neq 90°$ for $Fddd$.

prevails to around 270 GPa [11], whose vibrational spectra agree somewhat well with the experimental Raman data collected from the phase III. The $C2/c$ structure is predicted to transform into an orthorhombic $Cmca$ with a primitive cell containing 12 atoms ($Cmca$-12) at 385 GPa and into another $Cmca$ containing 4 atoms ($Cmca$-4) by further compression [11]. Subsequent transition into an exotic quasi-molecular mC24 structure is expected at around 470 GP, which has a space group of a monoclinic $C2/c$ [12]. Above 590 GPa an atomic metallic phase of a tetragonal $I4_1/amd$ is predicted to become the most stable structure [12, 13]. The tetragonal $I4_1/amd$ prevails over a wide range of pressure and transforms into another atomic phase with a space group of an orthorhombic $Cmcm$, called oC12 structure, above 2100 GPa [12]. A trigonal $R\bar{3}m$ structure with an ABC stacking is also predicted as another higher-pressure structure, which transforms into the face-centered cubic structure near 3500 GPa [13]. In the atomic metallic phase, the superconducting critical temperatures (T_c) have been calculated by the use of the Allen-Dynes formula [14], and T_c has been predicted to reach 340 K at around 1000 GPa in $I4_1/amd$ and 400 K at around 3500 GPa in $R\bar{3}m$ [2].

Recently, we have obtained an orthorhombic $Fddd$ structure in the metallic atomic phase using the genetic algorithm technique for crystal structure search [15]. The obtained structure is formed by a slight distortion in the ab plane of the tetragonal $I4_1/amd$, and the crystal symmetry is lowered. The structures are characterized by distortion angle θ as shown in Fig. 1: $\theta = 90°$ for $I4_1/amd$ and $\theta \neq 90°$ for $Fddd$. In pressure region of 1800-2000 GPa, where $Fddd$ mainly emerges as the most stable structure, the electron-phonon coupling is enhanced and T_c calculated by the Allen-Dynes formula reaches 350 K.

In the present study, we focus on the two atomic phases of solid hydrogen, $I4_1/amd$ and $Fddd$, at pressures of 500 and 600 GPa. We perform further investigations on the structural stability and the superconductivity of the two structures via another structure search technique, which we have developed and called the potential energy surface trekking (PEST) [16, 17].

2. Methods

In the PEST technique, the system is forced to ascend the potential energy surface to its ridges through various trekking paths following by the inversion of restoring force (ascent-run) and then to descend toward neighboring local minima by the release from the inversion (descent-run) [16, 17]. Since the trekking paths are independent from each other, many stable

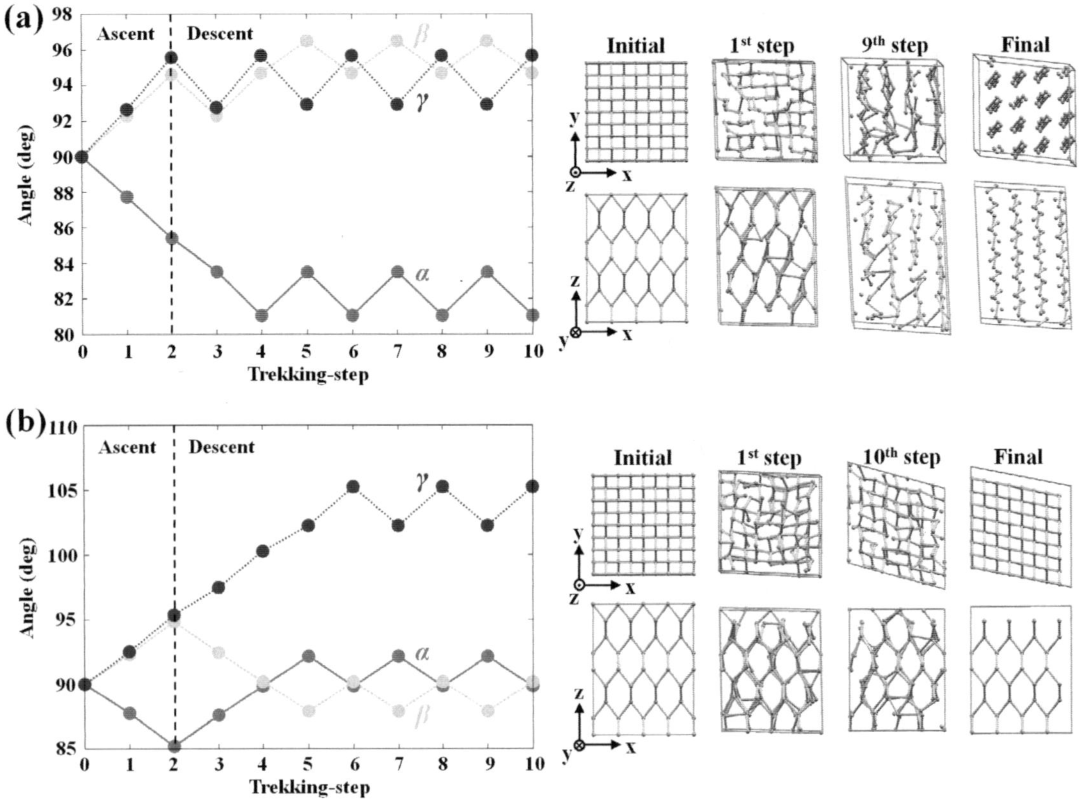

Figure 2. Evolution of cell angles in the PEST simulations: (a) a trekking path at 500 GPa and (b) a trekking path at 600 GPa. The structural transformations obtained through the trekking are shown on the right side.

structures existing around a target structure can be simultaneously and rapidly obtained. We have combined our PEST code with the Quantum ESPRESSO code [18], in which electronic structures are calculated within the density functional theory (DFT) by the use of plane waves and pseudopotentials.

In the DFT calculations, the generalized gradient approximation by Perdew, Burke and Ernzerhof [19] was used for the exchange-correlation functional, and the Rabe-Rappe-Kaxiras-Joannopoulos ultrasoft pseudopotential [20] was employed. The supercell including $4 \times 4 \times 2$ $I4_1/amd$ primitive cells, including 128 hydrogen atoms, was employed as the starting structure for the PEST simulations. The k-space integration over the Brillouin zone was performed on a $6 \times 6 \times 4$ grid, and the energy cutoff of the plane wave basis was set at 72 Ry. We created 64 starting directions for the trekking by adding the distortions to the $I4_1/amd$ supercell, *i.e.* adding either -0.2 or $+0.2$Å to six independent elements of a cell matrix $\mathbf{h} = (\mathbf{a}, \mathbf{b}, \mathbf{c})$, where $\mathbf{a}$, $\mathbf{b}$ and $\mathbf{c}$ are lattice vectors of the supercell. In the ascent-run, the matrix $\mathbf{h}$ is updated by the inversion of the restoring force F as follows: $h_{ij}^{t+1} = h_{ij}^{t} - (\delta h_{ij}) F_{ij}^{t} / |F_{ij}^{t}|$, where t is the number of trekking steps and δh_{ij} stepping parameters. The restoring force F is defined as a set of derivatives of the Gibbs free energy $\mathcal{G}(\mathbf{h})$ with respect to the six parameters of $\mathbf{h}$: $F_{ij} = -\partial \mathcal{G} / \partial h_{ij} = V\{[(\mathbf{p} - P\mathbf{I})\mathbf{h}^{-1}]_{ij} + [(\mathbf{p} - P\mathbf{I})\mathbf{h}^{-1}]_{ji}\}(1 - \delta_{ij}/2)$, where V is volume, $\mathbf{p}$ an averaged stress tensor, P a given pressure, $\mathbf{I}$ the unit tensor, and δ_{ij} the Kronecker delta [21].

Table 1. Results of the PEST simulations at pressures of 500 and 600 GPa. N_{tp} represents the number of the trekking paths leading to the corresponding structure.

500 GPa		600 GPa	
Structure	N_{tp}	Structure	N_{tp}
$I4_1/amd$ ($\theta = 90.0°$)	2	$Fddd$ ($\theta = 91.5°$)	5
$Fddd$ ($\theta = 91.2°$)	1	$Fddd$ ($\theta = 92.9°$)	1
$Fddd$ ($\theta = 91.8°$)	1	$Fddd$ ($\theta = 94.5°$)	1
$Fddd$ ($\theta = 92.6°$)	2	$Fddd$ ($\theta = 97.6°$)	2
$Fddd$ ($\theta = 95.4°$)	2	$Fddd$ ($\theta = 101.2°$)	2
$Fddd$ ($\theta = 96.8°$)	1		
$Fddd$ ($\theta = 102.4°$)	1		
$Cmca$-4	2		

For the stepping parameters, δh_{11}, δh_{22}, and δh_{33} were set at 0.4 Å, and δh_{12}, δh_{13}, and δh_{23} at 0.2 Å. An molecular dynamics (MD) run of 0.2 ps (200 MD steps) was carried out at each trekking step, and **p** was obtained by the average of the stress tensors for the last half of the MD steps. In the descent-run, the matrix **h** is updated by $h_{ij}^{t+1} = h_{ij}^{t} + (\delta h_{ij})F_{ij}^{t}/|F_{ij}^{t}|$. Pressure-temperature conditions for the simulations are 500 GPa and 300 K, and 600 GPa and 300 K.

The superconducting T_c was calculated by the use of the Allen-Dynes formula [14],

$$T_c = \frac{\omega_{\log}}{1.2} \exp\left[-\frac{1.04(1+\lambda)}{\lambda - \mu^*(1+0.62\lambda)}\right]. \tag{1}$$

In Eq. 1, the parameters of electron-phonon coupling constant λ and logarithmic-averaged phonon frequency $\omega_{\log}$ represent a set of characters for the phonon-mediated superconductivity. Using the QE code, we calculated these parameters with a $4 \times 4 \times 4$ q-point grid. The k-space integration was performed on a $48 \times 48 \times 24$ grid, and the energy cutoff of the plane wave basis was set at 80 Ry. The effective screened Coulomb repulsion constant μ^* was assumed to be 0.089, which is deduced as a reasonable value for atomic metallic hydrogen [22].

3. Results

Figure 2 (a) shows the evolution of three cell angles, α, β, and γ, obtained in a trekking path at 500 GPa. The ascent-run was switched to the descent-run at the second trekking step, and finally α decreased from 90° to about 82°, and β and γ increased to around 94°. We carried out the structural optimization at the 10th trekking step and obtained the structure consisting of H_2 molecules (see the rightmost structure in the figure). This structure is consistent with the earlier-predicted $Cmca$-4 structure, which has been considered as a candidate of the molecular phase in lower pressure region. The similar results obtained by another trekking path at 600 GPa are shown in Fig. 2 (b). This trekking path also shows the switching from the ascent-run to the descent-run at the second trekking step. The angles, α and β, remained at 90°, whereas γ increased to about 104° through 10 trekking steps. In contrast to the results shown in Fig. 2 (a), the atomic structure remained through the simulation, and the final structure became the $Fddd$ structure predicted as a candidate of the metallic phase.

Table 1 lists the obtained structures and the number of the trekking paths leading to them at 500 and 600 GPa. At 500 GPa, the structure returns to the starting $I4_1/amd$ with $\theta = 90°$ through two trekking paths, whereas it transforms into the $Fddd$ structure with $\theta \neq 90°$ through eight trekking paths. The $Fddd$ structure takes six stable states with respect to the change of

IOP Conf. Series: Journal of Physics: Conf. Series **950** (2017) 042009 doi:10.1088/1742-6596/950/4/042009

Table 2. Density of states at the Fermi level $N(E_F)$, electron-phonon coupling constant λ, logarithmic-averaged phonon frequency ω_{log}, and superconducting critical temperature T_c for $I4_1/amd$ and $Fddd$.

P (GPa)	Structure	$N(E_F)$ (states/Ry/proton)	λ	ω_{log} (K)	T_c (K)
500	$I4_1/amd$ ($\theta = 90°$)	0.2272	2.2332	1706	269.3
	$Fddd$ ($\theta = 92.88°$)	0.2301	2.0774	1743	263.2
600	$I4_1/amd$ ($\theta = 90°$)	0.2288	2.6920	1610	280.8
	$Fddd$ ($\theta = 91.42°$)	0.2241	2.5663	1477	251.5

the angle θ: 91.2°, 91.8°, 92.6°, 95.4°, 96.8°, and 102.4°. The enthalpy differences among them are less than 0.4 mRy/proton. In addition, two trekking paths lead to $Cmca$-4 with the molecular structure as shown in Fig. 2 (a). The other trekking paths lead to unstable structures obtained by disorderly freezing of H atoms in the course of the trekking, on which we have no further discussion in this paper. The transformation into multiple $Fddd$ structures was also obtained at 600 GPa. 11 trekking paths lead to the $Fddd$ structures, which show five stable θ values, 91.5°, 92.9°, 94.5°, 97.6°, and 101.2°. The number of the stable θ values including $\theta = 90°$ decreases from seven to five by pressurization from 500 to 600 GPa, which indicates that the potential wells get to be shallow with the pressurization and few local minima of the potential surface disappear. In addition, the potential barriers between $I4_1/amd$ and $Fddd$ are easily surmounted by few trekking steps. These results of the PEST simulations suggest that in metallic hydrogen crystal structure is softened with respect to the distortion of the angle θ and is easily fluctuated among the $I4_1/amd$ and $Fddd$ structures. The molecular $Cmca$-4 structure got to be unstable with pressurization and was not obtained by the structure search at 600 GPa.

We investigated the superconductivity of $I4_1/amd$ and $Fddd$ (Table 2). Though the $I4_1/amd$ structure was not obtained by the PEST simulations at 600 GPa, its superconducting parameters were calculated at the pressure for comparison. The two structures are metallic at the pressures, and the values of the density of states at the Fermi level $N(E_F)$ are in good agreement with each other. At 500 GPa, T_c shows the almost same value between the two structures: 269.3 K for $I4_1/amd$ and 263.2 K for $Fddd$. Both the structures are softened with pressurization to 600 GPa, and ω_{log} decreases by 5.6% for $I4_1/amd$ and by 15.3% for $Fddd$. This pressure-induced phonon softening is related to the results of PEST, the shallowing of the potential wells. The parameter λ increases by 0.46 for $I4_1/amd$ and 0.49 for $Fddd$ by the pressurization. However, there is a difference between the two structures with respect to the pressure dependence of T_c. The superconducting T_c increases by 11 K for $I4_1/amd$, whereas it decreases by 12 K for $Fddd$. This is caused by the difference of the phonon softening magnitude between $I4_1/amd$ and $Fddd$. The phonon softening involves the decrease of ω_{log} and the increase of λ. If the effect of the ω_{log} decreasing is smaller than that of the λ increasing, then T_c is enhanced by the softening. On the other hand, if the effect of the ω_{log} decreasing is larger, then T_c is decreased. Thus, $I4_1/amd$ corresponds to the former and $Fddd$ the latter.

4. Conclusion

We explored crystal structures in metallic hydrogen by applying PEST to the supercell of the tetragonal $I4_1/amd$ structure including 128 atoms. The angle θ, corresponding to a cell angle in the ab plane and characterizing the structure, was easily changed from 90° through few trekking steps and the following angles were obtained as stable states: $\theta = 91.2°$, 91.8°, 92.6°, 95.4°, 96.8°, and 102.4° at 500 GPa, and 91.5°, 92.9°, 94.5°, 97.6°, and 101.2° at 600 GPa. The number of the stable states decreases by the pressurization from 500 to 600 GPa, which indicates that the potential wells get to be shallow with the pressurization. These results suggest that crystal structure is softened with respect to the distortion and is easily fluctuated among the $I4_1/amd$ and $Fddd$ structures at the pressures. Bonev *et al.* predicted the decrease of melting temperature in hydrogen above 90 GPa [23], and Eremets and Trojan observed it by laser heating diamond anvil cell experiments, in which the melting temperature takes the maximum value of 1050 ± 60 K at 106 GPa and decreases to 880 ± 50 K by compression to 146 GPa [24]. In 2013, Chen *et al.* investigated the melting of hydrogen in higher pressure region using first-principles path-integral molecular dynamics simulations and predicted the transition from the metallic atomic solid phase into a low-temperature metallic atomic liquid phase at around 800 GPa [25]. Therefore, our results could be interpreted as the sign of the pressure-induced melting of the atomic solid hydrogen reported earlier.

The superconducting T_c shows 269.3 K for $I4_1/amd$ and 263.2 K for $Fddd$ at 500 GPa. These structures both show the pressure-induced phonon softening, and $\omega_{\log}$ is decreased by 5.6% for $I4_1/amd$ and by 15.3% for $Fddd$. On the other hand, λ is increased by 0.46 for $I4_1/amd$ and 0.49 for $Fddd$. As the results, T_c is enhanced to 280.8 K for $I4_1/amd$, whereas it is decreased to 251.5 K for $Fddd$ owing to the effect of the larger phonon softening than that in $I4_1/amd$.

Acknowledgments

This work was supported by JSPS KAKENHI, Grant-in-Aid for Specially Promoted Research (26000006) and Grant-in-Aid for Young Scientists (B) (15K17707).

References

[1] Ashcroft N W 1968 *Phys. Rev. Lett.* **21** 1748
[2] McMahon J M and Ceperley D M 2011 *Phys. Rev. B.* **84** 144515
[3] McMahon J M and Ceperley D M 2012 *Phys. Rev. B.* **85** 219902(E)
[4] Howie R T, Guillaume C L, Scheler T, Goncharov A F and Gregoryanz E 2012 *Phys. Rev. Lett.* **108** 125501
[5] Loubeyre P, LeToullec R, Hausermann D, Hanfland M, Hemley R J, Mao H K and Finger L W 1996 *Nature* **383** 702
[6] Cui L, Chen N H and Silvera I F 1995 *Phys. Rev. B* **51** 14987
[7] Hemley R J and Mao H K 1988 *Phys. Rev. Lett.* **61** 857
[8] Goncharov A F, Hemley R J, Mao H K and Shu J 1998 *Phys. Rev. Lett.* **80** 101
[9] Howie R T, Dalladay-Simpson P and Gregoryanz E 2015 *Nat. Mater.* **14** 495
[10] Eremets M I and Troyan I A 2011 *Nat. Mater.* **10** 927
[11] Pickard C J and Needs R J 2007 *Nature Physics* **3** 473
[12] Liu H, Wang H and Ma Y 2012 *J. Phys. Chem. C* **116** 9221
[13] McMahon J M and Ceperley D M 2011 *Phys. Rev. Lett.* **106** 165302
[14] Allen P B and Dynes R C 1975 *Phys. Rev. B* **12** 905
[15] Ishikawa T, Nagara H, Oda T, Suzuki N and Shimizu K 2014 *Phys. Rev. B* **90** 104102
[16] Ishikawa T, Suzuki N and Shimizu K 2014 *J. Phys.: Conf. Ser.* **500** 162003
[17] Ishikawa T 2014 *Comput. Mater. Sci.* **92** 36
[18] Giannozzi P *et al.* 2009 *J. Phys.: Condens. Matter* **21** 395502–395520
[19] Perdew J P, Burke K and Ernzefhof M 1996 *Phys. Rev. Lett.* **77** 3865
[20] Rappe A M, Rabe K M, Kaxiras E and Joannopoulos J D 1990 *Phys. Rev. B* **41** 1227–1230
[21] Allen M P and Tildesley D J (eds) 1989 *Computer Simulation of Liquids* (Oxford University Press)
[22] Richardson C F and Ashcroft N W 1997 *Phys. Rev. Lett.* **78** 118
[23] Bonev S A, Schwegler E, Ogitsu T and Galli G 2004 *Nature* **431** 669
[24] Eremets M I and Trojan I A 2009 *Lett. JETP* **89** 174
[25] Chen J, Li X Z, Zhang Q, Probert M I J, Pickard C J, Needs R J, Michaelides A and Wang E 2013 *Nat. Commun.* **4** 2064

New tetrahedral polymorphs of the group-14 elements

A Mujica[1], C J Pickard[2] and R J Needs[3]

[1] Departamento de Física, MALTA Consolíder Team, Universidad de La Laguna, La Laguna 38206, Tenerife, Spain

[2] Department of Materials Science & Metallurgy, University of Cambridge, 27 Charles Babbage Road, Cambridge CB3 0FS, United Kingdom

[3] Theory of Condensed Matter Group, Cavendish Laboratory, J J Thomson Avenue, Cambridge CB3 0HE, United Kingdom

E-mail: amujica@ull.es

Abstract. In this short communication we present the main results of an *ab initio* structural search for new polymorphs of the group-14 elements C, Si, and Ge, with a focus on low-density and energetically favorable sp^3-bonded tetrahedral polymorphs which could constitute candidate phases to be synthesized in the laboratory. We propose new orthorhombic and tetragonal polymorphs with enthalpies that are lower or comparable to those of other polymorphs of the group-14 elements that have either been experimentally synthesized or recently proposed as the structure of unknown phases obtained in experiments. The new phases have interesting structural and electronics properties. These include an orthorhombic phase that in Si is found to have a quasi-direct band gap at the zone center of about the right size for optimal photovoltaic applications, and a tetragonal phase with a chiral topology that displays an exotic pentagonal pattern and whose vibrational properties match those of an unknown phase of Si.

1. Introduction

The polymorphism of the technologically relevant group-14 elements carbon, silicon, and germanium has been intensively studied over the years and is still attracting much interest. A number of high pressure phases have been obtained by applying pressure to the cubic diamond ambient forms of silicon and germanium [1] whereas the versatility that stems from the ability of carbon to form sp, sp^2, and sp^3 hybrid bonds has resulted in a variety of synthetic carbon allotropes with a wide range of different structures and properties [2-6]. A number of zero-pressure metastable tetrahedral polymorphs of these materials, with both higher and lower densities than the corresponding diamond phases, have been experimentally observed after release of pressure from the stable high-pressure phases, in nanoindentation experiments or by other synthesis methods which involve the use of high pressure [1,6-8].

In this short conference communication (a fuller account of this study can be found in [9]) we present several energetically favorable and dynamically stable tetrahedral polymorphs of the group-14 elements that have been theoretically obtained by an *ab initio* random structure search performed within the framework of the density functional theory. These hypothetical structural forms make promising candidate metastable phases with the potential to be synthesized experimentally and have interesting crystallographic and electronics properties, including an

IOP Conf. Series: Journal of Physics: Conf. Series **950** (2017) 042010 doi:10.1088/1742-6596/950/4/042010

orthorhombic phase with 24 atoms per unit cell that in Si is found to have a quasi-direct band gap at the zone center of about the right size for optimal photovoltaic applications, and a tetragonal phase with 20 atoms per unit cell with a chiral topology that displays an exotic pentagonal pattern.

2. Details of the calculations

All total-energy calculations were performed within the framework of the density functional theory (DFT) whereas the computational searches for low-energy structures were performed using the *ab initio* random structure searching (AIRSS) approach developed by Pickard and Needs [10]. In AIRSS, ensembles of randomly chosen initial structures are relaxed to minima of the energy. This scheme has been successfully used in finding low-energy structures in many systems and the details of the present searches are similar to those reported in previous AIRSS studies [10]. We performed searches for structures with up to 24 atoms per unit cell, using the CASTEP [11] DFT code and the Perdew-Burke-Ernzerhof (PBE) [12] generalized gradient approximation (GGA) to the exchange-correlation (XC) functional in a plane-waves scheme that used ultrasoft pseudopotentials [13] with the outermost four valence electrons of each group-14 atom being treated explicitly and the innermost core electrons considered frozen. Default basis-set kinetic-energy cutoffs (as implemented in the CASTEP package) and convergence parameters were used for a first screening of candidate structures, further relaxing the configurations of interest to a higher level of accuracy, using a **k**-point grid of spacing $2\pi\times0.03$ Å^{-1} for the Brillouin zone integrations [14].

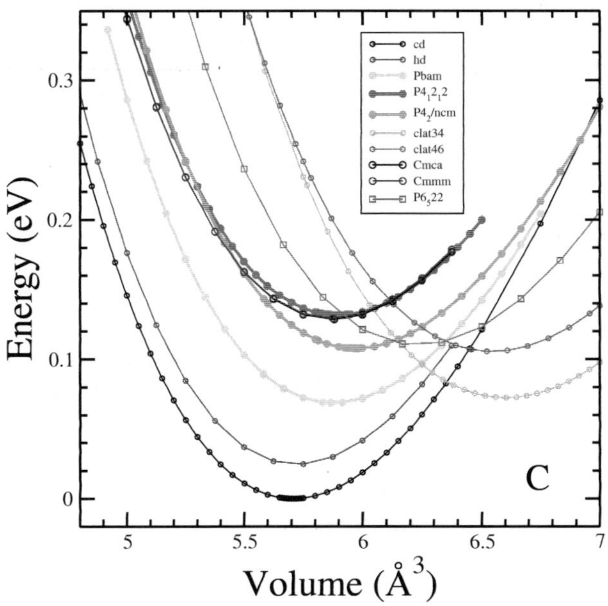

Figure 1. Energy-volume curves for several selected phases of C, with the different symbols corresponding to our calculated *E*-*V* values using VASP and the PBE. The novel phases discussed in the main text are shown in thicker line and solid symbols. The meaning of each curve's label is given in the main text.

These structures were then further relaxed to a very high accuracy and exhaustively studied (energetics, vibrational and electronic properties, pressure evolution, as well as the effect of different XC functionals, namely the LSDA and PBEsol [15]) using the Vienna *ab initio* simulation package (VASP) [16], which implements DFT within a projector augmented-wave (PAW) scheme [17], obtaining results in very good agreement with those obtained with CASTEP, which is a useful crosscheck. At this point we also studied the effect of the semicore *3d* electrons of Ge on the energetics of the tetrahedral phases that resulted from our study, which was found to be very small. The cutoff in the kinetic energies of the plane wave basis set used in the VASP

IOP Conf. Series: Journal of Physics: Conf. Series **950** (2017) 042010 doi:10.1088/1742-6596/950/4/042010

calculations wth the PBE were 520 eV (C), 320 eV (Si), and 230 eV (Ge, or 375 eV when the $3d$ electrons were explicitly included), with dense **k**-point meshes for the integrations within the Brillouin zones of each of the phases (e.g. $8 \times 8 \times 8$ for the Pbam and $P4_12_12$ phases, with 24 and 20 atoms per cell, respectively, see below).

All calculations shown here are for hydrostatic conditions and zero temperature, and the small zero-point energy was neglected. Under hydrostatic conditions, the structural optimization was conducted using the calculated forces on the atoms and the stress tensor, and full convergence was considered to have been achieved when the forces were less than 5 meV $Å^{-1}$ and the anisotropy in the diagonal components of the stress tensor was less than 0.1 GPa. For each phase, the calculated total energy, E, as a function of volume, V, was fitted using a fourth-order Birch-Murnaghan equation of state (EoS) [18] from which the pressure, p, and the enthalpy, $H = E + pV$, were obtained, as well as volumes and bulk moduli at equilibrium. Phonon frequencies were calculated within the finite displacement supercell method using the PHON code [19]. The phonon band structures were used to investigate the dynamical stability of the new phases.

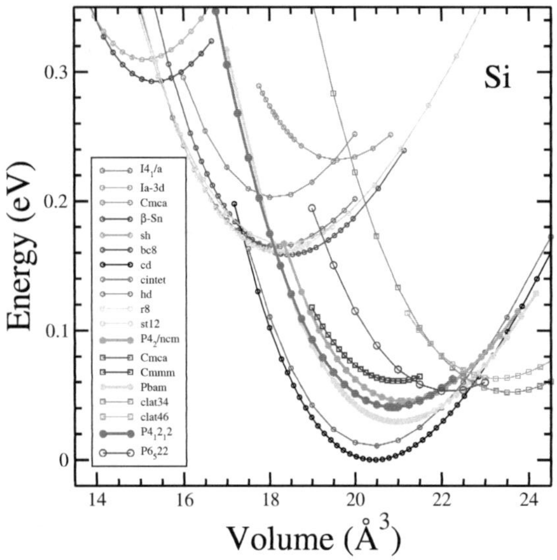

Figure 2. Energy-volume curves for several selected phases of Si, with the different symbols corresponding to our calculated E-V values using VASP and the PBE. The novel phases discussed in the main text are shown in thicker line and solid symbols. The meaning of each curve's label is given in the main text.

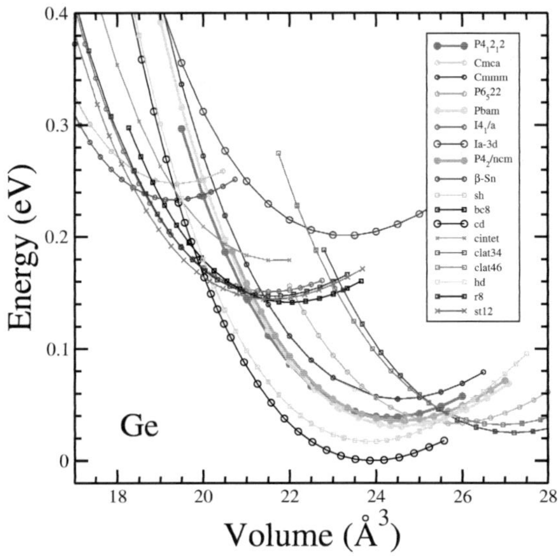

Figure 3. Energy-volume curves for several selected phases of Ge, with the different symbols corresponding to our calculated E-V values using VASP and the PBE. The novel phases discussed in the main text are shown in thicker line and solid symbols. The meaning of each curve's label is given in the main text.

3. Results and discussion

Figures 1-3 show the calculated energy-volume (E-V) curves for a selection of structural phases of C, Si, and Ge, respectively. In these figures we have used standard abbreviations for well known phases (see for example Ref. [1]) whereas we have followed the standard practice of using the space group symbol for denoting the novel tetrahedral phases. In this brief account we will summarily comment on several of these novel phases and direct the reader to Ref. [9] for a fuller report of our study. Among the well known phases included in Figures 1-3 we notice the cubic and hexagonal diamond phases (cd and hd, respectively), the metallic high-pressure β-Sn and simple hexagonal (sh) phases, and the high-density tetrahedral polymorphs known as bc8, r8, and st12 [1]. Ia$\bar{3}$d is the same structure reported for C in [20]. The structure labelled cintet in Figures 2 and 3 is the elemental analog of the tetrahedral variant of the cinnabar structure observed at high pressures in ZnTe and GaAs [1], whereas the high-density polymorph $I4_1/a$ was found to play an important role in the decompression kinetics from the high-pressure phases of Si and Ge [21]. All these structures came out during the search process.

Figures 4 and 5 show the structures of the low-density orthorhombic Pbam and tetragonal $P4_12_12$ phases, found with AIRSS, and which according to our calculations are dynamically stable and quite low in energy for each of the group-14 elements studied (see Fig. 1-3), and should thus be considered energetically competitive polymorphs for these materials at zero or low pressure. Both Pbam and $P4_12_12$ are slightly expanded with respect to the corresponding diamond phases, with equilibrium volumes larger than diamond by 2%-3%, and bulk moduli slightly below those of the diamond forms [9]. For comparison, the low-density clathrates of types I and II (referred to as clat46 and clat34, respectively) are also shown in Figures 1-3, as well as the $P6_522$ chiral framework structure previously obtained by Pickard and Needs [22], whose density is intermediate between that of the clathrates and that of Pbam and $P4_12_12$.

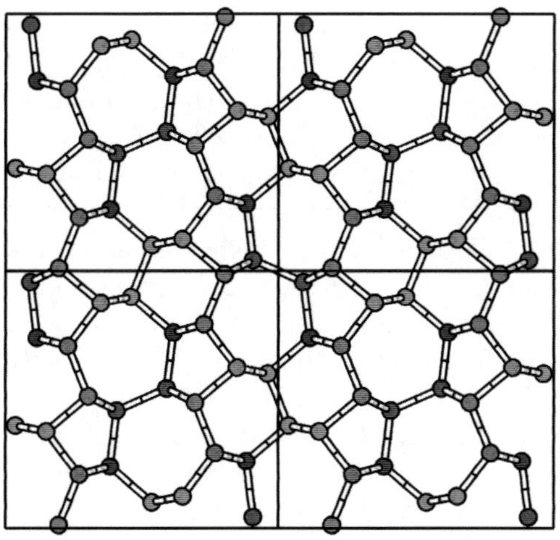

Figure 4. A plot of the Pbam structure viewed along the z axis, with its six different Wyckoff sites represented by different colors. Fivefold and sevenfold rings are clearly visible.

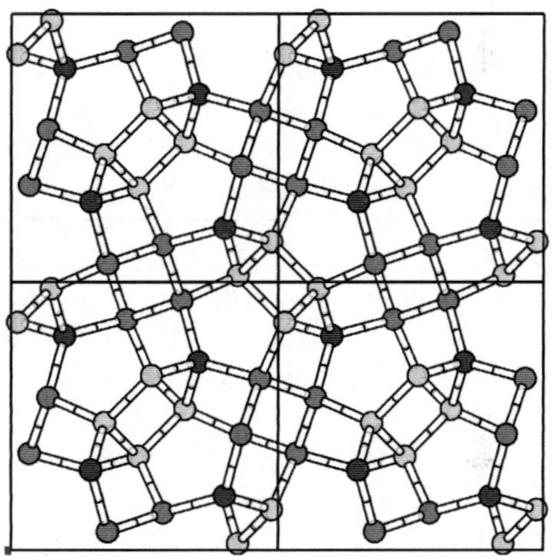

Figure 5. A plot of the $P4_12_12$ structure viewed along its fourfold axis, with the three different Wyckoff sites represented by different colors. The characteristic Cairo-type pattern is apparent in this plot.

The $P4_12_12$ form (20 atoms per unit cell at 2×8b and 4a Wyckoff sites, see [9]) is also a chiral framework structure, based on interconnected fivefold spirals chains. The presence of fivefold spirals is a rather unusual feature of this structure, as well as the large proportion of fivefold rings. Another exotic feature is that the two dimensional pattern of $P4_12_12$, when viewed along its fourfold main axis, corresponds to the so-called Cairo pentagonal tiling, reported in Islamic decoration and apparent in Figure 5. In C, $P4_12_12$ is close in energy to the Cmmm and Cmca phases (which are essentially mutually degenerate in energy for each of the three materials and indistinguishable in practice in Figures 1-3). Cmmm has been previously proposed for the structure of cold-compressed carbon nanotubes [6,23], and Cmca can be considered as a variant of this structure which shares the same structural motifs of fourfold, sixfold, and eightfold ring patterns as Cmmm, in a mutated arrangement. The tetragonal $P4_2/ncm$ structure reported by Zhao *et al.* [24] also appeared in our search. $P4_2/ncm$ (12 atoms per cell, at 4b and 8i sites) is made of slabs of corner-sharing diamondlike tetrahedra. Adjacent tetrahedra appear tilted through the direction of the fourfold axis in $P4_2/ncm$ with respect to the arrangement found in the diamond structure, but are otherwise remarkably close to being ideal (see Fig. 5 in Ref. [9]). The nearest-neighbor environment of each site in Pbam (24 atoms per cell at 3×4h and 3×4g sites [9]) is also tetrahedral, but in a sense much more distorted, with fivefold, sixfold, and sevenfold rings of atoms visible in the xy projection shown in Fig. 4. These can be interpreted as the result of strong rebonding in a diamond network, which however preserves a large number of sixfold chairlike diamond-type rings perpendicular to the xy plane. Angular distortion is consequently large. In both Pbam and $P4_12_12$, nearest-neighbor distances spread around the value for the perfectly tetrahedral network, viz. cd (diamond), with a similar response to compression and a large deviation in the orientation of the bonds.

The calculated electronic band structures of C-Pbam and C-$P4_12_12$ (at the DFT-PBE level) show them to be insulating with band gaps of 4.57 and 4.70 eV, respectively, whereas for C-cd our calculations yield 4.13 eV [9]. Band gaps are expected to be underestimated in DFT when calculated using a semilocal form for the XC density functional, as in the PBE approximation. Improving over the PBE values using a first-principles self-energy approach is very costly, but a measure of correction can nonetheless be attempted in order to discern an approximate trend for the true gaps. For example, our calculated PBE band gaps are 0.86 eV for Si-Pbam and 1.23 eV for Si-$P4_12_12$, which are larger than the calculated band gap for Si-cd of 0.63 eV. This value for Si-cd is in turn approximately half the experimental value of 1.17 eV. Roughly correcting the calculated band gaps of Si-Pbam and Si-$P4_12_12$ by adding the difference between the known experimental and theoretical values of the diamond phase (0.54 eV) results in an estimate of 1.4 eV for Si-Pbam and 1.77 eV for Si-$P4_12_12$. These values are close to the value considered optimal for photovoltaic conversion, ~1.5 eV. In particular, the estimated quasi-direct band gap of ~1.4 eV for Si-Pbam at the zone center makes it a very interesting phase in that respect. For Ge, the calculated band gap of the Pbam phase is close to zero (direct at the zone center) whereas for the $P4_12_12$ phase is indirect (1.00 eV).

Finally, in respect to vibrational properties, our simulation of the Raman spectrum of Si-$P4_12_12$ at equilibrium shows a strong Raman peak at a frequency around 480 cm^{-1}, significant Raman activity just below 200 cm^{-1} and weak Raman activity around 330 cm^{-1}, see Fig. 10 in Ref. [9]. Simulated Raman spectra can be a great advantage in the identification of phases obtained in the laboratory, and this can be the case with the hypothetical Si-$P4_12_12$ phase. In particular, the unknown structure of Si-XIII, which is formed in the laboratory under nanoindentation, has been examined in detail by Ruffell *et al.* [7] who concluded the existence of three signature peaks which are characteristic of the Raman spectrum of Si-XIII, at frequencies of 202, 333, and 478 cm^{-1}. These values agree quite well with the values calculated by us for the Raman peaks of the Si-$P4_12_12$ phase. The $P4_2/ncm$ structure had previously been proposed for Si-XIII [24], but the Raman spectrum of Si-$P4_2/ncm$ lacks the characteristic peak at 333 cm^{-1}. We propose that $P4_12_12$ is a better candidate for the structure of the Si-XIII phase.

4. Summary

In summary, we have outlined the results of our structural search for new polymorphs of the group-14 elements C, Si, and Ge, focusing mainly on low-density, energetically favorable polymorphs with tetrahedral coordination and sp^3 bonding. We have discussed the energetics, structure, and electronic and vibrational properties of several novel phases which can be compared with those of other observed and hypothetical phases of these materials. We have found a chiral structure with $P4_12_12$ symmetry and 20 atoms per cell which is low in energy and significantly different from any other structure previously proposed for these materials, exhibiting a characteristic Cairo-type two-dimensional pentagonal tiling. The calculated Raman data for the Si-$P4_12_12$ phase show good compatibility with those experimentally obtained for the unknown Si-XIII phase [7]. Of interest is also the orthorhombic structure with 24 atoms per cell and Pbam symmetry, which in the case of Si has an estimated quasi-direct band gap at the zone center optimal for photovoltaic applications. Recent experimental studies have demonstrated the possibility of synthesizing structures similar to Si-Pbam in the laboratory [8] and we hope that our results may entice further experimental research within this area, leading to new findings. The interested reader is referred to Ref. [9] and its supplemental material for an extended account of our survey on the polymorphism of the group-14 elements.

Acknowledgments
AM acknowledges support from the Ministerio de Educación, Cultura y Deporte (MECD, Spain), mobility Grant No. PRX12/00335, and from the Ministerio de Economía y Competitividad (MINECO), Projects No. MAT2010-21270-C04-03 and MAT2013-46649-C4-3-P). RJN and CJP were supported by the Engineering and Physical Sciences Research Council (EPSRC, UK), Grants No. EP/J017639/1 and EP/G007489/2. Access to the MALTA computer cluster (Universidad de Oviedo, MINECO Project No. CSD2007-00045) and the High Performance Computing Service of the University of Cambridge are also gratefully acknowledged.

References

[1] Nelmes R J and McMahon M I 1998 *Semicond. Semimetals* **54** 145; Mujica A, Rubio A, Muñoz A and Needs R J 2003 *Rev. Mod. Phys.* **75** 863; Mujica A, Radescu S, Muñoz A and Needs R J 2001 *Phys. Status Solidi* B **223** 379; 2001 *J. Phys.: Condens. Matter* **13** 35; 2002 *High Pressure Res.* **22** 455, and references therein.
[2] Geim A K and Novoselov K S 2007 *Nat. Mater.* **6** 183
[3] Mao W L, Mao H-K, Eng P J, Trainor T P, Newville M, Kao C-C, Heinz D L, Shu J, Meng Y and Hemley R J 2003 *Science* **302** 425
[4] Yamanaka S, Kubo A, Komaguchi K, Kini N S, Inoue T and Irifune T 2006 *Phys. Rev. Lett.* **96** 076602
[5] Popov M, Kyotani M, Nemanich R J and Koga Y 2002 *Phys. Rev.* B **65** 033408
[6] Wang Z, Zhao Y, Tait K, Liao X, Schiferl D, Zha C, Downs R T, Qian J, Zhu Y and Shen T 2004 *Proc. Natl. Acad. Sci. U.S.A.* **101** 136998
[7] Ruffell S, Haberl B, Koenig S, Bradby J E and Williams J S 2009 *J. Appl. Phys.* **105** 093513
[8] Kim D Y, Stefanoski S, Kurakevych O O and Strobel T A 2015 *Nat. Mater.* **14** 169
[9] Mujica A, Pickard C J and Needs R J 2015 *Phys. Rev.* B **91** 214104
[10] Pickard C J and Needs R J 2006 *Phys. Rev. Lett.* **97** 045504; 2011 *J. Phys.: Condens. Matter* **23** 053201
[11] Clark S J, Segall M D, Pickard C J, Hasnip P J, Probert M I J, Refson K and Payne M C 2005 *Z. Kristallogr.* **220** 567
[12] Perdew J P, Burke K and Ernzerhof M 1996 *Phys. Rev. Lett.* **77** 3865
[13] Vanderbilt D 1990 *Phys. Rev.* B **41** 7892
[14] Monkhorst H J and Pack J D 1976 *Phys. Rev.* B **13** 5188
[15] Perdew J P, Ruzsinszky A, Csonka G I, Vydrov O A, Scuseria G E, Constantin L A, Zhou X and Burke K 2008 *Phys. Rev. Lett.* **100** 136406
[16] Kresse G and Hafner J 1993 *Phys. Rev.* B **47** 558; Kresse G and Furthmüller J 1996 *Comput. Mater. Sci.* **6** 15; 1996 *Phys. Rev.* B **54** 11169
[17] Kresse G and Joubert D 1999 *Phys. Rev.* B **59** 1758; Blöchl P E 1994 *Phys. Rev.* B **50** 17953
[18] Birch F 1947 *Phys. Rev.* **71** 809
[19] Alfè D 2009 *Comput. Phys. Commun.* **180** 2622
[20] Li Z-Z, Lian C-S, Xu J, Xu L-F, Wang J-T and Chen C 2015 *Phys. Rev.* B **91** 214106
[21] Wang J-T, Chen C, Mizuseki H and Kawazoe Y 2013 *Phys. Rev. Lett.* **110** 165503
[22] Pickard C J and Needs R J 2010 *Phys. Rev.* B **81** 014106
[23] Zhao Z, Xu B, Zhou X-F, Wang L-M, Wen B, He J, Liu Z, Wang H-T and Tian Y 2011 *Phys. Rev. Lett.* **107** 215502
[24] Zhao Z, Tian F, Dong X, Li Q, Wang Q, Wang H, Zhong X, Xu B, Yu D, He J, Wang H-T, Ma Y and Tian Y 2012 *J. Am. Chem. Soc.* **134** 12362

First principles study on superconductivity of distorted bcc phase in phosphorus

A Nakanishi, T Ishikawa and K Shimizu

Center for Science and Technology under Extreme Conditions, Graduate School of Engineering Science, Osaka University, Japan

E-mail: nakanishi@hpr.stec.es.osaka-u.ac.jp

Abstract. We investigated structural stability and superconductivity of phosphorus under high-pressure using first-principles calculations. We found the structural transition from the simple hexagonal (sh) structure into the experimentally observed cI16 structure via the body-centered cubic (bcc) structure in pressure range of 262-286 GPa. The superconducting transition temperature T_c is largely increased from 0.5 K to 12 K by the transition from sh into bcc, which is the highest T_c in phosphorus. The T_c value is decreased to 6 K by the transition into the cI16 phase. The change of T_c is related to that of the density of states at the Fermi level.

1. Introduction

Phosphorus takes an orthorhombic *Cmca* (A17) (P-I) at ambient condition [1,2], and it transforms to a rhombohedral $R\overline{3}m$ (A7) (P-II) at 4.5 GPa and then to a simple cubic (sc) structure (P-III) at 10 GPa [3]. In the sc phase, the superconductivity is experimentally observed. The superconducting quantum interference device with vibrating coil clarified that the superconducting transition temperature T_c takes a maximum value of 9.5 K at 32 GPa and then gradually decreases to 4.3 K with increase of pressure to 100 GPa [4]. The sc structure takes the structural phase transition to the simple hexagonal (sh) structure (P-V) via an intermediated phase (P-IV) in the pressure range from 107 to 137 GPa [5]. The structure of the P-IV phase was theoretically predicted to be a modulated structure [6], and then it was experimentally identified to be an incommensurately modulated (mod) structure of an orthorhombic $Cmmm(00\gamma)s00$ with $\gamma = 0.267$ [7].

Phosphorus takes a further phase transition (P-VI) at 262 GPa. Sugimoto *et al.* experimentally identified the P-VI phase as a distorted bcc structure (cI16) using a synchrotron radiation x-ray powder diffraction measurements [8]. However, first-principles calculations predict that the sh structure first transforms into the bcc structure and then into an orthorhombic modulated superlattice of bcc (IM7), which is different from the experimentally observed structure [9]. Furthermore, the IM7 structure is predicted to transform into the hexagonal close-packed (hcp) structure at 346 GPa.

The experimental observations and first-principles calculations ensure the emergence of a distorted bcc phase in high pressure region, whereas the consistent results have not been obtained. In the present study, we carefully compared the enthalpies among the candidate structures in phosphorus and determined the sequence of the pressure-induced structural phase transitions using first-principles calculations based on density functional theory. Then, we calculated the superconducting T_c on the basis of the sequence to clarify the superconductivity of phosphorus in pressure region above 100 GPa, which has not been experimentally investigated.

2. Computational details

We used the exchange-correlation energy functional of Perdew-Burke-Ernzerhof [10] for the generalized gradient approximation, and the Vanderbilt ultrasoft pseudopotential [11]. We performed the calculations using the code package, Quantum ESPRESSO [12]. For the mod phase, we used the approximated structure with commensurate periodicity of $\gamma = 0.250$. The structural parameters for sc, mod, sh, bcc, cI16, IM7, and hcp at each pressure were obtained by the constant-pressure variable-cell structural optimization based on the Parrinello-Rahman method [13]. The $\boldsymbol{k}$-space integration over the Brillouin zone was performed by a $16 \times 16 \times 16$ grid for sc, sh, bcc, and hcp, a $16 \times 16 \times 4$ grid for mod, a $24 \times 24 \times 12$ grid for IM7, and an $8\times 8 \times 8$ grid for cI16 in the Monkhorst-Pack grid [14]. The cut-off kinetic energy for wave functions was set at 30 Ry. These $\boldsymbol{k}$-point grids and cut-off energies are enough to achieve convergence within 0.1 mRy/atom in the total energy.

Assuming that phosphorus shows the phonon-mediated superconductivity under high-pressure condition, we estimated T_c by the use of the Allen-Dynes formula [15]. We set effective screened Coulomb repulsion constant μ^* at 0.1. In the calculation of the phonon frequency, we employed the linear response method in which the first order corrections are calculated by means of the density functional perturbation theory [16,17]. The electron-phonon matrix element was calculated using a $32 \times 32 \times 32$ grid for sc, sh, bcc, cI16, and hcp and a $32 \times 32 \times 8$ grid for mod. These grids ensure the convergence of the $\boldsymbol{k}$-point sampling with Gaussian width of 20 mRy, which approximates the zero-width limits with respect to double-delta functions of the line width. To obtain the electron-phonon coupling constant λ and the logarithmic-averaged phonon frequency $\omega_{\log}$, we used an $8 \times 8 \times 8$ $\boldsymbol{q}$-point grid for sc, sh, bcc, and hcp, an $8 \times 8 \times 2$ $\boldsymbol{q}$-point grid for mod, and a $4 \times 4 \times 4$ $\boldsymbol{q}$-point grid for cI16.

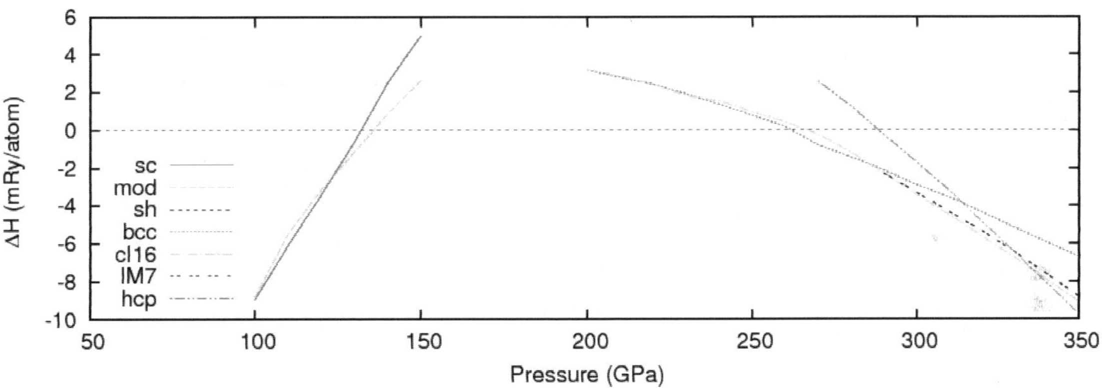

Figure 1. Enthalpies comparison in phosphorus. The relative enthalpies with respect to sh are plotted.

3. Results

3.1. Sequence of the structural phase transitions

First we investigated stable regions for the sc, mod, sh, bcc, cI16, IM7, and hcp phases. Figure 1 shows the comparison of the enthalpies among these seven phases in the pressure range from 50 to 350 GPa. The zero-point energies were not included in the calculations. The results indicate that the mod phase emerges in pressure range from 120 to 136 GPa, which shows a slight discrepancy with the experimentally observed one, 107-137 GPa, owing to the commensurate approximation. In our calculations, first sh transforms into bcc at 262 GPa and then into the experimentally observed cI16 at 286 GPa. The earlier-predicted IM7 has no stable pressure region up to at least 350 GPa. Sugimoto *et al.* reported that the sh and cI16 phases coexist over a wide pressure region of 262-322 GPa at room temperature [8]. Therefore, the predicted bcc phase could be experimentally obtained by careful investigation on the pressure region at low temperatures. By further compression, the hcp structure emerges as the most stable structure above 336 GPa.

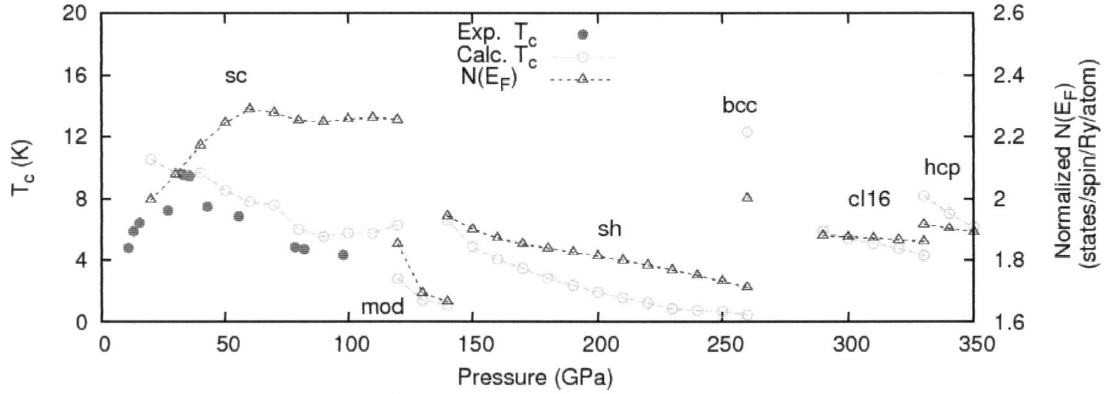

Figure 2. Changes of superconducting transition temperature T_c (open circle) and normalized density of states at the Fermi level $N(E_F)$ (open triangle) through the sequence of the structural phase transitions: sc → mod → sh → bcc → cI16 → hcp. Experimentally observed T_c values (solid circle) for sc [4] are also plotted together.

3.2. *Superconducting properties*

On the Basis of the sequence of the structural transitions shown above, we investigated the superconductivity of phosphorus under high-pressure. Figure 2 shows calculated T_c values as a function of pressure with the experimentally observed T_c data [4]. We eliminated the data at pressures where the phonon modes have imaginary frequency, i.e. the data at 10 GPa in sc and at 270 and 280 GPa in bcc. In 30-100 GPa, the calculated T_c values are in good agreement with the experimental values, which ensures the reliability of the results in pressure region above 100 GPa.

According to Figure 2, our calculations predict that T_c decreases from 6 to 3 K by the transformation from sc to mod, increases to 7 K by the transformation to sh, and gradually decreases to almost 0 K with further pressurization. However, T_c is largely enhanced to 12 K through the phase transition from sh to bcc at 260 GPa. This value exceeds that observed in the sc phase and sets a record for the highest T_c in phosphorus. The superconducting T_c decreases to 6 K through the phase transition to cI16, and again increases to 8 K by the transition to the hcp phase.

To clarify the origin of the T_c change through the structural phase transitions, we investigated the change of the density of states at the Fermi level $N(E_F)$ (open triangle in Figure 2). Since the unit cell of the mod, cI16, and hcp structures contain two or more atoms, $N(E_F)$ was normalized to the unit per atom. Above 120 GPa, T_c and normalized $N(E_F)$ have similar pressure dependency. This means that the T_c change is mainly involved by the $N(E_F)$ change. In the mod phase, the electronic states escape from the Fermi level by the formation of the modulated structure and pseudogap is formed, which causes the decrease of the normalized $N(E_F)$. The details have been discussed by Marqués *et al.*[18], Ishikawa *et al.* [19], and Nakanishi *et al.* [20]. The reason for the decrease of $N(E_F)$ in the mod phase is also applied to the case of the cI16 phase. The cI16 structure is formed by the distortion of bcc, which causes the creation of the pseudogap at the Fermi level. As the results, the $N(E_F)$ value of the cI16 phase is lower than those of the bcc and hcp phases.

4. Conclusion

In this study, we investigated the sequence of the pressure-induced structural phase transitions of phosphorus by comparing the enthalpies among the seven candidate structures. We found that the sh structure transforms to the bcc structure at 262 GPa and then to the experimentally observed cI16 structure at 286 GPa. The earlier predicted IM7 structure has no stable region for pressure.

IOP Conf. Series: Journal of Physics: Conf. Series **950** (2017) 042011 doi:10.1088/1742-6596/950/4/042011

On the Basis of the sequence of the structural phase transitions, we examined the pressure dependence of the superconducting transition temperature T_c. Our calculations predict that T_c largely increases from almost 0 K to 12 K by the structural transition from sh to bcc at 260 GPa, which sets a record for the highest T_c in phosphorus. The superconducting T_c was found to decrease by the transformation from bcc to cI16 and increase again by the transformation to hcp. The decrease of T_c in the cI16 phase is closely related to the decrease of DOS at the Fermi level $N(E_F)$, which is caused by the creation of the pseudogap at the Fermi level owing to the distortion from bcc.

References

[1] Hultgren R, Gingrich N S and Warren B E 1935 *J. Chem. Phys.* **3** 351.

[2] Brown A and Rundqvist S 1965 *Acta Crystallogr.* **19** 684.

[3] Kikegawa T and Iwasaki H 1983 *Acta Crystallogr. Sect.* B **39** 158.

[4] Karuzawa M, Ishizuka M and Endo S 2002 *J. Phys.: Condens. Matter* **14** 10759.

[5] Akahama Y, Kobayashi M and Kawamura H 1999 *Phys. Rev.* B **59** 8520.

[6] Ishikawa T, Nagara H, Kusakabe K and Suzuki N 2006 *Phys. Rev. Lett.* **96** 095502.

[7] Fujihisa H, Akahama Y, Kawamura H, Ohishi Y, Gotoh Y, Yamasaki H, Sakishima M, Takeya S and Honda K 2007 *Phys. Rev. Lett.* **98** 175501.

[8] Sugimoto T, Akahama Y, Fujihisa H, Ozawa Y, Fukui H, Hirao N and Ohishi Y 2012 *Phys. Rev.* B **86** 024109.

[9] Mikhaylushkin A S, Simak S I, Johansson B and Häussermann U 2007 *Phys. Rev.* B **76** 092103.

[10] Perdew J P, Burke K and Ernzerhof M 1996 *Phys. Rev. Lett.* **77** 3865.

[11] Vanderbilt D 1990 *Phys. Rev.* B **41** 7892.

[12] Giannozzi P *et al.* 2009 *J. Phys.: Condens. Matter* **21** 395502.

[13] Parrinello M and Rahman A 1980 *Phys. Rev. Lett.* **45** 1196.

[14] Monkhorst H J and Pack J D 1976 *Phys. Rev.* B **13** 5188.

[15] Allen P B and Dynes R C 1975 *Phys. Rev.* B **12** 905.

[16] Savrasov S Y 1996 *Phys. Rev.* B **54** 16470.

[17] Baroni S, de Gironcoli S, Corso A D and Giannozzi P 2001 *Rev. Mod. Phys.* **73** 515.

[18] Marqués M, Ackland G J, Lundegaard L F, Falconi S, Hejny C, McMahon M I, Contreras-García J and Hanfland M 2008 *Phys. Rev.* B **78** 054120.

[19] Ishikawa T, Nagara H, Mukose K, Kusakabe K, Miyagi H, and Suzuki N 2008 *High Pressure Res.* **28** 459.

[20] Nakanishi A, Ishikawa T, Nagara H, Shimizu K and Katayama-Yoshida H 2012 *High Pressure Res.* **32** 3.

IOP Conf. Series: Journal of Physics: Conf. Series **950** (2017) 042012 doi:10.1088/1742-6596/950/4/042012

Effect of pressure on optical properties of the transition metal dichalcogenide $MoSe_2$

S Caramazza[1], F Capitani[1,2], C Marini[3], A Mancini[4], L Malavasi[4], P Dore[5] and P Postorino[1]

[1] Dipartimento di Fisica, Università di Roma Sapienza, Piazzale Aldo Moro 2, 00185 Roma, Italy

[2] Synchrotron SOLEIL, L'Orme des Merisiers Saint-Aubin, BP 48, F-91192 Gif-sur-Yvette Cédex, France

[3] CELLS-ALBA, Carretera - BP 1413, Cerdanyola del Vallés, 08290, Barcellona, Spain, EU

[4] Dipartimento di Chimica, Università di Pavia, Via Taramelli 16, 27100 Pavia, Italy

[5] CNR-SPIN and Dipartimento di Fisica, Università di Roma Sapienza, P.le Aldo Moro 2, 00185 Roma, Italy

E-mail: paolo.postorino@roma1.infn.it

Abstract. Transition Metal Dichalcogenides TMDs MoX_2 (X=S, Se, Te) are an emergent class of layered materials displaying exciting optoelectronic properties which can be modified by varying the number of layers, by intercalation, or by applying an external strain/compression. In particular, these semiconducting materials can get a band gap closure under pressure and enter in a metallic phase. Here, we investigate the optical properties of $MoSe_2$ under high pressure by means of Raman spectroscopy over a wide pressure range (0-30 GPa). No anomaly has been observed in the pressure dependence of the frequencies of the vibrational modes A_{1g}, E^1_{2g} (Raman first order) and E_{1g} - E^2_{2g} (Raman second-order), in agreement with the absence previously reported of any pressure induced structural transition. Interestingly, our detailed line-shape analysis show a clear anomaly in the pressure behavior of the linewidth of the A_{1g} and E^1_{2g} phonons at the insurgent metallization process which was observed in previous infrared and transport experiments. Our results indicate that the linewidths of Raman phonons peaks can be sensitive to even subtle pressure-induced electronic rearrangements and can thus be used to monitor the insurgence of a pressure-induced semiconductor-metal transition.

1. Introduction

Graphene-like two-dimensional (2D) materials such as layered Transition Metal Dichalcogenides (TMD) attract a lot of interest because of their extraordinary physical and chemical properties, tunable in particular by varying number of layers, thermodynamic conditions (P,T) and doping [1]. Semiconducting TMDs belonging to the family of the MoX_2 compounds (X=S, Se, Te) are layered materials built up of a stacking of X-Mo-X layers [2], consisting of three hexagonal planes, two made by X atoms and one of Mo atoms in between. Within this three-layers structure, Mo is coordinated through covalent interactions with the X atoms in a trigonal prismatic arrangement (symmetry group: $P6_3/mmc$). The distance among the X-Mo-X layers, which is controlled by weak Van Der Waals (VdW) interactions, has a key role in determining the electronic properties of TMDs, such as semiconducting gap and transport properties [2]. Specifically, by strongly reducing the number of layers and thus the interlayer coupling, the optoelectronic properties of TMDs are significantly changed [1]. For example, in the $MoSe_2$ case, the bandgap (close to 1.1 eV) is indirect in the bulk system [3], while, in the single layer case, it becomes direct and shifts into the visible range (near to 1.55 eV). This is of great interest since optical properties of $MoSe_2$ can then be exploited for several potential applications in the visible range (solar cells, photodetectors and so on) [1,3].

In studying the structural and optical properties of TMDs, the high pressure technique is an alternative tool to modify the interlayer distances and thus the interlayer VdW coupling [4]. Indeed, by increasing pressure, TMD semiconducting gap and transport properties can be remarkably modified and, at very high pressure, a semiconducting-metal transition can occur. Recently, x-ray diffraction (XRD) measurements have shown that the MoS_2 system undergoes a subtle isostructural transformation at about 20 GPa (*layer sliding* transformation) with a collapse of the *c*-lattice parameter [5,6,7], in agreement with theoretical computations [8]. Experiments [6] and theoretical calculations [8] also show that the system enters a metallic phase within the same pressure range (19 GPa in ref. [6] and 20-25 GPa in ref. [8]). For $MoTe_2$, no x-ray diffraction (XRD) data are available at high pressures, but theoretical calculations show that this system apparently preserves its structure at least up to 100 GPa, and that a complete metallization occurs in the 13-19 GPa pressure range [9].

In the case of $MoSe_2$, Aksoy *et al.* [10] carried out XRD measurements up to 36 GPa and do not report any structural transition. More recently, experimental results obtained by Z. Zhao *et al.* [11] seem to confirm that $MoSe_2$ do not undergoes any structural transformation up to 60 GPa. In studying electronic and transport properties, the same authors also carried out infrared measurements, showing that, in $MoSe_2$ single crystals, remarkable changes occur with increasing pressure. In particular, they found the onset of a metallization process between 16 and 20 GPa, and within the 20-35 GPa range the energy gap rapidly decreases down to vanishingly small values. Transport measurements performed up to 42 GPa reveal indeed that the metallization process is complete at about 40 GPa [11]. From a theoretical point of view, DFT calculations performed by Riflikoví *et al.* [9] predict that $MoSe_2$ preserves its original structure up to 100 GPa, while it undergoes a complete metallization in the 28-40 GPa pressure range.

In conclusion, pressure dependent XRD structural studies of different MoX_2 systems do not show abrupt or well defined structural transitions, while infrared and transport properties reveal progressive but well detectable pressure-induced transformations in the electronic structure, such as metallization processes. The aim of the present work is to study pressure effects on the optical properties a MoX_2 system by employing Raman spectroscopy, which by probing the phonon spectrum is sensitive to structural properties of the system, but can also probe its electronic state and charge-lattice couplings by a detailed line-shape analysis [6,12,13]. We thus performed Raman measurements on a crystal powder of $MoSe_2$ in the 0-30 GPa range with the idea of confirming on one side the absence of any structural transition and, on the other side, of carrying out a detailed line-shape analysis looking for a spectroscopic signature of the insurgent metallization process.

2. Raman measurements on $MoSe_2$ at high pressure

The four Raman-active phonons predicted in MoX_2 systems are of A_{1g}, E_{1g}, E^{1}_{2g} and E^{2}_{2g} symmetry [2]. The first three modes are due to atomic vibrations within a X-Mo-X layer. A_{1g} is an out-of-plane mode (where the X atoms move outside the layer, along the *c*-axis). Note that the E_{1g} and the E^{1}_{2g} are in-plane modes: in the former, the X-atoms moves in opposite directions and the Mo-atoms do not move at all; in the latter, the X-atoms of adjacent layers move in opposite direction. The fourth mode, E^{2}_{2g}, is called "rigid-mode" and correspond to the relative motion of two adjacent X-Mo-X rigid layers. In the Raman spectrum of a $MoSe_2$ single crystal, Sekine *et al.* [14] observed the E^{2}_{2g} (25.4 $cm^{-1)}$ A_{1g} (242 cm^{-1}) and E^{1}_{2g} (286 cm^{-1}) phonons. For the E_{1g} mode, they found that it can be detected at 168 cm^{-1} only with polarized incident radiation. In the measured spectra, also second-order Raman contributions were observed. In particular, spectral features above 300 cm^{-1} were assigned to a two-phonon overtone or combination bands and the peak at 145 cm^{-1} to a two-phonon difference process [14], recently ascribed to the E_{1g} - E^{2}_{2g} combination band [15].

IOP Conf. Series: Journal of Physics: Conf. Series **950** (2017) 042012 doi:10.1088/1742-6596/950/4/042012

In the present work, we collected Raman spectra of $MoSe_2$ crystal powder (Sigma Aldrich: 325 mesh, 99.9%) as a function of pressure. Finely milling the sample, we obtained a powder with a micrometric/sub-micrometric average grain size.

A diamond-anvil cell (DAC), equipped with 400 μm culet II A diamonds, was employed in the 0-30 GPa pressure range. The gasket was made of a 250-μm-thick steel foil, with a sample chamber of 100 μm diameter and 50 μm height. A small quantity of powder of $MoSe_2$ was loaded inside the gasket hole together with NaCl powder used as hydrostatic medium. A ruby sphere was also put close to the sample into the sample chamber to measure the pressure through the standard ruby fluorescence techniques [16].

Raman measurements were carried out by means of a Horiba LabRAM HR Evolution microspectrometer using a 30 mW He-Ne laser (632.8 nm wavelength). The spectrometer is equipped with a notch filter used to reject the elastic contribution, a grating monochromator, and a Peltier-cooled charge-coupled device (CCD) detector. A confocal microscope with long working distance objectives is used to focus the laser beam on the sample and to collect the scattered signal. For the present measurements we used a 50x objective allowing a few-micrometer diameter laser spot on the sample. Raman spectra were collected in the backscattering geometry over the 120-500 cm^{-1}, with a spectral resolution better than 1 cm^{-1} thanks to a 1800 grooves/mm grating with a 800-mm focal length. Further details about the Raman experimental setup can be found in ref. [17].

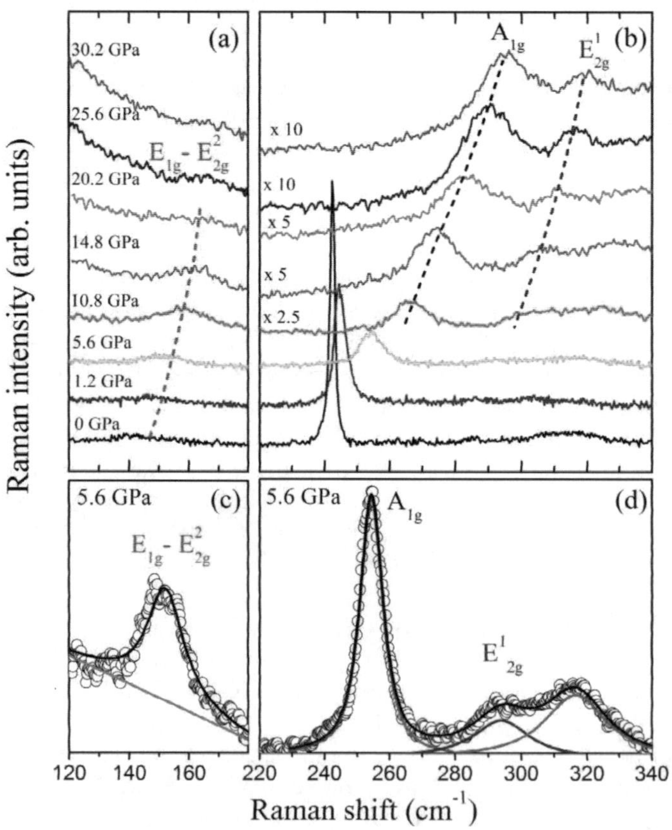

Figure 1. (a)-(b) Raman spectra of crystal powder of $MoSe_2$ as function of pressure at selected pressures. Dashed lines are guides for the eye. (c)-(d) the three vibrational modes (A_{1g}, E^{1}_{2g} and E_{1g} - E^{2}_{2g}) and best fit profiles are shown for the 5.6 GPa pressure.

The Raman spectra of $MoSe_2$ at selected pressures are shown in figure 1, over the 120-180 cm^{-1} range (panel a) and the 220-340 cm^{-1} range (panel b). We have not carried out any attempt of polarization analysis of the Raman signal since, as mentioned above, the sample consisted of fine crystal powder. We recall that the predicted Raman active modes, listed as a function of increasing frequency, are E^2_{2g}, E_{1g}, A_{1g}, and E^1_{2g}. The E^2_{2g} mode observed at about 25 cm^{-1} [14] is not detectable owing to the intense low-frequency elastic and quasi-elastic scattering, as well as the E_{1g} Raman mode (at around 170 cm^{-1} [14]) which can be observed only on a single crystal with a proper polarization configuration. The spectrum collected at ambient pressure with the sample already loaded in the DAC is characterized by the most intense and narrow A_{1g} mode at 243 cm^{-1}. The peak shows an abrupt and remarkable broadening as the sample is compressed. On this regard we notice that the A_{1g} is the only out-of-plane vibrational mode we observe. Indeed this is the vibrational mode inherently most sensitive to volume compression since the atomic vibration takes place along the *c-axis* which is the *soft* lattice direction [11,15]. The E^1_{2g} mode which is detectable when the sample is on a microscope slide, actually disappears when the powder is loaded in the DAC. As the powder is compressed and pressure is applied on the sample, the E^1_{2g} appears and it becomes increasingly evident.

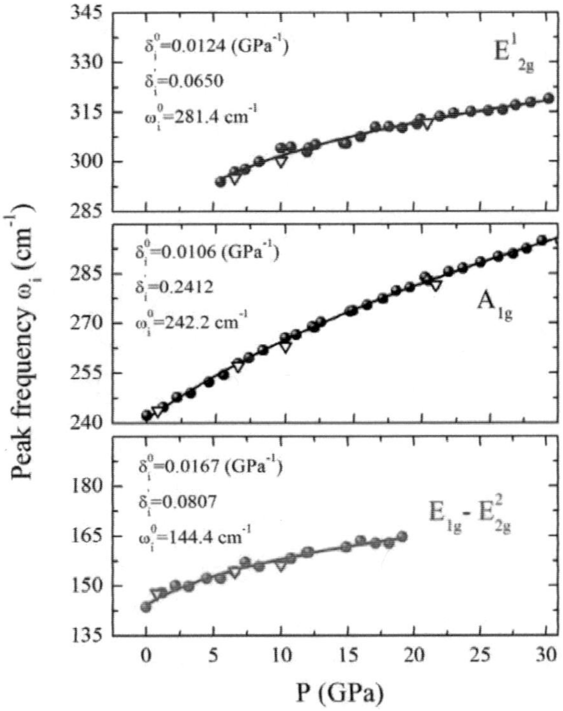

Figure 2. Pressure dependence of the three Raman peaks.
Full lines: best fit curves obtained from eq. (1).

At ambient pressure, the second order Raman peak ascribed to the E_{1g} - E^2_{2g} difference mode can be found at 143 cm^{-1} and it can be followed on increasing the pressure up to about 20 GPa, whereas the experimental noise masks the signal on further increasing the pressure. The weak broad peak above 300 cm^{-1}, ascribed to a combination band [14], appears at ambient pressure and it is still detectable on increasing the pressure.

As expected [11,15,18], the pressure dependences shown by all the frequencies of the measured Raman peaks show a clear hardening as the sample is compressed. In order to analyze the pressure dependence of the Raman peaks, a fitting procedure has been applied to extract the central peak frequency and the linewidth. Standard Lorentz functions have been used to model the Raman signal. In the low panels in figure 1, we report an example of best-fit at 5.6 GPa. The best fit parameters obtained for the A_{1g}, E^{1}_{2g} and the E_{1g} - E^{2}_{2g} second-order Raman modes are shown in figure 2 (peak frequency) and figure 3 (full-width at half-maximum, FWHM) as a function of pressure. Looking at figure 2 we notice that a remarkable frequency hardening is induced by pressure and that the continuous and regular trend observed is in agreement with the previously claimed [11,15,18] absence of any structural transition within the explored pressure range. We also collected several spectra on releasing the pressure at 21, 10, 6 and 0.8 GPa. The corresponding peak frequencies (see empty triangles in figure 2) perfectly follow the behavior observed on increasing pressure with no evidences of hysteretic processes, confirming again the absence of structural transitions. A close inspection of figure 2 shows a weak sublinear pressure dependence of the mode frequencies which have then been analyzed assuming a pressure dependence given by the widely used relation [19]:

$$\omega_i(P)/\omega_i^0 = [(\delta_i^0/\delta_i')P + 1)]^{\delta_i'} \quad (1)$$

where ω_i^0, δ_i^0, and δ_i', are taken as free fitting parameters. The best fit curves are also shown in figure 2, where a very good experimental/theoretical agreement can be appreciated. The best fitting parameters are reported in figure 2. For a direct comparison we also report the slopes obtained from a simple linear fit (i.e. by setting $\delta_i' = 1$) that is: 1.77 cm^{-1}/GPa, 0.92 cm^{-1}/GPa, 0.98 cm^{-1}/GPa for the A_{1g}, E^{1}_{2g} and the E_{1g} - E^{2}_{2g} modes respectively. The strong difference between the pressure sensitivity of the A_{1g} mode and the E^{1}_{2g} and E_{1g} - E^{2}_{2g} modes is explained by the larger lattice compression along the soft direction (*c*-axes) and the out-of-plane / in-plane nature of the vibrational modes. In particular the A_{1g} phonon, which involves the weak VdW forces along the *c*-axis, is more sensible to volume compression than the E^{1}_{2g}, which is a planar mode related to the strong covalent bond in the Se-Mo-Se planes [11,15].

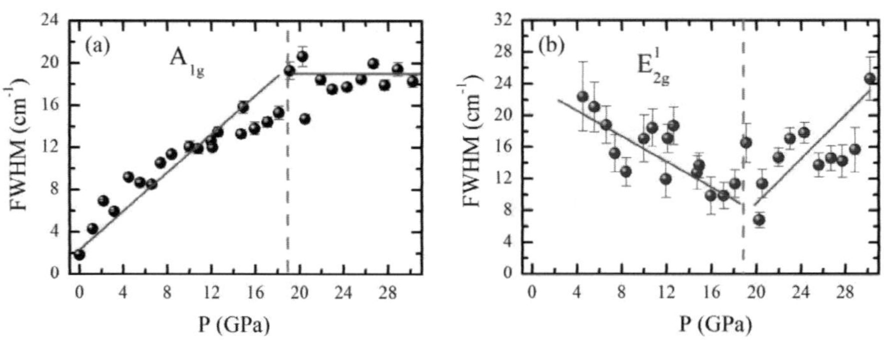

Figure 3. Pressure dependence of the FWHM of the A_{1g} and E^{1}_{2g} phonons. The vertical dashed lines and the solid red line are guide for the eye.

The full-widths at half-maximum, also obtained through the fitting procedure, are reported in figure 3 for the A_{1g} (panel a) and E^{1}_{2g} (panel b) peaks. A very different pressure dependence is evident, although in both cases an abrupt change occurs within the 18-20 GPa pressure range. As a function of the increasing pressure, the A_{1g} phonon shows an increasing FWHM (see panel a) up to ~19 GPa and on further increasing the pressure the linewidth keeps nearly constant. In the case of the E^{1}_{2g} (see panel b) a very different behavior is observed, but a turning point occurs at about the same pressure range. Bearing in mind that the high pressure infrared measurements show that a metallization process starts in the 16-20 GPa pressure range [11], the linewidth pressure behavior here observed establish a rather clear spectroscopic signature of the onset of the metallization process.

A similar anomaly has been indeed found in the pressure behavior MoS_2, where a sharp drop in the width of both the A_{1g} and the E^{1}_{2g} phonons was observed during the semiconducting-metal phase transition [6]. These results suggests an important role of the Raman FWHM of the A_{1g} and E^{1}_{2g} phonons in MoX_2 TMDs materials in identifying the insurgence of a metallization process.

We want to finally mention that such a strong sensitivity of the phonon peak linewidths to electronic modifications has been previously observed in a variety of systems (e.g. in pure Te [20]) and in many high pressure studies on strongly correlated electron systems (see for example [12]). The phonon peak width is indeed directly related to charge-lattice coupling and charge transfer processes which directly monitor the electronic state and the degree of charge delocalization of the system.

3. Conclusion

We carried out Raman measurements on $MoSe_2$ crystal powder as function of pressure up to 30 GPa. The evolution of the frequencies of the vibrational modes A_{1g}, E^{1}_{2g} (Raman first order) and E_{1g} - E^{2}_{2g} (Raman second-order) under applied pressure does not exhibit anomalies and can be very well reproduced by a sublinear curve. This confirms the absence of any structural transition within the 0-30 GPa pressure range, in agreement with experimental and theoretical results previously reported. More interestingly, a line-shape analysis of the Raman peaks reveals a clear anomaly in the A_{1g} and E^{1}_{2g} linewidth pressure behavior within the pressure range 18-20 GPa. This anomaly can be ascribed to the onset of the pressure induced semiconductor-metal transition observed in previous infrared and transport experiments. Present results suggest the use of a careful line-shape analysis of the Raman spectra to detect even subtle electronic rearrangement induced by the applied pressure.

References

[1] Wang Q H, Kalantar-Zadeh K, Kis A, Coleman J N and Strano M S 2013 *Nature Nanotech.* 7 699; Tang Q and Zhou Z 2013 *Progress in Materials Science* **58** 1244-1315
[2] Wilson J and Yoffe A 1969 *Adv. Phys.* **18** 193
[3] Tongay S, Zhou J, Ataca C, Lo K, Matthews T S, Li J, Grossman J C and Wu J 2012 *Nano Lett.* 12 5576-5580
[4] Capitani F, Hoeppner M, Joseph B, Malavasi L, Artioli G A, Baldassarre L, Perucchi A, Piccinini M, Lupi S, Dore P, Boeri L and Postorino P 2013 *Phys. Rev.* B **88** 144303
[5] Bandaru N, Kumar R S, Sneed D, Tschauner O, Baker J, Antonio D, Luo S-N, Hartmann T, Zhao Y and Venkat R 2014 *J. Phys. Chem.* C **118** 3230
[6] Nayak A P, Bhattacharyya S, Zhu J, Liu J, Wu X, Pandey T, Jin C, Singh A K, Akinwande D and Lin J F 2014 *Nat. Commun.* **5** 3731
[7] Chi Z-H, Zhao X-M, Zhang H, Goncharov A F, Lobanov S S, Kagayama T, Sakata M and Chen X-J 2014 *Phys. Rev. Lett.* **113** 036802
[8] Hromadová L, Martoňák R and Tosatti E 2013 *Phys. Rev.* B **87** 144105
[9] Riflikova M, Martoňák R and Tosatti E 2014 *Phys. Rev.* B **90** 035108
[10] Aksoy R, Selvi E and Ma Y 2008 *J. Phys. Chem. Solids* **69** 2138
[11] Zhao Z *et al.* 2015 *Nat. Commun.* **6** 7312
[12] Congeduti A, Postorino P, Caramagno E, Nardone M, Kumar A and Sarma D D 2001 *Phys. Rev. Lett.* **86** 1251
[13] Bera A, Pal K, Muthu D V S, Sen S, Guptasarma P, Waghmare U V, and Sood A K 2013 *Phys. Rev. Lett.* **110** 107401
[14] Sekine T, Izumi M, Nakashizu T, Uchinokura K and Matsuura E 1980 *J. Phys. Soc. Jpn.* **49** 1069-1077 (1980)
[15] Bhatt S V, Deshpande M P, Sathe V, Rao R and Chaki S H 2014 *J. Raman Spectrosc.* **45** 971-979
[16] Mao H K, Bell P M, Shaner J W and Steinberg D J 1978 *J. Appl. Phys.* **49** 3276
[17] Capitani F, Koval S, Fittipaldi R, Caramazza S, Paris E, Mohamed W S, Lorenzana J, Nucara A, Rocco L, Vecchione A, Postorino P and Calvani P 2015 *Phys. Rev.* B **91** 214308
[18] Sugai S and Ueda T 1982 *Phys. Rev.* B **26** 6554
[19] Postorino P, Congeduti A, Degiorgi E, Itié J P and Munsch P 2002 *Phys. Rev.* B **65** 224102
[20] Marini C, Chermisi D, Lavagnini M, Di Castro D, Petrillo C, Degiorgi L, Scandolo S and Postorino P 2012 *Phys. Rev.* B **86** 064103

IOP Conf. Series: Journal of Physics: Conf. Series **950** (2017) 042013 doi:10.1088/1742-6596/950/4/042013

Combined laser ultrasonics, and Raman scattering in diamond anvil cell system operating in the transmission configuration

P V Zinin[1,2], K Burgess[1], V Prakapenka[3], S K Sharma[1], I B Kutuza[2], N Chigarev[4], and V Gusev[4]

[1]Hawaii Institute of Geophysics and Planetology, University of Hawaii, Honolulu, USA
[2]Scientific-Technological Center of Unique Instrumentation, Moscow, Russia
[3]Center for Advanced Radiation Sources, University of Chicago, Chicago, USA
[4]LPEC, CNRS, PRES UNAM, Université du Maine, Av. O. Messiaen, 72085 Le Mans, France

E-mail: zosimpvz@mail.ru

Abstract. Recently it has been demonstrated that it is possible to detect both longitudinal and shear waves as well as measure their velocities for a non-transparent iron layer compressed in a diamond anvil cell (DAC) to pressures approaching 23 GPa by using laser ultrasonics (LU) technique. These experiments were conducted in the reflection configuration when point source and point receiver were located at the same side of the specimen. The purpose of the present study is to examine the generation and detection of the acoustical waves in the LU-DAC system operating in the transmission configuration. The experimental results obtained at 16.4 GPa presented in this article demonstrate that the LU-DAC technique can be applied for longitudinal wave velocity measurements in small specimens under high pressure in the transmission configuration.

1. Introduction

Understanding of the elastic behaviour of minerals under high pressure is a crucial factor for developing a model of the Earth's structure because most information about the Earth's interior comes from seismological data. More than fifty years ago, Birch proposed a simple empirical equation, Birch's law, that relates sound velocity to the density and mean atomic weight of the material the sound is passing through [1]. Because of this relation the velocity of sound can be measured in the Earth's interior as seismic waves travel from an earthquake on one side of the Earth to a seismograph on the other side. It can also be measured in the laboratory by sending ultrasound through samples at controlled pressures and temperatures [2]. Therefore, direct measurement of velocities and other elastic properties of minerals at elevated pressures and temperatures is the key to understanding the seismic information, allowing us to translate it into quantities such as chemical composition, mineralogy, temperature, and preferred orientation of minerals [3,4]. Recently it has been demonstrated that it is possible to detect both longitudinal and shear waves as well as measure their velocities for a non-transparent iron layer compressed in a DAC to pressures approaching 23 GPa by using LU in a point-source–point-receiver [5-7] or line-source–point-receiver [8] configurations. These experiments were conducted in the reflection configuration when point source and point receiver were located at the same side of the specimen. The purpose of the present study is to examine the generation and detection of the acoustical waves in the LU-DAC system operating in the transmission configuration.

2. Method

An experimental LU-DAC setup operating in the transmission configuration is shown in (figure 1). Schematics of the sample assemblage and of the sound wave trajectories in our LU experiments in a DAC are shown in figure 2.

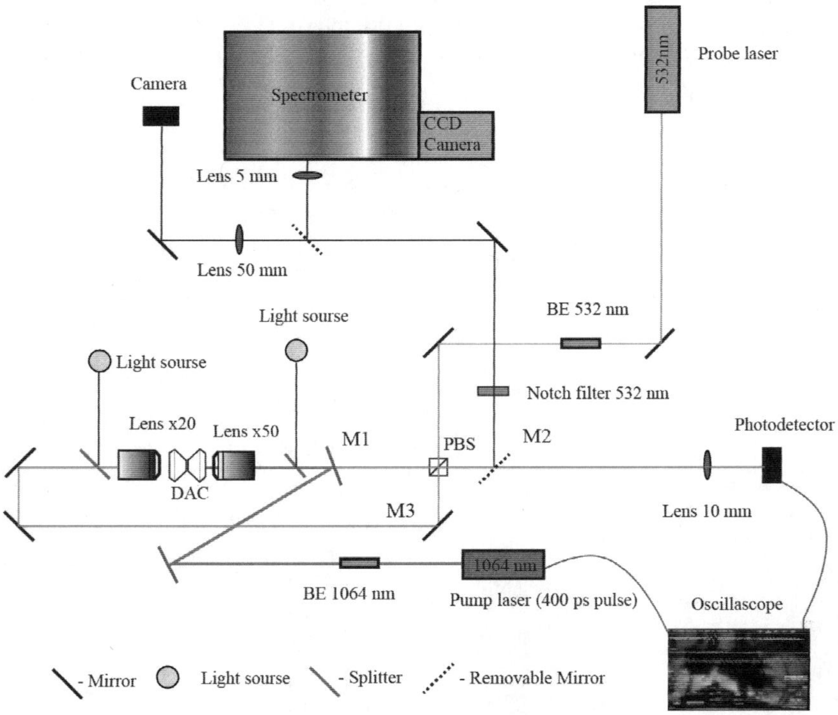

Figure 1. The sketch of the LU-DAC system operating in the transmission configuration. It consists of several components: (1) LU-DAC system (probe and pump lasers, photodetector, and oscilloscope); (2) the spectrometer for measuring Raman scattering for pressure determination; (3) the motorized sample stage; (5) double side high magnification imaging based on two long working distance infinity corrected objectives.

A flat iron specimen of about 150 μm in width is pressed between two diamond anvils (figure 2). The pump laser beam passes through a cylindrical lens in order to create a line-type source of material heating. The use of the beam expander (BE 1064 nm) allows shifting of the focal spot of the pump laser in z direction. The pump laser beam is focused by a Mitutoyo ×50 objective (Lensx50, figure1) with a working distance of 20 mm to a size of 3 μm on the sample surface. The position of the objective is fixed. The movement of the focal spot of the pump laser on the surface of the specimen in x direction in the x-z plane is achieved by rotating the M1 mirrors along the y axes. We do it remotely by using the compact piezo driven optical mount (NewPort, AG-M100L). The probe optical beam is from a continuous wave laser (Compass 315M, Diode-Pumped Laser, Coherent) with a maximum 150 mW power at 532 nm wavelength. For measurements of the surface acoustical waves, the photo-acoustic signals are detected by the photo deflection technique with a knife-screen, installed before the photo-detector [9]. We used a New Focus (model 1601) photo-detector with a 1 GHz bandwidth and a

2.5 GHz frequency bandwidth oscilloscope (WavePro 725ZI, LeCroy) to observe and record the signals. The use of the beam expander (BE 532 nm) allows shifting of the focal spot of the probe laser in z direction (perpendicular to the sample). The green light reflected from the specimen and passes through the polarized beam splitter (PBS). When the flip flop mirror (M2) is down, the reflected light after passing the PBS is focused by long distance focus lens (L5, f = 10 cm), on the sensor inside the Photodetector. For detecting acoustical waves inside the DAC only variation of the intensity of the reflected light is recorded. The LU-DAC system described above allows measuring velocities of the acoustical wave inside DAC in the transmission configuration, when the excitation and detection of the acoustical waves are on the opposite sides of the specimen in DAC (figure 2). In the transmission configuration, the probe laser beam is directed to the back side of the specimen in DAC through mirror (M3). The probe laser is focused on the sample by the objective, (Lens x20). Raman scattering system was used for determination of the pressure distribution over the specimen by measuring a shift of the diamond Raman line under pressure.

3. Results

The typical photoacoustic signal measured at 16.4 GPa is presented in figure 3. Following the notations introduced in [6,7], the L peak corresponds to the longitudinal wave (figure 2a). The peak marked as SL corresponds to skimming longitudinal wave [6]. As mentioned in [6], SL waves could combine, under certain conditions, with L waves leading to the so-called "head wave arrivals" (see SL-L peaks in figure 3).

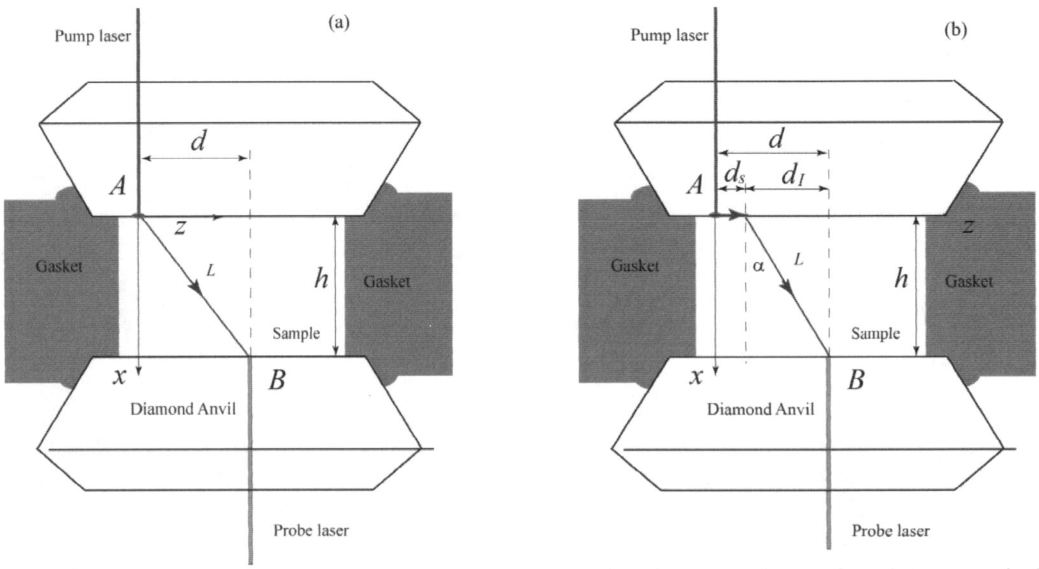

Figure 2. Sketch of the sound wave propagation in a flat iron specimen for the transmission configuration, where d is the distance between the points of the sound generation and detection; h is the thickness of the Fe-specimen. In the transmission case, the pump and probe lasers are incident to opposing side of the sample and the bulk wave travels through the material with no reflections. Excitation and detection of a number of different direct, reflected and refracted waves are possible using this technique.

As we can see (figure 3) only two waves have been detected (L and SL-L waves). Least square fits of two-parameters (sound velocity and sample thickness) to the experimental data $\tau(d)$ allowed us to determine the L-wave velocities in iron, and SL-L wave (figure 4). Time of flight of the L wave (τ_L) can be determined from a simple expression $\tau_L = 1/c_L\,(h^2 + d^2)^{1/2}$. Time of flight of *the SL-L* wave (τ_{SL})

has a simple form: $\tau_{SL} = \tau_I + \tau_S$, where $\tau_S = d_S/c_{SL}$, $\tau_I = h/\cos\alpha$, and α is the critical angle derived from the Snell's refraction law $\sin\alpha / c_L = 1/c_{SL}$.

Fitting of the L and SL-L wave arrivals at 16.4 GPa in the transmission configuration is shown in figure 4. Both the c_L and c_{SL} values of the correspondent waves in iron, determined in this work are in a good agreement with the results obtained earlier using the LU-DAC technique in the reflection configuration [6]. To our surprise the shear wave in iron was not detected in the transmission configuration.

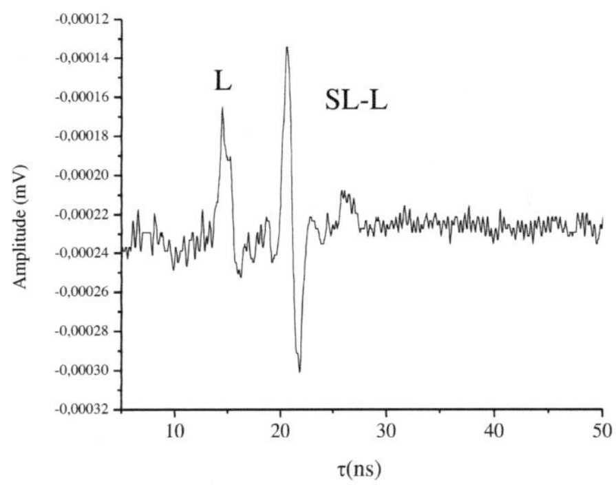

Figure 3. Photoacoustic signals measured at 16.4 GPa for $d = 41.5$.

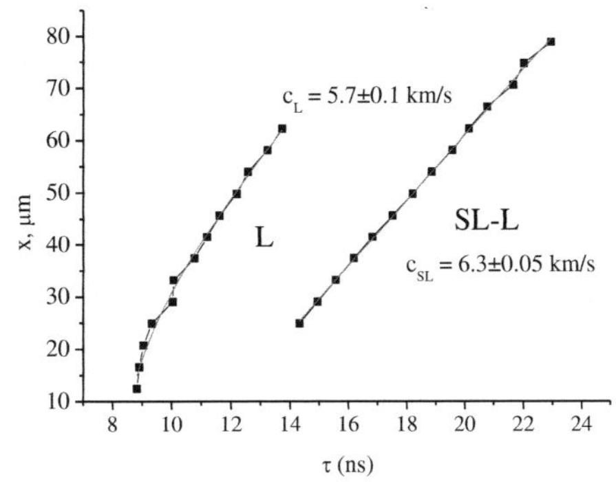

Figure 4. Fitting of the L and SL-L wave arrivals at 16.4 GPa.

4. Conclusion
The experimental results presented in this article demonstrate that the LU-DAC technique can be applied for sound velocity measurements in small iron specimens under high pressure in the transmission configuration. It was found that only longitudinal mode can be detected in this configuration.

Acknowledgement
This work was supported by NSF Grant NO. No EAR-1215796, and Russian Fund of Fundamental Research (grant 15-02-04994) and Program of the Presidium of RAS on Strongly Compressed Matter for financial support.

References
[1] Birch F 1952 *J. Geophys. Res.* **57** 227
[2] Bassett W A 1994 *Science* **266** 1662
[3] Bass D, Sinogikin S V, and Li B S 2008 *Elements* **4** 165
[4] Bass J D 2008 *Phys. Earth Planet. Inter.* **170** 207
[5] Chigarev N, Zinin P, Ming L C, Amulele G, Bulou A, and Gusev V 2008 *Appl. Phys. Lett.* **93** 181905
[6] Chigarev N, Zinin P, Mounier D, Bulou A, Ming L C, Acosta T, and Gusev V 2010 *High Pressure Res.* **30** 78
[7] Chigarev N, Zinin P, Mounier D, Bulou A, Zerr A, Ming L C, and Gusev V 2011 *J. Phys.: Conf. Ser.* **278** 017012
[8] Zinin P V, Chigarev N, Mounier D, Bulou A, Ming L C, Acosta T, and Gusev V 2010 *J. Phys.: Conf. Ser.* **215** 012053
[9] Rothenberg J E 1988 *Opt. Lett.* **13** 713

IOP Conf. Series: Journal of Physics: Conf. Series **950** (2017) 042014 doi:10.1088/1742-6596/950/4/042014

High pressure Raman spectroscopic study of low thermal expansion $AZr_2(PO_4)_3$, A=K, Rb and Cs

K Kamali and T R Ravindran

Materials Science Group, Indira Gandhi Centre for Atomic Research, Kalpakkam 603 102, India

E-mail: kamali@igcar.gov.in

Abstract. $AZr_2(PO_4)_3$ (A=K, Rb, Cs) are a set of flexible framework compounds that exhibit tunable ultralow thermal expansion over the wide temperature range of 293-1273 K. High pressure Raman spectroscopic measurements on $AZr_2(PO_4)_3$ reveals structural phase transformation above 2 GPa and amorphization above 20 GPa. On decompression from 20 GPa, the amorphous phase remains unchanged indicating irreversible nature of pressure induced amorphization (PIA). Various low frequency Raman modes corresponding to translation of Zr atoms and coupled rotation of PO_4 tetrahedra and ZrO_6 octahedra soften with increase in pressure which could be the driving force for the observed phase transition. Pressure induced phase transition could be due to collapse of Zr-O-P bond angles, and PIA due to kinetic hindrance of equilibrium decomposition. We have compared the phase transition and amorphization pressures of the various members of $AZr_2(PO_4)_3$ to account for the stability of the ambient rhombohedral phase.

1. Introduction

$NaZr_2(PO_4)_3$ (NZP) family of compounds are of great interest due to their remarkable property of ultralow thermal expansion [1]. In $NaZr_2(PO_4)_3$ substitution of Na by K, Rb and Cs does not change the parent rhombohedral structure of space group *R-3c* [2]. Thermal expansion coefficients can be tuned by substitution of interstitial cations of different sizes [3]. $NaZr_2(PO_4)_3$ (NZP), $KZr_2(PO_4)_3$ (KZP), $RbZr_2(PO_4)_3$ (RbZP) and $CsZr_2(PO_4)_3$ (CsZP) exhibit average thermal expansion coefficients of $\alpha_{av}\approx 4.5 \times 10^{-6}$ K^{-1}, -0.4×10^{-6} K^{-1}, -0.36×10^{-6} K^{-1} and 0.4×10^{-6} K^{-1} respectively from 293 to 1273 K [3]. NZP can also serve as a host for nuclear waste immobilization since it has the ability to accommodate most of the radioactive ions at the interstitial Na sites [4]. Alkali cation substituted NZP compounds are known to be stable up to 1273 K [5]. On the other hand, there are not many reports on the behavior of NZP family with respect to pressure: prototypic NZP undergoes structural phase transformations around 5 and 6 GPa maintaining structural integrity at least up to 30 GPa [6]. In the present work the influence of pressure on the phase stability of alkali cation substituted NZP compounds are investigated.

IOP Conf. Series: Journal of Physics: Conf. Series **950** (2017) 042014 doi:10.1088/1742-6596/950/4/042014

2. Experimental details

KZP, RbZP and CsZP were synthesised by sol gel technique described elsewhere [7]. The XRD patterns of all three samples match with the respective ICDD powder diffraction files confirming single phase.

Raman spectroscopic measurements were carried out using a micro Raman spectrometer (Renishaw, UK, model inVia) with 514.5nm laser excitation. For high pressure Raman measurements, AZP powder was loaded along with 4:1 Me:Et pressure medium and a few ruby specks for pressure measurement in a 250 μm hole drilled in a pre indented stainless steel gasket and mounted in a compact, symmetric diamond anvil cell with diamonds of 500 μm culet diameter. Spectra were recorded in both increasing and decreasing pressure cycles. The spectra were fitted to Lorentzian line shapes to determine the peak positions and full widths at half maxima (FWHM) using PeakFit software (JANDEL).

Table 1. Ambient Raman mode frequencies of KZP, RbZP and CsZP and their corresponding mode assignments.

KZP Raman frequencies (cm^{-1})	RbZP Raman frequencies (cm^{-1})	CsZP Raman frequencies (cm^{-1})	Mode assignment
73	85	105	E_g
120	126	134	E_g
155	159	168	A_{1g}
174	170	-	E_g
263	264	263	E_g
287	288	287	A_{1g}
317	315	313	E_g
403	401	-	A_{1g}
419	421	423	E_g
435	441	449	E_g
593	592	591	A_{1g}
636	636	634	A_{1g}
-	-	1007	E_g
-	1019	1013	E_g
1023	1028	1025	A_{1g}
1060	1059	1052	E_g
1086	1085	1079	A_{1g}

3. Results and discussions

At ambient conditions, $AZr_2(PO_4)_3$, A=K, Rb and Cs compounds with space group R-3c have six formula units in the crystallographic unit cell giving rise to 108 degrees of freedom. Factor group analysis results in the total irreducible representation of $\Gamma_{Total}=8A_{1g}+9A_{2g}+17E_g+9A_{1u}+10A_{2u}+19E_u$. The irreducible representations associated with the Raman and IR active phonons are $8A_{1g} + 17E_g$ and $9A_{2u} + 18E_u$ respectively. Detailed classification into internal modes of phosphate ion and external modes are described elsewhere [7].

Mode assignments for $AZr_2(PO_4)_3$, A= K, Rb and Cs are the same as those reported for $NaZr_2(PO_4)_3$ [7]. On the basis of the assignments reported earlier on similar compounds [8] we have labeled the internal vibrational modes ν_1, ν_2 and ν_3 and ν_4 of the tetrahedral phosphate group and external modes due to vibrations of A, Zr atoms. The bands in the range 900-1250 cm^{-1} arise due to stretching vibrations of PO_4 (ν_1 and ν_3), whereas 400-700 cm^{-1} corresponds to bending vibration of PO_4 (ν_2 and ν_4) and those below 400 cm^{-1} are due to translations of A, Zr ions and translations and librations of PO_4 units. Table 1 lists the ambient Raman mode frequencies of KZP, RbZP and CsZP and their corresponding mode assignments. Vibrational frequencies for the first three low energy Raman bands and bands around 420 cm^{-1} of all four compounds increase with increase in the size of alkali cations (Na-K-Rb-Cs) whereas other bands show a slight decrease in frequency. The anomalous behavior of the above four Raman modes was explained to be due to a contraction along the *a*-axis with increasing size of the A^I cations [8].

Figure 1 (a, b and c) show Raman spectra of $AZr_2(PO_4)_3$, A= K, Rb and Cs at different pressures. Pressure coefficients were obtained from mode frequency *vs.* pressure. KZP and RbZP show a phase transition around 3.4 and 2.2 GPa respectively. CsZP exhibits two phase transitions at 3.4 and 8.4 GPa which is comparable to the two transition pressures of 5 and 6.6 GPa of NZP [6]. Raman spectra after the first transition in KZP, RbZP and CsZP are observed to be similar to the Raman spectrum of NZP at 5.4 GPa indicating *R3* structure for the first high pressure phase as in NZP [6]. The transition is suggested to be due to a polyhedral Zr-O-P tilt mechanism [9]. Indeed, high pressure XRD studies on isostructural $RbTi_2(PO_4)_3$ have revealed a phase transition from *R-3c* to *R3* structure around 1.7 GPa [9]. Raman bands in NZP (72 and 112 cm^{-1}) [9], KZP (74 and 120 cm^{-1}), RbZP (85, 126, 159 and 170 cm^{-1}) and CsZP (105, 134 and 168 cm^{-1}) soften with increase in pressure and hence they are termed as soft modes. These soft modes may either be due to coupled rotation of PO_4 tetrahedra and ZrO_6 octahedra or Zr translation and PO_4 tetrahedra similar to that of prototype $NaZr_2(PO_4)_3$ [7]. Such modes are responsible for structural instabilities at high pressure prior to amorphisation.

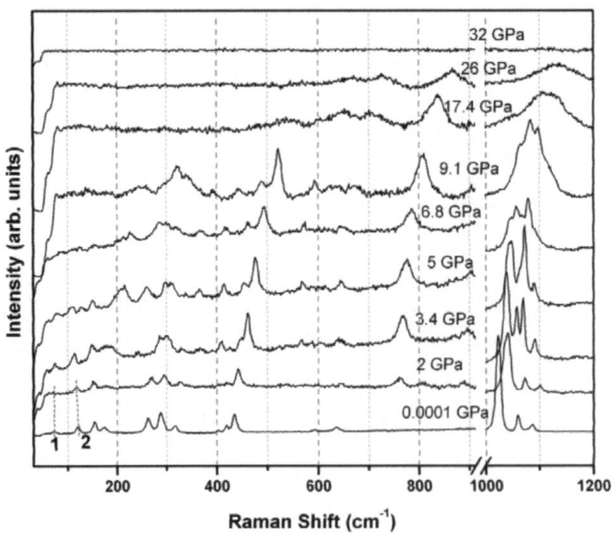

(a)

IOP Conf. Series: Journal of Physics: Conf. Series **950** (2017) 042014 doi:10.1088/1742-6596/950/4/042014

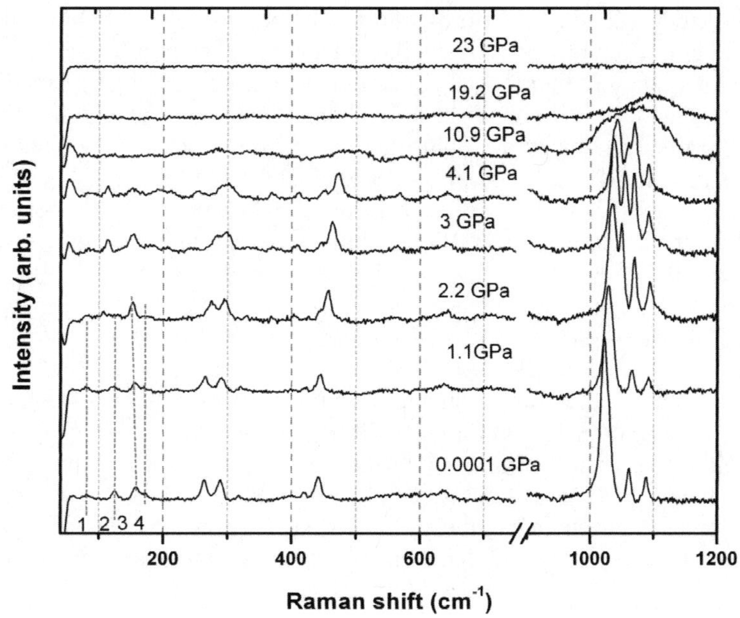

(b)

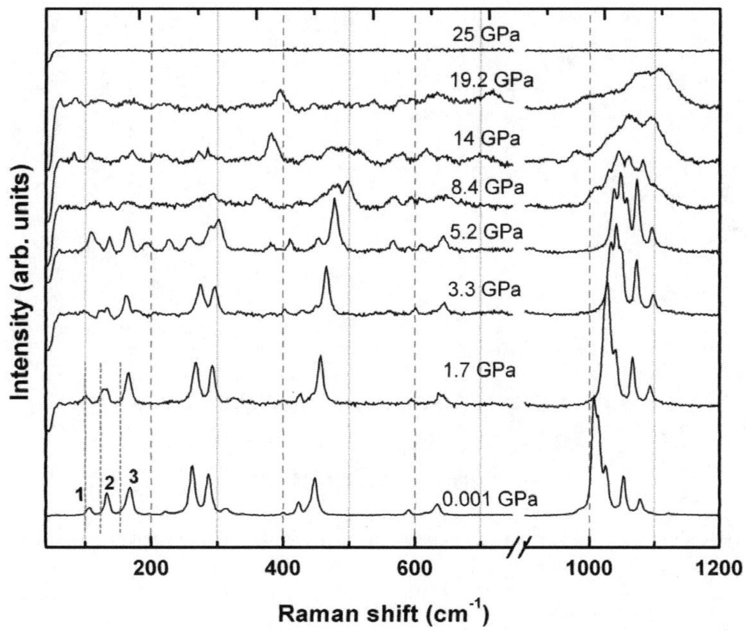

(c)

Figure 1. Raman spectra of (a) KZP, (b) RbZP and (c) CsZP at different pressures. KZP and RbZP show phase transition around 3.5 and 2.2 GPa and CsZP shows two transitions around 3.5 and 8.4 GPa. Pressure induced amorphisation for KZP, RbZP and CsZP occurs at higher pressures. Numbered dotted lines in *1*, *2* and *3* are identified as soft modes.

In KZP, RbZP and CsZP the intensities of the internal modes of PO_4 at high frequency as a function of pressure are different from one another. This suggests orientational disorder of polyhedral ions which is known to cause amorphization. Above 10 GPa broadening of modes is rapid; several weak bands lose their identity and only three broad bands centered around 900, 950 and 1020 cm^{-1} are present, indicating pressure induced amorphisation. NZP was found to be stable up to 32 GPa but compounds with larger cationic size amorphized at gradually lower pressures.

Pressure induced decomposition is expected to occur when the daughter compounds occupy less volume than the parent compound, if sufficient kinetics is provided to the system, i.e., at a high enough temperature [10].

$$3AZr_2(PO_4)_3 \longrightarrow A_3PO_4+2Zr_3(PO_4)_4 \quad (A= K, Rb, Cs)$$

The combined volume of the above daughter compounds ($A_3PO_4+2Zr_3(PO_4)_4$) is found to be less than the volume of parent compound ($3AZr_2(PO_4)_3$) indicating possible decomposition at high temperature and high pressure conditions. On the other hand all these compounds exhibit pressure induced amorphisation at ambient temperature. Amorphisation in these compounds might be due to kinetic hindrance of equilibrium decomposition. Figure 2 shows the variation of phase transition pressures and amorphisation pressure with increase in size of alkali cations in $AZr_2(PO_4)_3$ compounds.

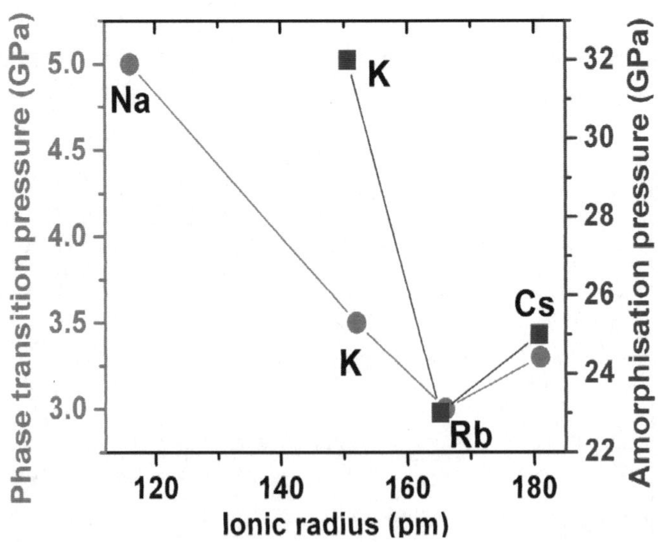

Figure 2. Variation of phase transition pressures and amorphisation pressures with increase in size of alkali cations (K, Rb and Cs).

Raman spectra of $AZr_2(PO_4)_3$, A= K, Rb, Cs were measured at pressures up to 25 GPa. Phase stability of these compounds at high pressure is discussed with respect to phase transformation and pressure induced amorphisation. Raman modes above the high pressure phase transition exhibit normal positive pressure coefficient that would result in positive thermal expansion.

IOP Conf. Series: Journal of Physics: Conf. Series **950** (2017) 042014 doi:10.1088/1742-6596/950/4/042014

References

[1] Miyazaki H, Ushiroda I, Itomura D, Hirashita T, Adachi N and Ota T 2008 *Jpn. J. Appl. Phys.* **47** 7262
[2] Ota T and Yamai I 1987 *J. Ceram. Soc. Jpn.* **95** 531
[3] Ota T and Yamai I 1986 *J. Am. Ceram. Soc.* **69** 1
[4] Scheetz E, Agarwal D K, Breval E and Roy R 1994 *Waste Manage.* **14** 489
[5] Petkov V I, Asabina E A, Markin A V, Smirnova N N and Kitaev D B 2005 *J. Therm. Anal. Calorim.* **80** 695
[6] Kamali K, Ravindran T R, Pandey K K, Sharma S M and Chandra Shekar N V 2015 *J. Solid State Chem.* **221** 225
[7] Kamali K, Ravindran T R, Ravi C, Sorb Y, Subramanian N and Arora A K 2012 *Phys. Rev. B* **86** 144301
[8] Tarte P, Rulmont A and Merckaert-Ansay C 1986 *Spectrochim. Acta A* **42** 10
[9] Hazen R M, Palmer D C, Finger L W and Stucky G D 1994 *J. Phys.:Condens.Matter* **6** 1333
[10] Arora A K 2000 *Solid State Commun.* **115** 665

IOP Conf. Series: Journal of Physics: Conf. Series **950** (2017) 042015 doi:10.1088/1742-6596/950/4/042015

Single-crystal X-ray diffraction study of $SrGeO_3$ high-pressure perovskite phase at 100 K

Akihiko Nakatsuka[1], Hiroshi Arima[2], Osamu Ohtaka[3], Keiko Fujiwara[1] and Akira Yoshiasa[4]

[1]Graduate School of Science and Engineering, Yamaguchi University, Ube 755-8611, Japan
[2]Institute for Materials Research, Tohoku University, Sendai 980-8577, Japan
[3]Graduate School of Science, Osaka University, Toyonaka 560-0043, Japan
[4]Graduate School of Science and Technology, Kumamoto University, Kumamoto 860-8555, Japan

E-mail: tuka@yamaguchi-u.ac.jp

Abstract. Single-crystal X-ray diffraction study of $SrGeO_3$ perovskite (cubic; space group $Pm\bar{3}m$) synthesized at 6 GPa and 1223 K was conducted at a low temperature of 100 K. The residual electron density revealed the presence of the bonding electron at the center of the Ge–O bond, in accordance with our previous conclusion that the Ge–O bond is strongly covalent. From comparison with our previous structure-refinement result at 296 K, the mean square displacement (MSD) of the O atom in the direction of the Ge–O bond is suggested to exhibit no significant temperature dependence, in contrast to that in the direction perpendicular to the bond. Thus, the strong covalency of the Ge–O bond can have a large influence on the temperature dependence of thermal vibration of the O atom.

1. Introduction

The perovskite-type compounds have widely been used for various industrial applications such as ferroelectric substances and ionic conductors. Recently, the high-pressure phase of $SrGeO_3$, with cubic perovskite-type structure (Fig. 1), was reported to be the first Ge-based transparent electronic conductor [1] in spite of its strongly covalent bonding nature. Detailed studies on the crystal structure of this compound are important for gaining insights into the correlation between its electronic conduction property and chemical bonding nature. The $SrGeO_3$ perovskite has also been investigated from interests as an analogue of $CaSiO_3$ and $MgSiO_3$ perovskites, major constituents in the Earth's lower mantle [2-4]. Its structural study can also provide geophysical imprecations in the Earth's interior. Recently, we successfully synthesized single crystals of $SrGeO_3$ perovskite phase at high pressure and high temperature and refined its crystal structure on the basis of single-crystal X-ray diffraction data at 296 K [5]. The structure refinement showed that the strong covalency of the Ge–O bond largely restricts the thermal vibration of the O atom. To further examine the effect of the bonding nature on the atomic thermal vibration, we here conduct single-crystal X-ray diffraction study of $SrGeO_3$ perovskite at a low temperature of 100 K and the obtained structure is compared with the reported one [5] at 296 K.

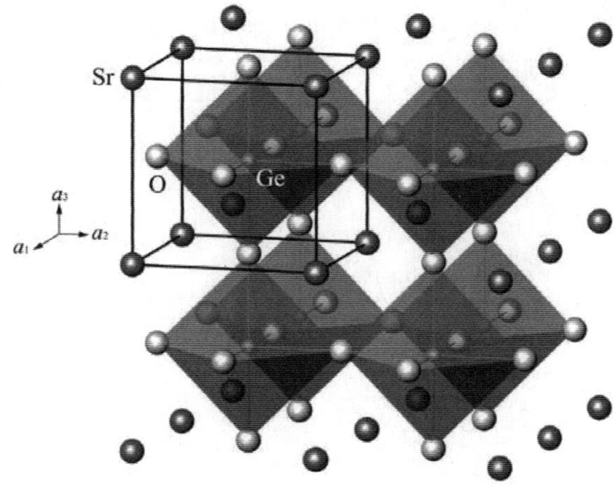

Figure 1. Polyhedral representation of the crystal structure of $SrGeO_3$ perovskite (after Ref. [5]).

2. Experiments and analysis

Single crystals of $SrGeO_3$ perovskite phase were grown under high pressure and high temperature of 6 GPa and 1223 K using a 700 ton cubic anvil-type press. The detailed procedure of the high pressure experiment is provided in our previous study [5]. A single crystal with a size of 0.10 × 0.09 × 0.09 mm^3 was selected from the recovered sample and was put into a silica-glass capillary for the low-temperature X-ray diffraction experiment. Intensity data were collected at 100 K using a Rigaku AFC-7R four-circle diffractometer with a graphite-monochromatized Mo Kα radiation (λ = 0.71069 Å) at the operating condition of 60 kV and 250 mA. The data collections at the low-temperature were conducted by directly cooling the crystal through a continuous cold N_2 gas flow generated using a Rigaku GN2 low-temperature apparatus. Prior to the data collections, the sample temperature was calibrated using a K-type thermocouple placed exactly at the sample position. Temperature fluctuation during the data collections was kept within ±0.2 K. The unit-cell parameters were determined by the least-squares method from a set of 25 reflections within the range of $43° \leq 2\theta \leq 50°$. Intensities of 521 reflections within $2° \leq 2\theta \leq 120°$ were measured using the continuous ω–2θ scan mode. After the intensity data were corrected for Lorentz-polarization factors and absorption effects (ψ-scan method [6]), they were averaged in Laue symmetry $m\bar{3}m$ to give 116 independent reflections. Of these, independent reflections with $F_o \leq 3\sigma(F_o)$ were eliminated. Even if independent reflections had intensities of $F_o > 3\sigma(F_o)$ after averaging, those averaged from data set of equivalent reflections including reflection(s) with $F_o \leq 3\sigma(F_o)$ were also discarded since these reflections were potentially affected by multiple scattering. Moreover, independent reflections with $(\sin\theta)/\lambda < 0.229$ Å^{-1} were eliminated to reduce secondary extinction effects and to avoid dependence on atomic charge as far as possible in the choice of atomic scattering factors. Finally, 61 independent reflections were used in the present refinement. Internal residual of the equivalent reflections (R_{int}) was 0.0176.

The structure refinement was carried out by minimizing the function $\sum w(F_o - F_c)^2$ using a full matrix least-squares program RADY [7]. Scattering factors of Sr^{2+} [8], Ge^{4+} [8] and O^{2-} [9] were used. Anomalous dispersion coefficients for each scattering factor were taken from *International Tables for Crystallography* [8]. Several correction models for the secondary extinction effects were attempted during the refinement, and the isotropic correction of Type I [10, 11] with a Gaussian mosaic spread distribution model yielded the best fit. The structure refinement converged smoothly to R = 0.0073 and wR = 0.0053. Crystal data, data collection and structure refinement details are summarized in Table 1. The refined structural parameters are given in Tables 2 and 3.

Table 1. Crystal data, data collection and structure refinement details.

Temperature (K)	100
Crystal system	cubic
Space group	$Pm\bar{3}m$
a (Å)	3.7921(6)
V (Å^3)	54.53(2)
No. of measured reflections	521
No. of independent reflections	116
R_{int}	0.0176
R	0.0073
wR	0.0053
Weighting scheme	$1/\sigma^2(F_o)$
No. of observed independent reflections used in refinement[a]	61
No. of parameters	6

[a] Reflections with $F_o > 3\sigma(F_o)$ and $(\sin\theta)/\lambda \geq 0.229$ Å^{-1} were used.

Table 2. Positional parameters and equivalent isotropic displacement parameters (Å^2) at 100 K.

Atom	SOF[a]	x	y	z	U_{eq}
Sr	1	0	0	0	0.00204(6)
Ge	1	0.5	0.5	0.5	0.00103(6)
O	1	0	0.5	0.5	0.0033(4)

[a] Site occupancy factor.

Table 3. Anisotropic displacement parameters (Å^2) at 100 K.

Atom	U_{11}	U_{22}	U_{33}	U_{12}	U_{13}	U_{23}
Sr	0.0020(1)	0.0020	0.0020	0	0	0
Ge	0.0010(1)	0.0010	0.0010	0	0	0
O	0.0021(5)	0.0038(3)	0.0038	0	0	0

3. Results and discussion

Our previous study [5] reported that the degree of covalency included in the Ge–O bond calculated from the bond valence is 48.9% at 296 K and thus showed that the sixfold-coordinated Ge–O bond in the $SrGeO_3$ perovskite has the strong covalency comparable to the fourfold-coordinated Si–O bonds in silicates with about 50% covalency. We here show in Figure 2 the difference Fourier map around a Ge position at 100 K. Positive residual electron density peaks with a height of 0.43 eÅ^{-3}, which can be regarded as bonding electrons, are clearly observed at the centers of the Ge–O bonds. This observation supports the presence of the strong covalency of the Ge–O bond.

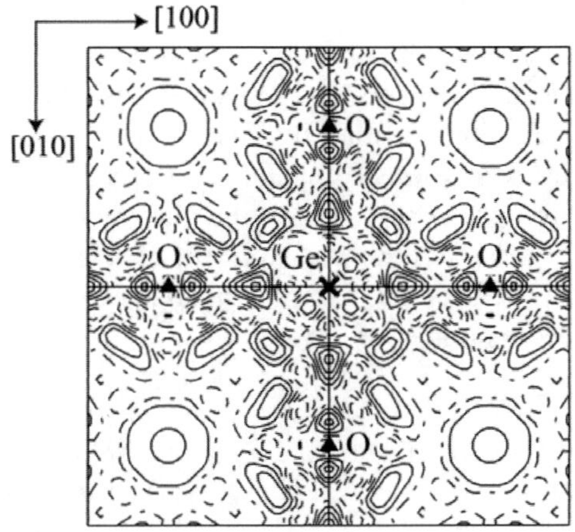

Figure 2. Difference Fourier map at 100 K on the section passing through a Ge position and parallel to (001) plane. The contour interval is 0.1 e$Å^{-3}$. Positive, negative and zero contours are solid, dashed and dashed-and-dotted lines, respectively. The Ge and O positions are denoted with cross and solid triangles, respectively.

The site-symmetry constraints require that the displacement ellipsoids of the Sr and Ge atoms are always spherical and that of the O atom is a uniaxial ellipsoid with one determinable ellipsoid-axis in the Ge–O direction and two undeterminable ones in the directions perpendicular to it. Our previous study [5] showed from the structure refinement at 296 K that the thermal vibration of the O atom is remarkably anisotropic and its displacement ellipsoid is considerably suppressed in the direction of the Ge–O bond. The reported mean square displacements (MSDs) of the O atom at this temperature are as follows: 0.0011(8) $Å^2$ in the direction of the Ge–O bond and 0.0077(7) $Å^2$ in the direction perpendicular to the bond. On the other hand, the present results at 100 K show that the ellipticity of the displacement ellipsoid of the O atom (Fig. 3) is much smaller than that at 296 K; the present values of its MSDs at 100 K are 0.0021(5) $Å^2$ in the direction of the Ge–O bond and 0.0038(3) $Å^2$ in the direction perpendicular to the bond. It is suggested, thus, that the MSD of the O atom in the direction of the Ge–O bond exhibits no significant temperature dependence, whereas that in the direction perpendicular to the bond largely increases with increasing temperature. This thermal vibration behavior of the O atom can be attributed to the strong covalency of the Ge–O bond.

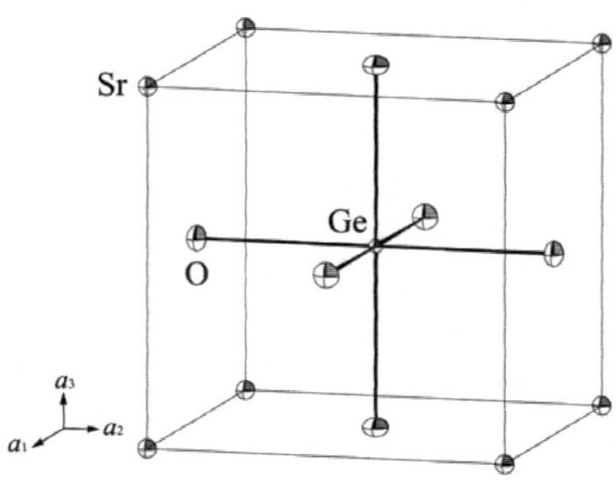

Figure 3. Unit cell of the cubic $SrGeO_3$ perovskite at 100 K, with displacement ellipsoids drawn at the 80% probability level.

IOP Conf. Series: Journal of Physics: Conf. Series **950** (2017) 042015 doi:10.1088/1742-6596/950/4/042015

4. Conclusion

From single-crystal X-ray diffraction analysis of high-pressure perovskite phase of $SrGeO_3$ at 100 K, we have observed the residual electron density peaks originating in the bonding electrons of the Ge–O bonds. This is a supportive observation for the presence of the strong covalency of the Ge–O bond suggested by our previous study [5]. The MSDs of the O atom at 100 K are 0.0021(5) Å^2 in the direction of the Ge–O bond and 0.0038(3) Å^2 in the direction perpendicular to the bond. The comparison of these values with the reported ones at 296 K leads to the suggestion that the thermal vibration of the O atom in the Ge–O direction exhibits no significant temperature dependence because of the strong covalency of the Ge–O bond.

Acknowledgments

This work was partly supported by a Grant-in-Aid for Scientific Research (C) (No. 15K05344) from the Japan Society for the Promotion of Science. The characterization of crystallinity of synthesized crystals was performed under the research project (No. 2010G608) at Photon Factory BL-10A, High Energy Accelerator Research Organization, Tsukuba, Japan. We also thank Mr. R. Takagaki and Mr. M. Ogawa of Yamaguchi University for their experimental assistance.

References

[1] Mizoguchi H, Kamiya T, Matsuishi S and Hosono H 2011 *Nature Comm.* **2** 470 doi: 10.1038/ncomms1484
[2] Akaogi M, Kojitani H, Yusa H, Yamamoto R, Kido M and Koyama K 2005 *Phys. Chem. Minerals* **32** 603–613
[3] Hashimoto T, Yoshiasa A, Sugahara M, Arima H, Fukui H, Murai K and Okube M 2008 *J. Phys.: Conf. Series* **121** 102002
[4] Andrault D, Itié J -P and Farges F 1996 *Am. Mineral.* **81** 822–832
[5] Nakatsuka A, Arima H, Ohtaka O, Fujiwara K and Yoshiasa A 2015 *Acta Cryst.* E**71** 502–504
[6] North A C T, Phillips D C and Mathews F S 1968 *Acta Cryst.* A**24** 351–359
[7] Sasaki S 1987 *RADY*. National Laboratory for High Energy Physics, Japan
[8] *International Tables for Crystallography* (vol C) 1992 ed A J C Wilson (Dordrecht: Kluwer)
[9] Tokonami M 1965 *Acta Cryst.* **19** 486
[10] Becker P J and Coppens P 1974*a* *Acta Cryst.* A**30** 129–147
[11] Becker P J and Coppens P 1974*b* *Acta Cryst.* A**30** 148–153

Volume and bond length dependences of the electronic structure of 6-fold and 8-fold coordinated Co^{2+} in pressure transformed CoF_2

J.A. Barreda-Argüeso [1], F. Aguado [1], J. González [1], R. Valiente [2], L. Nataf [3], M.N. Sanz-Ortiz [1,4], F. Rodríguez [1,*]

[1] MALTA Consolider Team, DCITIMAC, Facultad de Ciencias, Universidad de Cantabria, 39005 Santander, Spain

[2] MALTA Consolider Team, Dpt. Física Aplicada, Facultad de Ciencias, Universidad de Cantabria, Santander, Spain

[3] Synchrotron SOLEIL, L'Orme des Merisiers, Saint-Aubin, BP 48, 91192 Gif-sur-Yvette Cedex, France

[4] SCIC biomaGUNE, Parque tecnológico de San Sebastián, Guipúzcoa, Spain

* Corresponding author's e-mail: fernando.rodriguez@unican.es

Abstract. This work reports an optical absorption study on CoF_2 under high-pressure conditions in the 0-80 GPa range. A recent structural study[1] performed on the pressure-induced phase-transition sequence of CoF_2 reveals that this compound undergoes a structural transformation associated with a change of Co^{2+} coordination from 6 (rutile type) to 8 (fluorite type) at 15 GPa, what implies a profound change in the Co^{2+} electronic ground state associated with the inversion of d orbitals from ($t_{2g}^5e_g^2$) to ($e_g^4t_{2g}^3$). This allows us to explore the Co^{2+} electronic structure, and hence $10Dq$, as a function of the density and the Co-F bond length in both CoF_6 (rutile phase) and CoF_8 (fluorite phase) coordinations. The results are compared with those obtained in $KCoF_3$ (perovskite) in a wide range of Co-F distances (from 2.04 to 1.83 Å).

1. Introduction

There is a lack of experimental studies devoted to investigate the volume/density and bond length (R) dependencies of the crystal-field d-orbital splitting e_g-t_{2g} ($10Dq$) in transition metal ions (TM) occupying high-symmetry crystallographic sites. The situation is even harder if different coordination geometries are involved. Although experimental studies on $10Dq(R)$ have been usually performed at ambient pressure on series of TM compounds with different crystal structure and composition [1-4], rest-of-lattice and compositional effects can significantly mask the actual bond-length dependence of the TM electronic structure [5]. High-pressure spectroscopy solves out this problem since the electronic structure of a given compound can be measured as a function of pressure and eventually of V and/or R provided that its structure as a function of pressure is known.

Although pioneering works by Drickamer *et al.* [6] dealt with that problem in a lot of TM oxides and halides, correlations between $10Dq$ and V were performed on TM-doped Al_2O_3 or MgO (TM: Ti,

Cr, V) since the equation of state was known only for few materials at that time [6, 7]. However impurity systems pose a scale problem of whether the *TM*-ligand bond length and its variation with pressure are the same for the host (Al-O) and impurity (Cr-O) as for example in ruby (Al_2O_3: Cr^{3+}) [4,8-10]. Recent high-pressure studies by optical spectroscopy provided such correlations in pure *TM*-ion compounds: NiO [11] and $BiFeO_3$ [11, 12]. These studies gave evidence of $10Dq(V)$, and thus of $10Dq(R)$ dependences close to $V^{-5/3}$ and R^{-5}, respectively, given that in these compounds $R(P)$ scales as $V^{1/3}(P)$, and therefore both dependences obey the same power law, in agreement with crystal-field theory (CFT) predictions. This result is quite surprising since CFT in the point charge approximation is very simple and does not take explicitly into account the *TM*-ligand bonding [13-15]. Although efforts to justify the R^{-5} law have been done on the basis of ligand field theory [16,17] and by means of LCAO-type calculations, there is a lack of experimental studies aiming to validate whether CFT predictions on $10Dq$ apply for 1) different *TM* coordinations in the same crystal, *i.e.* 6-fold and 8-fold cubic coordinations in CoF_2; 2) the same coordination in different crystals, *i.e.* 6-fold coordination in CoF_2 and $KCoF_3$; and 3) a coordination change, *i.e.* $10Dq$(6-fold)/$10Dq$(8-fold) for the same R value.

This work reports an optical absorption study on CoF_2 under high-pressure conditions in the 0-80 GPa range. A recent structural study [18] performed on the pressure-induced phase-transition sequence of CoF_2 reveals that this compound undergoes a structural transformation associated with a change of Co^{2+} coordination from 6 (rutile type) to 8 (fluorite type) at 15 GPa, what implies a profound change in the Co^{2+} electronic ground state associated with inversion of d orbitals splitting from ($t_{2g}^{5}\,e_g^{2}$) to ($e_g^{4}\,t_{2g}^{3}$). This different ground state electronic configuration allows us to explore clearly the Co^{2+} electronic structure, and hence $10Dq$, as a function of the density and $R_{Co\text{-}F}$ in the two coordinations: CoF_6 (rutile phase) and CoF_8 (fluorite phase), taking the structural data reported elsewhere [18]. The results will be compared with those obtained in $KCoF_3$ where the 6-fold coordination of Co^{2+} in the perovskite structure is stable up to 80 GPa, thus providing the widest $10Dq(R)$ range ever measured in *TM* ions.

2. Experimental

2.1. *Crystal structure and x-ray diffraction*

Single crystals of CoF_2 and $KCoF_3$ grown by the Bridgman method according to methods described elsewhere [18]. At ambient conditions CoF_2 crystallizes in the tetragonal space group $P4_2/mnm$ (rutile phase) [18] and $KCoF_3$ in the cubic $Pm3m$ (perovskite phase) [19,20]. The evolution of the crystal structure of $KCoF_3$ with pressure was studied by angle dispersive x-ray diffraction (AXRD) on powdered samples using the 12.2.2 beamline at the *Advance Light Source* (ALS). Pressure was applied by means of a symmetric piston cylinder Diamond Anvil Cell (DAC). High pressure experiments on polycrystalline CoF_2 were performed in the *Materials Science and Powder Diffraction* beamline (BL04) at *ALBA* synchrotron using a Boehler-Almax DAC. In all experiments samples were loaded with several Ruby spheres (10 μm diameter) as pressure gauge [21] using methanol-ethanol-water and silicone oil as pressure transmitting media.

2.2. *Optical Absorption and Raman Spectroscopy*

Optical absorption and Raman experiments were performed on single-crystal plates of CoF_2 (90x80x14 μm^3) and $KCoF_3$ (100x80x20 μm^3) were used for high-pressure optical absorption experiments. The optical spectroscopy experiments were carried out in membrane and Almax- Boehler DACs. 200-μm-thick Inconel gaskets were preindented to 40 μm. 200-μm-diameter holes were perforated with a BETSA motorized electrical discharge machine. The DAC was loaded with a single crystal plate of CoF_2 or $KCoF_3$ and ruby microspheres (10 μm diameter) as pressure probes [21] using silicon oil as pressure-transmitting medium.

Optical absorption under high-pressure conditions was performed on a prototype fiber-optics microscope equipped with two 25× reflecting objectives mounted on two independent x -y -z translational stages for the microfocus beam and the collector objective, and a third independent x -y -z

translational stage for the DAC holder. Optical absorption data and images were obtained simultaneously with the same device [22]. Spectra in the UV-VIS and NIR were recorded with Ocean Optics USB 2000 and NIRQUEST 512 monochromators using Si- and InGaAs-CCD detectors, respectively.

Unpolarized micro-Raman scattering measurements were performed with a triple monochromator Horiba-Jobin-Yvon T64000 spectrometer in subtractive mode backscattering configuration, equipped with an Horiba Symphony liquid-nitrogen-cooled CCD detector. The 514.5-nm and 647-nm lines of an Coherent Innova 70 Ar^+-Kr^+ laser were focused on the sample with a 20× objective for micro-Raman, and the laser power was kept below 4 mW in order to avoid heating effects. The laser spot was 20 μm in diameter and the spectral resolution was better than 1 cm^{-1}. The Raman technique was used in combination with X-ray diffraction to check the sample structure through the characteristic first-order modes in CoF_2 [18] and absence of first-order modes in $KCoF_3$ as well as to determine structural phase-transition pressures. The Raman high-pressure experiments were performed on the same CoF_2 and $KCoF_3$ single crystals employed in the optical absorption measurements.

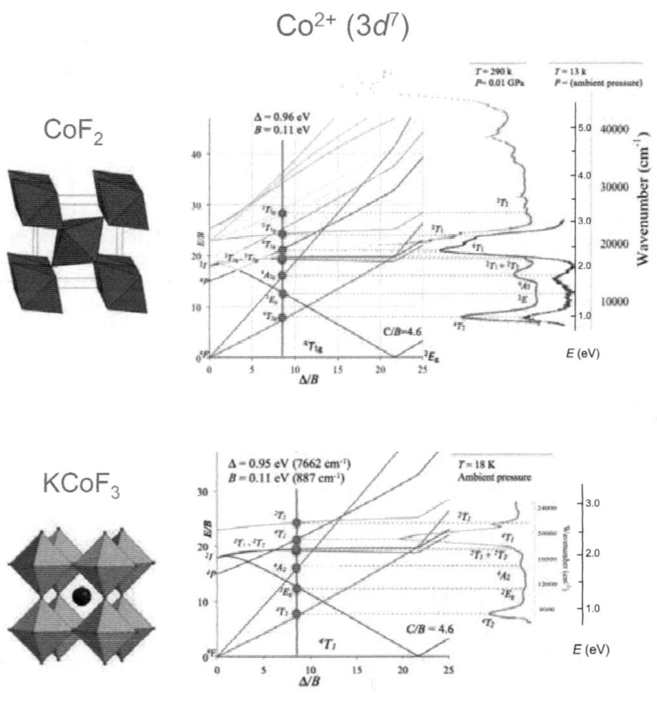

Figure 1. Optical absorption spectra of CoF_2 and $KCoF_3$ at ambient pressure and 290 K and low temperature (13 and 18 K, respectively). Spectra are normalized to the absorption coefficient. The peak energies are compared with those predicted by the Tanabe-Sugano diagram for octahedral Co^{2+} ($3d^7$): Racah (B) and crystal-field ($10Dq$) parameters obtained by fitting the experimental energies to the calculated energies are: $\Delta = 0.950 \pm 0.005$ eV and $B = 0.105 \pm 0.003$ eV for $KCoF_3$; and $\Delta = 0.962 \pm 0.005$ eV and $B = 0.107 \pm 0.003$ eV for CoF_2 ($\Delta/B = 9$).

3. Results and discussion

Figure 1 shows the electronic absorption spectrum of CoF_2 and $KCoF_3$ at ambient pressure, as well as the Tanabe-Sugano and corresponding B and $10Dq$ parameters, which account for the Co^{2+} transition energies. Both compounds exhibit a similar spectrum as it mainly corresponds to the CoF_6 coordination unit with similar Co-F bond distances: R = 2.042 and 2.035 Å [23,24]; and associated CF parameters: B = 0.107 and 0.105 eV; and $10Dq$ = 0.962 and 0.950 eV, for CoF_2 and $KCoF_3$, respectively. Interestingly, the similarity of spectra strengthen the relevance of CoF_6 to account for the CF electronic structure of Co^{2+} beyond other ionic shells of the crystal, the influence of which seems to be much weaker than the first F^- shell in spite of the distinct crystal structure: rutile (CoF_2) and perovskite ($KCoF_3$). However, the slightly bigger $10Dq$ of CoF_2 with respect to $KCoF_3$ contrasts with the shorter Co-F distance in $KCoF_3$ must be ascribed to rest-of-the-lattice CF effects beyond CoF_6, which is estimated to be less than 4% of $10Dq$.

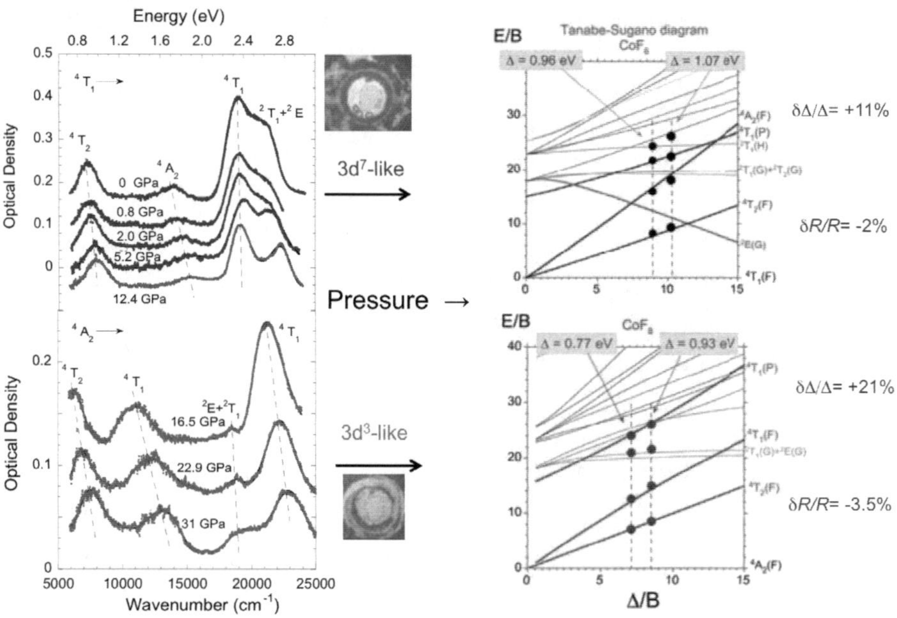

Figure 2. Pressure dependence of the optical absorption spectrum of CoF_2. Spectra correspond to rutile-type structure (blue), orthorhombic (green) with 6-fold coordinated Co^{2+}, and to fluorite-type phase (red) with 8-fold coordinated Co^{2+} [18]. The bands correspond to crystal-field transitions from the electronic ground state $^4T_1(F)$ in CoF_6 and $^4A_2(F)$ in CoF_8 to different excited states Γ_i according to the assignment given in the corresponding Tanabe-Sugano diagrams for d^7 and d^3, respectively. The energy of the Γ_i state is represented as a function of the crystal-field splitting and are both given in units of the Racah parameter B [24]. Colour representing Γ_i state has been maintained in both figures. The two sets of experimental points in the Tanabe-Sugano diagram denote the low pressure and high-pressure limits in both 6-fold and 8-fold coordinations. The relative crystal-field, $\Delta 10Dq$ / $10Dq$ (or $\delta\Delta/\Delta$), and corresponding $\delta R/R$ variations are given right side.

The variation of the absorption spectra with pressure for the two compounds is shown in figures 2 and 3. The extreme pressure points are represented in the Tanabe-Sugano diagram to illustrate the $10Dq/B$ range swept by pressure in each coordination: CoF_6 (d^7-like) or CoF_8 (d^3-like). The variations of $10Dq$ and R in the corresponding pressure range are also indicated. Interestingly, the CoF_6 (*rutile*) to CoF_8 (*fluorite*) change of coordination taking place in CoF_2 at 15 GPa induces a drastic change in the absorption spectrum which is associated with the corresponding change of electronic configuration (from d^7-like to d^3-like) affecting the ground and excited states, and hence the absorption spectra. However such an abrupt change is not observed in $KCoF_3$ since the CoF_6 octahedral coordination is stable in the 0 - 80 GPa range as indicated by AXRD experiments whose *V(P)* data are represented in figure 3 (right). Consistently, the absorption spectrum of $KCoF_3$ in the whole pressure range is concomitant with an electronic configuration d^7 and thus the experiment provides the widest $10Dq(R)$ range explored ever in Co^{2+}.

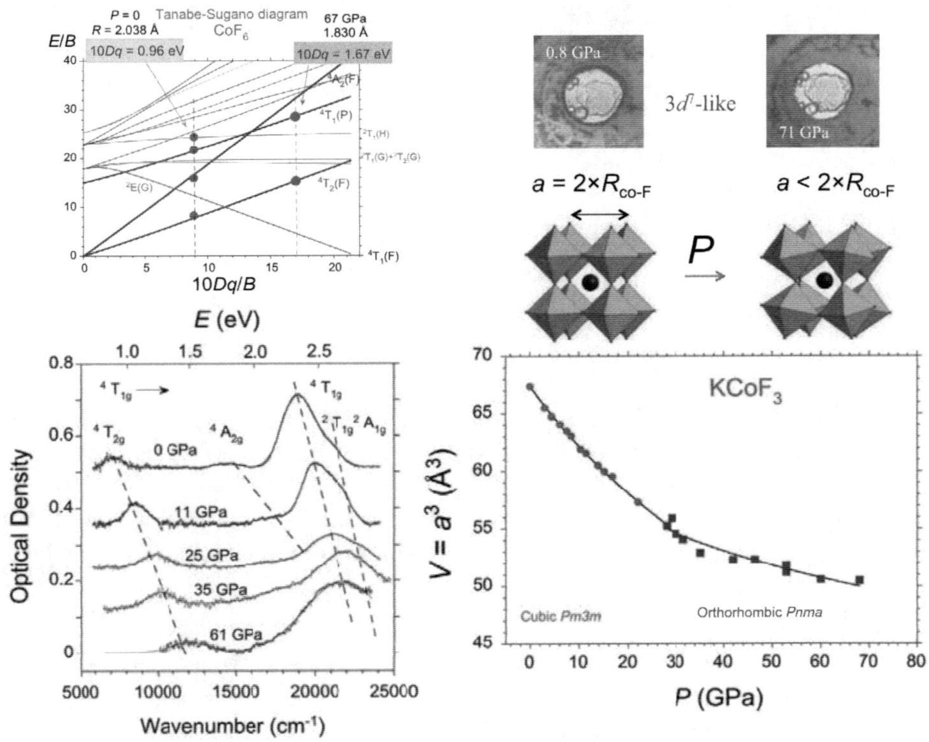

Figure 3. Variation of the absorption spectrum of $KCoF_3$ with pressure (left) and of the perovskite cell volume *V(P)* (right). The peak energies are compared with those predicted by the Tanabe-Sugano diagram. The *B* and $10Dq$ parameters obtained by fitting the experimental energies *E(P)* to the calculated crystal-field energies are: $10Dq = 0.960 \pm 0.005$ eV and 1.67 ± 0.01 eV, and $B = 0.105 \pm 0.003$ and 0.100 ± 0.003 eV for $KCoF_3$ at 0 and 67 GPa, respectively. The Tanabe Sugano-diagram compares the $10Dq/B$ range swept by pressure for CoF_6 in $KCoF_3$ (red dots): $10Dq/B = 10 \rightarrow 18$. The corresponding *R* variation obtained from the Equation of State *V(P)* is indicated.

In general, pressure-induced phase transitions in both CoF_2 and $KCoF_3$ affect the electronic spectra as evidenced through their $10Dq(P)$ behaviour (figure 4). The different slope of $10Dq(P)$ exhibited by the perovskite $KCoF_3$ unravels the effect of a cubic (*Pm-3m*) to orthorhombic (*Pnma*) second-order phase transition at 24 GPa. Such a phase transition, which is associated with rotations of the CoF_6 octahedra, keeps the 6-fold octahedral coordination of Co^{2+} in the high-pressure orthorhombic phase, and thus $10Dq$ varies continuously at the phase transition, although its pressure dependence is different in each phase. This pressure behaviour contrasts with that exhibited by CoF_2, whose $10Dq(P)$ values show a peculiar pressure dependence through which the corresponding $10Dq(R)$ have been derived using the available XRD and XAS structural data reported elsewhere [18]. In CoF_2 pressure induces an increase of $10Dq$ from 0.96 to 1.06 eV in the 0-6 GPa (rutile phase), whereas at 15 GPa, it abruptly decreases by -0.3 eV along with the 6- to 8-fold coordination change. Above 15 GPa, $10Dq$ increases at a rate of 9 meV GPa^{-1} in the fluorite phase, and at 4.4 meV GPa^{-1} beyond 44 GPa in the cotunnite phase [18].

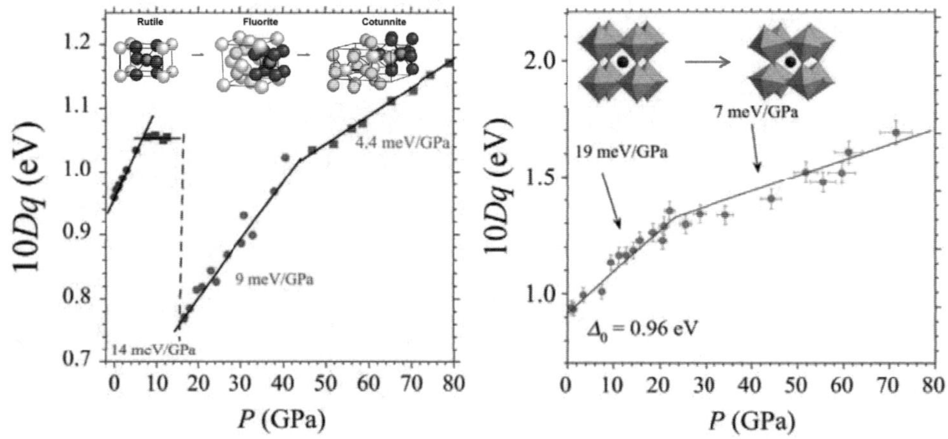

Figure 4. Variations of the Co^{2+} crystal-field splitting energy between e_g and t_{2g} d-orbitals with pressure in CoF_2 (left) and $KCoF_3$ (right). The coordination of Co^{2+} and the $10Dq(P)$ pressure shift are indicated for each crystal phase. Anomalies in $10Dq(P)$ observed at 6, 15 and 44 GPa in CoF_2, and at 24 GPa in $KCoF_3$, correspond to structural phase transitions $P4_2/mnm$ (rutile) ↔ $Pbca$ ↔ $Fm3m$ (fluorite) ↔ $Pnma$ (cotunnite) in CoF_2, and $Pm3m$ ↔ $Pnma$ (perovskite) in $KCoF_3$, respectively.

4. Concluding remarks

The knowledge of EOS and $R(P)$ in the two investigated compounds from XRD and XAS (Figure 3) [18,20] allows us to derive $10Dq(R)$ from the corresponding $10Dq(P)$. A noteworthy result concerns the regular variation of $10Dq(R)$ in CoF_6 and CoF_8 with respect to variations of $10Dq(P)$ that show anomalies at the phase transitions in both CoF_2 and $KCoF_3$. Such anomalies are concurrent with similar anomalies in $R(P)$ thus highlighting the suitability of R (and neither V nor density) to properly describe pressure-induced variations of $10Dq$. Furthermore $10Dq(R)$ obeys a potential law as R^{-n} with n close to 5 in both compounds. In particular, we get an exponent $n = 5.1\pm0.3$ in $KCoF_3$ in an R range of 12% whereas in CoF_2 $n = 5.0\pm0.1$ (rutile phase) and $n = 6.0\pm0.2$ (fluorite phase) as deduced from their relative changes of R and $10Dq$ in the explored R-ranges (figures 2 and 3).

The reliability of CFT to fairly account for R-dependence of $10Dq$ is also confirmed through the relative $10Dq$ values obtained for CoF_6 and CoF_8 in CoF_2 at the same Co-F distance, $R = 2.04$ Å, as $10Dq(oct)/10Dq(cub) = 1.10$. Its similitude with the theoretical value of 9/8 supports CFT predictions to accurately describe experimental $10Dq(R)$.

Finally, this study constitutes a reference system to validate *ab initio* methods aiming to calculate electronic structure (including excited states) of *TM* ions besides of well-suited DFT methods to calculate crystallographic and vibrational structures.

Acknowledgments
Financial support from the Spanish Ministerio de Economia y Competitividad (Project No. MAT2015-69508-P) and MALTA INGENIO-CONSOLIDER 2010 (Ref. No. CDS2007-0045) is acknowledged.

The ALS is supported by the Director, Office of Science, Office of Basic Energy Sciences of the US Department of Energy, under Contract No. DE-AC02-05CH11231.

References

[1] Rodríguez F and Moreno M 1986 *J. Phys. Chem.* **84** 692

[2] Grinberg M 2006 *Opt. Mat.* **28** 26

[3] Rodríguez-Lazcano Y, Nataf L, Rodríguez F 2009 *Phys. Rev. B* **80** 085115-1

[4] Marco-de-Lucas MC, Rodríguez F, Güdel HU, and Furer N 1994 *J. Lumin.* **60/61** 581

[5] García-Lastra J M, Barriuso M T, Aramburu J A, and Moreno M 2005 *Phys. Rev. B* **72** 113104

[6] Drickamer H G and Frank C W 1973 *Electronic structure, electronic transitions and the high pressure chemistry and physics of solids* (Chapman and Hall: London)

[7] Tonkov E Y 1992 *High Pressure Phase Transformations: A Handbook* (Gordon & Breach: Amsterdam)

[8] Duclos S D, Vohra Y K and Ruoff A L 1990 *Phys. Rev. B* **41** 5372

[9] Millot M, Broto J M, Gonzalez J and Rodríguez F 2010 *Phys. Rev. B* **81** 075120

[10] Duan W, Paiva G, Wentzcovitch R M and Fazzio A 1998 *Phys. Rev. Lett.* **81** 3267

[11] Gavriliuk A G, Trojan I A and Struzhkin V V 2012 *Phys. Rev. Lett.* **109** 086402

[12] Gomez-Salces S, Aguado F, Rodriguez F, Valiente R, Gonzalez J, Haumont R and Kreisel J 2012 *Phys. Rev. B* **85** 144109

[13] Lever A B P 1984 *Inorganic Electronic Spectroscopy* (Elsevier: Amsterdam)

[14] Burns R G 1993 *Mineralogical Applications of Crystal Field Theory* (Cambridge University Press: Cambridge)

[15] Griffith J S 1908 *The theory of Transition-Metal Ions* (Cambridge University Press: Cambridge)

[16] Curie D, Barthou C and Canny B 1974 *J. Chem. Phys.* **61** 3048

[17] Liu L, Li X D, Li Y C, Tang L Y, Liu J and Ban Y 2009 *Chin. J. High Press. Phys.* **23** 209

[18] Barreda-Argüeso J A, López-Moreno S, Sanz-Ortiz M N, Aguado F, Valiente R, González J, Rodríguez F, Romero A H, Muñoz A, Nataf L, and Baudelet F 2013 *Phys. Rev. B* **88** 214108

[19] Julliard J and Nouet J 1975 *Rev. Phys. Appl.* **10** 325

[20] Aguado F, Rodriguez F and Redfern S A T 2009 *High Pressure Res.* **29** 525

[21] Syassen K 2008 *High Pressure Res.* **28** 75

[22] Barreda-Argüeso J A and Rodríguez F 2014 *Microscopio para la caracterización espectroscópica de una muestra* (Patent PCT/ES2014/000049)

[23] Costa N M R, Paixao J A, De Almeida M J M and Andrade L C R 1993 *Acta Crystallogr. B* **49** 591

[24] Sugano S and Tanabe Y 1958 *J. Phys. Soc. Jpn.* **13** 880

IOP Conf. Series: Journal of Physics: Conf. Series **950** (2017) 042017 doi:10.1088/1742-6596/950/4/042017

The effect of high pressure on the lattice structure and dynamics of phenacenes

F Capitani[1,2] M Höppner[3], L Malavasi[4], C Marini[5], P Dore[1], L Boeri[3,6] and Paolo Postorino[1]

[1]Dipartimento di Fisica, Università di Roma Sapienza, P.le Aldo Moro 2, 00185 Roma, Italy
[2]Present Address: Ligne AILES - Synchrotron SOLEIL, L'Orme des Merisiers, F-91192 Gif-sur-Yvette, France
[3]Max Planck Institute for Solid State Research, Heisenbergstrasse 1, D-70569 Stuttgart, Germany
[4]Dipartimento di Chimica and INSTM, Università di Pavia, Via Taramelli 16, I-27100 Pavia, Italy
[5]CELLS-ALBA, Carretera B.P. 1413, Cerdanyola del Vallès, 08290 Barcelona, Spain.
[6]Institute of Theoretical and Computational Physics, TU Graz, Petersgasse 16, A-8010, Graz, Austria

E-mail: `francesco.capitani@synchrotron-soleil.fr`

Abstract. We studied the effect of high pressure on three phenacenes, aromatic molecules with a zig-zag configuration of the benzene rings. The lattice structure and vibrational dynamics of crystalline phenanthrene ($C_{14}H_{10}$, three benzene rings), chrysene ($C_{18}H_{12}$, four), and picene ($C_{22}H_{14}$, five) were investigated by means of X-ray diffraction and Raman measurements. Raman spectra were compared with theoretical ones obtained from *ab-initio* Density Functional Theory calculations. Experimental and theoretical results allowed to identify the onset of a structural transition in phenanthrene at 7.8 GPa under hydrostatic conditions and at 5.7 GPa under non-hydrostatic conditions. We found that this transition is related to a reorientantion of the molecules in the *ab* plane. On the contrary, chrysene and picene do not undergo any phase transition in the investigated pressure range, thus suggesting that molecular size plays an important role in the occurence of pressure induced structural modifications in aromatic compounds.

1. Introduction

Phenacenes are a class of polycyclic aromatic hydrocarbons (PAHs) characterized by a zig-zag configuration of the aromatic rings in the molecular structure. Like other PAHs, they are prototypical organic systems where molecules are held together in the crystalline state by weak intermolecular interactions. Intermolecular distances can thus be conveniently reduced by the application of an external pressure, which finely tunes structural [1–3], electronic [4, 5] and vibrational properties [3, 6] of the system. When the pressure exceeds a critical threshold, usually a few tens of GPa, it also induces irreversible reactions leading to new and peculiar structural phases [7–9]. Moreover, experimental investigations of aromatic hydrocarbons provide new useful benchmarks for theoretical calculations [10, 11].

In this respect, experimental studies on phenacenes under pressure are still low in number and not systematic. The first three members of phenacene are phenanthrene $C_{14}H_{10}$ which is

IOP Conf. Series: Journal of Physics: Conf. Series **950** (2017) 042017 doi:10.1088/1742-6596/950/4/042017

made of three fused benzene rings, chrysene $C_{18}H_{12}$ (four rings) and picene $C_{22}H_{14}$ (five rings). Phenanthrene has been recently investigated under pressure by means of X-ray diffraction (XRD) and Raman spectroscopy [3]. In this work, two structural phase transitions at 2.3 and 5.6 GPa were claimed. Concerning chrysene, preliminary spectroscopic works have been performed in the past at very low pressure [12, 13], whereas high pressure structural studies have never been performed. Picene has been studied only by means of optical measurements [5, 6, 14] showing the absence of any structural phase transition up to 23 GPa, where an irreversible transformation to hydrogenated amorphous carbon occurs [5].

Here, we report a first step towards a systematic investigation of the effect of high pressure (HP) on phenacenes by studying the structural and vibrational properties of phenanthrene, chrysene and picene up to 20 GPa. Our results show that only phenanthrene undergoes a phase transition whereas chrysene and picene keep their original structure up to the maximum pressure. The comparison among the three compounds points out the key role of the molecular size in determining the occurrence of a structural transition.

2. Methods

2.1. Experimental

Phenanthrene and chrysene powder samples were purchased from Sigma Aldrich. Picene powder was obtained through a newly developed synthesis procedure [15].

High pressure XRD measurements were performed at ID-09 beamline of the ESRF synchrotron (Grenoble, France) with a fixed wavelength of 0.4127 Å. The sample-to-detector distance and the image plate orientation angles were calibrated using CeO_2 standard. The two-dimensional diffraction images, acquired from a MAR555 detector, were converted to one dimensional 2θ diffraction patterns using the FIT2D software [16]. Pressure on the sample was generated by a membrane diamond anvil cell (DAC) equipped with IA diamonds (culet size 400 μm) and a stainless steel gasket. A $\sim$ 50μm sample was loaded, together with Helium as hydrostatic medium, inside a 150 μm hole which was previously drilled in the gasket.

High pressure Raman measurements on phenanthrene and chrysene were carried out with a LabRam Infinity and a LabRam HR Evo micro-spectrometers, respectively. Both instruments are by Horiba Jobyn-Yvon and are installed at the Department of Physics of the Sapienza University of Rome. Samples were excited by a He-Ne laser through a 50x long working distance objective which also collected the scattered radiation from the samples (backscattering geometry). Both the spectrometers allowed a spectral resolution better than 3 cm^{-1}. A screw driven DAC equipped with 700 μm culet IIA diamonds was used to pressurize the sample. A 250 μm hole was drilled in gasket and filled with sample without any pressure transmitting medium. A ruby sphere was added near to the culet center (no farther than 50 μm) for measuring the pressure *in situ* by the ruby fluorescence technique [17]. At each pressure step, Raman spectra were collected from three different points almost equidistant from the ruby and separated by $\sim$ 50 μm one from the other. This allowed to check the homogeneity of the pressure response over the sample [18]. The comparison among these spectra showed that peak frequencies never exceeded a 2 cm^{-1} difference among the three acquisition points, thus witnessing the absence of strong pressure gradients.

2.2. Theoretical

Raman and IR cross sections of crystalline phenanthrene were calculated at selected pressures using Density Functional Perturbation Theory (DFPT) [19, 20], as implemented in the *quantum-espresso* package [21]. We used norm-conserving pseudopotentials [22] together with the local density approximation. The plane-wave (charge density) cutoff energy was set to 100 Ry (400 Ry) for all calculations. We used a uniform $3\times4\times3$ grid for **k**-space integration. We checked the choice of the plain wave cutoff by performing calculations under pressure keeping constant the number of plane waves by increasing/decreasing the cutoff according to $E_{\mathrm{cutoff}} \propto V^{-\frac{2}{3}}$.

3. Results and Discussion

At ambient conditions, phenanthrene [23], chrysene [24] and picene [25] crystallize in a monoclinic unit cell where molecules are arranged in a herringbone configuration in the *ab* plane and are stacked along the *c* axis. In particular, phenanthrene and picene have the same space group $P2_1$ and two molecules per unit cell ($Z = 2$), whereas the space group of chrysene is $I2/c$ with $Z = 4$.

The XRD patterns of the three compounds have been collected in the 0-20 GPa pressure range. Data at selected pressures are shown in figure 1. At ambient conditions, a Le Bail fit of the diffraction patterns provides the following lattice parameters: $a = 8.458(2)$ Å, $b = 6.164(2)$ Å, $c = 9.482(5)$ Å, $\beta = 98.11(2)°$ for phenathrene; $a = 8.352(3)$ Å, $b = 6.205(2)$ Å, $c = 25.16(2)$ Å, $\beta = 115.84(4)°$ for chrysene; $a = 8.196(3)$ Å, $b = 6.324(2)$ Å, $c = 13.82(5)$ Å, $\beta = 93.75(6)°$ for picene. All these parameters are in agreement with the literature [23–25]. On increasing pressure, Bragg peaks shift towards higher angular values (lower interplanar distances) as a consequence of the volume reduction. This shift is continuous for chrysene, figure 1(b), and picene, figure 1(c), where also the overall shape of the diffraction patterns keeps constant over the whole pressure range. This indicates that the ambient pressure crystalline lattice of chrysene and picene is maintained, although compressed, up to 20 GPa. In other words, chrysene and picene do not undergo any structural transition in the investigated pressure range.

On the contrary, phenanthrene undergoes a phase transition at about 7.8 GPa. As a matter of fact, around this pressure value, the intensity of the lowest angle Bragg peak decreases and a new peak appears around $2\theta \sim 2.8°$ (look at the 8.7 GPa pattern in figure 1). Other subtle changes are observed for the reflections around $2\theta \sim 8°$. It is worth noting that our data does not show any other phase transition, at variance with the results reported by Huang *et al.* in Ref.[3], where a transition at 2.3 GPa and another one at 5.6 GPa were claimed. The comparison between our XRD patterns, figure 1(a), and those from Ref.[3] shows that the transition found by Huang *et al.* at 5.6 GPa is the same we found at 7.8 GPa. The two different values of the transition pressure are likely due to different hydrostatic conditions since here we employed helium as hydrostatic medium, whereas no medium was used by Huang and coworkers [3]. As to the phase transition found at 2.3 GPa by Huang *et al.*, our diffraction patterns do not show any anomaly around this pressure value. The absence of the phase transition at 2.3 GPa and the lowering of the pressure value under non-hydrostatic conditions for the other transition are further confirmed by our Raman measurements, which will be discussed in the following.

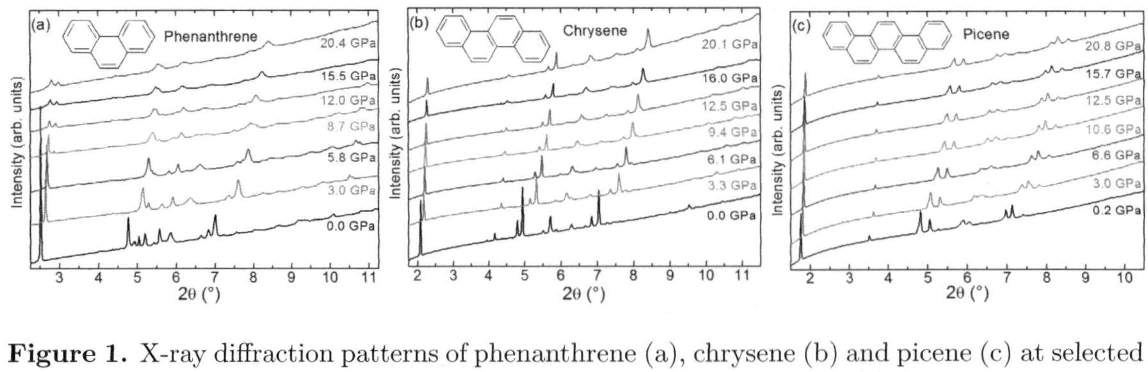

Figure 1. X-ray diffraction patterns of phenanthrene (a), chrysene (b) and picene (c) at selected pressures. Sketches of the molecular structures are also shown.

IOP Conf. Series: Journal of Physics: Conf. Series **950** (2017) 042017 doi:10.1088/1742-6596/950/4/042017

In Ref.[26], we analysed the diffraction patterns of phenanthrene using a structural model generated from an algorithm for crystal structure prediction and we identified a triclinic $P1$ unit cell with Z=3 formula units as the best structural candidate for the HP phenanthrene phase [26]. The lattice parameters of this phase obtained from a Le Bail fit of experimental data at 7.8 GPa are $a = 6.951(10)$ Å, $b = 8.967(10)$ Å, $c = 9.522(5)$ Å, $\alpha = 100.42(11)°$, $\beta = 106.86(9)°$, $\gamma = 106.03(9)°$. In the HP phase, phenanthrene molecules are arranged in a parallel fashion within the ab plane, thus pointing out that the herringbone arrangement typical of phenacenes is no longer energetically convenient at HP (see Ref.[26]). This *motif* transition in the ab plane implies a rotation of the molecular units under pressure. This reorientation is easier when the steric hindrance of the molecules is smaller, especially when the available volume is reduced by pressure. This considerations strongly suggest a prominent role of the small molecular size of phenanthrene (three benzene rings) for the occurence of the HP structural transition. Therefore, the bigger steric hindrance of chrysene (four benzene rings) and picene (five benzene rings) is likely responsible for the structural stability of these systems at least up to 20 GPa.

The occurence of a phase transition can be also revealed by a vibrational spectroscopy. In fact, when the symmetry of the system changes across a phase transition, as in the case of phenanthrene (from $P2_1$ to $P1$), the vibrational selection rules change and a more or less pronounced variation can be observed in the vibrational spectrum of the system.

We thus performed Raman measurements on the three phenacenes as a function of pressure in the 0-8 GPa range. The pressure dependence of the Raman spectra of phenanthrene and chrysene are shown in figure 2. The spectra of picene are instead reported in Ref.[6] and do not show any indication of a phase transition, in agreement with the XRD data shown in figure 1 and with the infrared measurements under pressure reported in Ref.[5].

In any case, the investigated pressure range is limited by the appeareance of a strong pressure induced luminescence background which prevents to collect Raman data above 8.5 GPa for phenanthrene, 7.7 GPa for chrysene and 6 GPa for picene [6]. This phenomenon has been observed in several aromatic hydrocarbons under pressure and it is generated by crystalline defects which increase under pressure [4].

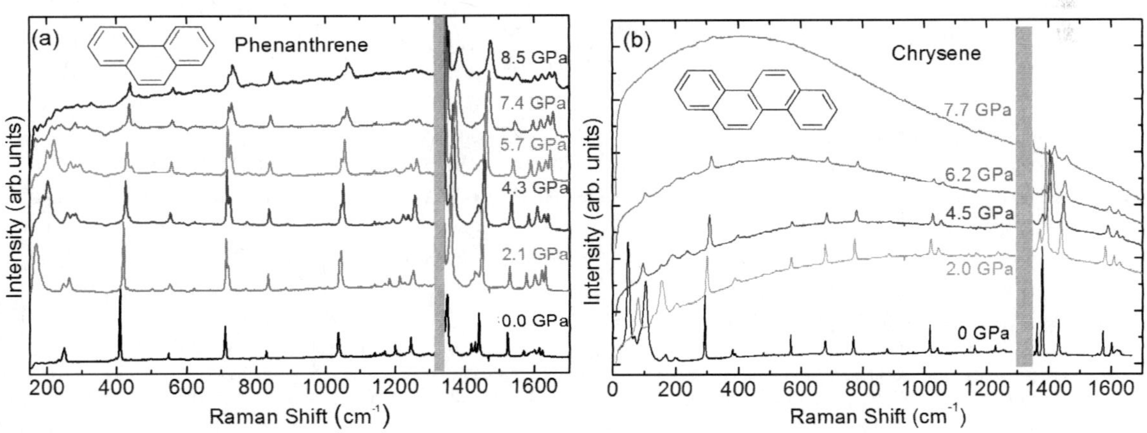

Figure 2. Raman spectra of phenanthrene (a) and chrysene (b) at selected pressures. The grey area around 1330 cm^{-1} masks the Raman peak of the diamond.

As to figure 2, a hardening of the vibrational frequencies and a broadening of the peaks is observed on increasing pressure. In particular, the Raman peaks below 200 cm^{-1}, ascribed to intermolecular vibrations [27, 28], show the strongest hardening. Indeed, pressure can easily modify the intermolecular distances owing to the weakness of the intermolecular interactions.

The shift of the Raman peaks of chrysene under pressure is continuous (see figure 2(b)), without signs of any phase transition in agreement with XRD data.

On the other hand, some pressure induced spectral changes (such as peaks splitting, intensity enhancement/reduction, appeareance of new peaks) can be observed in the Raman spectra of phenanthrene (see figure 2(a)). In the 0-6 GPa pressure range, these spectral changes can be nicely reproduced by DFT calculations based on the low pressure $P2_1$ phase of phenanthrene. Above 6 GPa, measured spectra are no longer well reproduced by calculations, especially some new peaks appearing in the experimental data, thus suggesting the occurrence of a phase transition in agreement with XRD data.

We report in figure 3 the comparison of the measured and calculated Raman spectra for two frequency regions: around 730 cm^{-1}, panels (a)-(c), and 1050 cm^{-1}, (d)-(f). The lineshape analysis of the experimental spectra is based on a standard least-square fitting procedure with Lorentzian or Gaussian functions. The theoretical spectra shown in panels (c) and (f) are calculated using the monoclinic $P2_1$ structure of phenanthrene at 0, 4, 8 GPa with relaxed internal positions. At 0 GPa, the experimental spectra can be fitted with a two component

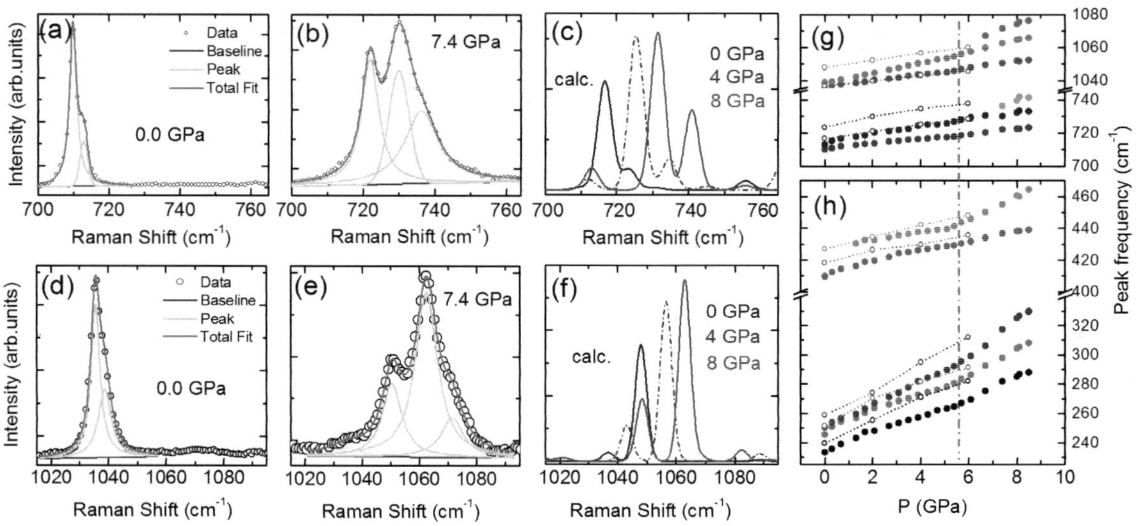

Figure 3. Lineshape analysis of the Raman spectra of phenanthrene within the 700-760 cm^{-1} and 1000-1080 cm^{-1} ranges at 0 GPa, panels (a)-(d), and 7.4 GPa, (b)-(e). Panels (c)-(f): DFT calculated spectra at 0, 4, 8 GPa for the $P2_1$ phase of phenanthrene. Panels (g)-(h): Measured (full points) and calculated (empty points) vibrational frequencies of some selected modes as a function of pressure (g)-(h). The vertical grey dotted line marks the pressure value where a slope change is observed.

model in both spectral regions (see panels (a),(d)), in agreement with the calculated spectra (see panels (c),(f)). Above 5.7 GPa, a considerable broadening is observed in both regions. Moreover, there is an inversion of the intensity of the two components of the spectral feature around 710 cm^{-1} (compare panels (a) and (b)). The spectral analysis at 7.4 GPa, panels (b) and (e), reveals a change in the relative Raman cross section of the two components and the appeareance of a shoulder on the high frequency side in both the spectral regions. In contrast, the Raman spectra calculated in the $P2_1$ structure are characterized by only two components at all pressures and do not show the appeareance of any new peak (see panels (c),(f)). Therefore, the appeareance of the high frequency shoulders in the experimental spectra indicates a structural change from the $P2_1$ structure to a high pressure phase of phenanthrene with reduced symmetry. This agrees with the results of our XRD analysis under hydrostatic compression which shows a transition to a structure with $P1$ symmetry and Z=3 molecules in the unit cell.

The analysis of experimental spectra also provided the vibrational frequencies of the observed peaks. The pressure dependence of some of these frequencies is shown in figure 3(g)-(h), where the calculated values of the corresponding vibrational modes of the $P2_1$ phase is also shown for comparison. Both calculated and measured frequencies increase as a function of pressure, with similar slopes. The theoretical values are systematically $\sim$ 10 cm^{-1} higher than the experimental ones. This amounts to 1 % relative error, which is within the typical accuracy of LDA-DFPT calculations for phonon frequencies [6, 29].

More interestingly, above 5.7 GPa, a slope change of the experimental frequencies occurs, *i.e.* they increase more rapidly under pressure. Such a slope change demonstrates that pressure has a different and stronger effect on the vibrational frequencies above 5.7 GPa. This behavior is often observed when the system enters a new phase [30]. On further increasing pressure, two new peaks, already shown in panels (b) and (e), appear around 740 and 1070 cm^{-1} and persist up to the maximum pressure investigated (orange and dark cyan points). Therefore, both the slope change and the appeareance of new peaks indicates the occurrence of a phase transition in phenanthrene at high pressure.

Yet, the phase transition in the XRD data is at 7.8 GPa whereas the phase transition deduced from the Raman spectra occurs at 5.7 GPa. This disagreement can be ascribed to the different experimental conditions. In fact, the XRD data were collected with helium as a pressure transmitting medium (hydrostatic conditions), whereas the Raman spectra were collected without a medium. The non-hydrostaticity thus explains the lower transition pressure of about 5.7 GPa, which is indeed in agreement with the value obtained from the non-hydrostatic XRD measurements reported in Ref.[3].

4. Conclusions

In this paper, we studied the effect of high pressure on the lattice structure and vibrational spectra of crystalline phenanthrene, chrysene and picene, by means of X-ray diffraction, Raman spectroscopic measurements and DFT calculations. We showed that the onset of a structural transition occurs in phenanthrene at 7.8 GPa under hydrostatic conditions, whereas chrysene and picene keep the same structure up to the maximum pressure here investigated of about 20 GPa. Bearing in mind the shorter length of phenanthrene (three aromatic rings) molecules compared to chrysene (four) and picene (five), we conclude that the molecular size strongly affects the pressure induced modifications of the structure of phenacenes.

Acknowledgements

Authors acknowledge ESRF for beamtime and M. Hanfland for the assistance provided on ID-09 beamline. This work is supported by Fondazione Cariplo (grant no. 2013-0632).

References

[1] Oehzelt M, Aichholzer A, Resel R, Heimel G, Venuti E and Della Valle R G 2006 *Phys. Rev. B* **74** 104103
[2] Fabbiani F P A, Allan D R, Parsons S and Pulham C R 2006 *Acta Crystallogr. B* **62** 826–42
[3] Huang Q W *et al.* 2013 *J. Chem. Phys.* **139** 104302
[4] Dreger Z A, Lucas H and Gupta Y M 2003 *J. Phys. Chem. B* **107** 9268–74
[5] Fanetti S, Citroni M, Malavasi L, Artioli G A, Postorino P and Bini R 2013 *J. Phys. Chem. C* **117** 5343–51
[6] Capitani F *et al.* 2013 *Phys. Rev. B* **88** 144303
[7] Ciabini L, Santoro M, Gorelli F A, Bini R, Schettino V and Raugei S 2007 *Nat. Mater.* **6** 39–43
[8] Wen X D, Hoffmann R and Ashcroft N W 2011 *J. Am. Chem. Soc.* **133** 9023–35
[9] Fitzgibbons T C, Guthrie M, Xu E s, Crespi V H, Davidowski S K, Cody G D, Alem N and Badding J V 2015 *Nat. Mater.* **14** 43–47

[10] Schatschneider B, Monaco S, Liang J J and Tkatchenko A 2014 *J. Phys. Chem. C* **118** 19964–74
[11] Schatschneider B, Monaco S, Tkatchenko A and Liang J J 2013 *J. Phys. Chem. A* **117** 8323–31
[12] Offen H W 1966 *J. Chem. Phys.* **44** 699–703
[13] Baldwin B A and Offen H W 1967 *J. Chem. Phys.* **46** 4509–14
[14] Joseph B *et al.* 2012 *J. Phys. Condens. Matter* **24** 252203
[15] Protti S, Artioli G A, Capitani F, Marini C, Dore P, Postorino P, Malavasi L and Fagnoni M 2015 *RSC Adv.* **5** 27470–5
[16] Hammersley A P, Svensson S O, Hanfland M, Fitch A N and Hausermann D 1996 *High Pressure Res.* **14** 235–48
[17] Mao H K, Bell P, Shaner J W and Steinberg D J 1978 *J. Appl. Phys.* **49** 3276
[18] Capitani F, Fasolato C, Mangialardo S, Signorelli S, Gontrani L and Postorino P 2015 *J. Phys. Chem. Solids* **84** 13–6
[19] Baroni S, de Gironcoli S, Dal Corso A and Giannozzi P 2001 *Rev. Mod. Phys.* **73** 515–62
[20] Lazzeri M and Mauri F 2003 *Phys. Rev. Lett.* **90** 036401
[21] Giannozzi P *et al.* 2009 *J. Phys. Condens. Matter* **21** 395502
[22] We used the pseudopotentials C.pz-vbc.UPF and H.pz-vbc.UPF from `http://www.quantum-espresso.org`
[23] Trotter J 1963 *Acta Crystallogr.* **16** 605–8
[24] Burns D M and Iball J 1960 *Proc. R. Soc. London A* **257** 491–514
[25] De A, Ghosh R, Roychowdhury S and Roychowdhury P 1985 *Acta Crystallogr. C* **41** 907–9
[26] Capitani F, Höppner M, Malavasi L, Marini C, Artioli G A, Hanfland M, Dore P, Boeri L and Postorino P 2016 *J. Phys. Chem. C* **120** 14310–6
[27] Godec J and Colombo L 1976 *J. Chem. Phys.* **65** 4693–700
[28] Colangeli L, Mennella V, Baratta G A, Bussoletti E and Strazzulla G 1992 *Astrophys. J.* **396** 369–77
[29] Brown-Altvater F, Rangel T and Neaton J B 2016 *Phys. Rev. B* **93**(19) 195206
[30] Capitani F, Trequattrini F, Palumbo O, Paolone A and Postorino P 2016 *J. Phys. Chem. B* **120** 2921–28

IOP Conf. Series: Journal of Physics: Conf. Series **950** (2017) 042018 doi:10.1088/1742-6596/950/4/042018

Study of the orpiment and anorpiment phases of As_2S_3 under pressure

S Radescu[1], A Mujica[1], P Rodríguez Hernández[1], A Muñoz[1], J Ibáñez[2], J A Sans[3], V P Cuenca-Gotor[3] and F J Manjón[3]

[1] Departamento de Física and Instituto Universitario de Materiales y Nanotecnología, MALTA Consolider Team, Universidad de La Laguna, La Laguna 38200, Tenerife, Spain

[2] Consejo Superior de Investigaciones Científicas, Instituto Jaume Almera, Barcelona 08028, Spain

[3] Instituto de Diseño para la Fabricación y Producción Automatizada, and MALTA Consolider Team, Universitat Politècnica de València, Camí de Vera s/n, 46022 València, Spain

E-mail: `sradescu@ull.es`

Abstract. In this work we study the pressure behaviour of the orpiment (monoclinic) and anorpiment (triclinic) layered structures of As_2S_3 by means of *ab initio* calculations performed within the density functional theory, as part of an ongoing theoretical and experimental joint effort to provide a comprehensive picture of the bonding of this interesting material and the evolution of its structural, electronic, and vibrational properties under pressure.

1. Introduction

Arsenic sulfide binary compounds with different stoichiometries have been researched in the past following interest in their optical, electronic, and structural properties, and in particular As_2S_3 has been found to be an excellent glass former at high temperatures, with some significant properties [1-5]. Under normal conditions, the stable phase of As_2S_3 is a semiconductor named orpiment with a crystalline layered structure in which each atom of As is surrounded by three atoms of S whereas each atom of S is surrounded by two atoms of As. Despite efforts to synthesize new polymorphs of this material no other crystalline metastable phases at normal conditions could be artificially produced in laboratories for a long time. Quite recently however, a low symmetry (triclinic) new phase of As_2S_3 named anorpiment has been naturally discovered analyzing the mineral specimens of an old peruvian mine [3]. The structure of this anorpiment phase is similar to that of the long known orpiment phase in that it is made of covalently bonded layers of As and Se atoms with a similar coordination (though the interlayer pattern is not exactly the same) arranged in a different stacking sequence, with interlayer van der Waals forces playing a significant role in the structural properties. In this work we study the orpiment and anorpiment phases of As_2S_3 theoretically (within the first-principles framework of the density functional theory) as part of an ongoing theoretical and experimental joint effort to provide a comprehensive picture of the bonding of this interesting material and the evolution of its structural, electronic, and vibrational properties under pressure.

IOP Conf. Series: Journal of Physics: Conf. Series **950** (2017) 042018 doi:10.1088/1742-6596/950/4/042018

2. Details of the calculations

All the calculations were done within the first-principles theoretical framework of the density functional theory (DFT) implemented in the computational scheme of pseudopotentials and plane waves. The Vienna Ab Initio Simulation Package (VASP) code was used [6-8]. Both the Perdew-Burke-Ernzerhof (PBE) [9] and its PBEsol modified version [10] of the generalized gradient approximations to the exchange and correlation (XC) energy were used as well as the projector-augmented wave scheme [11,12]. The $4s$ and $4p$ outermost electrons of As and the $3s$ and $3p$ electrons of S were dealt with explicitly in the valence, the rest of the electrons being considered frozen at the core. The cutoff in the kinetic energy of the plane-waves basis set was 360 eV which together with the use of dense Monkhorst-Pack grids of k-points for integrations in reciprocal space [13] (6×4×2 for orpiment and 4×2×2 for anorpiment), ensured a convergence in the energy differences between phases of around 1 meV. The van der Waals (vdW) corrections to the total energy were taken into account within the DFT using Grimme's so-called D2 method [14].

All the results that we present in this short communication correspond to zero temperature, with zero-point energy neglected, and hydrostatic pressure conditions (that is, diagonal and isotropical stress tensor). For each phase, a full structural optimization at fixed volumes (of both internal parameters and cell shape) was performed through the calculation of the forces on atoms and the components of the stress tensor. The structural relaxation was deemed complete when the forces on atoms were less than 0.005 eV $Å^{-1}$ and the anisotropy in the diagonal components of the stress tensor was less than 0.1 GPa. The calculated total energy E as a function of volume V for the fully relaxed configurations were fitted using a Birch-Murnaghan equation of state (EoS) [15] from which the pressure p, enthalpy $H = E + pV$, and properties at zero pressure (equilibrium volume V_0, bulk modulus B_0 etc) were obtained [16].

For the calculation of phonons at the zone center we used the small-displacements method as implemented in the PHON code [17] followed by the symmetry analysis of the eigenvectors for each mode in order to identify its possible Raman or infrared activity. These calculations were done at a range of volumes (and respective pressures) allowing the study of the pressure evolution of the frequencies of the different modes.

3. Results and discussion

Both structural evolution and the change in energetics, lattice vibrations, and electronic structure have been studied under compression, of which we will give here a very brief account.

The orpiment structure is monoclinic with space group (SG) $P2_1/c$, No. 14, with Z=4 and As and S atoms at 2e positions whereas, the recently discovered anorpiment structure is triclinic, SG P-1, No. 2, Z=4 and As and S atoms at 2i positions. The orpiment and anorpiment structures are plotted in projection in Figures 1 and 2, respectively, and Figures 3 and 4 show the calculated evolution of the first few neighbours distances under compression.

Figure 5 shows the calculated energy-volume (E-V) curves of the orpiment and anorpiment phases of As_2S_3 obtained using different XC functionals with and without van der Waals corrections, which shows the effects of the different approximations. Thus, for example, for the orpiment phase the zero-pressure equilibrium volumes V_0 (per cell) obtained using the PBE functional are 477.8 $Å^3$ and 584.8 $Å^3$, with and without van der Waals corrections, respectively, whereas using the PBEsol functional the values obtained are 418.6 $Å^3$ (with vdW) and 445.1 $Å^3$ (without vdW), cf. the experimental value of 467.7(4) $Å^3$ [3]. For the anorpiment phase the results for the equilibrium volume are 502.0 $Å^3$ (PBE with vdW), 582.3 $Å^3$ (PBE without vdW), 454.5 $Å^3$ (PBEsol with vdW), and 485.2 $Å^3$ (PBEsol without vdW) - cf. the experimental value of 488.4 $Å^3$. [3]. In general, including vdW corrections decreases the equilibrium volume and results in E-V curves with a better defined minimum, in comparison to those E-V curves obtained without vdW, that is, the vdW corrections enhance the stability of the zero-pressure

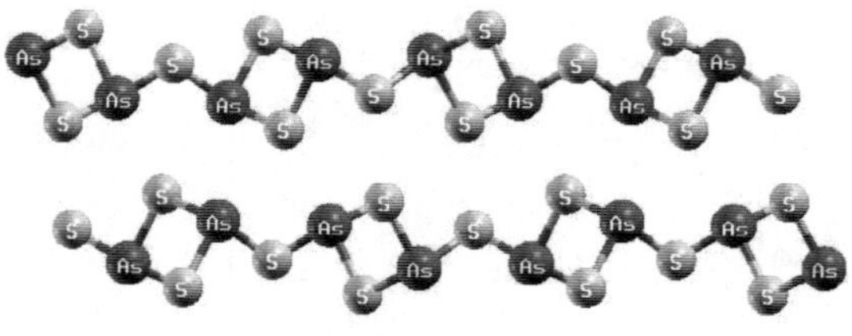

Figure 1. Plot of the orpiment structure, showing the positions of As and S atoms in projection, perpendicularly to the layers.

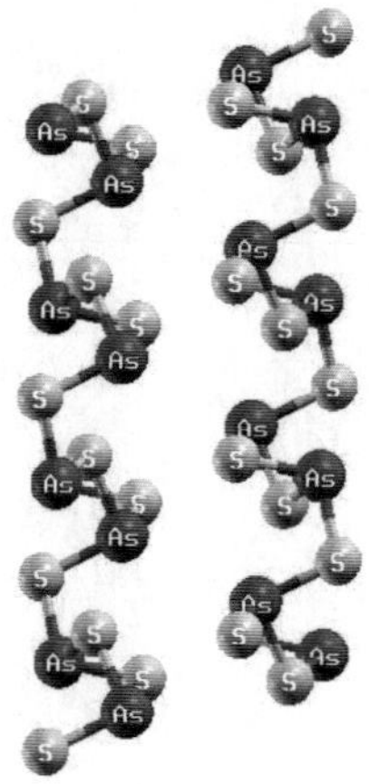

Figure 2. Plot of the anorpiment structure, showing the positions of As and S atoms in projection, perpendicularly to the layers.

state. On the other hand, the PBEsol XC functional and the PBE XC functional with vdW corrections provide the results that are closer to the experimental values of V_0, and for this last functional the calculated bulk moduli at zero pressure are 14 GPa and 19 GPa for the orpiment and anorpiment phases, respectively.

Finally, in this brief account of our theoretical results we show in Figures 6-8 the evolution of the calculated frequencies of the phonon modes at Γ under pressure for the two phases. The mechanical representation of the orpiment structure yields the following multiplicities for the symmetry of the modes at the zone center: $15A_g + 15B_g + 15A_u + 15B_u$, of which the A_g and B_g modes are active in Raman and the A_u and B_u modes are infrared-active (two B_u and one A_u acoustic modes excepted). For the triclinic anorpiment phase, the symmetry analysis results in $30A_g + 30A_u$ modes, of which the A_g modes are Raman active and the 27 non-acoustic A_u modes are infrared active. The results of our assignment will be useful for the experimental

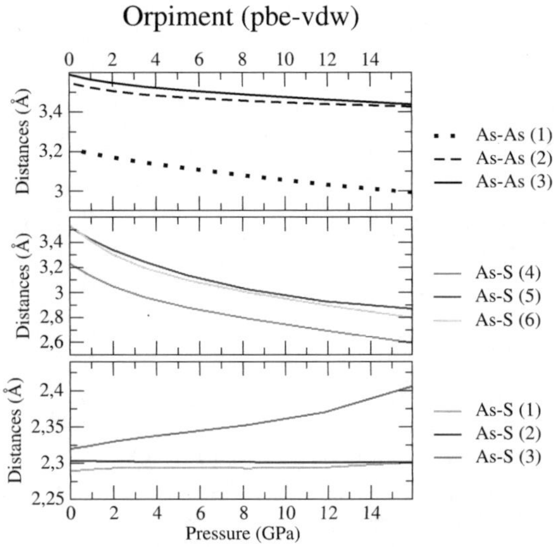

Figure 3. Calculated evolution of the interatomic distances between near neighbours of the orpiment phase obtained using the PBE funtional and van der Waals corrections. The species for each neighbour is specified in the labels to each curve.

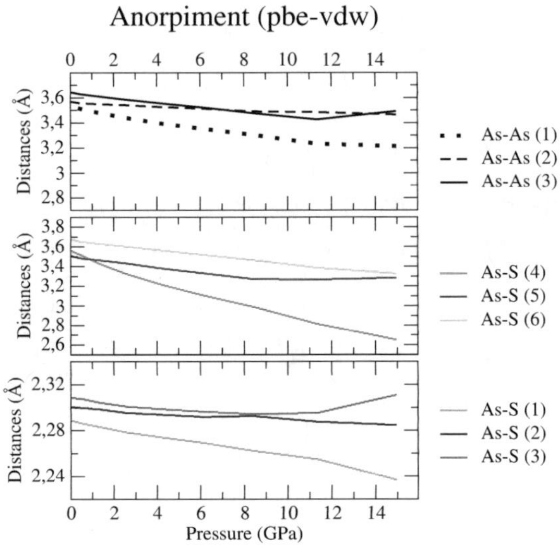

Figure 4. Calculated evolution of the interatomic distances between near neighbours of the anorpiment phase obtained using the PBE funtional and van der Waals corrections. The species for each neighbour is specified in the labels to each curve.

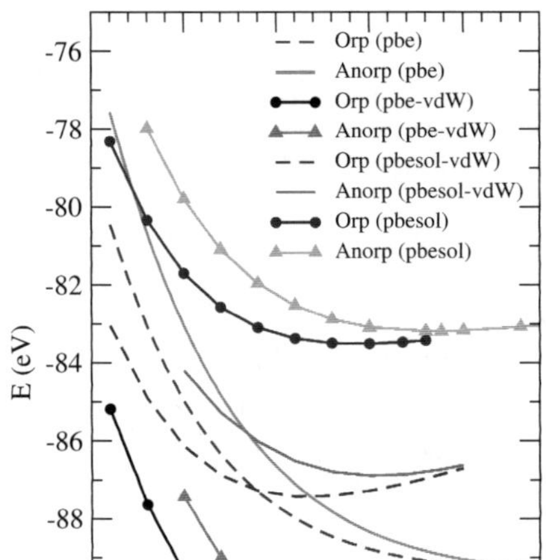

Figure 5. Calculated energy-volume (E-V) curves for the orpiment (Orp) and anorpiment (Anorp) phases of As_2S_3 obtained using the PBE and PBEsol XC functionals, with and without van der Waals (vdW) corrections. Both E and V are given per cell.

study of the modes, as this information is not easy to obtain experimentally.

4. Summary

The pressure behaviour of the orpiment and anorpiment layered structures of As_2S_3 has been investigated by means of *ab initio* calculations performed within the density functional theory.

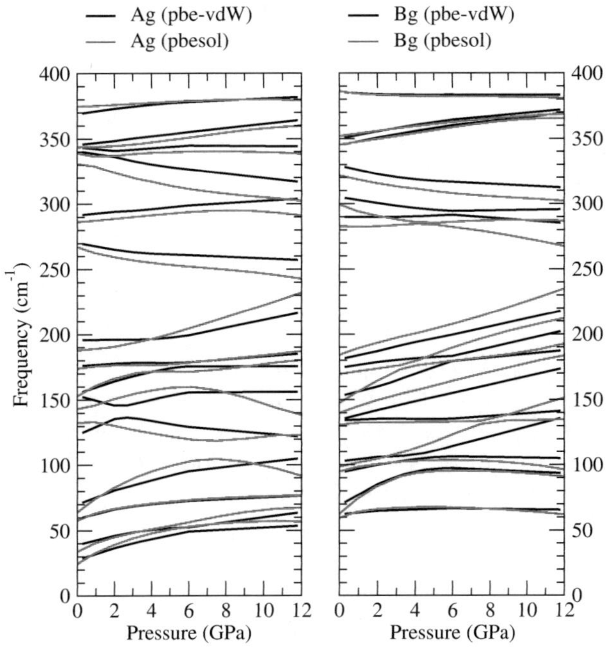

Figure 6. Calculated pressure evolution of the phonon frequencies of the Raman-active modes of the orpiment phase (Ag and Bg modes), obtained using the PBE XC functional with vdW corrections (pbe-vdW) and the PBEsol XC functional (pbesol).

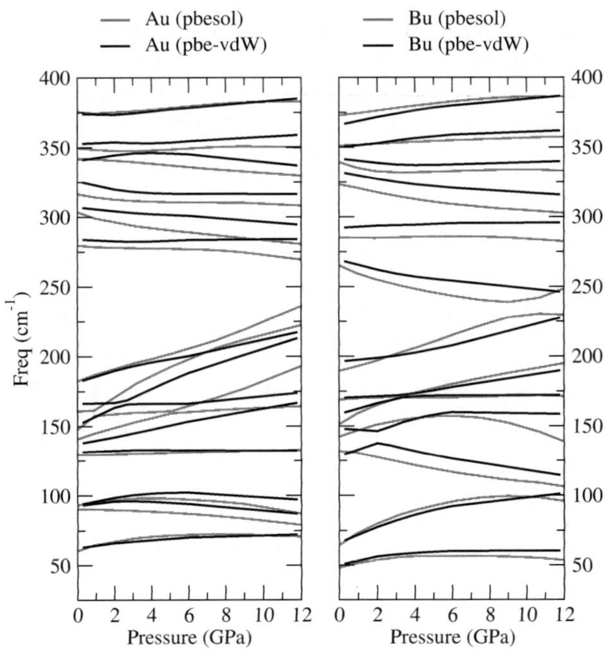

Figure 7. Calculated pressure evolution of the phonon frequencies of the infrared-active modes of the orpiment phase (Au and Bu modes), obtained using the PBE XC functional with vdW corrections (pbe-vdW) and the PBEsol XC functional (pbesol).

The phonon frequencies at the zone center were calculated using the finite displacements method, and the symmetry of the modes were assigned, which will help in future Raman and infrared experimental studies of these phases under compression. The effect of van der Waals interactions was especifically investigated and found to be important for the structural response of these layered structures, enhancing their stability. These results are part of an ongoing comprehensive effort to shed further light on the properties of As_2S_3, with a particular emphasis on their pressure evolution.

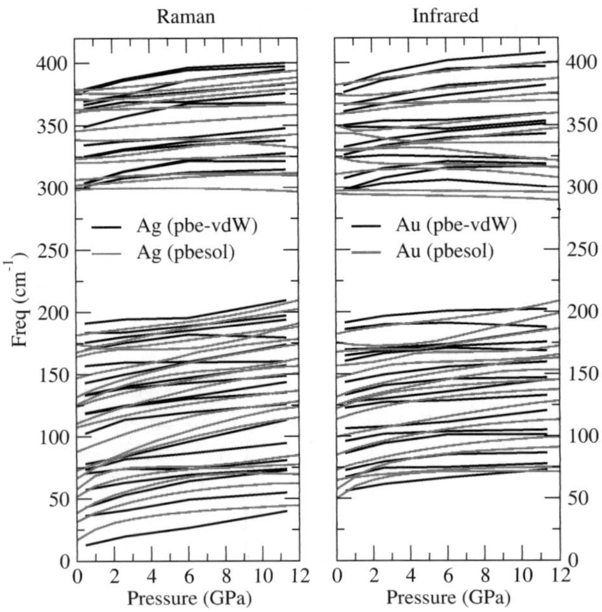

Figure 8. Calculated pressure evolution of the phonon frequencies of the Raman-active (Ag) and infrared-active (Au) modes of the anorpiment phase, obtained using the PBE XC functional with vdW corrections (pbe-vdW) and the PBEsol XC functional (pbesol).

Acknowledgments
The authors acknowledge the financial support from the Ministerio de Economía y Competitividad (MINECO) of Spain through Projects No. MAT2013-46649-C04-02-P and MAT2013-46649-C04-03-P. Computer time in the MALTA computer cluster at the University of Oviedo, Spain, is also gratefully acknowledged (MINECO Project No. CSD2007-00045).

References
[1] Brazhkin V V, Katayama Y, Kondrin M V, Lyapin A G and Saitoh H 2010 *Phys. Rev.* B **82** 140202(R)
[2] Gibbs G V, Wallace A F, Zallen R, Downs R T, Ross N L, Cox D F and Rosso K M 2010 *J. Phys. Chem.* A **114** 6550
[3] Kampf A R, Downs R T, Housley R M, Jenkins R A and Hyršl J 2011 *Mineral. Mag.* **75** 2857
[4] Bolotina N B, Brazhkin V V, Dyuzheva T I, Katayama Y, Kulikova L F, Lityagina L V and Nikolaev N A 2013 *JETP Letters* **98** 539
[5] Bolotina N B, Brazhkin V V, Dyuzheva T I, Lityagina L M, Kulikova L F, Nikolaev N A and Verin I A 2013 *Crystallogr. Rep.* **58** 61
[6] Kresse G and Hafner J 1993 *Phys. Rev.* B **47**, 558
[7] Kresse G and Furthmüller J 1996 *Comput. Mater. Sci.* **6** 15
[8] Kresse G and Furthmüller J 1996 *Phys. Rev.* B **54** 11169
[9] Perdew J P, Burke K and Ernzerhof M 1996 *Phys. Rev. Lett.* **77** 3865
[10] Perdew J P, Ruzsinszky A, Csonka G I, Vydrov O A, Scuseria G E, Constantin L A, Zhou X and Burke K 2008 *Phys. Rev. Lett.* **100** 136406
[11] Kresse G and Joubert D 1999 *Phys. Rev.* B **59** 1758
[12] Blöchl P E 1994 *Phys. Rev.* B **50** 17953
[13] Monkhorst H J and Pack J D 1976 *Phys. Rev.* B **13** 5188
[14] Grimme S 2006 *J. Comp. Chem.* **27** 1787; Grimme S, Antony J, Ehrlich S and Krieg S 2010 *J. Chem. Phys.* **132** 154104
[15] Birch F 1947 *Phys. Rev.* **71** 809
[16] Mujica A, Rubio A, Muñoz A and Needs R J 2003 *Rev. Mod. Phys.* **75** 863 and references therein.
[17] Alfè D 2009 *Comput. Phys. Commun.* **180** 2622

IOP Conf. Series: Journal of Physics: Conf. Series **950** (2017) 042019 doi:10.1088/1742-6596/950/4/042019

Three-Phase Melting Curves in the Binary System of Carbon Dioxide and Water

E H Abramson

Department of Earth and Space Sciences, University of Washington, Seattle, WA 98195, USA

E-mail: evan@ess.washington.edu

Abstract. Invariant, three-phase melting curves, of ice VI in equilibrium with solid CO_2, of ice VII in equilibrium with solid CO_2, and of solid CO_2 in simultaneous equilibrium with a majority aqueous and a majority CO_2 fluid, were explored in the binary system of carbon dioxide and water. Diamond-anvil cells were used to develop pressures of 5 GPa. Water exhibits a large melting temperature depression (73°C less than its pure melting temperature of 253°C at 5 GPa) indicative of large concentrations of CO_2 in the aqueous solution. The melting point of water-saturated CO_2 does not show a measureable departure from that of the pure system at temperatures lower than ~200°C and only 10°C at 5 GPa (from 327°C).

The binary system of carbon dioxide and water, despite its importance to planetary sciences, has received little attention at high pressures and temperatures. A recent paper [1] follows the melting curve of ice VI under conditions of CO_2 saturation up to 2.3 GPa, while the extent of fluid/fluid miscibility has been reported to a maximum pressure of only 0.35 GPa [2]. Preliminary to studying the fluid/fluid solubility surface we explored its boundaries given by the invariant three-phase melting curves of ice VI in equilibrium with solid CO_2, ice VII in equilibrium with solid CO_2, and solid CO_2 in simultaneous equilibrium with a majority aqueous and a majority CO_2 fluid.

Diamond-anvil cells of a modified Merrill-Bassett [3] design were first loaded with water and a bubble of air. They were subsequently immersed in liquid CO_2 at ~10°C and 58 bar and re-opened, allowing liquid CO_2 to displace the air, then re-sealed. Gaskets of Inconel 718 were found to corrode above ~300°C in the mixed fluid and were used only for temperatures less than 200°C. For most of the loads, and all brought above 200°C, rhenium was used. The two gasket materials yielded comparable data at the lower temperatures.

The cells were placed in an oven with optical access, heated, and the contents observed in transmitted light. Temperature was determined by way of thermocouples situated adjacent to the diamonds. The system was calibrated at the melting point of Pb and is believed to be accurate to 0.1°C at room temperature and better than 1°C at the highest temperatures. Pressures were determined with either ruby fluorescence [4] or Raman scattering from cubic boron nitride (cBN), each load containing one or the other of the markers. Ruby fluorescence was used only below 200°C and pressure was measured by the difference in the R1 line positions of a marker contained within the cell and another in contact with the outer surface of one of the diamonds and thus at a similar temperature; any temperature drift occurring between acquisitions of the two spectra was compensated by an added shift [5] of 0.02 GPa/°C.

cBN was purchased from Diamond Innovations (Ohio, U.S.A) as commercial grinding grit (BMP-400 30-40). The crystals produced a tolerable amount of fluorescence, the power per wavenumber being roughly comparable to that of the Raman signal. The crystals were used without annealing and, among those used, the center frequencies of their signals were offset by as much as 1 cm^{-1} (equivalent to 0.3 GPa); the offsets were found to be independent of pressure and temperature, and once measured could be subtracted out. Raman shifts were measured for both the marker in the cell and a reference crystal held at room temperature and 1 bar.

Variations of the Raman frequencies, ν, of cBN with pressure, *P*, and absolute temperature, *T*, have been reported [6]. In order to use cBN as a pressure marker the data of reference 6 were fit to a polynomial:

$$\nu = c_1P + c_2PT + c_3P^2 + c_4T + c_5T^2 + c_6 \quad (1)$$

with c_1= 3.303 cm^{-1}/GPa, c_2= 1.85x 10^{-4} cm^{-1}/GPa/K, c_3= -9.72 x 10^{-3} cm^{-1}/GPa^2, c_4=5.28 x 10^{-3} cm^{-1}/K, c_5= -2.94 x 10^{-5} cm^{-1}/K^2, and c_6=1054.96 cm^{-1}. The resulting fit gives a root-mean-square deviation from the data of 0.06 GPa and shows no systematic deviations, either in pressure or temperature. In use, the nominal Raman frequency at the pressure and temperature of the reference crystal (0 GPa, ~294 K) is calculated from equation (1) and then added to the measured difference between the high pressure and reference crystals; this corrected frequency is used in equation (1) to solve for pressure.

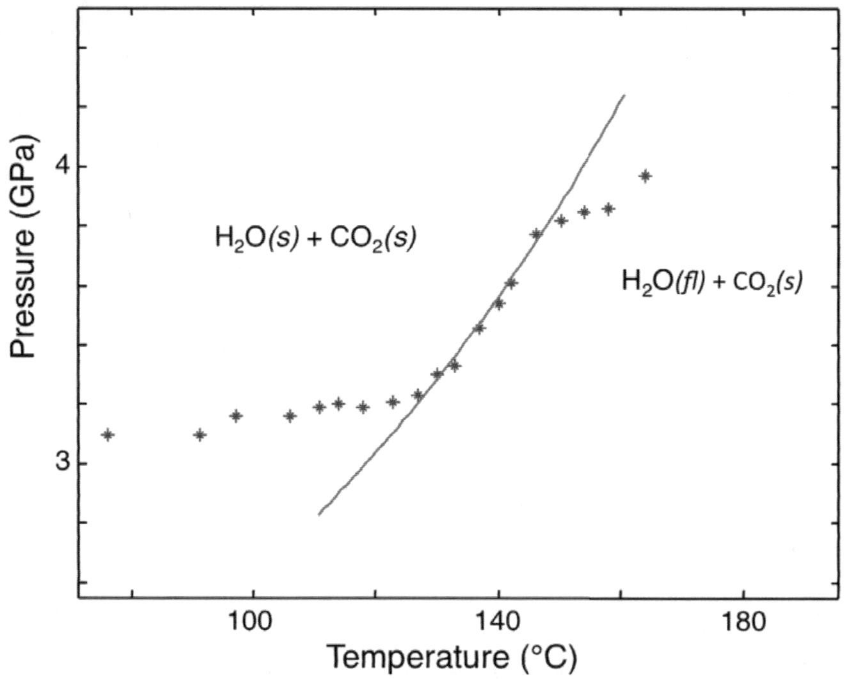

Figure 1. Pressure rises as ice VII melts isochorically, in equilibrium with solid CO_2 and an aqueous solution of the same. A section of the final, fitted curve is shown.

During the melting of water it is difficult to determine visually whether both water and CO_2 solids are present. We therefor monitor the (approximately) isochoric pressure changes as temperature is increased; melting on the three-phase line is signaled by a rapidly rising pressure, along a path which is independent of starting pressure (figure 1). The melting of CO_2 can be observed by the same method but also, often, visually.

Data are shown in figure 2. Up to pressures of ~5 GPa the water ices exhibit significant and increasing melting point depression indicative of increasing concentrations of CO_2 in the aqueous phase. In contrast, the melting point of water-saturated CO_2 does not show a measureable departure

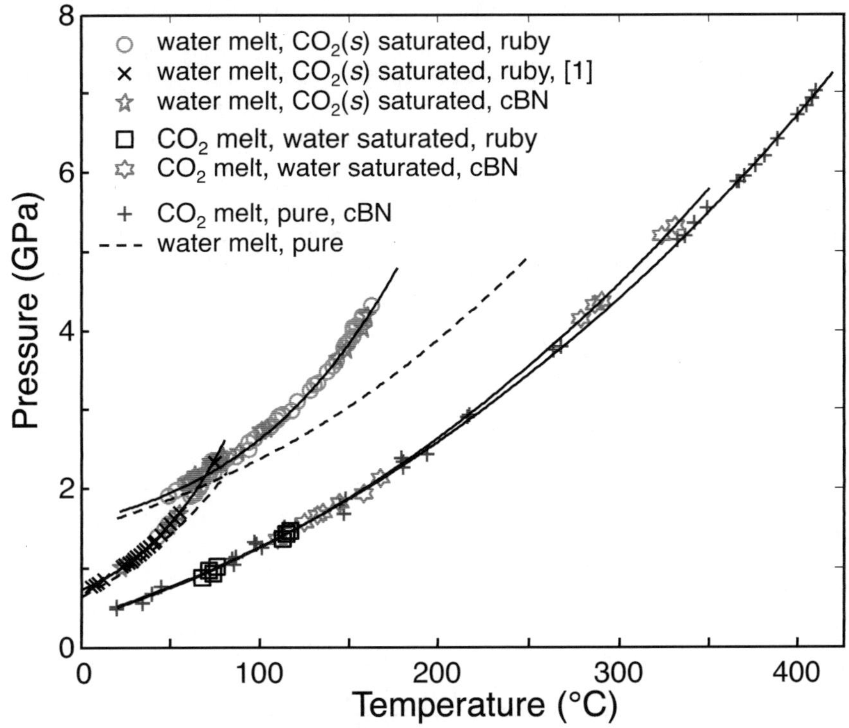

Figure 2. Melting pressures of water (ices VI and VII) in equilibrium with solid carbon dioxide, carbon dioxide in equilibrium with saturated water, and pure carbon dioxide, are plotted against temperature. Points are designated as taken with either ruby or cBN as a pressure marker. The dashed curves denote pure water melting lines of ice VI [7] and ice VII [8]. Solid curves are fits to the data.

from that of the pure system at temperatures lower than ~200°C, indicating that up to these temperatures the solubility of water in fluid CO_2 remains small.

Figure 3 shows an enlarged section of figure 2. Both water ices can be seen to melt from the field of stability of the other. The current data for ice VI agree nicely with those of reference 1. The

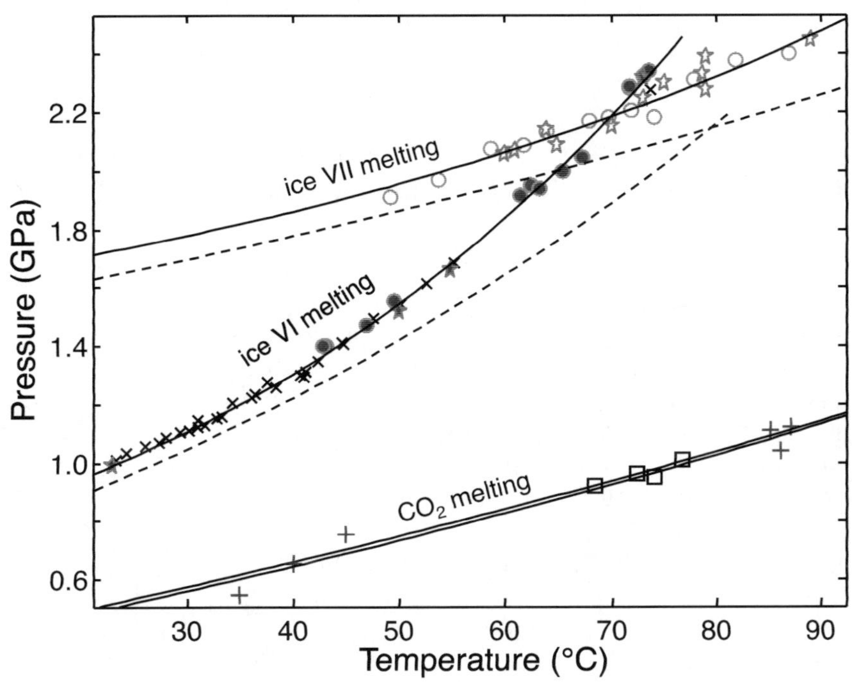

Figure 3. An enlarged section of figure 2. Symbols have the same meanings, however those representing data taken along the ice VI melt are now distinguished by being filled. The current data agree nicely with those of Bollengier *et al.* [1] (black "x").

invariant, four-phase point at which ice VI, ice VII, CO_2(s), and aqueous solution co-exist at equilibrium is found to be at 69.9°C and 2.19 GPa.

Curves in the figures are fits of the data to the Simon-Glatzel relation, equation 2, with parameters given in table 1. The melting curves of ices VI and VII were fit only to points for which (the more precise) ruby served as the pressure marker. A combination of the current data and those of reference 1 were used to define the ice VI curve. Rms misfits of the current (ruby-derived) pressures were 0.02 and 0.03 GPa, for ices VI and VII, respectively. In order to better determine the small departure of the melting curve of water-saturated CO_2 from that of the pure compound, the latter was also measured; uncertainties in the calibration of the cBN standard should thus be negligible for this purpose. Pressures determined with cBN had a rms misfit of 0.05 GPa.

$$P = P_0 + a((T/T_0)^b - 1) \tag{2}$$

Table 1. Parameters for fits to equation (2).

	P_0 (GPa)	T_0 (K)	*a* (GPa)	*b*
ice VI, CO_2(s) saturated	0.77	279	0.412	7.201
ice VII, CO_2(s) saturated	2.21	345	0.875	5.277
pure CO_2	0.00	216.59	0.451	2.444
CO_2, water saturated	0.00	216.59	0.404	2.588

In conclusion, the melting curves of both water (saturated in carbon dioxide) and of carbon dioxide (saturated in water) have been measured to a pressure of 5 GPa. Simon-Glatzel expressions have been fit to the data and the parameters given. The four-phase point at which ice VI, ice VII, solid carbon dioxide and fluid are in joint equilibrium has been determined. A commercial grinding grit of cubic boron nitride has proven to provide an adequate, and easily obtainable, pressure standard.

Acknowledgments
I would like thank Scott Temple for his work in the laboratory, including a substantial contribution to the collected data. This work was supported by the Department of Energy, Contract No. DE-NA0001843.

References

[1] Bollengier O, Choukroun M, Grasset O, Le Menn E, Bellino G, Morizet Y, Bezacier L, Oancea A, Taffin C and Tobie G 2013 *Geochim. Cosmochim. Acta* **119** 322-39
[2] Todheide K and Franck E U 1963 *Z. Phys. Chem.* (Neue Folge) **37** 387-401
Takenouchi S and Kennedy G C 1964 *Am. J. Sci.* **262** 1055-74
[3] Merrill L and Bassett W A 1974 *Rev. Sci. Instrum.* **45** 290-4
[4] Mao H-K and Bell P M 1986 *J. Geophys. Res.* **91** 4673-6
[5] Ragan D D, Gustavsen R and Schiferl D 1992 *J. Appl. Phys.* **72** 5539-44
[6] Datchi F and Canny B 2004 *Phys. Rev.* B **69** 144106
[7] Wagner W, Saul A and Pruss A 1994 *J. Phys. Chem. Ref. Data* **23** 515-27
[8] Datchi F, Loubeyre P and LeToullec R 2000 *Phys. Rev.* B **61** 6535-46

IOP Conf. Series: Journal of Physics: Conf. Series **950** (2017) 042020 doi:10.1088/1742-6596/950/4/042020

Structural phase transition of potassium under high-pressure and low-temperature condition

Masafumi Sakata[1], Yuki Nakamoto[1], Takahiro Matsuoka[1], Yasuo Ohishi[2], Katsuya Shimizu[1]

[1] KYOKUGEN, Graduate School of Engineering Science, Osaka University, 1-3 Machikaneyama-cho, Toyonaka, Osaka 560-8531, Japan

[2] Japan Synchrotron Radiation Research Institute / SPring-8, 1-1-1 Kouto, Sayo-cho Sayo-gun, Hyogo 679-5148, Japan

sakata@stec.es.osaka-u.ac.jp

Abstract. We performed electrical resistance and powder X-ray diffraction measurements of potassium under high-pressure and low-temperature condition. In the temperature dependence of resistance under several pressure, no superconducting transition was observed below 1 K. From the powder X-ray diffraction measurement, the K-III phase, it has a host-guest structure, is stable in wide pressure range. Furthermore, the K-IV phase disappeared and the amorphous-like region with no X-ray diffraction peaks appeared at low temperature below 150 K.

1. Introduction

Superconductivity in elements under high pressure has been investigated intensively. At present, 53 elements have been reported as the superconductor at low temperature. Especially, 23 of superconducting elements shows the superconductivity only under high pressure. Among superconductive elements, relatively high superconducting transition temperature (T_c) of around 20 K has been observed in light alkali and alkali earth metals, lithium and calcium [1,2]. The alkali and alkali earth metals show the successive pressure-induced structural phase transitions. The phase transition causes to the drastic change of electronic properties of elements under high pressure. For example, the lithium indicates the re-entrant metal-insulator-metal transition related to the borders between each structural phases. In the case of the calcium, superconductivity emerges in the Ca-III phase and the T_c rising tendency is observed below Ca-VII phase. It is noted that the T_c shows the maximum in the Ca-VII which has a host-guest structure [3]. The host-guest structure is commonly appeared under high pressure among alkali earth metal elements (calcium, strontium, and barium) and the maximum of T_c is in the high pressure phase with this structure.

Among alkaline metal elements, lithium and cesium show superconductivity under high pressure. On the other hand, other alkaline metals reported no superconductivity yet. At room temperature, potassium shows successive pressure-induced structural phase transition as body-centered cubic (bcc) → face-centered cubic (fcc, 12 GPa) → K-III (22 GPa) → K-VI (*oP*8, 54 GPa) → K-V (90 GPa) [4]. K-III phase is a host-guest structural phase similar to Ca-VII phase showing relatively high-T_c superconductivity under high pressure. Therefore, K-III phase has a possibility of superconductivity under high pressure. However, the no superconducting transition is reported in the *T-P* region below 1.5 K up to 17 GPa and below 4 K up to 44 GPa [5]. In contrast to experimental absence of

IOP Conf. Series: Journal of Physics: Conf. Series **950** (2017) 042020 doi:10.1088/1742-6596/950/4/042020

superconductivity on potassium even if under high pressure, several theoretical calculation suggests that the possibility of superconductivity on potassium. For example, Profeta *et al.* calculated that the T_c in fcc phase is estimated about 3 K and with increasing pressure, T_c rise to about 10 K in K-III phase from a first principle investigation [6].

The investigation of phase transition at low temperature is important to understand the ground state such as superconducting state because of difference structure between room and low temperatures. For example, it is reported that the simple cubic structure of Ca-III phase is distorted at low temperature by several research groups including us [2, 7]. Additionally, tetragonal Ca phase with $I4_1/amd$ structure is observed only at low temperature below 10 K under the pressure at 30~40 GPa, which is close to the border between Ca-II (fcc) and Ca-III (sc) phases [8]. The superconductivity of calcium is observed as a small resistance drop at 2 K and 44 GPa [9], this pressure is just above the $I4_1/amd$ phase of calcium.

In this study, we investigate about pressure-induced structural phase transition at low temperature and superconductivity of potassium from electrical resistance and powder X-ray diffraction measurements using diamond anvil cell (DAC).

2. Experimental

A potassium block with a stated purity of 99.95 % was purchased from Aldrich Chemical Company. The sample settings to a DAC for electrical resistance and powder X-ray diffraction measurements were performed in an argon atmosphere to prevent sample oxidization or any other chemical reaction. No pressure medium was used. For the electrical resistance measurement, the copper film deposited on the diamond culet was used as an electrode with 4-probe arrangement. Aluminum oxide was used as an insulating layer on a rhenium metal gasket. The ac-resistance of potassium sample in the DAC was measured by using the ac resistance bridge LR-700 (Linear Research Inc.). The pressure was determined by using ruby fluorescence method. The X-ray diffraction experiments were carried out by an angle-dispersive method using an image-plate detector. High-quality X-ray diffraction patterns of potassium under high-pressure and low-temperature condition were obtained by the synchrotron radiation BL10XU in SPring-8. In the case of the X-ray diffraction measurement, the pressure determined by using the frequency shift of diamond Raman peak by pressurization.

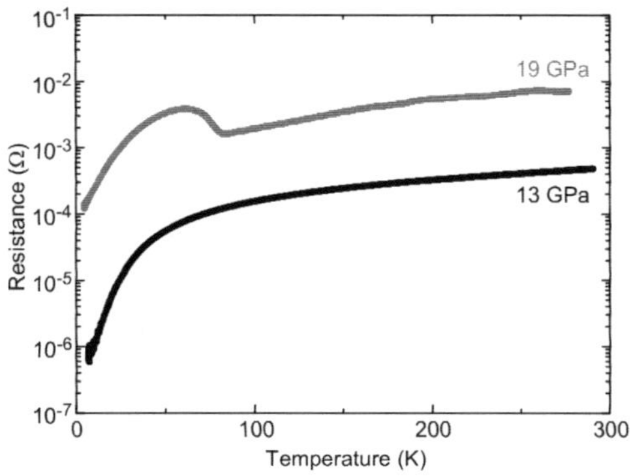

Figure 1. Temperature dependence of electrical resistance under pressure at 13 and 19 GPa.

3. Results and discussions

3.1. Temperature dependence of resistance under high pressure.

Figure 1 shows the temperature dependence of resistance of potassium under several pressures as 13 and 19 GPa in fcc phase. In this phase, metallic temperature dependence of resistance and no superconducting transition was observed below 1 K. It coincides with previous report about no superconductivity in fcc phase. Additionally, the partially insulating behaviour appeared from 80 to 70 K. The pressure was determined only at room temperature and the DAC has a possibility of pressure changing during a cooling-heating cycle. One of possible reason of the anomaly in resistance around

70 K is a crossing the boundary between fcc and K-III phase at low temperature because the pressure at 19 GPa is close to the boundary (22 GPa). Other is any electronic transition such as a charge density wave (CDW). In order to reveal the reason of this anomaly, the pressure dependence of anomaly such as a shift of characteristic temperature is important. However, the diamond anvils were broken and there is no result about pressure dependence of resistance just above 19 GPa. At present, there are no more information about this anomaly.

3.2. Phase diagram from the powder X-ray diffraction measurement under high pressure.

Figure 2(a) shows several powder X-ray diffraction patterns under some high-pressure and low-temperature conditions. At around 30 K, the potassium sample shows pressure-induced structural phase transitions to fcc and then K-III phases under similar pressure to that at room temperature. Therefore, phase boundaries between bcc and fcc, fcc and K-III phases are almost temperature-independent. On the other hand, the transition from K-III phase was observed above 80 GPa at low temperature. Furthermore, the K-IV phase disappeared at low temperature and no X-ray diffraction peaks observed above about 100 GPa, it is amorphous-like behavior. With increasing pressure, the X-ray diffraction peaks re-appeared above about 130 GPa by phase transition to higher-pressure phase.

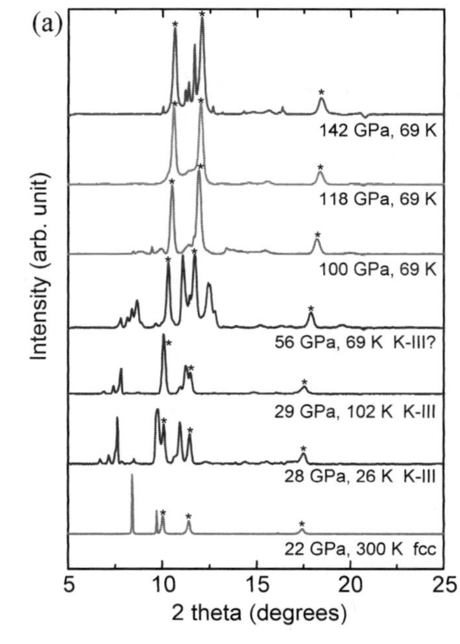

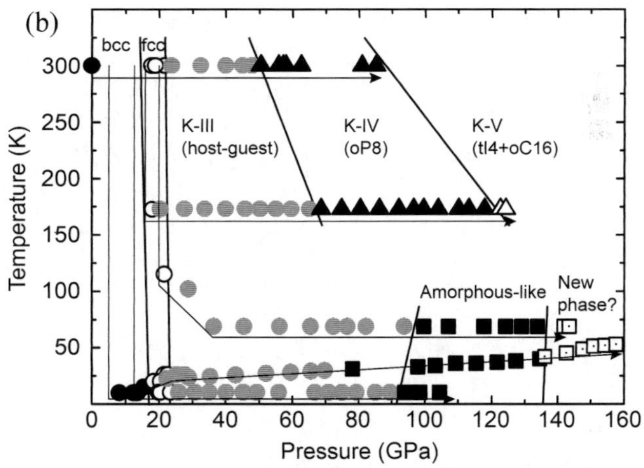

Figure 2. (a) powder X-ray diffraction patterns under several *P-T* conditions using the X-ray with λ = 0.41268 Å. Asterisks indicate diffraction peaks of Re gasket. (b) *P-T* phase diagram determined from powder X-ray diffraction measurements.

Figure 2(b) shows the *P-T* phase diagram determined from powder X-ray diffraction. The arrows indicate the pressure-temperature pathway during measurements. The pressure region of K-III phase with host-guest structure at low temperature is extended to higher pressure than at room temperature. At present, the superconductivity has been researched only up to 44 GPa. The extension of pressure region of K-III phase suggests the possibility of superconducting phase above 44 GPa. The K-IV phase is observed only above 150 K and amorphous-like region without long-range structural ordering appears at low temperature. It possibly indicates that several crystal structures have similar stability and thermal energy is needed to the structural transition from K-III to higher-pressure structural phase. Under higher pressure than amorphous-like region, structural high-pressure phase appeared. In this study, we could not determine the crystal structure of this high pressure phase at low temperature because of hinderance by large peaks from Re gasket. It seems that the diffraction pattern of this high pressure phase is inconsistent with reported pattern of K-V phase at room temperature [4]. However, reported structural analysis determines the crystal structure of K-V phase as a mixture of two difference structures (*tI*4 and *oC*16) [4]. Therefore, it is difficult to compare with our measured pattern.

IOP Conf. Series: Journal of Physics: Conf. Series **950** (2017) 042020 doi:10.1088/1742-6596/950/4/042020

Acknowledgments
This research is financially supported by the Japan Society for the Promotion of Science (JSPS) through its "Funding Program for Next Generation World-Leading Researchers (GR068)", and the Scientific Research (S) 19104009 and Global COE Program (Core Research and Engineering of Advanced Materials-Interdisciplinary Education Center for Materials Science), MEXT, Japan. This work was concluded under proposal No. 2011B1357 and 2012A1112 of the SPring-8.

References

[1] Shimizu K, Ishikawa H, Takao D, Yagi T and Amaya K 2002 *Nature* **419** 597
[2] Sakata M, Nakamoto Y, Shimizu K, Matsuoka T and Ohishi Y 2011 *Phys. Rev. B* **83** 220512(R)
[3] Fujihisa H, Nakamoto Y, Sakata M, Shimizu K, Matsuoka T, Ohishi Y, Yamawaki H, Takeya S and Gotoh Y 2013 *Phys. Rev. Lett.* **110** 235501
[4] Lundegaard L F, Marqués M, Stinton G, Ackland G J, Nelmes R J and McMahon M I 2009 *Phys. Rev. B* **80** 020101(R)
[5] Schilling J S 2006 *High Pressure Research* **26** 145
[6] Profeta G, Franchini C, Lathiotakis N N, Floris A, Sanna A, Marques M A L, Lüders M, Massidda S, Gross E K U and Continenza A 2006 *Phys. Rev. Lett.* **96** 047003
[7] Mao W L, Wangd L, Ding Y, Yang W, Liu W, Kim D Y, Luo W, Ahuja R, Meng Y, Sinogeikin S, Shu J and Mao H-k 2010 *Proc. Natl. Acad. Sci. USA* **107** 9965
[8] Li B, Ding Y, Yang W, Wang L, Zou B, Shu J, Sinogeikin S, Park C, Zou G, and Mao H-k 2012 *Proc. Natl. Acad. Sci. USA* **109** 16459
[9] Dunn K J and Bundy F P 1981 *Phys. Rev. B* **24** 1643

IOP Conf. Series: Journal of Physics: Conf. Series **950** (2017) 042021 doi:10.1088/1742-6596/950/4/042021

Recent ab initio phase diagram studies: Iridium

L Burakovsky[1], M J Cawkwell[1], D L Preston[2], D Errandonea[3] and S I Simak[4]

[1] Theoretical and [2] Computational Physics Divisions, Los Alamos National Laboratory, Los Alamos, New Mexico 87545. USA
[3] Departamento de Física Aplicada-ICMUV, MALTA Consolider Team, Universidad de Valencia, Edificio de Investigación, C/Dr. Moliner 50, Burjassot, 46100 Valencia, Spain
[4] Department of Physics, Chemistry and Biology, Linköping University, 58183 Linköping, Sweden

E-mail: burakov@lanl.gov

Abstract. The phase diagram of iridium is investigated using the Z methodology in conjunction with the VASP *ab initio* molecular dynamics package. The Z methodology is a novel technique for phase diagram studies which combines the direct Z method for the computation of melting curves and the inverse Z method for the calculation of solid-solid phase boundaries. We compare our results to the available experimental data on iridium. We offer explanation for the 14-layer hexagonal structure observed in experiments by Cerenius and Dubrovinsky.

1. Introduction

Iridium (Ir) is one of the very special materials in nature. It is the only refractory face-centered cubic (fcc) metal with a melting point of 2719 K [1]. Its shear modulus, $G = 210$ GPa [2], is the largest among such fcc metals, and the second largest among all the elemental metals behind hexagonal close-packed (hcp) metal osmium (Os) with $G = 222$ GPa [2]. Ir is also one of the two most dense elemental metals having a density of 22.65 g/cc at ambient pressure (P) and T=0 or 22.56 g/cc at T=293.15 K [1] along with Os for which the corresponding values are essentially the same: 22.66 and 22.59 [3]. Because it exhibits excellent mechanical properties and high resistance to oxidation and corrosion at elevated T, Ir is used in numerous applications as a static component at high T and/or in aggressive environments. Despite the technological advances, the current understanding of the mechanical properties of Ir is rather poor, and the knowledge of its phase diagram is virtually nonexistent.

The similarity of the mechanical properties of Ir to those of the refractory hcp metals, specifically Os, and the fact that in a number of binary systems Ir forms solid solutions with a hcp structure which are stable in a wide range of concentrations, raised the issue of a potential phase transformation of Ir into the hcp structure. Such a fcc-hcp phase transformation in Ir was found in [4], and the transition P at room T was estimated to be 9 GPa and the fcc-hcp phase boundary was predictied to be a straight line with a positive slope of 150 K/GPa.

The subsequent experimental studies could not confirm a fcc-hcp transition in Ir at room T (except perhaps for a study by Cerenius and Dubrovinsky [5] discussed in what follows). It was however noted in [6] that the earlier data on the shock Hugoniot of Ir suggest the existence of a structural transformation above 1.14 Mbar. With the more recent shock Hugoniot data

IOP Conf. Series: Journal of Physics: Conf. Series **950** (2017) 042021 doi:10.1088/1742-6596/950/4/042021

available in the online Shock Wave Database (SWD) [7], it is now possible to firmly establish the existence of a solid-solid transformation on the Ir Hugoniot. Figure 1 shows two parallel segments shifted from each other by ~ 0.25 km/s that describe U_s-U_p shock Hugoniot data on Ir. This situation is typical of a solid-solid phase transition on the Hugoniot; for instance, figure 3 of ref. [8] demonstrates α-ω-β transition sequence in zirconium described by 3 parallel segments shifted from each other by ~ 0.2 and ~ 0.3 km/s, respectively [8]. The transition in Ir occurs at $1.10 < U_p < 1.36$ km/s or, if converted from U_s-U_p to P-U_p, at $140 < P < 180$ GPa [7], so we take the transition P to be 160 ± 20 GPa.

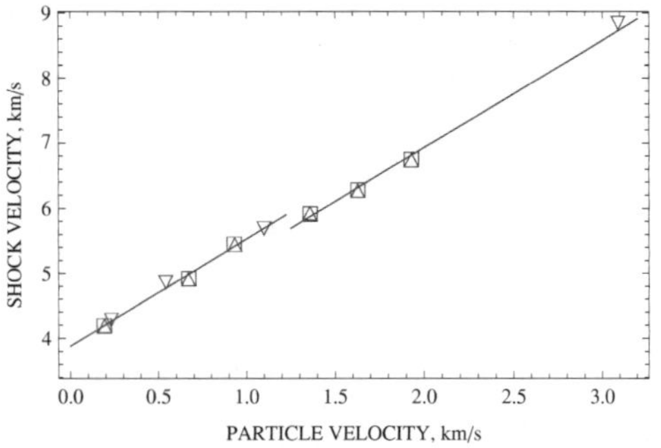

Figure 1. Shock velocity, U_s, vs. particle velovity, U_p, along the principal Hugoniot in Ir; data from [7]. A solid-solid phase transition occurs at $1.10 < U_p < 1.36$ km/s.

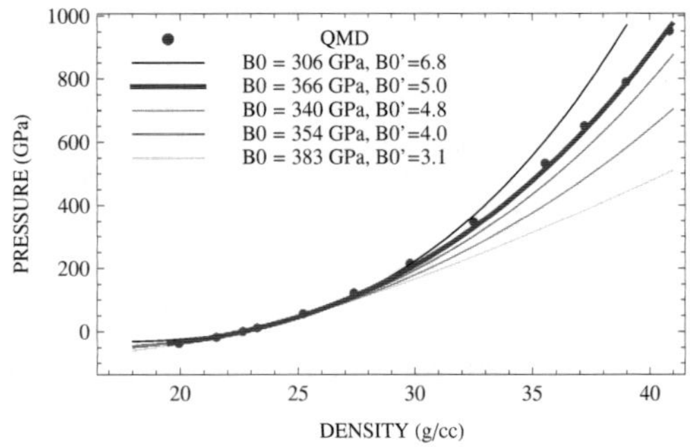

Figure 2. The T=0 Ir EOS from VASP compared to the experimental and theoretical 300 K Birch-Murnaghan EOSs.

In relatively recent experimental studies of Ir by Cerenius and Dubrovinsky [5], an unusual transition from fcc to a 14-layer hexagonal (hex) supercell was detected, at ~ 60 GPa. X-ray diffraction spectra showed a substantial increase in the intensity of the (111) peak of Ir accompanied by the appearance of a distinct saw-tooth pattern. Despite their low intensity, sufficient signal-to-noise ratio made it possible to resolve the peaks. Cerenius and Dubrovinsky noted that these peaks could not be explained by the formation of stacking faults or other types of defects (e.g., twinning) but could rather correspond to a distortion of the fcc lattice of Ir and the formation of a superlattice structure. These peaks were properly indexed and associated with a 14-layer hex supercell with cell parameters $a = 2.60$ Å and $c = 29.7$ Å at 65 GPa. Hence, the supercell is ideal to within 0.1%: $c/a = 29.7/2.60 = 11.423$ vs. $14\sqrt{2/3} = 11.431$. We are not aware of any additional experiments on Ir to pressures comparable to those of [5]. However, subsequent theoretical studies based on *ab initio* approaches failed to confirm the existence of a transformation of fcc-Ir into another solid structure [9, 10]. Specifically, in [9] all possible atomic-layer arrangements for a 14-layer hex cell were considered and their T=0 enthalpies were compared to that of fcc-Ir as functions of atomic volume. The enthalpies of these different 14-layer structures form a family of quasi-parallel curves (figure 3 of ref. [9]; only sixty such distinct curves exist, out of a total of $\sim 2^{14}$, because of a high level of degeneracy [9]) such that the enthalpy differences are virtually independent of volume, yet all of these curves remain higher and never cross the fcc one. Hence, no transition from fcc-Ir can be inferred based on thermodynamic considerations. However, according to [9], the enthalpy difference between fcc and the most stable of the sixty 14-layer structures is only 12 meV/atom, or ~ 140 K; such a small difference can be easily overcome by contribution from the corresponding entropy term. Thus, at least some of the hex structures considered in [9] should be expected to become thermodynamically more stable than fcc with increasing T.

In the present work we determine the melting curve of Ir to ~ 600 GPa using the Z method, which we briefly describe in the following section. Our Z method calculations are carried out using the quantum molecular dynamics (QMD) code VASP (Vienna Ab initio Simulation Package), which is based on density functional theory (DFT). We use the generalized gradient approximation (GGA) with the Perdew-Burke-Ernzerhof (PBE) exchange-correlation functional. We model Ir using the electron core-valence representation $[^{54}\mathrm{Xe}\ 4f^{14}]\,5d^7\,6s^2$, i.e. we assign the 9 outermost electrons of Ir to the valence. The valence electrons are represented with a plane-wave basis set with a cutoff energy of 370 eV, while the core electrons are represented by projector augmented-wave (PAW) pseudopotentials.

We first calculated the T=0 isotherm of fcc-Ir. This was done by performing a short QMD run and extracting the corresponding value of P. We used a $5 \times 5 \times 5$ (500-atom) supercell with a single Γ-point. With such a large supercell, full energy convergence (to $\lesssim 1$ meV/atom) is already achieved, which was verified by performing short runs with $2 \times 2 \times 2$ and $3 \times 3 \times 3$ k-point meshes and comparing their output with that of the 500-atom run with a single Γ-point.

Our results on the T=0 isotherm are shown in figure 2. We note that each of the papers [5, 11, 12] that discuss Ir EOS data uses the 3rd-order Birch-Murnaghan (BM) EOS $P(\rho) = \frac{3}{2} B_0 \left((\rho/\rho_0)^{7/3} - (\rho/\rho_0)^{5/3} \right) \left[1 + \frac{3}{4}(B_0' - 4) \left((\rho/\rho_0)^{2/3} - 1 \right) \right]$, where B_0 and B_0' are the values of the bulk modulus and its pressure derivative at the reference point $\rho = \rho_0$. Since the P=0 values of the density of Ir at T=0 and 300 K differ by $\sim 0.4\%$ (22.65 vs. 22.56 g/cc [1]), and T=300 K introduces a negligibly small thermal pressure correction, the T=0 and T=300 K isotherms can be described by the same values of B_0 and B_0'. Consequently, we can compare room-T isotherm data to our T=0 isotherm as determined from QMD. This comparison is shown in figure 2. The experimental results from two groups, each of which accessed $P \sim 65$ GPa, were presented as BM fits with different choices for B_0 and B_0': $B_0 = 306$ GPa and $B_0' = 6.8$ and $B_0 = 354$ GPa and $B_0' = 4.0$ from [5] (the former set results from fitting with both B_0 and B_0' as free parameters, while the latter from fitting with $B_0' = 4.0$ kept fixed), and $B_0 = 383$ GPa and $B_0' = 3.1$ from [11]. Another set of parameters, $B_0 = 340$ and $B_0' = 4.8$ comes from theoretical modeling [12]. It is seen that although excellent agreement of the QMD points with the available experimental data (up to $\rho \sim 26$ g/cc) is found for each of the (B_0, B_0') sets, it is the 3rd-order BM isotherm with $B_0' = 5$ that best fits the QMD data at densities above the experimental range.

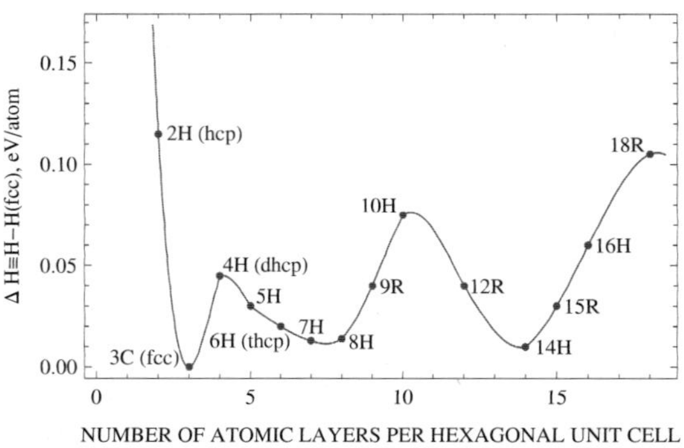

Figure 3. Differences between the T=0 enthalpies of hex structures of Ir and that of fcc-Ir as a function of the number of atomic layers per unit hex cell, at a fixed unit cell volume.

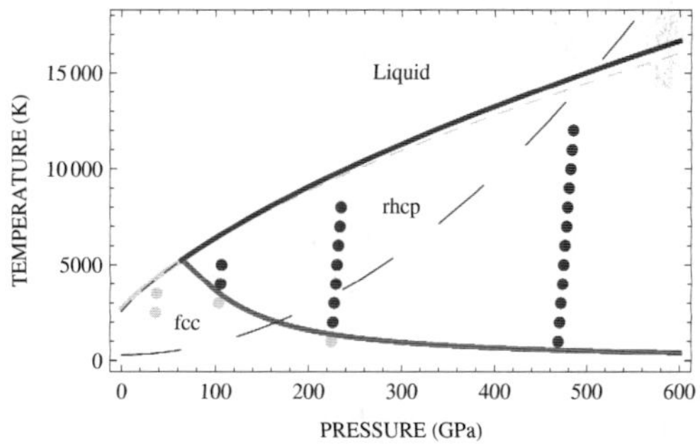

Figure 4. Phase diagram of Ir obtained from the Z methodology: fcc-Ir melting curve (green line), liquid Ir solidified into solid fcc (green bullets), 9R-Ir melting curve (blue line), liquid Ir solidified into solid rhcp (blue bullets), and the (tentative) fcc-rhcp solid-solid phase boundary (violet).

We then turned to hex structures of Ir. We calculated the T=0 enthalpies of different hex structures of Ir having the number of atoms per unit cell from 2 (hcp) to 18 (18R). This was done by optimizing the value of c/a, i.e. determining the c/a that minimizes the energy, at a fixed volume of a hex supercell, and then performing a short QMD run and extracting the corresponding value of P. We used supercells with the number of atoms from 540 (for $6 \times 6 \times 1$ 15R) to 686 (for both $7 \times 7 \times 2$ 7H and $7 \times 7 \times 1$ 14H) with a single Γ-point. With such large supercells, just like for fcc-Ir, full energy convergence (to $\lesssim 1$ meV/atom) is already achieved, which allowed us to resolve the enthalpy differences to high accuracy. We found out that all the hex structures of Ir have the values of c/a equal to the ideal ones to within 1%, so that basically $c/a = N\sqrt{2/3}$, where N is the number of atoms per unit cell, for all the unit cell volumes considered (P from 0 to $\sim$ 900 GPa). The hex enthalpies form a system of quasi-parallel curves similar to that shown in figure 3 of ref. [9]. Figure 3 shows different hex enthalpies (including that of fcc-Ir as 3C hex structure) as a function of the number of atoms or, equivalently, the number of atomic layers, per hex unit cell at some fixed unit cell volume (since the enthalpy curves are quasi-parallel, their differences do not depend on volume, thus for any other fixed unit cell volume the plot will look similar to that in figure 3). In figure 3, only the structures having the lowest enthalpy among those with the same number of layers per unit cell are shown, in each case that there are more than one such structure (for $N \geq 6$). Specifically, for the hexagonal structures listed below, points in figure 3 indicate the following (periodic) stacking sequences, 2H (hcp): AB, 3C (fcc): ABC, 4H (double-hcp, or dhcp): ABAC, 5H: ABCAC, 6H (triple-hcp, or thcp): ABCACB, 7H: ABCABAC, 8H: ABACBABC, 9R (α-Sm): ACABABCBC, 10H: ABACBACABC, 12R: ABABCACABCBC, 14H: ABACBCBACABCBC, 15R: ABABACACACBCBCB, 18R: ABACACBCBACABABCBC.

The plot clearly exhibits two distinct minima, for 7H and 8H, and for 14H the enthalpy of which is the closest one to that of fcc-Ir: the 14H-fcc enthalpy difference is 11 meV/atom, in agreement with [9]. Since none of the hex structures becomes more stable than fcc at T=0 (or T=300 K, according to the Ir phase diagram from Section 3), we offer the following explanation for the appearance of the 14-layer superstructure in experiments by Cerenius and Dubrovinsky. It is not clear from [5] how the hydrostaticity of the pressurized Ir was controlled (whether or not a pressure medium was used, etc.) It is quite possible that in their experiments nonhydrostatic stresses were present under the action of which fcc-Ir loses its thermodynamic stability and the hex structure that is energetically the closest one to fcc-Ir gets thermodynamically stabilized. Such a structure is 14H, with 14 atomic layers per unit cell, just like the one found in ref. [5].

2. Z method applied to the melting curve of Ir

Now we switch to the calculation of the melting curves of different solid structures of Ir using the Z method. The Z method was developed to calculate melting curves using first-principles based software, specifically VASP. The Z method was introduced for the first time in our paper on the *ab initio* melting curve of Mo [13]. The method has since been applied to the study of a large number of melting curves of different materials, and comparisons with experimental data on Pb, Ta, Fe and Pt at ESRF show good agreement [14]. In contrast to previous melting curve calculations based on the Z method, in our recent work on Os [15], the method was utilized as closely as possible to the original concept, but at the expense of an extensive suite of QMD simulations. Specifically, we directly observe melting in the computational cell, and then determine the corresponding melting temperature (T_m) and pressure (P_m) (at a given density); see [15] for more detail. Earlier on, T_m and P_m would be determined from bracketing the true melting point by the highest solid (P_s, T_s) and the lowest liquid (P_l, T_l) states on the Z-shaped isochore, and using $(P_m, T_m) \approx ((P_s + P_l)/2, (T_s + T_l)/2)$. Here we use the same idea of directly observing melting and directly calculating T_m and P_m for the melting curve of Ir. In the case of Ir, it is important to note that, if a material has more than one thermodynamically stable crystal structure, the Z method yields the solid-liquid equilibrium boundaries of those structures.

The phase having the highest solid-liquid equilibrium T over some pressure range is the most stable, thus the physical melting curve, including triple points, is the envelope of the solid-liquid equilibrium boundaries.

Here we applied this strategy to the study of the melting curve of Ir. We calculated the melting curves of the following 8 solid structures: fcc, hcp, dhcp, 7H, 8H, 9R, 14H, and 15R. We used the same supercells as for the $T=0$ enthalpy calculation described above. We simulated 4 melting points for each of the 4 structures fcc, hcp, 9R and 14H, and then, after these 4 melting curves were obtained, we simulated 2 melting points for each of the remaining 4 structures, dhcp, 7H, 8H and 15R, to see if they lie close to any of those 4 curves. At a given density we performed a sequence of very long runs, each up to 25000 time steps or 25 ps, with initial Ts separated by relatively small increments of 375 K. We performed 13 such runs for each of the 24 melting points on the 8 melting curves; that is, our simulations covered a range of initial T of 4500 K in each case. We carried out a total of 312 runs which, with an average of $\sim$ 17500 time steps per run, amounted to a total of $\sim$ 5.5 million time steps of our melting simulations.

The melting curve of fcc-Ir, as the best fit to the corresponding four QMD melting points, is described by (T_m in K, P in GPa) $T_m^{\rm fcc}(P) = 2719\,(1 + P/31.2)^{0.59}$. The melting curves of hcp, 14H and 15R are close or slightly below the fcc one. However, the melting curves of dhcp, 7H, 8H and 9R are all close to each other and cross that of fcc; in other words, another solid phase becomes thermodynamically more stable than fcc-Ir at high PT conditions. We chose 9R to be the prototype solid structure at high PT; its melting curve, as the best fit to the corresponding four QMD melting points, is described by $T_m^{\rm 9R}(P) = 2553\,(1 + P/27.5)^{0.60}$. The fcc and 9R melting curves cross each other at (P, T) = (64, 5250), the fcc-9R-liquid triple point. Both melting curves are shown in figure 4.

Since for a number of solid structures of Ir the solid-liquid equilibrium boundaries (melting curves) are indistinguishable within mutual error bars, these structures have very close free energies. Consequently, the most stable structure cannot be determined using the Z method, and an alternative approach is needed.

3. Inverse Z method applied to the phase diagram of Ir

To cope with this difficulty, and to locate the solid-solid phase boundary in Ir, we use the inverse Z method introduced in our earlier work [14] where it was applied to the study of the phase diagrams of Pt and Ta; see [14] for more detail. The method consists in solidifying the initially liquid state into a final solid structure at a given fixed T. The subsequent identification of the crystal structure of the final state can be done by means of a number of techniques: (i) comparison of radial distribution functions (RDFs); (ii) comparison of X-ray diffraction patterns in reciprocal momentum space; (iii) geometric structure analysis (coordination number, angles between interatomic bonds, etc.). Two different final-state crystal structures at (P_1, T_1) and (P_2, T_2) then bracket the corresponding solid-solid phase boundary.

Here we applied the inverse Z method to locate the solid-solid boundary in Ir. We used a computational cell of 686 atoms prepared by melting a $7\times7\times7$ solid bcc supercell which should avoid bias towards solidifying either into fcc or any of the hex structures. We carried out NVT simulations using the Nosé-Hoover thermostat with a timestep of 1 fs. Complete solidification typically required from 15 to 25 ps, or $15000 - 25000$ timesteps, which is comparable with the duration of a typical direct-Z NVE run; the 25 inverse-Z runs (figure 4) contributed to a total of $\sim$ 0.5 million time steps, in addition to $\sim$ 5.5 million from direct-Z melting simulations.

We found that Ir solidifies into fcc below the violet line in figure 4, while above this line it solidifies into a hex structure. The line is described by the equation (T in K, P in GPa) $T(P) = 10500/(1+(P/64)^{1.5})$. The Ir principal Hugoniot described by [16] $T(P) = 293+0.087653\,P^{1.9332}$ crosses this line at (P, T) = (166.5, 2020), the Hugoniot transition point, such that the transition pressure of 166.5 GPa is consistent with 160 ± 20 GPa from SWD.

The RDFs did not allow us to discriminate between different hex structures: *hcp*, *dhcp*, *thcp*, $9R$, etc. Upon fast quenching of the hex structures to low T, where RDFs are more discriminating, we observed all of the above four hex structures, and perhaps some other hex structures as well. Hence, the inverse Z method indicates that there may be a number of energetically competitive hex structures of Ir at high PT. Our results indicate that structures with different stacking sequences are energetically very close. Hence, the energy cost of forming a stacking fault between two such structures is virtually zero. Consequently, the actual layer stacking could be non-periodic and, in principle, random. A random-stacking hex close-packed (*rhcp*) structure was first introduced for hard-sphere colloids [17], and has since been the subject of literature discussions [18, 19]. In general, when different stacking sequences become energetically degenerate, that is, the energy difference between any two such structures is $\sim 1-10$ meV per atom, or $\sim 10-100$ K, then in the resulting structure any two adjacent layers can occur with equal probability.

We are not yet aware of any direct reference to elements with a *rhcp* structure, except for our own suggestion that Pt may be such an element (having *rhcp* as its high PT phase [14]). In view of the fact that a *rhcp* phase may have been discovered experimentally in Au above 250 GPa by Dubrovinsky *et al.* [20] (see [14] for more detail), all the three third-row fcc noble metals Ir, Pt and Au become good candidates for being *rhcp* elements.

4. Conclusions

To conclude, we have applied the Z methodology, a novel technique for the computation of both melting curves and solid-solid phase boundaries, to the study of the phase diagram of Ir, and our results compare favorably with the existing experimental data. This study demonstrates that the Z methodology is a powerful utility for the calculation of phase diagrams. With the knowledge of the T=0 solid-solid transition points from the cold free-energy calculations, and of the solid-solid-liquid triple points from the direct Z method, complete solid-solid phase boundaries can be constructed, as both the previous examples of Pt and Ta [14] and the present example of Ir clearly demonstrate. We have offered the following explanation for the 14-layer hex structure observed by Cerenius and Dubrovinsky: because of possible nonhydrostaticity in their experiments, *fcc*-Ir loses its stability and converts into the energetically closest hex structure, 14H, with the periodic layer stacking ABACBCBACABCBC.

References

[1] Arblaster J W 2010 *Platinum Metals Rev.* **54** 93-102
[2] http://www.periodictable.com/Properties/A/ShearModulus.v.html
[3] Arblaster J W 2013 *Platinum Metals Rev.* **57** 177-85
[4] Zil'bershteyn V A and Estrin E I 1968 *Fiz. Metal. Metalloved.* **28** 369-70 and references therein
[5] Cerenius Y and Dubrovinsky L 2000 *J. Alloys Compounds* **306** 26-9
[6] Shock R N and Johnson K 1971 *Fiz. Metal. Metalloved.* **31** 1100-1
[7] http://www.ihed.ras.ru/rusbank/
[8] Greeff C W 2005 *Model. Simul. Mater. Sci. Eng.* **13** 1015-27
[9] Grussendorff S, Chetty N and Dreysse H 2003 *J. Phys. Cond. Mat.* **15** 4127-34
[10] Cawkwell M J *et al.* 2006 *Phys. Rev. B* **73** 064104; Fang H *et al.* 2010 *Physica B* **405** 732-7
[11] Cynn H, Klepeis J E, Yoo C-S, and Young D A 2002 *Phys. Rev. Lett.* **88** 135701
[12] Kuchhal P and Dass N 2003 *Pramana* **61** 753-7; Kapoor K, Kumar A and Dass N 2014 *ibid.* **82** 549-61
[13] Belonoshko A B *et al.* 2008 *Phys. Rev. Lett.* **100** 135701
[14] Burakovsky L *et al.* 2014 *J. Phys. Conf. Ser.* **500** 162001 and references therein
[15] Burakovsky L, Burakovsky N and Preston D L 2015 *Phys. Rev. B* **92** 174105
[16] Kinslow R, editor 1970 *High-Velocity Impact Phenomena* Academic Press, New York and London, p. 547
[17] Auer S and Frenkel D 2001 *Nature* **409** 1020-3
[18] Dolbnya I P, Petukhov A V, Aarts D G, Vroege G J and Lekkerkerker H N W 2005 *Europhys. Lett.* **72** 962-8
[19] Byelov D V *et al.* 2010 *Phase Transitions* **83** 107-14
[20] Dubrovinsky L *et al.* 2007 *Phys. Rev. Lett.* **98** 045503

IOP Conf. Series: Journal of Physics: Conf. Series **950** (2017) 042022 doi:10.1088/1742-6596/950/4/042022

Improved AC-calorimeter for measuring the absolute value of specific heat up to 10 GPa

K Umeo

N-BARD, Hiroshima University,Kagamiyama 1-3-1, Higashi-Hiroshima, 739-8526, Japan

kumeo@sci.hiroshima-u.ac.jp

Abstract. We developed a low-temperature AC calorimeter for measuring the absolute value of specific heat under pressures up to 10 GPa. We used Bridgman anvil cell made of WC with the top of 3 mm in diameter. A hollow at the top prevents expansion of the sample space over the anvil top. As the thermometer and the heater, two chip resistors were mounted on the outer part of the Cu-Be gasket in a frypan like shape. Thereby, the thermometer is free from pressure. In order to thermally isolate the gasket from the anvil, diamond powder of 0.25 μm was laid on the anvil top. A jump of specific heat at the superconducting transition of In is observed under pressures up to 8.4 GPa as clearly as at ambient pressure.

1. Introduction

The measurement of specific heat C of a solid under pressures provides important information on phase transitions and different types of internal excitations and their dependence on the volume. For such a measurement, an AC method is suitable, because the sample is not need to be isolated thermally from the heat bath [1, 2]. The AC calorimeter using a piston-cylinder pressure cell is widely used [3-5]. In this type of cell, however, the maximum pressure is limited to about 4 GPa. More recently, the pressure range has been extended above 10 GPa by using Bridgman and diamond anvil cells [6-8]. In such methods, the temperature is usually measured by a thermocouple inside the pressure cell. The pressure effect on the electromotive force of the thermocouple makes it difficult to determine the absolute value of C. Therefore, we have developed an AC calorimeter using the Bridgman anvil cell, in which the thermometer is not pressurized.

In our previous paper [9], we have used the Bridgman anvil with the top of the diameter of 5 mm. The sample has been packed in a Cu-Be gasket of 143 mg which is heavier than the sample weight of a few mg. We have measured the $C(T)$ of Yb-based heavy-fermion antiferromagnets and β-pyrochlore oxide superconductor under pressures [10-12]. Because the heat capacity of addenda without sample became larger than 50 % of total heat capacity, it was difficult to obtain the $C(T)$ of the sample with rather small specific heat. Furthermore, applying pressure above 4 GPa has broken the Cu-Be gasket. In this paper, we present improved AC calorimeter to measure the sample with small heat capacity under higher pressures up to 10 GPa. As a test of our calorimeter, the specific heat of superconducting Indium has been measured. The detail of measurements on superconducting Lead up to 9 GPa will be reported in a forthcoming paper [13].

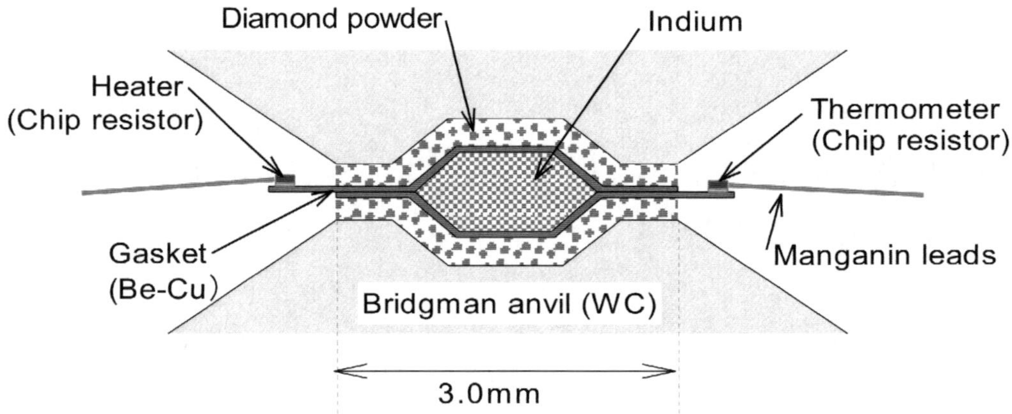

Figure 1. Cross-section of the Bridgman anvil cell of the AC calorimeter.

2. Experimental section

2.1. Design of calorimeter

Figure 1 shows the cross-section of the assembly for our calorimeter. We used Bridgman anvil cell made of WC with the top of 3 mm in diameter. A hollow at the top prevents expansion of the sample space over the anvil top. In order to test our calorimeter, we chose an Indium (In) piece of 3~4 mg as a sample. Two RuO_2-chip resistors ($0.6\times0.3\times0.2$ mm^3) as the thermometer and the heater were mounted on the fringe (width: 0.7 mm, length: 0.8 mm) of the Cu-Be gasket of 6 mg in the shape of a frypan by GE varnish. Thereby, the thermometer is free from pressure. The resistances of thermometer and heater at room temperature are 3.9 kΩ and 120 Ω, respectively. To isolate thermally the gasket from the anvil, diamond powder of 0.25 μm was laid on the anvil top, and manganin wires (diameter 50 μm, length 5 mm) were used as electrical leads. In this way, the weight of Cu-Be gasket was reduced down to 4% of that for the previous calorimeter [9].

2.2. Model and conditions

The methods of the AC calorimetry have been reported in a number of publications [1, 2]. We consider the one-dimensional heat flow problem according to the model introduced by P. F. Sullivan and G. Seidel [1]. Here, some necessary equations to analyze our data are presented. Detailed introduction of these formulas will be shown in another paper [13].

Let us consider a bar shaped sample with the length of L, density of ρ, thermal conductivity of κ, and heat capacity with C. An oscillatory heat $\dot{Q}(t) = Q_0\{1-\sin(\omega t)\}$ is put into the sample from one face (x=0). The opposite face (x=L) is coupled to a heat bath at a constant temperature T_b through a thermal conductance K_b. The thermal diffusivity D and reciprocal k of the thermal diffusion length are difined by $D = \kappa/(\rho C)$ and $k = \{\omega/(2D)\}^{1/2}$, respectively. Under a steady-state condition, the amplitude ΔT_{ac} of the oscillatory temperature T_{ac} can be represented by the following form [1, 2],

$$\Delta T_{ac} = Q_0(\omega C)^{-1}\delta(\omega, \tau_{ext}, \tau_{int}). \quad (1)$$

Here, P. F. Sullivan and G. Seidel have introduced two characteristic relaxation times, τ_{ext} and τ_{int},

$$\tau_{ext} = CK_b^{-1}, \quad (2)$$

$$\tau_{int} = L^2\left(\sqrt{90}D\right)^{-1}. \quad (3)$$

When the frequency ω satisfies,

$$\omega\tau_{ext} >> 1 \text{ and } \omega\tau_{int} << 1, \quad (4)$$

$\delta(\omega, \tau_{ext}, \tau_{int})$ becomes [1,2]

$$\delta(\omega, \tau_{ext}, \tau_{int}) = \left\{1 + (\omega\tau_{ext})^{-2} + (\omega\tau_{int})^2 + \sqrt{40}\tau_{int}\tau_{ext}^{-1}\right\}^{-1/2}. \quad (5)$$

However, in order to analyze our data, we must consider the condition,

$$\omega\tau_{ext} \geq 1 \text{ and } \omega\tau_{int} \leq 1. \quad (6)$$

Therefore, we represent the amplitude ΔT_{ac} from the formula of (1) using C, Q, ω, τ_{ext} and τ_{int} without the condition of (4) [13].

$$\delta(\omega, \tau_{ext}, \tau_{int}) = \left[G^2 + H^2\right]^{1/2}, \quad (7)$$

$$G(\omega, \tau_{ext}, \tau_{int}) = (\omega\tau_{ext})^{-1}\sinh kL \sin kL + (2kL)^{-1}(\sinh kL \cos kL + \cosh kL \sin kL), \quad (8)$$

$$H(\omega, \tau_{ext}, \tau_{int}) = -(\omega\tau_{ext})^{-1}\cosh kL \cos kL + (2kL)^{-1}(\cosh kL \sin kL - \sinh kL \cos kL), \quad (9)$$

$$kL = (45/2)^{1/2}(\omega\tau_{int})^{1/2}. \quad (10)$$

The phase between the oscillatory heat flux and T_{ac} is shifted by ϕ as

$$\phi(\omega, \tau_{ext}, \tau_{int}) = \pi/2 - \arctan|H/G|. \quad (11)$$

2.3. Measurements procedures

The amplitude and phase of the temperature oscillation at the sample were measured by a lock-in amplifier (NF Corporation, LI 5640). An AC current applying to the heater was generated by the transformation of the AC voltage supplied by an internal oscillator in the lock-in amplifier by a *V/I* converter (NF Corporation, As 913). The current supplied to the thermometer by a home-made current source was set as 0.3 μA for T<1.5 K and 1.0 μA for T>1.5 K. The use of a battery for the current source prevents the noise from the power source. The KEYTHLEY 2182 digital nano-volt meter measured the resistance of the thermometer to give the temperature T_s on the sample. The sample heating power was controlled to hold the ΔT_{ac} to be less than 3 % of T_s for T_s<10 K, and 10 % for T_s>10 K.

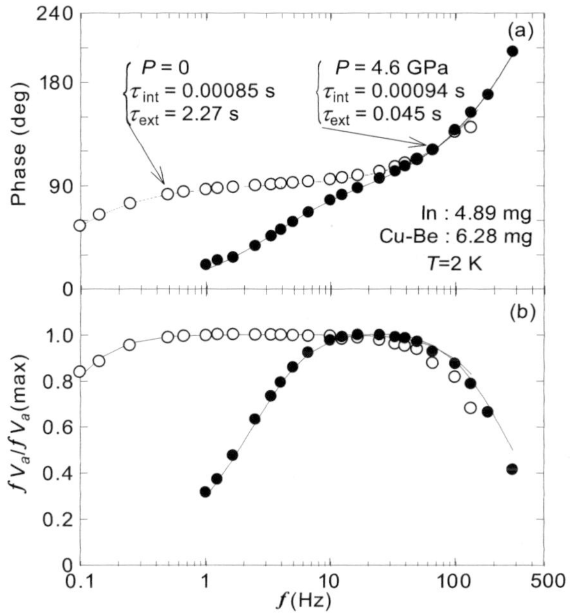

Figure 2. (a) Frequency dependence of phase shift ϕ between the sinusoidal heat input and the oscillatory sample temperature. The experimental data points are fitted by the equation (11). The time constants τ_{ext} and τ_{int} have been determined by the fits. (b) Normalized frequency dependence of fV_a, where f is the frequency of the heat flow and V_a is the output voltage of the Lock-in Amplifier which is proportional to the amplitude of the AC temperature. The solid lines are reproduced by the equation (7)-(10) using τ_{ext} and τ_{int} determined by the above fits.

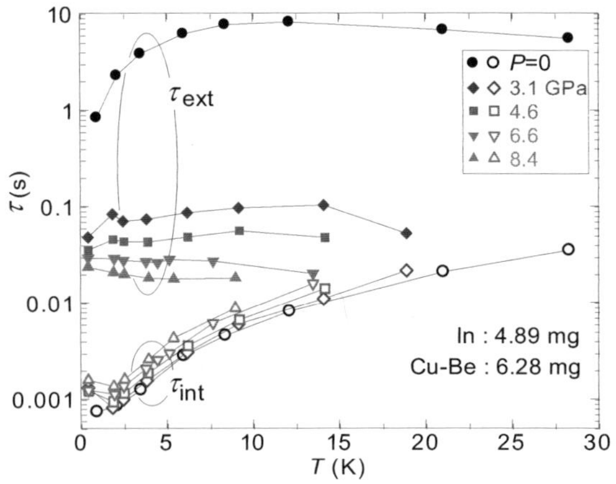

Figure 3. Temperature dependences of τ_{ext} and τ_{int} under various constant pressures.

Before the measurement of the specific heat, we must estimate two relaxation times, τ_{ext} and τ_{int}, to determine the optimum frequency f (=$2\pi/\omega$) of the heating signal. Both τ_{ext} and τ_{int} at constant pressure and temperature are determined to satisfy the equation (6) concerning the frequency dependence of the phase shift ϕ between the oscillatory heat flux and T_{ac}. As an example, frequency dependences of ϕ at the pressures of 0 and 4.3 GPa and T=2 K are shown in Fig. 2 (a). Furthermore, as shown in Fig. 2 (b), the normalized experimental values fV_a/fV_a(max), where V_a ($\propto \Delta T_{ac}$) is an output voltage from the Lock-in amplifier, can be calculated by the function of $\delta(\omega, \tau_{ext}, \tau_{int})$ using the parameters of τ_{ext} and τ_{int} in Fig. 2 (a). The good agreement between the experimental (○ and ●) and calculated values (solid lines) assures the validity of the one dimensional heat flow model. In the frequency range of $1 \leq f \leq 30$ Hz for P=0 and $10 \leq f \leq 30$ Hz for P = 4.3 GPa, ϕ is almost $\pi/2$, and fV_a/fV_a(max) is approximately 1.0. The lowest optimum frequency increases with increasing pressure as manifested by the decrease of τ_{ext}.

The temperature dependences of τ_{ext} and τ_{int} at various constant pressures are shown in Fig. 3. For P=0, τ_{ext} has a broad maximum at 10 K, whereas τ_{int} has a minimum at around 2 K. Thus, the condition (6) is satisfied in the wide temperature range from 0.5 K to 20 K. With increasing pressure up to 8.4 GPa, however, τ_{ext} decreases down to about 1/100 of the value for P=0, while τ_{int} increases slightly. This strong decrease in τ_{ext} suggests the increase of thermal conductance K_b of the diamond powder. As a result, the optimum temperature region becomes narrower at high pressures. However, at even 8.4 GPa the condition (6) is satisfied below 5 K. We have chosen the frequencies in $1 \leq f \leq 5$ Hz for P = 0 and $3 \leq f \leq 25$ Hz for $P > 0$.

3. Results and discussions

The temperature dependences of specific heat divided by temperature C/T of In (4.89 mg) and Cu-Be plate under various pressures are shown in Fig. 4. The absolute value of C at P=0 is slightly larger than the value (open circles) estimated from the reference data of In [14] and previous date of Cu-Be [7]. This difference may be the contributions from the thermometer and heater. The value of pressure was estimated by the pressure dependence of the superconducting transition temperature T_c. The maximum pressure of 8.4 GPa was obtained in the load of 66 kN at room temperature. With increasing pressure, the decrease of C for $T>T_c$ originates from the increase in the Debye temperatures of In. The jump of C at the superconducting transition of In is observed under pressures as clearly as at ambient pressure as shown in the inset. The transition temperature width of 0.005 K at 8.4 GPa gives the estimation of the pressure distribution less than 0.05 GPa.

The sets of data of T_c and jump ΔC of specific heat for In at P=0 are summarized in Table 1.

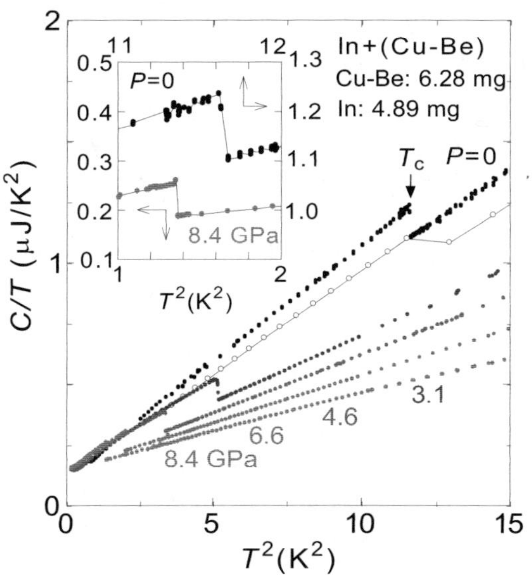

Figure 4. Specific heat C of In and Cu-Be divided by temperature T, C/T, vs T^2 under various constant pressures. The open circles are the data estimated from the reference data of In [14] and previous date of Cu-Be [9]. The inset shows the specific heat jump of In at the superconducting transition temperature T_c for P=0 and 8.4 GPa.

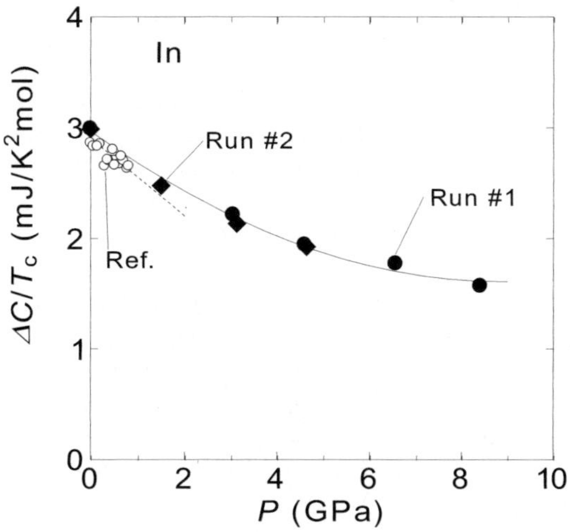

Figure 5. Pressure dependence of the specific heat jump divided by T_c, $\Delta C/T_c$, of In. Solid line is a fit to our data. The dotted line is the average of reference data [17].

These values of ΔC are in agreement with reference data [14-16] within 10 %. Figure 5 shows the pressure dependence of $\Delta C/T_c$. To the best of our knowledge, no reference data of $\Delta C/T_c$ have been reported for P>1 GPa. The value of d($\Delta C/T_c$)/dP at P=0 is -0.31 (mJ/K^{2}mol)/GPa, which is in good agreement with the reference data [17] of -0.35 (mJ/K^{2}mol)/GPa. These agreements ensure the measurements of absolute value of C for both P=0 and P>0 by our calorimeter.

Table 1. Superconducting transition temperature T_c and specific heat jump divided by T_c, $\Delta C/T_c$, for In under P=0.

	Weight of Cu-Be (mg)	Weight of In (mg)	T_c (K)	$\Delta C/T_c$ (mJ/K^{2}mol)
Run #1	6.28	4.89	3.414	2.99
Run #2	6.25	3.89	3.433	2.99
Ref. 14			3.405	2.79
Ref. 15			3.396	2.87
Ref. 16			3.403	3.03

4. Summary

We improved the AC-calorimeter for measuring the absolute value of specific heat up to 10 GPa using Bridgman anvil cell made of WC with the top of 3 mm in diameter. A hollow at the top prevents expansion of the sample space over the anvil top. As for the thermometer and the heater, two chip resistors were mounted on the outer part of the Cu-Be gasket in a frypan like shape. In order to thermally isolate the gasket from the anvil, diamond powder of 0.25 μm was laid on the anvil top. By this methods, the absolute value of specific heat of superconducting In was precisely measured up to 8.4 GPa.

Acknowledgments
We thank T. Takabatake, T. Onimaru and K. Suekuni for fruitful discussions. The specific heat measurements under pressures were performed at N-BARD, Hiroshima University. This work was supported by the Scientific Research (C) (21540363, 25400375) from MEXT, Japan.

References

[1] Sullivan P F, and Seidel G 1968 *Phys. Rev.* **173** 679
[2] Hatta I, and Ikushima A J 1981 *Japan. J. Appl. Phys.* **20** 1995
[3] Eichler A, and Gey W 1979 *Rev. Sci. Instrum.* **50** 1445
[4] Eichler A, Bohn H, and Gey W 1980 *Z. Physik B - Condensed Matter* **38** 21
[5] Bohn H, and Eichler A 1991 *Z. Physik B - Condensed Matter* **83** 105
[6] Bouquet F, Wang Y, Wilhelm H, Jaccard D, and Junod A 2000 *Solid State Commun.* **113** 367
[7] Demuer A, Marcenat C, Thomasson J, Calemczuk R, Salce B, Lejay P, Braithwaite D, and Flouquet J 2000 *J. Low Temp. Phys.* **120** 245
[8] Jaccard D, and Sengupta K 2010 *Rev. Sci. Instrum.* **81** 043908
[9] Kubo H, Umeo K, and Takabatake T 2007 *J. Phys. Soc. Jpn.* **76** Suppl. A 221
[10] Kubo H, Umeo K, Katho K, Ochiai A, and Takabatake T 2008 *J. Phys. Soc. Jpn.* **77** 023704
[11] Umeo K, Kubo H, Muro Y, Nakamura F, Suzuki T, and Takabatake T 2009 *J. Phys. Conf. Ser.* **150** 042223
[12] Umeo K, Kubo H, Yamaura J, Hiroi Z and Takabatake T 2009 *J. Phys. Soc. Jpn.* **78** 123602
[13] Umeo K to be submitted to *Rev. Sci. Instrum.*
[14] O'Neal H R, and Phillips N E 1965 *Phys. Rev.* **137** A748
[15] Clement J R, and Quinnell E H 1953 *Phys. Rev.* **92** 258
[16] Bryant C A, and Keeson P H 1961 *Phys. Rev.* **123** 491
[17] Eichler A, Cieslik J, and Gey W 1981 *Physica* **108B** 1005

IOP Conf. Series: Journal of Physics: Conf. Series **950** (2017) 042023 doi:10.1088/1742-6596/950/4/042023

Integrated form of the Anderson-Grüneisen equation of state for *p-V-T* data fit: Application to the compounds of boron, carbon, silicon and some metals.

O O Kurakevych[1], Y Le Godec[1], and V L Solozhenko[2]
[1]IMPMC, UPMC Sorbonne Universités, UMR CNRS 7590, Muséum National d'Histoire Naturelle, IRD UMR 206, 75005 Paris, France
[2]LSPM–CNRS, Université Paris Nord, 93430 Villetaneuse, France

E-mail: oleksandr.kurakevych@impmc.jussieu.fr

Abstract. The pressure-volume-temperature equations of state have been constructed by combining experimental data and semiempirical estimations for a number of compounds recently synthesized under extreme pressure-temperature conditions. The solids with various bonding types were considered: covalent hard and superhard boron-rich and diamond-like compounds (B_6O, $B_{13}N_2$, BP, cBC_5, and nano-cBN), ionic semiconductors (Mg_2C and Mg_2C_3), as well as intercalation compounds (clathrates Na_4Si_{24} and $Na_{24+x}Si_{136}$), and simple substances (γ-B_{28} and t'-B_{52} boron allotropes, Na and Mg metals, and open-framework silicon allotrope o-Si_{24}). We also showed how the reliable *p-V-T* equations of state may be constructed using different types of available data.

1. Introduction

Recent studies of phase transformations and chemical interactions in various systems under high pressure and high temperature (HPHT) conditions have led to the discovery of a number of novel materials [1-4]. We need to explore the HPHT thermodynamics for understanding the syntheses of these new materials for new challenging applications as superhard [5,6], advanced electronic [7] and photovoltaic [8], superconductors [9] as well as clathrate thermoelectric materials [10,11]: (1) boron allotropes [12] (orthorhombic γ-B_{28} [13-16], pseudo-cubic t'-B_{52} [17]) and boron-rich compounds (subnitride $B_{13}N_2$ [18,19]), (2) superhard compounds with diamond structure (nanostructured cBN [20] and c-BC_5 [21,22]); (3) covalent clathrates of new stoichiometries ($Na_{4-x}Si_{24}$ [8,10] & $Na_{24+x}Si_{136}$ [10,11]) and even (4) new unexpected semiconductors, like antifluorite Mg_2C [23,24], dense Mg_2C_3 [25] and pure silicon allotrope with quasi-direct bandgap, Si_{24} [8]. Although a part of the lacking data can be replaced by fitted parameters of common models [26-28] or with *ab initio* calculations [24], the reliable *p-V-T* equations of state (EOS) data are crucial for that.

In the present paper we describe the method of construction of such equations of state using integrated form of the Anderson-Grüneisen equation [29,30]. The method is efficient even in the case of small number of experimental data [31] and may be easily combined with *ab initio*, semiempirical and even empirical modeling [32].

2. Theoretical background

In our previous works [31,33] we have shown that the Anderson-Grüneisen equation [29,30], which takes into account the pressure dependence of thermal expansion through the volume change, i.e.

$$\alpha(p,T) = \alpha(0,T)\left[\frac{V(p,T)}{V(0,T)}\right]^{\delta_T}, \quad \mathbf{(1)}$$

can be integrated (under the assumption that δ_T is constant) to

$$V(p,T) = \left[\, V(0,T)^{-\delta_T} + V(p,300)^{-\delta_T} - V(0,300)^{-\delta_T}\,\right]^{-1/\delta_T}, \quad \mathbf{(2)}$$

where thermal expansion (i.e. $V(0,T)$ at 0.1 MPa) and isothermal compression (i.e. $V(p,300)$ at 300 K) can be presented in any analytical form, e.g. polynomial

$$V(0,T) = V(0,300)\,[1+a\,(T\text{-}300)+b\,(T\text{-}300)^2] \quad \mathbf{(3)}$$

and Murnaghan (or any other) equation of state

$$V(p,300) = V(0,300)\left(1 + B'_0\, p / B_0\right)^{-1/B_0}. \quad \mathbf{(4)}$$

Finally, a set of parameters is needed to describe an EOS using equations (2-4) is $V_0 \equiv V(0,300) \equiv M/\rho_0$, B_0, B_0', a, b and δ_T (usually ~4-6). Such form of the EOS, (2), allows one easily approximate the $V(p,T)$ in the vicinity of a new compound formation, which is often the principal domain of interest in the terms of HPHT thermodynamics. Here the unit cell volume can be estimated *in situ* using X-ray diffraction (300-K EOS may be measured on decompression and recovery).

In some other cases, the knowledge of phase equilibrium curves may allow evaluating the HPHT EOS of unknown phase through the known one. For example, we succeeded to fit the experimental melting curves (*p-T* coordinates) of α- and β-B_2O_3 and to find the parameters determining the *p-V-T* EOS for liquid B_2O_3 [34]. The melting was observed experimentally, while the bulk modulus and thermal expansion parameters were adjusted so that the experimental curve fits the theoretical one.

3. Boron-rich solids.

In the case of boron-rich solids the only unknown bulk modulus, that of pseudo-cubic t'-B_{52} phase was estimated using the density data by the method described elsewhere for various elastic characteristics of covalent and ionic materials [5,32,35,36] and previously justified for boron allotropes [37,38]. At the same time, the thermal expansion parameters are known just for B_6O [39,40] and β-B_{106} [41]. For both dense allotropes, γ-B_{28} and t'-B_{52}, we propose to take, in the first approximation, values of β-B_{106} [41], while for boron subnitride $B_{13}N_2$ one can take a value of suboxide B_6O having similar crystal structure . Except for B_6O, the δ_T parameter – linking 300-K *p-V* data with 0.1-MPa thermal expansion data – was fixed to 5.5. Figure shows that the parameters well agree with *in situ* experimental observations for $B_{13}N_2$. Better fit may be obtained with $\delta_T = B_0' = 4$ (just like in the case of B_6O, $\delta_T = B_0' = 6$) or by adjusting the a and b thermal expansion parameters, or even by suggestion of a pressure drop from 5 to 4 GPa. So, the lack of experimental data does not allow making a choice, and so far we suggest a value of $\delta_T = 5.5$.

4. Diamond-like solids.

For diamond-like phases – nano-cBN, cBC_5 and BP – all bulk moduli were established experimentally. Only for BP the thermal expansion data at 0.1 MPa have been available in literature. The a and b parameters were established from the literature data (for nano-cBN – its conventional counterpart [42], for c-BC_5 – the linear combination of diamond and boron). δ_T parameter was chosen as 5.5 for all compounds. Figure shows that the parameters well agree with experimental *in situ* observations for cBC_5. Just like in the case of $B_{13}N_2$ discussed above, the "ideal" match of an experimental point to the theoretical curve may be achieved, but it is not clear which parameter should be used for such adjustment (e.g. *a* or pressure drop during the transformation).

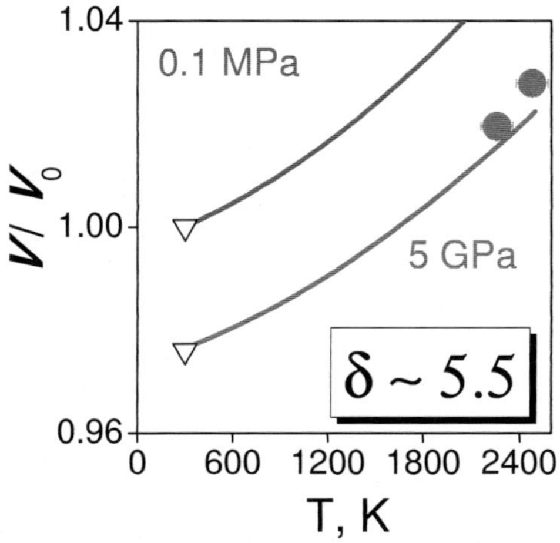

Figure 1. *p-V-T* equation of state of $B_{13}N_2$. Experimental points [27] were obtained at HASYLAB (multianvil press MAX80, resistive heating).

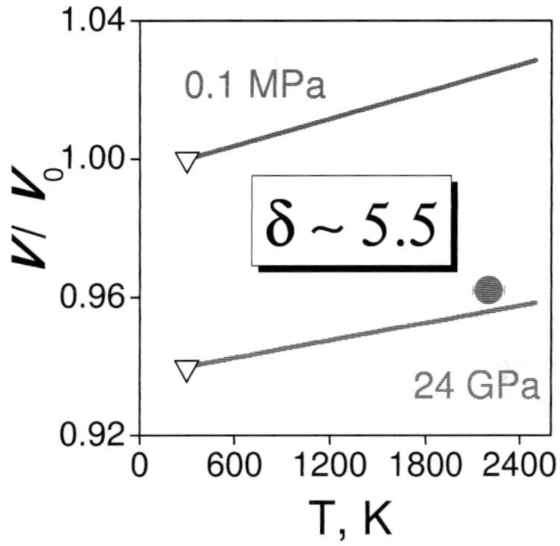

Figure 2. *p-V-T* equation of state of c-BC_5. Experimental points [22] were obtained at ESRF (diamond anvil cell, laser heating).

5. Na-Si system.
The equations of state for sodium and silicon have been previously measured. Our special interest to Na is due to the fact that Na is highly compressible solid (i.e. the reliable *p-V-T* data can be obtained in the region where pressure and temperatures can be easily measured). The data down to $V/V_0 \sim 0.85$ [43] can be perfectly fitted with proposed equation (2) (Figure 3). The value of $\delta_T = 6$.

In the case of clathrate compounds of the Na-Si system the situation with the data is the most complicated. High-temperature data, especially, at low pressure can hardly be obtained, since the compounds easily decompose. From another side, at high pressure these compounds often have only narrow domains of stability. However, it has been established that elastic properties mainly depend on the rigid silicon framework and are close to those of diamond silicon [44,45]. For $Na_{24+x}Si_{136}$ and Na_4Si_{24} the values of bulk moduli were fixed to that of $Na_{24}Si_{136}$, while for open framework "high-pressure" clathrate silicon, to the values of Si_{136} [44,46]. The experimental results on these compounds will be published elsewhere.

6. Mg-C system.
The equation of state of magnesium has been previously studied in ref. [47]. Figure 4 shows the results of the fit to equation (2). One can see that thermal expansion at ambient pressure and 300-K EOS data allowed us to predict the *p-V-T* data quite reasonably by adjusting only one parameter, i.e. $\delta_T = 3.8$.

Magnesium and carbon, both of which form numerous compounds with other elements, have remarkably low affinity for one another. At ambient pressure, only the reaction of Mg or MgO with hydrocarbons leads to the formation of metastable carbides with reasonable yields, and pure Mg and carbon do not react to form stable compounds at any temperature. After exploring pressure as additional dimension for chemistry of the Mg–C system, four magnesium carbides are now known: (1) tetragonal MgC_2, (2) orthorhombic α-Mg_2C_3, (3) monoclinic β-Mg_2C_3, and (4) cubic Mg_2C.

For Mg-C compounds stable at HP (Mg_2C [23,24] and β-Mg_2C_3 [25]), bulk moduli were established experimentally. In the case of Mg_2C the *a* and *b* parameters for thermal expansion were established by fitting the *p-V* data at high temperature (~1500 K) [24], while in other cases they were estimated from the literature data. δ_T parameter was chosen as 5.5 for all compounds except Mg_2C. Figure 5 shows that the parameters well agree with experimental *in situ* observations for Mg_2C.

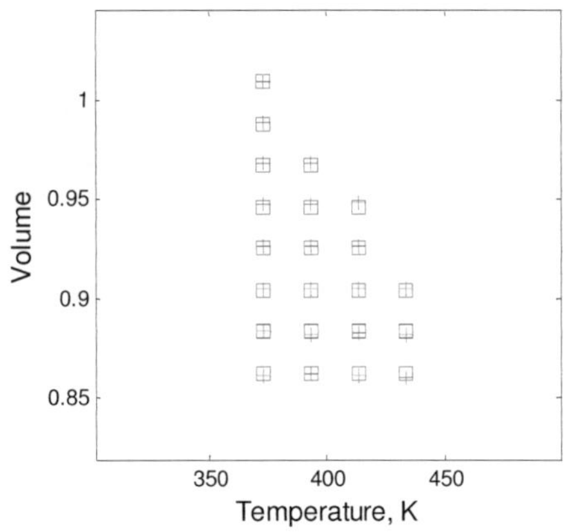

Figure 3. *p-V-T* equation of state of Na. Experimental points from ref. [43].

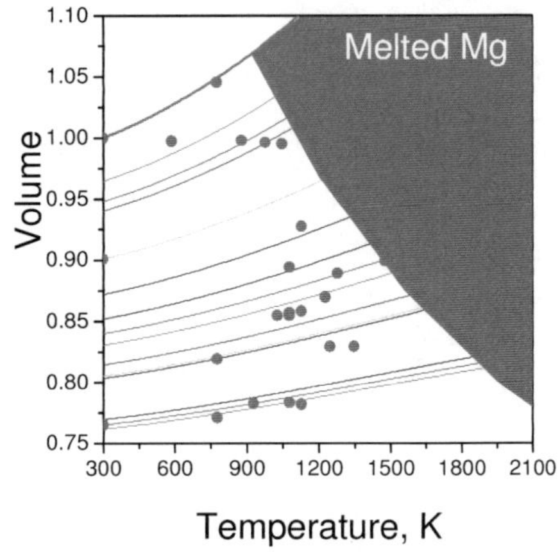

Figure 4. *p-V-T* equation of state of Mg. Experimental points from ref. [47].

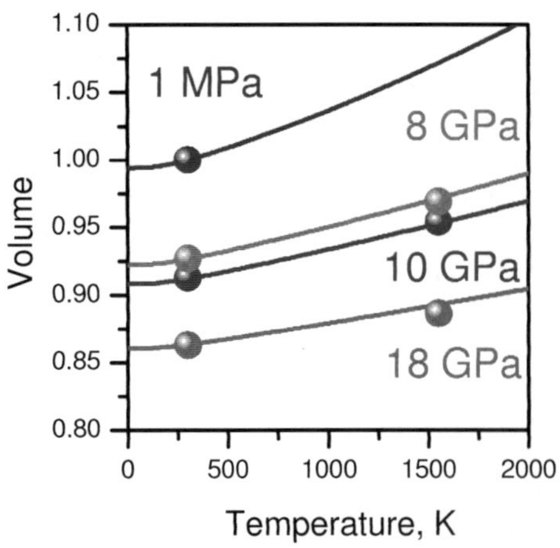

Figure 5. *p-V-T* equation of state of Mg_2C. Experimental points [24] were obtained at ESRF (multianvil press, resistive heating).

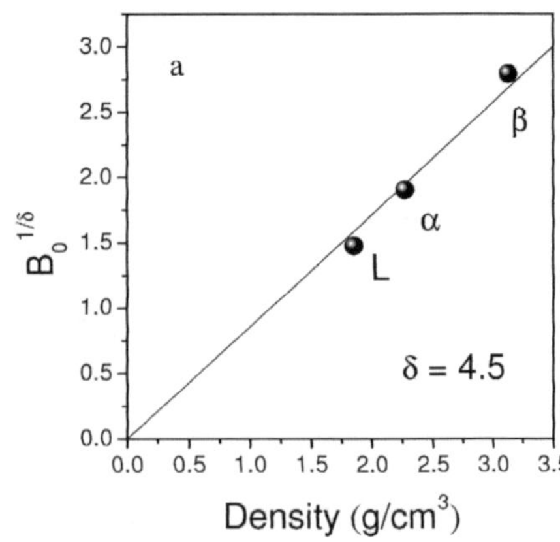

Figure 6. Bulk moduli of crystalline and liquid forms of boron oxide B_2O_3 [34].

7. Crystalline and liquid boron oxide B_2O_3.

Equations of state of α-, β-, and liquid boron oxide were obtained by fitting our experimental *p-V-T* data, while for self-consistency some parameters were adjusted to be compatible with experimental equilibria at high pressure, in the *p-T* domain where the kinetic effects become negligible [34]. In this way we have estimated both the 300-K equation of state (*p-V* coordinates) and coefficient δ_T describing the *p-T* evolution of thermal expansion and compressibility through the volume dependence. Only thermal expansion of liquid phase has been known from experimental measurements [48]. It is interesting to note that for a given composition the bulk modulus vs density show the power dependence.

Table 1. ***p-V-T*** **equation of state data for compounds synthesized at high pressure.**

	Phase	Parameters of EOS
Boron-rich solids	γ-B_{28} [37,38]	$\rho_0 = 2.544$, $B_0 = 237$, $B_0' = 2.7$, $a = 6^*$, $b = 0^*$ and $\delta_T = 5.5^*$
	t'-B_{52} [17]	$\rho_0 = 2.493$, $B_0 = 232^*$, $B_0' = 2.7^*$, $a = 6^*$, $b = 0^*$ and $\delta_T = 5.5^*$
	B_6O [31,33,49]	$\rho_0 = 2.601$, $B_0 = 180$, $B_0' = 6$, $a = 14^*$, $b = 5^*$ and $\delta_T = 6$
	$B_{13}N_2$ [19,27,50]	$\rho_0 = 2.666$, $B_0 = 200$, $B_0' = 4.0$, $a = 14^*$, $b = 5^*$ and $\delta_T = 5.5^*$
Diamond-like phases	Nano-cBN [42,51]	$\rho_0 = 3.615$, $B_0 = 375$, $B_0' = 2.3$, $a = 15^*$, $b = 0^*$ and $\delta_T = 5.5^*$
	c-BC_5 [22]	$\rho_0 = 3.267$, $B_0 = 335$, $B_0' = 4.5$, $a = 13^*$, $b = 0^*$ and $\delta_T = 5.5^*$
	BP [52-54]	$\rho_0 = 2.966$, $B_0 = 174$, $B_0' = 3.2$, $a = 16.5$, $b = 0$ and $\delta_T = 5.5^*$
Mg-C system	Mg_2C [23,24]	$\rho_0 = 2.503$, $B_0 = 87$, $B_0' = 5.1$, $a = 48$, $b = 7.1$ and $\delta_T = 4.3$
	β-Mg_2C_3 [25]	$\rho_0 = 2.580$, $B_0 = 103$, $B_0' = 4.0$, $a = 48^*$, $b = 7.1^*$ and $\delta_T = 5.5^*$
Na-Si system	$Na_{24+x}Si_{136}$ [11]	$\rho_0 = 2.318$, $B_0 = 90^*$, $B_0' = 4^*$, $a = 17$, $b = 0$ and $\delta_T = 5.5^*$
	Na_4Si_{24} [10]	$\rho_0 = 2.395$, $B_0 = 90^*$, $B_0' = 4^*$, $a = 17^*$, $b = 0$and $\delta_T = 5.5^*$
	Si_{24} [8]	$\rho_0 = 2.163$, $B_0 = 90^*$, $B_0' = 4^*$, $a = 12^*$, $b = 0^*$ and $\delta_T = 5.5^*$

Units: ρ_0 in g/cm^3; B_0 in GPa; B_0 is dimensionless; a in 10^6 K^{-1}; b in 10^9 K^{-2}; δ_T is dimensionless.

8. Conclusion

Finally, we have illustrated the applicability of a simple integrated form of Anderson-Grüneisen equation for description of the *p-V-T* data for a wide range of substances. A number of methods to construct the equations of state with limited number of available data has been shown and applied to the recently discovered high-pressure materials.

9. Acknowledgements

The EOS measurements using *in situ* X-ray diffraction were performed at ID06 & ID27 beamlines at the European Synchrotron Radiation Facility (Grenoble, France) and at F2.1 & P02.2 beamlines at

HASYLAB-DESY (Hamburg, Germany). We are grateful to W. Crichton, J. Guignard, M. Mezouar, C. Lathe and Z. Konôpková for providing assistance in using these beamlines. This work was financially supported by the Agence Nationale de la Recherche (grant ANR-2011-BS08-018).

10. References

[1] McMillan P F 1997 *Europ. J. Solid State Inorg. Chem.* **34** 621-6.
[2] McMillan P F 2002 *Nat. Mater.* **1** 19-25.
[3] McMillan P F 2003 *Chem. Comm.* 919-23.
[4] McMillan P F 2004 *High Press. Res.* **24** 67-86.
[5] Mukhanov V A, Kurakevych O O and Solozhenko V L 2010 *J. Superhard Mater.* **32** 167-76.
[6] Kurakevych O O and Solozhenko V L 2011 *High Press. Res.* **31** 48-52.
[7] Yamanaka S, Kubo A, Inumaru K, Komaguchi K, Kini N S, Inoue T and Irifune T 2006 *Phys. Rev. Lett.* **96** 076602.
[8] Kim D Y, Stefanoski S, Kurakevych O O and Strobel T A 2015 *Nat. Mater.* **14** 169–73.
[9] Tanigaki K, Shimizu T, Itoh K M, Teraoka J, Moritomo Y and Yamanaka S 2003 *Nat. Mater.* **2** 653-5.
[10] Kurakevych O O, Strobel T A, Kim D Y, Muramatsu T and Struzhkin V V 2013 *Cryst. Growth Des.* **13** 303-307.
[11] Yamanaka S, Komatsu M, Tanaka M, Sawa H and Inumaru K 2014 *J. Amer. Chem. Soc.* **136** 7717-25.
[12] Solozhenko V L and Kurakevych O O 2013 *Sci. Rep.* **3** 2351, doi:10.1038/srep02351.
[13] Solozhenko V L, Kurakevych O O and Oganov A R 2008 *J. Superhard Mater.* **30** 428-9.
[14] Oganov A R, Chen J, Gatti C, Ma Y, Ma Y, Glass C W, Liu Z, Yu T, Kurakevych O O and Solozhenko V L 2009 *Nature* **457** 863-7.
[15] Oganov A R, Chen J, Gatti C, Ma Y, Ma Y, Glass C W, Liu Z, Yu T, Kurakevych O O and Solozhenko V L 2009 *Nature* **460** 292.
[16] Oganov A R, Solozhenko V L, Gatti C, Kurakevych O O and Le Godec Y 2011 *J. Superhard Mater.* **33** 363-79.
[17] Kurakevych O O and Solozhenko V L 2013 *J. Superhard Mater.* **35** 60-3.
[18] Kurakevych O O and Solozhenko V L 2007 *Acta Crystallogr. C* **63** i80-2.
[19] Solozhenko V L and Kurakevych O O 2009 *J. Solid State Chem.* **182** 1359-64.
[20] Solozhenko V L, Kurakevych O O and Le Godec Y 2012 *Adv. Mater.* **24** 1540-4.
[21] Solozhenko V L, Kurakevych O O, Andrault D, Godec Y L and Mezouar M 2009 *Phys. Rev. Lett.* **102** 179901.
[22] Solozhenko V L, Kurakevych O O, Andrault D, Le Godec Y and Mezouar M 2009 *Phys. Rev. Lett.* **102** 015506.
[23] Kurakevych O O, Strobel T A, Kim D Y and Cody G D 2013 *Angew. Chem. Int. Ed.* **52** 8930-3.
[24] Kurakevych O O, Le Godec Y, Strobel T A, Kim D Y, Crichton W A and Guignard J 2014 *J. Phys. Chem. C* **118** 8128-33.
[25] Strobel T A, Kurakevych O O, Kim D Y, Le Godec Y, Crichton W, Guignard J, Guignot N, Cody G D and Oganov A R 2014 *Inorg. Chem.* **53** 7020-7.
[26] Solozhenko V L, Kurakevych O O, Turkevich V Z and Turkevich D V 2008 *J. Phys. Chem. B* **112** 6683-7.
[27] Solozhenko V L, Kurakevych O O, Turkevich V Z and Turkevich D V 2010 *J. Phys. Chem. B* **114** 5819-22.
[28] Solozhenko V L, Turkevich V Z, Kurakevych O O, Turkevich D V and Taniguchi T 2013 *J. Phys. Chem. C* **117** 18642-7.
[29] Anderson O, Oda H, Chopelas A and Isaak D 1993 *Phys. Chem. Miner.* **19** 369-80.
[30] Anderson O L and Isaak D G 1993 *J. Phys. Chem. Solids* **54** 221-7.
[31] Kurakevych O O and Solozhenko V L 2011 *J. Superhard Mater.* **33** 421-8.

[32] Mukhanov V A, Kurakevych O O and Solozhenko V L 2008 *J. Superhard Mater.* **30** 368-78.
[33] Kurakevych O O and Solozhenko V L 2014 *J. Superhard Mater.* **36** 270-8.
[34] Solozhenko V L, Kurakevych O O, Le Godec Y and Brazhkin V V 2015 *J. Phys. Chem. C* **119** 20600-5.
[35] Mukhanov V A, Kurakevych O O and Solozhenko V L 2008 *High Press. Res.* **28** 531-7.
[36] Mukhanov V A, Kurakevych O O and Solozhenko V L 2009 *Phil. Mag.* **89** 2117 - 27.
[37] Le Godec Y, Kurakevych O O, Munsch P, Garbarino G and Solozhenko V L 2009 *Solid State Comm.* **149** 1356-8
[38] Le Godec Y 2011 *J. Superhard Mater.* **33** 388-93.
[39] Tsagareishvili D S, Tushishvili M C and Tsagareishvili G V 1991 *AIP Conf. Proc.* **231** 392-5.
[40] Tushishvili M C, Tsagareishvili G V and Tsagareishvili D S 1992 *J. Hard Mater.* **3** 225-33.
[41] Lundstrom T, Lonnberg B and Bauer J 1998 *J. Alloy. Comp.* **267** 54-8.
[42] Datchi F, Dewaele A, Le Godec Y and Loubeyre P 2007 *Phys. Rev. B* **75** 214104.
[43] Makarenko I N, Nikolaenko A M, Ivanov V A and Stishov S M 1975 *JETP* **42** 875.
[44] San-Miguel A, Keghelian P, Blase X, Melinon P, Perez A, Itie J P, Polian A, Reny E, Cros C and Pouchard M 1999 *Phys. Rev. Lett.* **83** 5290-3.
[45] San Miguel A and Toulemonde P 2005 *High Press. Res.* **25** 159-85.
[46] Tang X, Dong J, Hutchins P, Shebanova O, Gryko J, Barnes P, Cockcroft J K, Vickers M and McMillan P F 2006 *Phys. Rev. B* **74** 014109.
[47] Errandonea D, Meng Y, Hausermann D and Uchida T 2003 *J. Phys.: Cond. Matter* **15** 1277.
[48] Napolitano A, Macedo P B and Hawkins E G 1965 *J. Amer. Ceram. Soc.* **48** 613-6.
[49] Solozhenko V L, Kurakevych O O and Bouvier P 2009 *J. Raman Spectr.* **40** 1078-81.
[50] Kurakevych O O and Solozhenko V L 2009 *Solid State Comm.* **149** 2169-71.
[51] Le Godec Y, Kurakevych O O, Munsch P, Garbarino G, Mezouar M and Solozhenko V L 2012 *J. Superhard Mater.* **34** 336-8.
[52] Slack G A and Bartram S F 1975 *J. Appl. Phys.* **46** 89-98.
[53] Le Godec Y, Mezouar M, Kurakevych O O, Munsch P, Nwagwu U, Edgar J H and Solozhenko V L 2014 *J. Superhard Mater.* **36** 61-4.
[54] Solozhenko V L, Kurakevych O O, Le Godec Y, Kurnosov A V and Oganov A R 2014 *J. Appl. Phys.* **116** 033501.

IOP Conf. Series: Journal of Physics: Conf. Series **950** (2017) 042024 doi:10.1088/1742-6596/950/4/042024

Electrical resistivity of FeAs, $FeAs_2$ and Fe_2As at homogeneous high pressures

Nao Takeshita*, Akira Iyo, Shigeyuki Ishida, Hiroshi Eisaki, Yoshiyuki Yoshida

National Institute of Advanced Industrial Science and Technology (AIST), Tsukuba, Ibaraki 305-8568, Japan

*takeshita.n@aist.go.jp

Abstract. Electrical resistivity of binary compounds containing iron and arsenic, FeAs, $FeAs_2$ and Fe_2As, is measured at homogeneous high pressures up to 15 GPa. The obtained results indicate the absence of pressure-induced superconductivity for these compounds in the specified pressure range. A new ground state possibly emerges in FeAs above 3 GPa.

1. Introduction

Since the discovery of iron pnictide high-T_c superconductors [1], many iron- and arsenic-containing superconductors have been developed. They belong to the group of superconductors with the second highest critical temperature, for which the pressure studies led to the discovery of a new phenomenon called pressure-induced superconductivity.

Earlier, we investigated some 1111-type (LnFeAsO$_{1-y}$, Ln = La, Ce, Pr, Nd...) [2] and 122-type compounds (AEFe$_2$As$_2$, AE = Ba, Sr, Ca) [3], and observed that tiny amounts of superconducting impurities exhibit large resistive anomalies. In some cases, a drop in resistivity was mistakenly identified as a bulk superconducting transition. In a similar way, several studies reported "pressure-induced superconductivity" for some materials even without reaching zero resistivity[4,5]. In order to avoid such confusions, special efforts were made to investigate bulk superconductivity by using dense homogeneous high pressure samples of superior quality for our studies. Thus, the resistivity at high pressures for binary compounds FeAs, $FeAs_2$, and Fe_2As, which are often obtained for iron pnictide superconductors during sample preparation, are studied in this work for the first time.

Pressure-induced superconductivity studies were recently reported for monopnictides of the 3d transition metals, CrAs and MnP [6-8]. These compounds and FeAs have the same crystal structure (MnP type) and exhibit a helimagnetic ground state at atmospheric pressure [9]. In addition, pressure-induced superconductivity can potentially affect the transport properties of FeAs at high pressures.

2. Experimental

High quality FeAs, $FeAs_2$, and Fe_2As polycrystalline samples were used in this study. They were synthesized by a high pressure method (HP) using a cubic type multi-anvil apparatus at pressure $P \sim 2$ GPa and temperature $T \sim 1200$ K. The HP synthesis produced very dense and homogeneous samples, which is important for electrical resistivity measurements at high pressures. In order to observe the helimagnetic transition of FeAs in detail, a single crystal FeAs sample was prepared by a solid-state reaction method.

Electrical resistivity measurements at high pressures were performed in a cubic-type setup for low temperatures [10]. As mentioned above, tiny impurity amounts or crystallographic randomness in a sample can lead to misinterpretation of pressure-induced phenomena. In order to avoid this outcome, high pressure has to be isotropically supplied to the sample, which makes the cubic system the best apparatus for this study.

3. Results

3.1. FeAs

Figure 1 shows the temperature dependence of the electrical resistivity at high pressures for the FeAs polycrystalline sample. The general trend reveals a decrease in resistivity with pressure; which, however, does not go down significantly even at the high pressures. This indicates no increase in randomness of the crystal even at the highest pressure value of 15 GPa. A small anomaly was detected at around 65 K and 1 GPa (Figure 1), which can be possibly related to the helimagnetic transition in FeAs observed in an earlier study at around 77 K [9]. However, at higher pressures this anomaly becomes virtually undetectable.

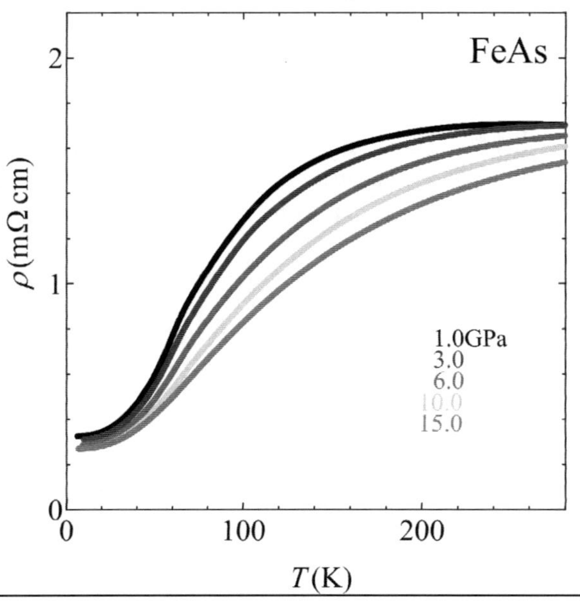

Figure 1. Temperature dependence of the electrical resistivity of the FeAs polycrystalline sample at different pressures.

Another experiment was done with the FeAs single crystal. The results are shown in Figure 2(a). At 1 GPa, a resistive anomaly at T_N = 69 K is clearly seen. The temperature corresponding to this anomaly shifts when more pressure is applied. In order to determine T_N, the resistivity must be differentiated with respect to temperature as shown in Figure 2(b). Below 3 GPa, T_N decreases with pressure; this trend becomes reversed at higher pressure values. At 3 GPa, the differential curve becomes broad and appears to be split into two separate peaks; however, it contains no characteristic differences from the resistivity curves below or above 3 GPa. At the highest pressure of 11 GPa, it is difficult to even determine T_N from the figure. By plotting T_N (defined by the maximum of the differential curve) versus pressure it is possible to construct a pressure phase diagram for FeAs (Figure 3). Contrary to our expectations, no pressure-induced superconductivity was observed in the experimental temperature range (T > 4K) at given pressures. A possible phase boundary appears to exist at around 3 GPa; however, no more details about this can be provided at this time. Structural

IOP Conf. Series: Journal of Physics: Conf. Series **950** (2017) 042024 doi:10.1088/1742-6596/950/4/042024

studies or a neutron diffraction experiment at high pressures would help to determine the ground state of FeAs above 3 GPa.

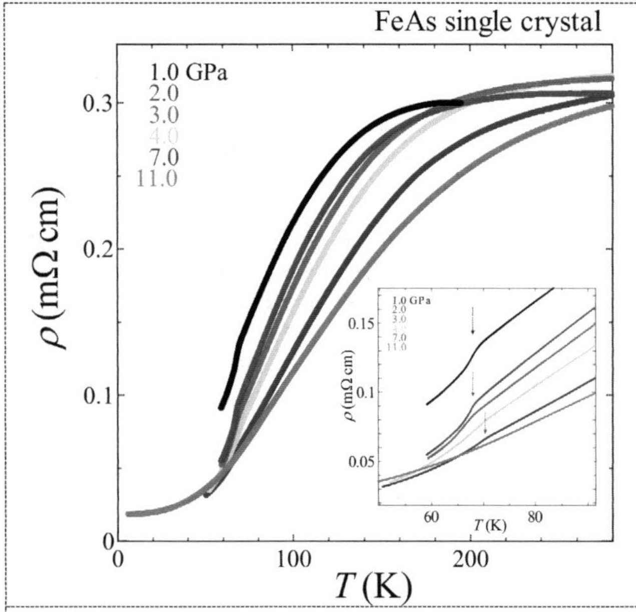

Figure 2(a). Temperature dependence of the electrical resistivity of the FeAs single crystal at different pressures.

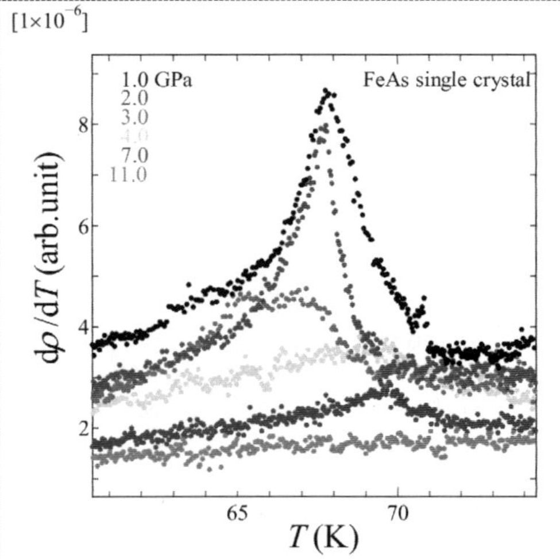

Figure 2(b). Temperature derivative of the electrical resistivity of the FeAs single crystal. The peak broadens at 3 GPa and shifts to the right at higher pressures.

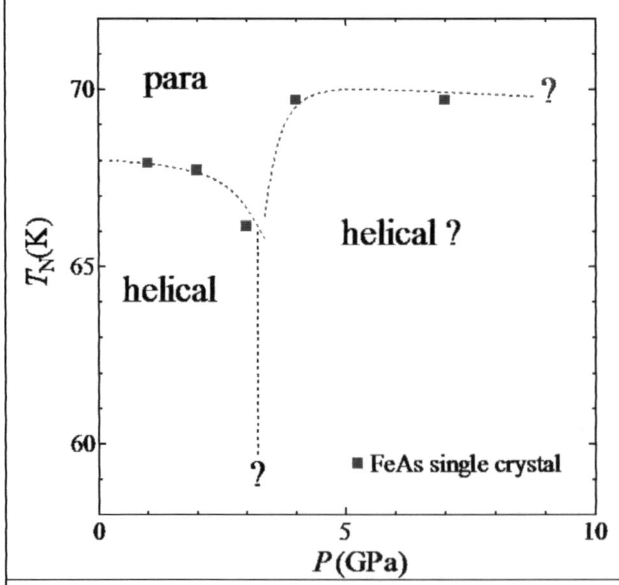

Figure 3. A proposed phase diagram for FeAs defined by the resistivity measurements. The detail on the right remains undefined.

IOP Conf. Series: Journal of Physics: Conf. Series **950** (2017) 042024 doi:10.1088/1742-6596/950/4/042024

3.2. $FeAs_2$

$FeAs_2$ was reported to be a diamagnetic semiconductor with a band gap of 0.22 eV at room temperature [11]. Its resistivity at high pressures as a function of temperature is shown in Figure 4. Following the general trend of resistivity decrease, it becomes metallic at high pressures. Note that the residual resistivity at the highest pressure of 13 GPa is very small, almost one-tenth of that at atmospheric pressure; nevertheless, $FeAs_2$ remains semiconductive below 130 K. There is still a possibility for pressure-induced superconductivity at pressures higher than 13 GPa.

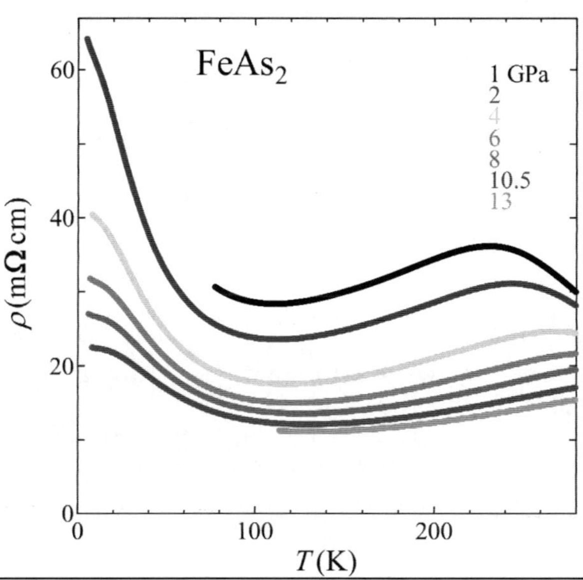

Figure 4. Electrical resistivity of $FeAs_2$ as a function of temperature at different pressures. The resistivity exhibits a metallic behavior at 13 GPa and starts to increase at around 130 K.

3.3. Fe_2As

Fe_2As has a Cu_2Sb-type crystal structure. Its unit cell contains two Fe sites and one As site. LiFeAs, which is a typical 111-type iron-pnictide superconductor, has the same crystal structure. An antiferromagnetic transition occurs in Fe_2As at around T_N = 353 K [12]. Unfortunately, our high pressure apparatus was not capable of heating samples to such a temperature because of the plastic materials near the anvils; therefore, no T_N values could be confirmed for Fe_2As at any pressure in this study.

Figure 5 shows the temperature dependence of the electrical resistivity for the polycrystalline Fe_2As. At 1 GPa, a poorly resolved anomaly appears between 50 and 70 K, which becomes more unclear at higher pressures (thus making it impossible to determine its critical temperature). Zocco *et al.* measured the electrical resistivity of Fe_2As at high pressures to calculate the heat capacity of Fe_2As at ambient pressure [13]. In contrast to the results presented in this work, no anomaly was detected around 50 ~ 70 K.

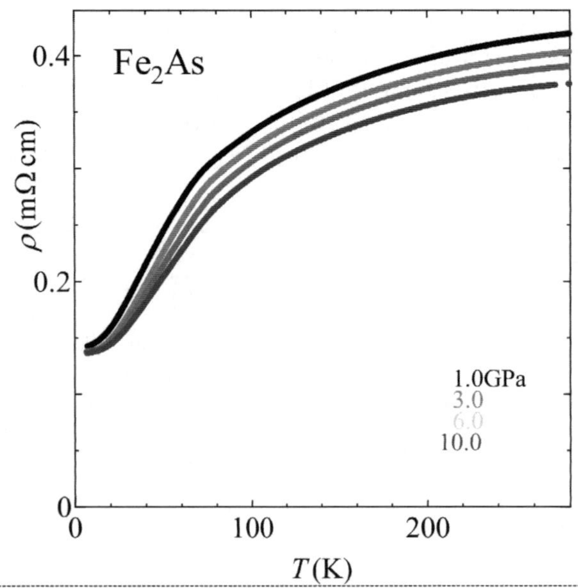

Figure 5(a). Electrical resistivity of the polycrystalline Fe_2As as a function of temperature at different pressures. Out of the three binary polycrystalline compounds considered in this work, it displays the smallest resistivity.

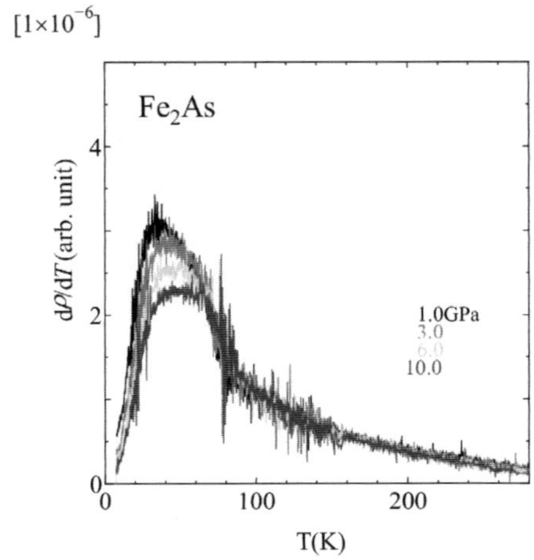

Figure 5(b). Temperature derivative of the resistivity in Fe_2As at different pressures. The anomaly, observed at around 50 K and 1 GPa, disappears at higher pressures.

4. Summary

Electrical resistivity of three binary compounds, FeAs, $FeAs_2$, and Fe_2As, were measured at homogeneous high pressures. The obtained results reveal a non-superconducting behavior for each of the studied compounds regardless of the pressure, thus eliminating a possibility of pressure-induced superconductivity for these materials (often detected as impurities in iron pnictide compounds). A pressure phase diagram for FeAs was successfully constructed from the obtained data. The T_N values initially decrease with pressure in the low pressure range; however, T_N starts increasing again above the critical pressure of 3 GPa. It appears that a ground state emerges in FeAs at this point. To characterize this new state, further structural studies or neutron diffraction measurements are required.

References

[1] Y. Kamihara, T. Watanabe, M. Hirano, H. Hosono, *J. Am. Chem. Soc.* **130**, 3296 (2008).
[2] N. Takeshita, A. Iyo, H. Eisaki, H. Kito, T. Ito, *J. Phys. Soc. Jpn.* **77**, 075003 (2008).
[3] T. Yamazaki, N. Takeshita, R. Kobayashi, H. Fukazawa, Y. Kohori, K. Kihou, Chul-Ho Lee, H. Kito, A. Iyo, H. Eisaki, *Phys. Rev. B* **81**, 224511 (2010).
[4] L. Sun, X.-J. Chen, J. Guo, P. Gao, Q.-Z. Huang, H. Wang, M. Fang, X. Chen, G. Chen, Qi Wu, C. Zhang, D. Gu, X. Dong, L. Wang, K. Yang, A. Li, X. Dai, H. K. Mao, Z. Zhao, *Nature* **483**, 67–69 (2012).
[5] N. Takeshita, T. Yamazaki, A. Iyo, H. Eisaki, H. Kito, T. Ito, K. Hirayama, H. Fukazawa, Y. Kohori, *J. Phys. Soc. Jpn.* **77SC**, 131-133 (2008).
[6] W. Wu, J. Cheng, K. Matsubayashi, P. Kong, F. Lin, C. Jin, N. Wang, Y. Uwatoko, J. Luo, *Nat. Commun.* 5:5508 doi: 10.1038/ncomms 6508 (2014).

[7] H. Kotegawa, S. Nakahara, H. Tou, H. Sugawara, *J. Phys. Soc. Jpn.* **83**, 093702 (2014).
[8] J.-G. Cheng, K. Matsubayashi, W. Wu, J. P. Sun, F. K. Lin, J. L. Luo, Y. Uwatoko, *Phys. Rev. Lett.* **114**, 117001 (2015).
[9] K. Selte, A. Kjekshus, A. F. Andresen, *Acta Chemica Scandinavia* **26**, 3101 (1972).
[10] N. Môri, H. Takahashi, N. Takeshita, *High Pressure Research* **24**, 225-232 (2004).
[11] A. K. L. Fan, G. H. Rosenthal, H. L. McKinzie, A. Wold, *J. Solid State Chem.* **5**, 136 (1972).
[12] T. Kanomata, T. Goto, H. Ido, *J. Phys. Soc. Jpn.* **43,** 1178 (1977).
[13] D. A. Zocco, D. Y. Tütün, J. J. Hamlin, J. R. Jeffries, S. T. Weir, Y. K. Vohra, M. B. Maple, *Supercond. Sci. Technol.* **25,** 084018 (2012).

IOP Conf. Series: Journal of Physics: Conf. Series **950** (2017) 042025 doi:10.1088/1742-6596/950/4/042025

Record high magnetic ordering temperature in a lanthanide at extreme pressure

J Lim[1,2], G Fabbris[1,3,4], D Haskel[3] and J S Schilling[1]

[1] Department of Physics, Washington University, St. Louis, Missouri 63130, USA
[3] Advanced Photon Source, Argonne National Laboratory, Argonne, Illinois 60439, USA

E-mail: `jss@wuphys.wustl.edu`

Abstract. Today's best permanent magnet materials, $SmCo_5$ and $Nd_2Fe_{14}B$, could likely be made significantly more powerful were it not necessary to dilute the strong magnetism of the rare earth ions (Sm, Nd) with the $3d$ transition elements (Fe, Co). Since the rare-earth metals order magnetically at relatively low temperatures $T_o \leq 292$ K, transition elements must be added to bring T_o to temperatures well above ambient. Under pressure $T_o(P)$ for the neighboring lanthanides Gd, Tb, and Dy follows a notably nonmonotonic, but nearly identical, dependence to ~60 GPa. At higher pressures, however, Tb and Dy exhibit highly anomalous behavior, T_o for Dy soaring to temperatures well above ambient. We suggest that this anomalously high magnetic ordering temperature is an heretofore unrecognized feature of the Kondo lattice state.

Measurements of the magnetic and superconducting properties of matter can now be carried out to multimegabar pressures where the increase in energy (1 - 10 eV/atom) is sufficient to significantly alter these properties and possibly result in novel many-body states of matter. Stable magnetic systems can destabilize under extreme pressure, leading perhaps to surprising new forms of magnetism and/or superconductivity.

Rare-earth elements play a central role in many modern technologies, including permanent magnets, computer memories, and applications requiring giant magnetostriction. The high degree of localization of their magnetic $4f$ electrons results in relatively weak interatomic exchange interactions and low magnetic ordering temperatures, the highest being $T_o \simeq 292$ K for Gd metal [1]. This is unfortunate since the strongest permanent magnets, $SmCo_5$ and $Nd_2Fe_{14}B$, require the addition of a substantial concentration of a magnetic transition metal, Co or Fe, in order to raise T_o to a temperature well above ambient. Unfortunately, this addition dilutes the strong magnetism of the rare-earth component. A significant increase in T_o for rare-earth materials devoid of transition elements could lead to many new applications.

Here we report the results of resistivity studies on the neighboring rare earths Gd, Tb, and Dy to extreme pressures [2, 3]. Above 60 GPa the magnetic ordering temperatures of Tb and Dy show a significant increase, T_o for Dy soaring to ~400 K at 157 GPa. We suggest that at extreme pressures, in contrast to Gd, both Tb and Dy enter an anomalous Kondo lattice state that is responsible for the significant increase in T_o. This scenario receives support from the fact

[2] Present address: Institute for Shock Physics, Washington State University, Pullman, Washington 99164, USA.
[4] Present address: Department of Condensed Matter Physics and Materials Science, Brookhaven National Laboratory, Upton, NY 11973, USA.

that in the pressure region above 60 GPa dilute magnetic impurities of Tb and Dy are able to drastically suppress the superconducting transition temperature of Y.

Resistivity samples are cut from 99.9% pure elements Dy and Gd (Alfa Aesar) and Tb and Y (Ames Lab [4]). The dilute magnetic alloys are prepared by argon arc-melting stoichiometric amounts of Gd, Tb, and Dy dopant with Y. To generate pressures to 1-2 Mbar, a diamond anvil cell (DAC) made of CuBe alloy [5] is fitted with two opposing diamond anvils (1/6-carat, type Ia) with 0.35 mm diameter culets beveled at 7° to 0.18 mm central flats. A Re gasket (6-7 mm diameter, 250 μm thick) is preindented to 30 μm and a 80 μm diameter hole electro-spark drilled through the center. The center section of the preindented gasket surface is filled with a 4:1 cBN-epoxy mixture to insulate the gasket and serve as pressure medium (see Fig. 1). The thin sample (dimensions $\sim 30\times30\times5$ μm^3) is then placed on top of four thin Pt leads to allow a four-point dc electrical resistivity measurement. Further details of the non-hydrostatic high pressure resistivity technique are given in a paper by Shimizu *et al.* [6].

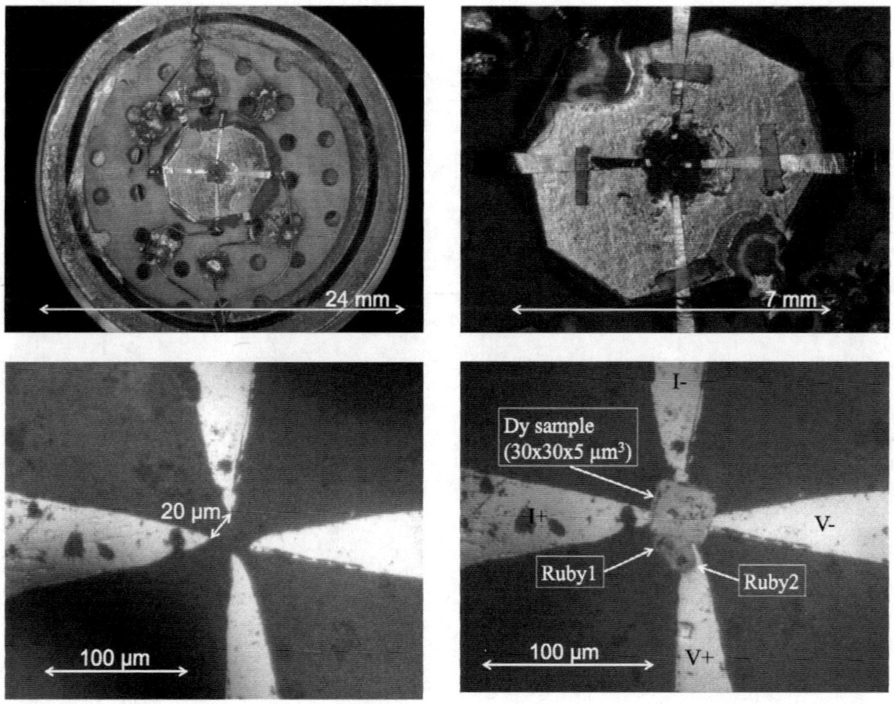

Figure 1. Images at three magnification levels of the non-hydrostatic pressure cell used with an insulated Re gasket mounted on a diamond anvil for four-probe electrical resistivity measurements. Two images at bottom show four (4μm thick) Pt leads before and after a Dy sample is placed on top.

A He-gas-driven membrane allows the pressure to be changed at any temperature above 3 K [7]. Pressure is determined using both ruby fluorescence [8] and Raman spectroscopy from the frequency shift of the diamond vibron [9]. Temperatures as low as 1.3 K can be reached in an Oxford flow cryostat. Further experimental details of the DAC and cryostat are given elsewhere [2, 5, 10, 11].

In Fig. 2 the electrical resistance $R(T)$ of Dy is plotted versus temperature at 8 different pressures to 157 GPa; for $P > 2.1$ GPa the data have been shifted vertically so as not to cross.

IOP Conf. Series: Journal of Physics: Conf. Series **950** (2017) 042025 doi:10.1088/1742-6596/950/4/042025

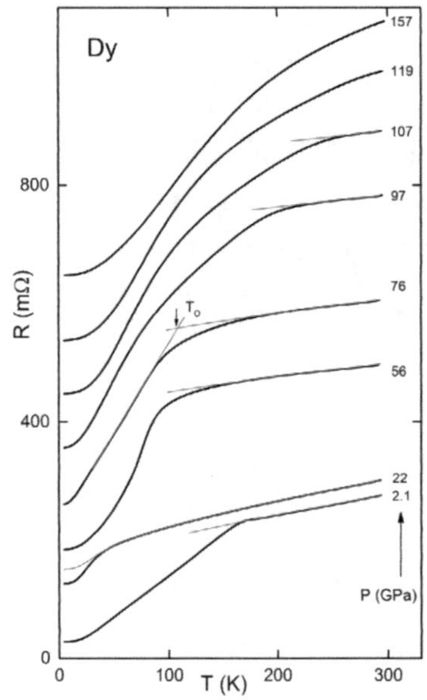

Figure 2. Resistance versus temperature for Dy at eight different pressures. Except at 2.1 GPa, the curves are shifted vertically for clarity. Red lines with small positive slope give temperature dependence of phonon resistance. See Refs. [2, 3] for details.

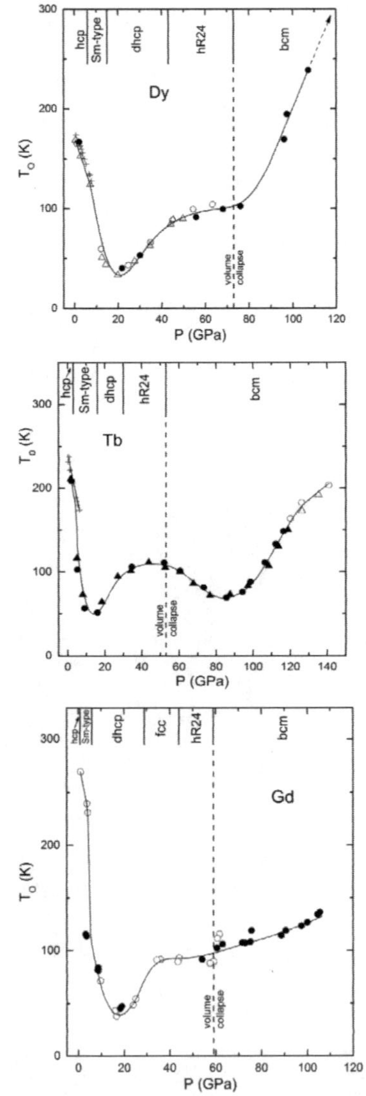

Figure 3. Magnetic ordering temperature T_o of Dy [2], Tb [3], and Gd [2] versus pressure. (+) earlier studies to ~8 GPa on Dy [13] and Tb [16]. References for crystal structures at top of graphs Dy [13], Tb [14], Gd [19]; vertical dashed line gives pressure where 5-6% volume collapse occurs. In all plots the extended solid line through data points is guide to the eye.

The magnetic ordering temperature T_o is defined by the kink in the $R(T)$ dependence clearly seen near 170 K at 2.1 GPa, the lowest pressure of the experiment. At higher pressures this kink broadens somewhat into a knee due to pressure gradients across the sample, but remains clearly visible to 107 GPa. We define T_o by the intersection point of two straight lines, as illustrated for the data at 76 GPa in Fig. 2. The observed kink/knee is due to the suppression of spin-disorder scattering upon cooling as magnetic ordering begins to set in at T_o [12]. For pressures above 107 GPa the knee in $R(T)$ is no longer visible, signalling that T_o for Dy has shifted above 300 K. Similar resistivity data have also been obtained for Tb [3] and Gd [2].

In Fig. 3 T_o is plotted to Mbar pressures for Dy, Tb, and Gd. The results are in reasonable agreement both with earlier results to ∼8 GPa on Dy [13] and Tb [14] as well as resistivity studies on Dy to 69 GPa [15]. The pressure dependence $T_o(P)$ is seen to be highly non-monotonic, presumably in response to changes in crystal structure (see top of graphs). It is notable that to ∼50 GPa the $T_o(P)$-dependences are nearly identical for all three rare earths. Above 60 GPa, however, the dependences begin to differ markedly: whereas in Gd T_o increases only slowly, $T_o(P)$ in Dy displays a rapid increase that continues unabated to the highest pressure obtained 157 GPa. Above 60 GPa $T_o(P)$ for Tb first decreases but then begins to increase rapidly. One should note that above 60 GPa the compressibility of rare earths is approximately 10-times less than that at ambient pressure [13, 14]. Plotted versus relative volume, the increase in T_o for Dy above 80 GPa is far more rapid than the initial decrease at ambient pressure; in fact, on such a plot $T_o(P)$ extrapolates to ∼400 K at 157 GPa [2].

We now address the question as to the mechanism(s) responsible for the highly non-monotonic dependence of the magnetic ordering temperature T_o of Dy, Tb, and Gd on pressure. The striking similarities in $T_o(P)$ to ∼60 GPa for all three point to a common mechanism. The magnetic ordering of these metals at ambient pressure is well explained [16] by a conduction-band-driven indirect Ruderman-Kittel-Kasuya-Yosida (RKKY) exchange interaction [17]. Within this model the magnetic ordering temperature scales with the local moment, the magnetic susceptibility of the conduction band, and the square of the exchange interaction J between the $4f$ local moment and the conduction electrons. In fact, Fleming and Liu [18] show that the initial decrease in T_o with pressure for Dy, Tb, and Gd is a direct consequence of shifts in the valence band. Furthermore, Dy [13], Tb [14], and Gd [19] undergo a very similar set of structural phase transitions driven by increasing $5d$ electron occupation with pressure [20]. Thus changes in the susceptibility of the conduction band associated with the multiple structural phase transitions are the likely cause for the highly non-monotonic dependences $T_o(P)$ shared by Dy, Tb, and Gd to ∼60 GPa.

For pressures above ∼60 GPa, however, the pressure dependences of $T_o(P)$ for Dy, Tb, and Gd are seen to differ significantly, as pointed out above. Since their structural phase diagrams are so similar, these significant differences in $T_o(P)$ almost certainly do not originate from shifts in the energy bands but likely signal differing changes in the magnetic state of these rare earths. A long-standing strategy [21, 22] to probe the magnetic state of a given ion is to alloy the magnetic ion in dilute concentration with a superconductor and determine to what extent the superconducting transition temperature is suppressed ΔT_c. As seen in the inset in Fig. 4, the pressure dependence $T_c(P)$ for Y(0.5 at.% Gd) is seen to faithfully track that for Y to the maximum pressure 126 GPa [23]. This gives evidence that over this pressure range Gd remains a conventional magnetically stable lanthanide. The absence of magnetic instabilities in Gd, even at extreme pressures, is not surprising since the magnetic state of Gd with its half-filled $4f^7$ shell is the most stable of all elements, its $4f^7$ level lying ∼ 9 eV below the Fermi level [24].

These findings for Y(0.5 at.% Gd) are in sharp contrast to those for Y(1 at.% Dy) shown in Fig. 4. As for T_o, the pressure dependence of ΔT_c is seen to show a sharp increase above 73 GPa, reaching the value $\Delta T_c \approx 9$ K at the highest pressure. Such a dramatic suppression of superconductivity strongly suggests that Kondo pair breaking is involved. This can occur when the magnetic state of a rare earth nears a magnetic instability, a precursor to a change in valence where an electron jumps out of the $4f$ orbital into the conduction band. Very recent x-ray absorption near edge structure (XANES) studies confirm that no change in valence occurs in Dy to 115 GPa [25]. As the magnetic $4f$ level approaches the Fermi level, the exchange interaction with the conduction electrons takes on a negative sign, signalling the onset of strong Kondo resonance phenomena. This enhancement of the magnitude of $|J|$ with pressure causes the magnetic ordering temperature $T_o \propto |J|^2$ to increase until $|J|$ becomes so large that the local magnetic moment begins to be compensated through the exponentially increasing Kondo spin screening, as anticipated in the simple Kondo-lattice or Doniach model [26, 27] illustrated in Fig. 5. This could then lead to an anomalously high value of T_o, such as observed for Dy and Tb at extreme pressure, a value surpassing that possible for normal positive exchange interactions.

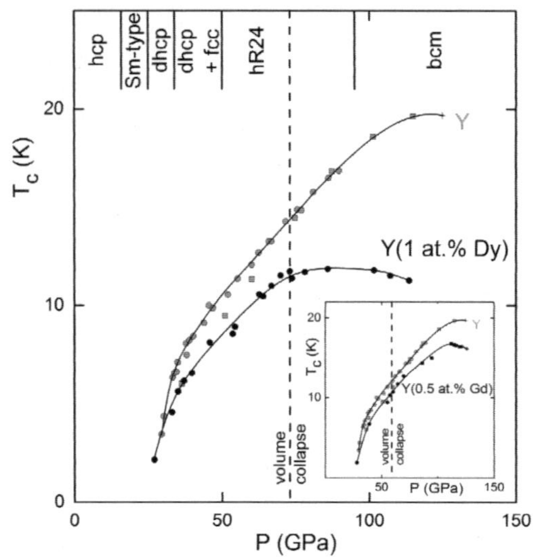

Figure 4. Superconducting transition temperature T_c versus pressure for Y(1 at.% Dy) compared to that for Y [2]. Inset shows similar graph for Y(0.5 at.% Gd) [23]. Vertical dashed line marks pressure of volume collapse for Dy at 73 GPa [13] and in inset for Gd at 59 GPa [19]. At top of graph are crystal structures taken on by Y [15]. Extended solid line through data points is guide to the eye.

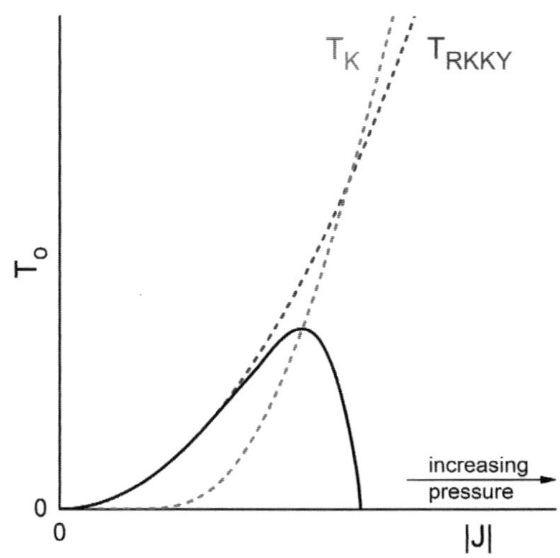

Figure 5. Magnetic ordering temperature T_o plotted versus the absolute value of the negative exchange parameter $|J|$ according the Doniach model [26, 27]. Since T_o increases only as $|J|^2$, but the Kondo temperature T_K increases exponentially with $|J|$, the latter ultimately dominates and quenches magnetic ordering.

At pressures significantly higher than those in the present experiment, the simple Doniach model in Fig. 5 would lead one to anticipate that T_o for Dy and Tb would pass through a maximum and fall rapidly to 0 K at a quantum critical point [26, 27]. Future experiments will search for this behavior either by a significant expansion of the present pressure range or by identifying a rare earth where the instabilities in the magnetic state occur at lower pressures.

In summary, measurements of the electrical resistivity of Dy, Tb, and Gd to extreme pressures show that the magnetic ordering temperature T_o exhibits a highly non-monotonic pressure dependence, rising dramatically for $P > 80$ GPa to unprecedentedly high values approaching $\sim$400 K for Dy with a similarly rapid increase for Tb. It is suggested that these anomalously high values of T_o in Dy and Tb signal that these rare earths have entered an anomalous magnetic state, one likely related to a Kondo lattice. If so, the anomalously high magnetic ordering temperatures observed here for Dy and Tb would be an heretofore overlooked feature of the Kondo lattice state itself. A search for further lanthanide and actinide systems with anomalously high magnetic ordering temperatures would be of considerable interest and is underway.

If the mechanism(s) responsible for these anomalously high values of T_o under extreme pressure can be clearly identified, one can hope to reproduce these conditions in a suitable compound at ambient pressure. This hope receives some support from the fact that the Curie temperature of the ternary compound $CeRh_3B_2$ is 115 K [28], a value exceeding even that of $GdRh_3B_2$ and 100-times higher than that anticipated from simple de Gennes scaling, surpassing even that of $GdRh_3B_2$. Perhaps Dy under exreme pressure and $CeRh_3B_2$ at ambient pressure share a common mechanism for their anomalous magnetic properties.

Acknowledgments

The authors would like to thank T. Matsuoka and K. Shimizu for sharing information on their high-pressure electrical resistivity techniques used in this study. Thanks are due A. Gangopadhyay for his help in preparing the dilute magnetic alloys. This work was supported by the National Science Foundation (NSF) through Grant No. DMR-1104742 and by the Carnegie/DOE Alliance Center (CDAC) through NNSA/DOE Grant No. DE-FC52-08NA28554. Work at Argonne National Laboratory is supported by the U.S. Department of Energy, Office of Science, under contract No. DE-AC02-06CH11357.

References

[1] Colvin R V, Legvold S and Spedding F H 1960 *Phys. Rev.* **120** 741
[2] Lim J, Fabbris G, Haskel D and Schilling J S 2015 *Phys. Rev. B* **91** 045116
[3] Lim J, Fabbris G, Haskel D and Schilling J S 2015 *Phys. Rev. B* **91** 174428
[4] Material Preparation Center, Ames Lab, US DOE, Ames, Iowa http:/www. mpc.ameslab.gov.
[5] Schilling J S 1984 *Mat. Res. Soc. Symp. Proc.* **22** 79
[6] Shimizu K, Amaya K and Suzuki N 2005 *J. Phys. Soc. Jpn.* **74** 1345
[7] Daniels W B and Ryschkewitsch W 1983 *Rev. Sci. Instr.* **54** 115
[8] Chijioke A D, Nellis W J, Soldatov A and Silvera I F 2005 *J. Appl. Phys.* **98** 114905
[9] Akahama Y and Kawamura H 2006 *J. Appl. Phys.* **100** 043516
[10] Klotz S, Schilling J S and Müller P 1991 *Frontiers of High Pressure Research*, ed H D Hochheimer and R D Etters (New York: Plenum) p 473
[11] Debessai M, Hamlin J J and Schilling J S 2008 *Phys. Rev. B* **78** 064519
[12] Coles B R 1958 *Adv. Phys.* **7** 40
[13] Patterson R, Saw C K and Akella J 2004 *J. Appl. Phys.* **95** 5443
[14] Cunningham N C, Qiu W, Hope K M, Liermann H-P and Vohra Y K 2007 *Phys. Rev. B* **76** 212101
[15] Samudala G K, Tsoi G M, Weir S T and Vohra Y K 2014 *High Press. Res.* **34** 266
[16] Jackson D D, Malba V, Weir S T, Baker P A and Vohra Y K 2005 *Phys. Rev. B* **71** 184416
[17] Ruderman M A and Kittel C 1954 *Phys. Rev.* **96** 99
[18] Fleming G S and Liu S H 1970 *Phys. Rev. B* **2** 164; Liu S H 1962 *Phys. Rev.* **127** 1889
[19] Hua H, Vohra V K, Akella J, Weir S T, Ahuja R and Johansson B 1998 *Rev. High Pressure Sci. Technol.* **7** 233
[20] Duthie J and Pettifor D 1977 *Phys. Rev. Lett.* **38** 564
[21] Matthias B T, Suhl H and Corenzwit E 1958 *Phys. Rev. Lett.* **1** 92
[22] Maple M B 1976 *Appl. Phys.* **9** 179
[23] Fabbris G, Matsuoka T, Lim J, Mardegan J R L, Shimizu K, Haskel D and Schilling J S 2013 *Phys. Rev. B* **88** 245103
[24] Yin Z P and Pickett W E 2006 *Phys. Rev. B* **74** 205106
[25] Bi W, Song J, Soni V, Haskel D and Schilling J S (unpublished).
[26] Doniach S 1977 *Valence Instabilities and Related Narrow-Band Phenomena*, ed R D Parks (New York: Plenum) p 169
[27] Yang Y-f, Fisk Z, Lee H-O, Thompson J D and Pines D 2008 *Nature* **454** 611
[28] Dhar S K, Malik S K and Vijayaraghavan R 1981 *J. Phys. C* **14** L321

IOP Conf. Series: Journal of Physics: Conf. Series **950** (2017) 042026 doi:10.1088/1742-6596/950/4/042026

Magnetic and Structural Phase Transitions in Thulium under High Pressures and Low Temperatures

Yogesh K Vohra, Georgiy M Tsoi, Gopi K Samudrala

Department of Physics, University of Alabama at Birmingham, Birmingham. AL 35294, USA

Abstract. The nature of *4f* electrons in many rare earth metals and compounds may be broadly characterized as being either "localized" or "itinerant", and is held responsible for a wide range of physical and chemical properties. The pressure variable has a very dramatic effect on the electronic structure of rare earth metals which in turn drives a sequence of structural and magnetic transitions. We have carried out four-probe electrical resistance measurements on rare earth metal Thulium (*Tm*) under high pressures to 33 GPa and low temperatures to 10 K to monitor the magnetic ordering transition. These studies are complemented by angle dispersive x-ray diffraction studies to monitor crystallographic phase transitions at high pressures and low temperatures. We observe an abrupt increase in magnetic ordering temperature in *Tm* at a pressure of 17 GPa on phase transition from ambient pressure hcp-phase to α-Sm phase transition. In addition, measured equation of state (EOS) at low temperatures show anomalously low thermal expansion coefficients likely linked to magnetic transitions.

1. Introduction

The localized state in rare earth metals is marked by tightly bound shells or narrow bands of highly correlated electrons near the Fermi level with localized magnetic moments for all of the rare earth elements. The regular trivalent rare earth structural sequence, observed at low pressures for La through Lu excepting Eu and Yb as a function of increasing pressure or decreasing atomic number, typifies this behaviour [1]. The structures are all close packed and of high symmetry, and the sequence hexagonal close packed (hcp) → α-Sm → double hexagonal close packed (dhcp) → face centered cubic (fcc) may be reproduced solely by transfer of *sp* electrons to the *d*-band. The f-delocalization transition is of first order, and may be accompanied by a discontinuous drop in volume, loss of magnetic moment on the 4f-shell, a lowering of electrical resistivity, and a stiffening of the crystalline lattice. The magnetic ordering temperatures at high pressures and low temperatures in heavy lanthanides have been investigated using magnetic susceptibility measurements and it was reported that magnetic transition disappear at pressures above 5-10 GPa [2]. Recently, it has been shown that electrical transport measurements is a very sensitive technique in measuring magnetic ordering temperatures in rare earth metal terbium and these measurements have been validated by direct observation of magnetic ordering by neutron diffraction studies [3].

2. Experimental Methods

High pressure electrical resistance measurements were made by employing an eight-probe designer diamond anvil cell in a four-probe configuration to pressures up to 33 GPa and temperatures to 10 K. The eight tungsten microprobes were encapsulated in a homoepitaxial diamond film and were exposed only near the tip of the diamond in order to make electrical contact with the Thulium (*Tm*) sample at high pressures and low temperatures. Two electrical leads were used to pass a dc current through the sample and two additional leads were used to monitor the voltage across the sample. The sample was cut from a 0.1-mm thick foil of polycrystalline *Tm* and loaded into the cell along with a ruby pressure marker. The pressure was monitored by the ruby fluorescence technique and care was taken to calibrate the ruby R_1 emission at low temperatures as described in an earlier publication [4]. The electrical resistance of the *Tm* sample was measured with increasing pressure, and the sample was then decompressed, and measured again during a second compression in order to verify if the magnetic transition could be recovered. The electrical resistance measurements were complemented by angle-dispersive image plate x-ray diffraction on *Tm* to high pressures and liquid helium temperatures (4 K). X-ray diffraction was then performed on sample pressurized at the HPCAT 16 ID-B beamline at the Advanced Photon Source at Argonne National Laboratories in Chicago, Illinois. A 30.494 keV beam of x-rays was collimated to a spot size with Full Width at Half Maximum (FWHM) of 5x7 μm and scanned across the sample area while spectra were collected to pressure up to 36 GPa and low temperatures to 4 K.

3. Results and Discussion

The following sequence of phase transitions are known in *Tm* from *hcp* to *Sm-type* at 17 GPa, *Sm-type* to *dhcp* phase at 32 GPa, and *dhcp* to *hR24* phase at 61 GPa, and *hR24* phase to monoclinic *C2/m* phase at 124 GPa [5]. At ambient pressure, *Tm* shows a c-axis sinusoidal Anti-Ferromagnetic (AFM) phase at Néel Temperature T_N = 56 K followed by a modulated Ferromagnetic (FM) phase at T_C = 25 K. The AFM transition temperature T_N was tracked in an earlier magnetic susceptibility measurement on *Tm* to a pressure 12 GPa and T_N was found to show a decrease with increasing pressure [2]. Figure 1 shows the four probe electrical resistance (R) for *Tm* as a function of temperature at a pressure of 7 GPa exhibiting anomalous behaviour near 50 K. The temperature derivative of electrical resistance (dR/dT) is also shown in figure 1 where the magnetic transition T_N is clearly evident as a minimum at 52.5 K for 7 GPa. The procedure shown in figure 1 was utilized for all data points in the hcp-phase below 17 GPa. In pressure range above 17 GPa for the α-Sm phase, the electrical resistance changes are more gradual as shown in figure 2. In these cases, the temperature derivative of electrical resistance (dR/dT) shows a change in slope at the magnetic transition T_N of 131.8 K at a pressure of 17.6 GPa (figure 2).

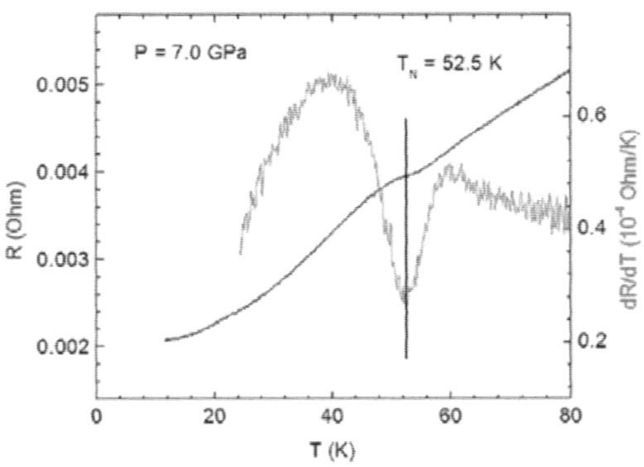

Figure 1. Electrical resistance data at a pressure of 7 GPa for *Tm* indicating T_N.

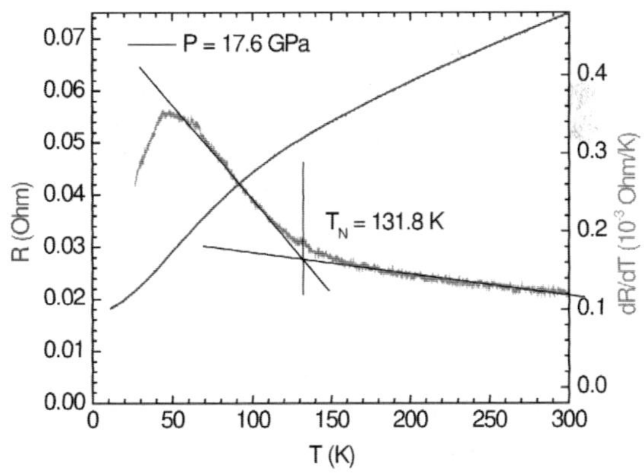

Figure 2. Electrical resistance data at a pressure of 17.6 GPa for *Tm* indicating T_N.

The magnetic transition temperature T_N data for both the hcp-phase and α-Sm phases are combined and are shown as a function of pressure in figure 3. The three independent experiments that were performed on *Tm* show remarkable consistency with each other. The magnetic transition temperature T_N for the hcp-phase was observed to decrease gradually first to 7 GPa and then at a faster rate till transition to α-Sm Phase at 17 GPa. There is an abrupt increase in T_N by as much as 100 K on transition from the hcp-phase to α-Sm phase at 17 GPa (figure 3). On increasing pressure beyond 17 GPa till 33 GPa, the magnetic transition temperature T_N was found to be relatively insensitive to changes in pressure. It is to be noted that additional changes in T_N are anticipated when *Tm* transforms finally to dhcp and other high phases above 30 GPa [5].

IOP Conf. Series: Journal of Physics: Conf. Series **950** (2017) 042026 doi:10.1088/1742-6596/950/4/042026

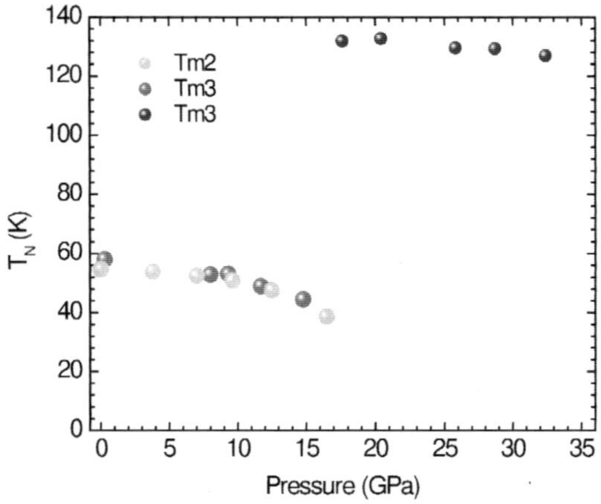

Figure 3. This shows the experimentally measured magnetic ordering temperature T_N for Thulium as a function of pressure. This plot is a combination of data from three separate experiments. An abrupt increase in T_N is observed near 17 GPa.

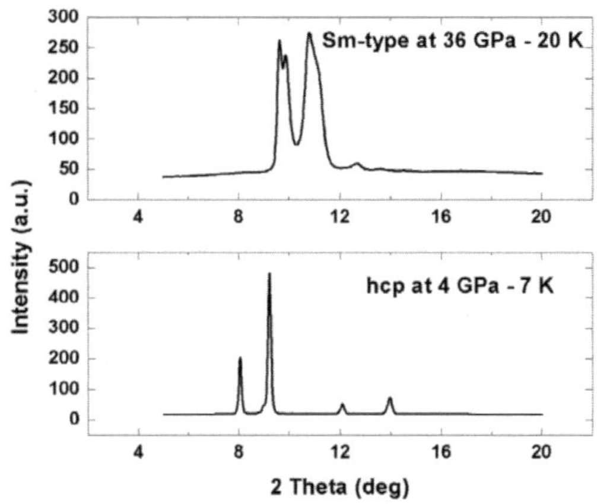

Figure 4. The angle-dispersive x-ray diffraction patterns for *Tm* at high pressures and low temperatures.

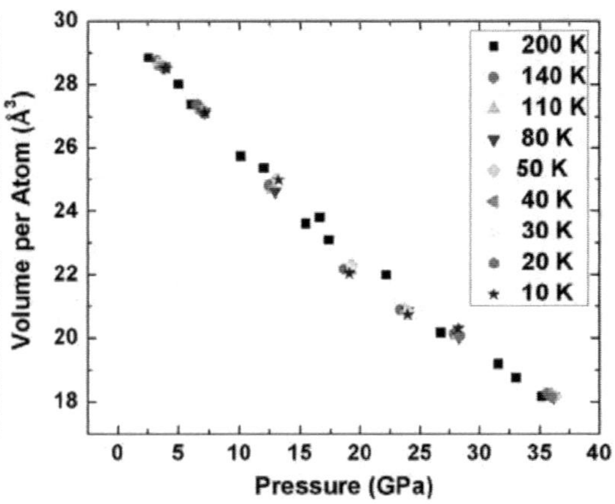

Figure 5. The measured Pressure-Volume (P-V) curves for *Tm* at various temperatures from 10 K to 200 K.

The integrated image plate x-ray diffraction data is shown in figure 4 for the hcp phase at 4 GPa and 7 K as well as for the α-Sm Phase at 36 GPa and 20 K. The wavelength for x-ray beam is λ = 0.4066 Å for both spectra shown in figure 4. The pressure was measured by ruby-fluorescence at low temperatures. The measured Pressure-Volume (P-V) data or equation of state is shown in figure 5 at various temperatures to a pressure of 36 GPa. There are two important conclusions to be drawn from figure 5; the first is that there is no discernible volume change on phase transition from hcp-Sm-type phase transition at all temperatures. This is consistent with the EOS data obtained at ambient temperature. The second important conclusion is that there is no discernible volume contraction measured on cooling at various pressures. The measured EOS at various temperatures between 10 K to 200 K for *Tm* basically overlap with each other indicating that any volume contraction on cooling is basically countered by slight volume increase on magnetic ordering.

4. Conclusions

The four probe electrical resistance measurements have been carried out on heavy rare earth metal Thulium to 33 GPa and 10 K using a designer diamond anvil cell. The electrical resistance measurements are complemented by angle dispersive x-ray diffraction measurements at high pressures and low temperatures using a synchrotron source. The paramagnetic to anti-ferromagnetic transition is readily detected in electrical resistance measurements and the Néel Temperature T_N is observed to decrease gradually in the hcp-phase till the transition pressure of 17 GPa to the α-Sm phase. The transformation to the α-Sm phase results in an abrupt increase in T_N by as much as 100 K. The magnetic ordering temperature for the α-Sm phase is found to be relatively insensitive to pressure between the pressure range of 17 GPa and 33 GPa. The experimentally determined equation of state of Thulium to 36 GPa at various temperatures between 10 K and 200 K show anomalously low thermal expansion coefficient due to the presence of magnetic ordering at low temperatures.

5. References

[1] Samudrala G and Vohra Y 2013, *Handbook on the Physics and Chemistry of Rare Earths* vol 43 ed J G. Bünzli and V K Pecharsky (Amsterdam: Elsevier), p 275.
[2] Jackson D, Malba V, Weir S, Baker P, and Vohra Y 2005 *Phys. Rev.* B **71**184416.
[3] Thomas S, Montgomery J, Tsoi G, Vohra Y, Chesnut G, Weir S, Tulk C, dos Santos A 2013 *High Pressure Research* **33** 555.
[4] Tsoi G, Stemshorn A, Vohra Y, Wu P, Hsu F, Huang Y, Wu M, Yeh K, and Weir S 2009 *J. Phys.: Condens. Matter* **21**, 232201.
[5] Montgomery M, Samudrala G, Tsoi G, and Vohra Y 2011 *J. Phys.: Condens. Matter* **23** 155701.

Acknowledgments

This material is based upon work supported by the Department of Energy-National Nuclear Security Administration under Award Number DE-NA0002014. Portions of this work were performed at HPCAT (Sector 16), Advanced Photon Source (APS), Argonne National Laboratory. HPCAT operations are supported by DOE-NNSA under Award No. DE-NA0001974 and DOE-BES under Award No. DE-FG02-99ER45775, with partial instrumentation funding by NSF.

IOP Conf. Series: Journal of Physics: Conf. Series **950** (2017) 042027 doi:10.1088/1742-6596/950/4/042027

Extreme instability of magnetic Fe-moment in amorphous alloys under pressure

J Kamarád, M Míšek and Z Arnold

Institute of Physics, AS CR, v.v.i., Na Slovance 2, Prague 8, Czech Republic

E-mail: kamarad@fzu.cz

Abstract. The effect of high hydrostatic pressure on magnetization of the selected Fe-, FeNi-, FeCr- and FeW-based amorphous alloys has been studied in pressure range up to 1.2 GPa. The very strong non-linear decrease of magnetization under pressure was observed in all the studied alloys, with the extreme value of $d\ln M/dP$ = - $780*10^{-3}$ GPa^{-1} in the case of the amorphous $Fe_{93}Zr_7$ alloy. Moreover, the spin glass behaviour of the alloy at temperatures below 50 K has been strongly changed by pressure. The received results support the theoretical concepts based on a dominant role of the pressure-induced widening of the 3d-band of transition metals under high pressure.

1. Introduction

The volume instability of magnetic moment of the 3d-elements is a source of all the huge (and useful) magneto-volume anomalies observed in transition metals and alloys, particularly in the Invar alloys. The strong magneto-volume instabilities were verified by the enormous decrease of both, the magnetization and the Curie temperature of Invar-like materials with increasing pressure. The changes of magnetic Fe-moment induced by external pressure has been studied theoretically and experimentally in the Fe-based alloys, intermetallics and amorphous materials for a long time and few models have been proposed in the past to understand it [1-5]. Despite of this, an origin of the anomalies is discussed up to now. [3,4].

Recently we have presented the direct measurement of magnetization of the crystalline $Fe_{64}Ni_{36}$ alloy under high pressure in a diamond anvil cell [6]. A non-linear steep decrease of saturated magnetization of the alloy has been observed at the pressure range from 3 GPa to 4 GPa, where a step-like decrease of Fe-moment was also determined by the selective "local" probes (XES, XMCD, Mössbauer spectra) [7]. As a consequence of both, the structural disorder and the presence of metalloids, magnetism of amorphous transition metal (TM) alloys is more complex than ferromagnetism of crystalline alloys, including the spin glass (SG) state observed at ambient pressure at low temperatures [1,2]. To test effects of the local and the long range atomic order on the Fe-moment instability, we have measured magnetization of selected amorphous Fe-based alloys under hydrostatic pressure up to 1.2 GPa.

2. Experimental

The amorphous Fe-based alloys (Table I) were received in form of ribbons that were prepared by a melt spinning method. Composition of the alloys was determined using EDAX. Magnetization M (per 3d-TM-atom), of the selected alloys was measured under hydrostatic pressure up to 1.2 GPa at magnetic field up to 6 T using the miniature CuBe pressure cell [8] and a SQUID magnetometer. The ribbons were fixed in the pressure cell in direction parallel with magnetic field in SQUID.

3. Results and Discussion

The strongly non-linear pressure dependence of magnetization with a sharp decrease of magnetization at low pressure range has been observed in the most of the studied amorphous alloy, see Fig.1 and Fig.2. The only $Fe_{84}B_{16}$ alloy with the linear pressure dependence of M represents an exception among the studied alloys (see Table I). The extreme pressure induced decrease of magnetization, M(5K,0T), has been observed in the amorphous $Fe_{93}Zr_7$ alloy (see Fig.1), on the other hand, the amorphous alloy $Fe_{47}Ni_{25}Si_{18}B_{10}$ exhibit a very slight sensitivity of magnetization to pressure not only at low temperatures, but, even at temperature 200 K, see Fig.2. All results are summarized in Table I, where for comparison, the pressure data of the crystalline pure Fe and the Invar alloy are presented too.

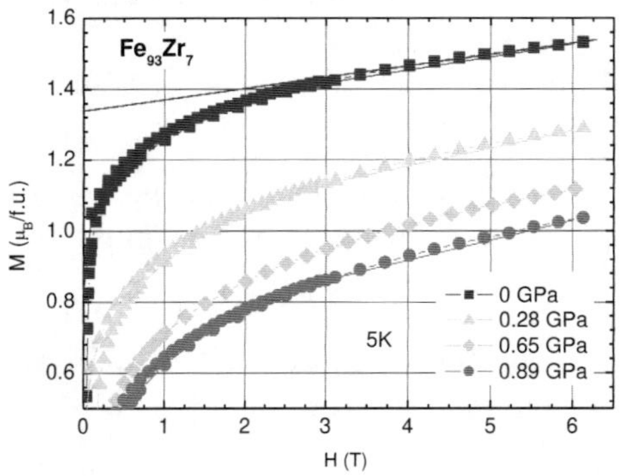

Fig.1: Pressure dependence of magnetization isotherms of the $Fe_{93}Zr_7$ alloy at 5 K.

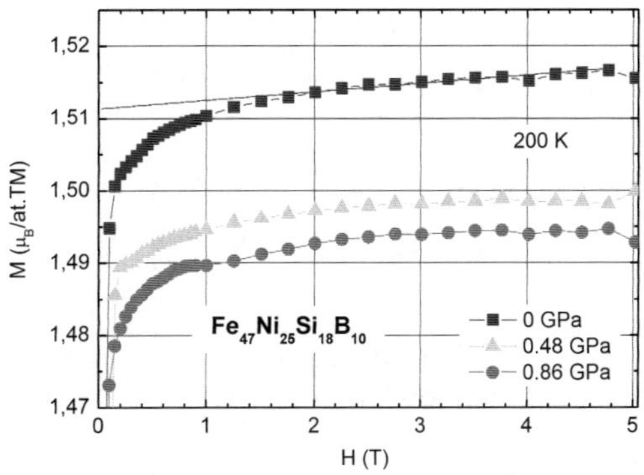

Fig.2: Magnetization isotherms of $Fe_{47}Ni_{29}Si_{18}B_{10}$ alloy under different pressures.

In the case of crystalline TM-based alloys, the high values of d$\ln M$/dP were observed in the FeNi Invar-like alloys with face centered cubic (fcc) crystal structure in contrast to very low value of d$\ln M$/dP in the pure crystalline Fe with body centered cubic (bcc) structure, see Table I. Significantly different densities of states of the bcc-Fe and the fcc-Fe at vicinity of the Fermi energy $N(E_F)$ are considered in theoretical works as an origin of these pressure effects [1,3,5]. The different sensitivity of magnetization to external pressure in the amorphous alloys points to a possible difference in short range order (SRO) of their atoms, i.e., fcc-like and bcc-like SRO, in the $Fe_{93}Zr_7$ and $Fe_{84}B_{16}$ alloys, respectively. The Mössbauer spectra and the X-ray absoption K-edge spectra of the Fe in the amorphous $Fe_{100-x}B_x$ alloys exhibited clearly similar features as the spectra of crystalline bcc-Fe alloys [9]. Using the Maxwell thermodynamic relation, $dM/dP = - dV/dH$, the pronounced decrease of M under pressure can be compared with forced volume magnetostriction $d\omega/dH$, where $\omega = dV/V$, and with high-field susceptibility χ_{hf} related to $d\omega/dH \approx 2\kappa C\chi_{hf}M$, where κ is compressibility and C is magneto-volume coupling constant [10]. The experimental data received for the amorphous $Fe_{92}Zr_8$ and $Fe_{83}B_{17}$ alloys, $d\omega/dH = 385*10^{-10}$ Oe^{-1} and $50*10^{-10}$ Oe^{-1}, respectively [11,12], and data of χ_{hf} presented in Table I, perfectly reflect the observed giant difference in pressure behaviour of magnetization of the amorphous (and crystalline) 3-d transition metals alloys with different SRO and consequently with different electronic structure.

Due to a similarity of SRO in amorphous and crystalline alloys, itinerant electron model of magnetism is used to the description of magnetism of the amorphous 3d-TM alloys. Hence, the known Stoner criterion for the itinerant ferromagnets, $N(E_F)J_f > 1$, can be used to assess the pressure effects in the amorphous alloys. Dominant increase of effective interaction J_f with decreasing interatomic distances should lead to increase of T_C under pressure. On the other hand, a broadening of 3d-band (a decrease of $N(E_F)$ under pressure) leads to a pressure-induced decrease of both, T_C and magnetization, that can be described by Wohlfarth's relations: $dT_C/dP \sim - A/T_C$ and $d\ln M/dP \sim d\ln T_C/dP + \kappa B$, where

IOP Conf. Series: Journal of Physics: Conf. Series **950** (2017) 042027 doi:10.1088/1742-6596/950/4/042027

κ is compressibility and A,B are positive constants [10]. The recent systematic theoretical study of pressure effects on electronic structure and magnetism of 3d-transition metals showed that the dominant

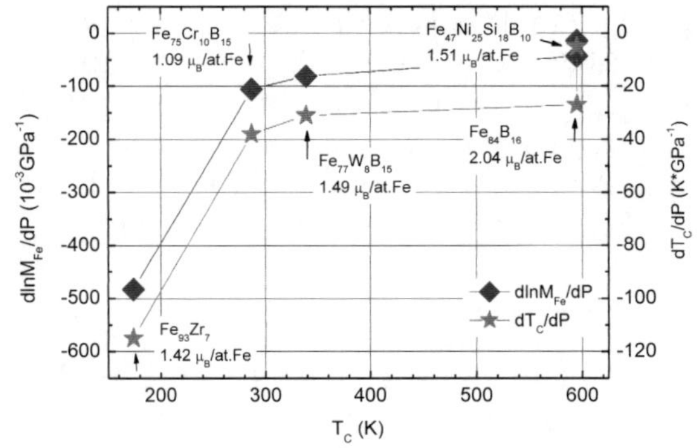

Fig.3: Dependences of dT_C/dP and $d\ln M/dP$ on T_C reflect the above mentioned relations

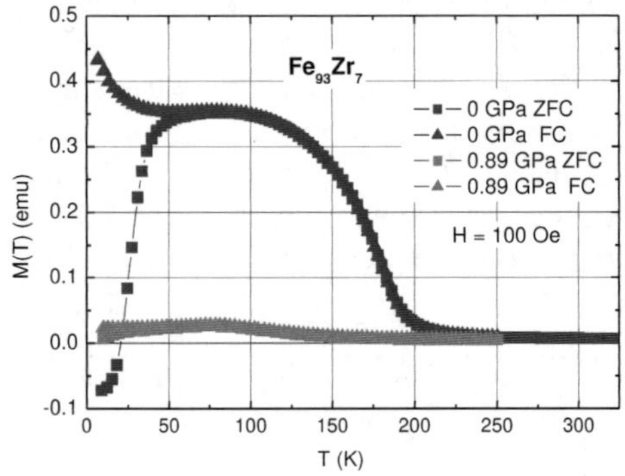

Fig.4: Spin glass behaviour of the $Fe_{93}Zr_7$ alloy is strongly changed by high pressure

Table I: Magnetization M and the Curie temperature T_C of the studied alloys and their pressure derivatives within the low and the high pressure range and the high field susceptibility χ_{hf}.

Composition	M(5K,0T) μ_B/at.TM	d ln$M$(5K)/dP 10^{-3} GPa^{-1} 0÷0.3 GPa	d ln$M$(5K)/dP 10^{-3} GPa^{-1} 0.3÷1 GPa	T_C K	d lnT_C/dP 10^{-3} GPa^{-1}	χ_{hf}(5K) 10^{-5}
Fe – bcc [7]	2.18	- 4.5 ± 0.2		1043	-	4.3
$Fe_{64}Ni_{36}$ – fcc [7]	1.66	- 110 ± 5		520	- 85±4	120
$Fe_{93}Zr_7$ – a	1.42	- 780	- 520 ± 30	175	- 560±50	205
$Fe_{84}B_{16}$ – a	2.04	- 54 ± 2		595	- 45±5	7.2
$Fe_{77}W_8B_{15}$ – a	1.49	- 150	- 103 ± 7	339	- 90±8	30
$Fe_{75}Cr_{10}B_{15}$ – a	1.09	- 163	- 137 ± 5	287	- 135±10	32
$Fe_{47}Ni_{25}Si_{18}B_{10}$ – a	1.51	- 74	- 9 ± 1	595	- 9±1	~ 1

pressure–induced broadening of the 3d-band leads to the suppression of their magnetism, namely in the case of Fe [5]. The results presented in Fig.3 well agree with theoretical concepts. Values of T_C on the x-axis of Fig.3 represent different compositions of the alloys.

The more pronounced decrease of magnetization at low pressure range can be connected with low density (higher compressibility) of amorphous alloys at ambient pressure. Pycnometric measurements showed lower density (of about -6% ÷ -14%) than one determined by Vegard's law. However at low pressure range, detailed data of compressibility of the amorphous alloys are missing.

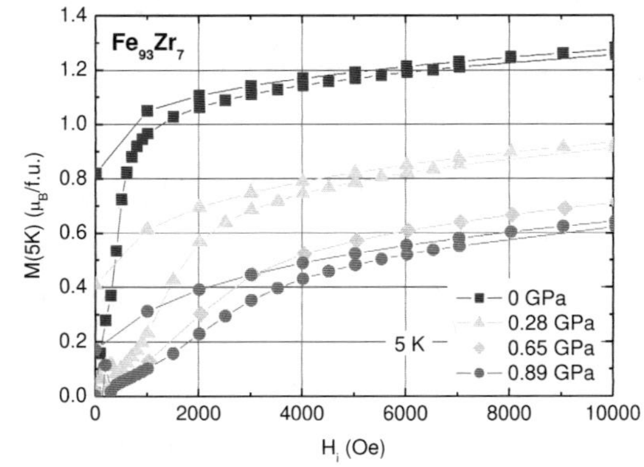

Fig.5: Effect of pressure and magnetic field on SG-ferromagnet transition in $Fe_{93}Zr_7$ alloy

The pronounced difference between ZFC and FC curves of temperature dependence of magnetization of the $Fe_{93}Zr_7$ alloy is presented in Fig.4. The earlier experimental and theoretical studies [2] showed that the amorphous Fe-rich alloys form spin glasses (SG) at lowered temperatures and the SG was described as randomly oriented ferromagnetic clusters. The transition of the alloys from SG to ferromagnetic state with increasing temperature was explained by an enhancement of the short-range ferromagnetic interactions with respect to the long-range antiferromagnetic interactions [2]. The observed effect of pressure and magnetic field on SG in the $Fe_{93}Zr_7$ alloy is compatible with the above mentioned concept as can be seen in Fig.5. In contrast to temperature, pressure stabilizes the random orientation of ferromagnetic clusters and the critical field of the transition from SG to long-range ferromagnetism increases with increasing pressure. The SG-ferromagnetic transition at temperature 5 K is induced by field of 400 Oe at ambient pressure or by field 2500 Oe under pressure 0.9 GPa. However, magnetization of the clusters is strongly suppressed by pressure.

Acknowledgement
This work was supported by the grant No.: 15-03777S of the Grant Agency of CR. M.M. appreciates a care within the AS CR Program of post-docs support.

References
[1] Wasserman E F 1990 *Ferro-magnetic Materials* vol 5, ed K H J Buschow and E P Wohlfarth (New York: Elsevier) p 237 and refs. therein
[2] Kakehashi Y 1991 *Phys. Rev.* B **43** 10820 and refs. therein
[3] Entel P, Hoffmann E, Mohn P, Schwarz K and Moruzzi V L 1993 *Phys. Rev.* B **47** 8706
[4] van Schilfgaarde M, Abrikosov I A and Johansson B 1999 *Nature* **400** 46
[5] Iota V, Klepeis J-H P, Yoo Ch-S, Lang J, Haskel D and Srajer G 2007 *Appl. Phys. Lett.* **90** 042505
[6] Kamarád J, Míšek M and Arnold Z 2014 *High Press. Res.* **34** 365
[7] Nataf L *et al* 2009 *Phys. Rev.* B **80** 134404 and refs. therein
[8] Kamarád J, Machátová Z and Arnold Z 2004 *Rev. Sci. Instrum.* **75** 5022
[9] Taniwaki M and Maeda M 1988 *Materials Science and Engineering* **99** 47
[10] Wohlfarth E P 1969 *J. Phys.* C **2** 68
[11] Tange H, Tanaka Y, Goto M and Fukamichi K 1989 *J. Magn. Magn. Mats.* **81** L243
[12] Tange H, Goto M and Ishio S 1983 *Physica* B **119** 188

IOP Conf. Series: Journal of Physics: Conf. Series **950** (2017) 042028 doi:10.1088/1742-6596/950/4/042028

High-pressure Synthesis and Bulk Modulus of Non-centrosymmetric Superconductor Mo_3Al_2C

C Sekine[1], U Sai[1], J Hayashi[1], Y Kawamura[1] and E Bauer[2]

[1] Muroran Institute of Technology, 27-1 Mizumoto-cho, Muroran 050-8585, Japan
[2] Vienna University of Technology, A-1040 Wien, Austria

E-mail: sekine@mmm.muroran-it.ac.jp

Abstract. Mo_3Al_2C is a superconductor without inversion symmetry (the transition temperature $T_C \sim 9$K). We have succeeded in preparing high-quality samples of Mo_3Al_2C using the high-pressure synthesis technique. The samples were characterized by powder x-ray diffraction (XRD) analysis. Furthermore, powder XRD patterns for the samples with synchrotron radiation have been studied under high pressures up to around 10 GPa and the volume versus pressure curve for the compound has been investigated. A bulk modulus was estimated to be 221 GPa.

1. Introduction

The Mo-based superconductor Mo_3Al_2C crystallizes with the β-Mn type structure of space group $P4_132$ with a transition temperature $T_C \sim 9$K [1, 2]. The structure of the compound does not have a center of inversion. Superconductors without inversion symmetry such as $CePt_3Si$ recently received special attention due to their exotic properties [3]. A mixture of spin-singlet and spin-triplet pairs condensate is proposed to explain superconductivity in the non-centrosymmetric compounds. The absence of Hebbel-Slichter coherence peak, a power-law behavior of spin-lattice relaxation rate, specific heat deviating from BCS model and positive pressure dependence of T_C suggest that Mo_3Al_2C is an unconventional superconductor [2]. However, the sample reported ref. 2 includes Mo_2C as an impurity phase. In order to study the superconductivity of Mo_3Al_2C in detail, high-quality samples without Mo_2C, which is a superconductor with $T_C \sim 3$K [4], are required.

We have succeeded in preparing high-quality samples of Mo_3Al_2C without Mo_2C under high pressure using a multi-anvil type high-pressure press. In this study, we report the results of powder x-ray diffraction (XRD) measurements of Mo_3Al_2C under high pressure to investigate the structural stability of the compound.

2. Experimental

Single-phase polycrystalline samples of Mo_3Al_2C were synthesized at high temperatures and high pressures using a double-stage multi-anvil high-pressure apparatus. Mo_3Al_2C was prepared by reaction of stoichiometric amounts of 3N (99.9% pure) Mo, 3N-Al and 5N-C powders at 4 GPa. The reaction temperature was 1200°C. The sample cell assembly is similar to that used for the high-pressure synthesis of $CeRu_4As_{12}$ [5]. The prepared samples were characterized by powder x-ray diffraction (XRD) using Co $K\alpha_1$ radiation and silicon as a standard.

Using synchrotron radiation, powder XRD patterns of Mo_3Al_2C were measured with a diamond-anvil cell (DAC) and an imaging plate at room temperature and high pressures [6]. High-pressure

XRD experiments with $\lambda = 0.619$ Å were performed at the beam line 18C of the High Energy Accelerator Research Organization (KEK) Photon Factory in Tsukuba. We employed the DAC with diamond culet diameters of 500 μm. The sample was finely ground and loaded in the 100 μm diameter hole drilled in a stainless steel (T301) gasket. The pressure in the DAC was determined by a pressure shifts in the R-line fluorescence spectrum of ruby. A 4:1 methanol-ethanol solution was used as the pressure medium.

3. Results and discussion

Figure 1 shows a powder XRD pattern of Mo_3Al_2C prepared under high pressure. Most of the observed diffraction peaks were indexable using the β-Mn type structure. We were unable to detect other impurity phases including Mo_2C at the experimental resolution. The lattice constant determined by a least-squares fit to the data was 6.8623 Å, in reasonable agreement with 6.8630 Å as reported previously [2]. Specific heat and magnetic susceptibility measurements of Mo_3Al_2C prepared under high pressure reveal superconductivity of the samples at around 9K, in agreement with the data reported earlier [2].

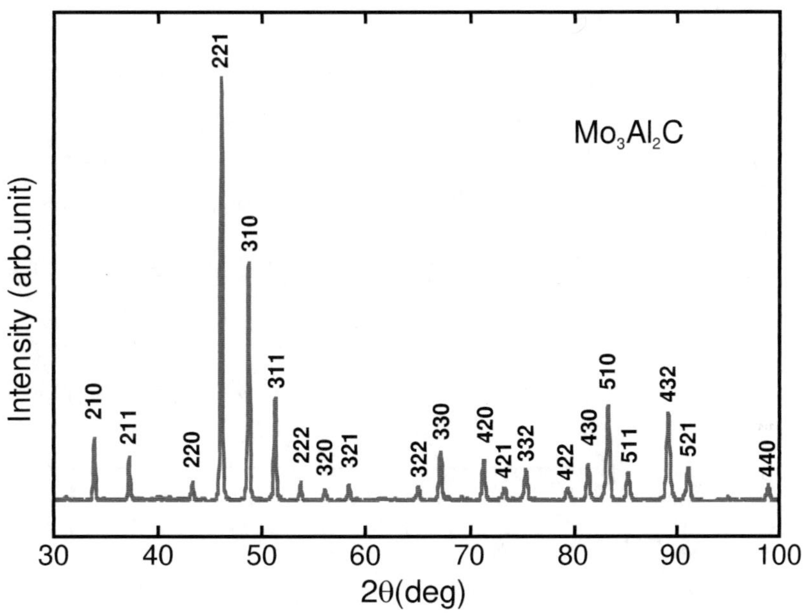

Figure 1. Powder x-ray diffraction pattern of Mo_3Al_2C prepared under high pressure

Using a synchrotron radiation, the powder XRD of Mo_3Al_2C has been studied under high pressures up to 9.74 GPa with the diamond anvil cell at room temperature. The d-values of the diffraction peaks decrease with increasing pressure. The lattice constant of Mo_3Al_2C decreases to 6.7270 Å at 9.74 GPa. No obvious change, such as the appearance of any new diffraction peaks, was observed. The result indicates that there is no change in crystal structure type with in this pressure range. When pressure is reduced from 9.74 GPa to the ambient pressure, the d-values of the diffraction peaks return to those at ambient pressure. This behavior is completely reversible.

Figure 3 shows the relative cell volume (V/V_0) versus pressure curve for Mo_3Al_2C. Here, V_0 is the unit cell volume at ambient pressure. The cell volume of Mo_3Al_2C monotonically decreases with increasing pressure up to around 10 GPa. This experimental data can be fitted by a Birch equation of state [7]:

IOP Conf. Series: Journal of Physics: Conf. Series **950** (2017) 042028 doi:10.1088/1742-6596/950/4/042028

$$P = \frac{3}{2}B_0\left\{\left(\frac{V}{V_0}\right)^{-\frac{7}{3}} - \left(\frac{V}{V_0}\right)^{-\frac{5}{3}}\right\}\cdot\left[1 - \frac{3}{4}(4 - B_0')\left\{\left(\frac{V}{V_0}\right)^{-\frac{2}{3}} - 1\right\}\right] \tag{1}$$

where B_0 is the bulk modulus, B_0' its first pressure derivative, V the volume, and P the pressure. A least-square fit to the data of Mo_3Al_2C gives the following values: $B_0 = 221$ GPa and $B_0' = 2.3$.

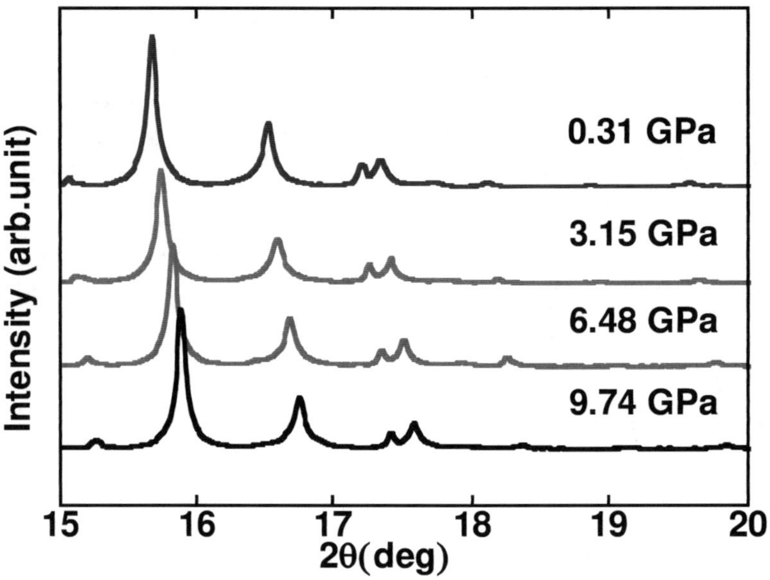

Figure 2. Powder x-ray diffraction patterns of Mo_3Al_2C at each high pressure

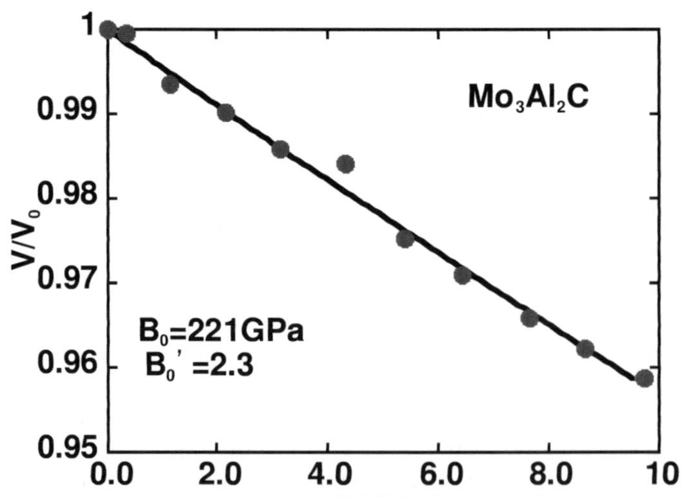

Figure 3. Pressure dependence of the relative unit cell volume V/V_0 of Mo_3Al_2C in the pressure range up to around 10 GPa. The line shows the compression curve of the bulk modulus B_0=221 GPa, which was obtained by fitting of a Birch equation of state to data.

4. Summary

We have synthesized high-quality samples of Mo_3Al_2C without Mo_2C using the high-pressure synthesis technique. The crystal structure of Mo_3Al_2C is found to be stable at high pressures up to 9.74 GPa. The bulk modulus of Mo_3Al_2C has been estimated from the XRD data under high pressure. The value of B_0 for the compound is 221 GPa. In order to investigate the superconducting properties of Mo_3Al_2C in detail, photoelectron spectroscopy and NMR measurements are currently in progress.

Acknowledgments
This work was performed under the approval of the Photon Factory Program Advisory Committee (Proposal No. 2013G211, 2015G098). This work was supported by JSPS KAKENHI Grant Number 23340092.

References

[1] Toth L E and Zbasnik J 1968 *Acta Metall.* **16** 1177
[2] Bauer E, Rogl G, Chen X-Q, Khan R T, Michor H, Hilscher G, Royanian E, Kumagai K, Li D Z, Li Y Y, Podloucky R and Rogl P 2010 *Phys. Rev.* B **82** 064511
[3] Bauer E, Hilsher G, Michor M, Paul C, Scheidt C E W, Gribanov A, Seropegin Y, Noel H, Sigrist M and Rogl P 2004 *Phys. Rev. Lett.* **92** 027003
[4] Matthias B T and Hulm J K 1952 *Phys. Rev.* **87** 799
[5] Sekine C, Kawata T, Kawamura Y and Yagi T 2013 *J. Korean Phys. Soc.* **63** 359
[6] Shirotani I, Noro T, Hayashi J, Sekine C, Giri R and Kikegawa T 2004 *J. Phys. Condesns. Matter* **16** 7853
[7] Birch F 1947 *Phys. Rev.* **71** 809

IOP Conf. Series: Journal of Physics: Conf. Series **950** (2017) 042029 doi:10.1088/1742-6596/950/4/042029

Stress computation and optimizing the shape of a miniature opposed anvil through finite element analysis

Nobuya Kishii[1], Shota Tateno[1] and Masashi Ohashi[1,2]

[1]Graduate School of Natural Science and Technology, Kanazawa University, Kakuma-machi, Kanazawa, 920-1192, Japan

[2]Institute of Science and Engineering, Kanazawa University, Kakuma-machi, Kanazawa, 920-1192, Japan

E-mail: ohashi@se.kanazawa-u.ac.jp

Abstract. We have demonstrated stress computation of high-pressure apparatus to obtain effective pressure with a miniature opposed anvil cell. The anvils and the gasket were assumed to be made of tungsten carbide (WC) and of stainless steel (SUS 304), respectively. The pressure medium was presumed to be liquid to generate hydrostatic pressure. Considering cell thickness, we set the anvil radius to approximately 6 mmϕ. In order to keep a large space with a diameter of 1 mmϕ, the diameter of 3 mmϕ was fixed on the tip of anvils. Using the stresses found in the analysis, we evaluated the anvil in terms of pressure efficiency and fragility. Calculating pressure in the sample space between anvils and the maximum differential stress of an anvil, we found that the best pressurization efficiency and lowest fragility was achieved with an anvil structure with an angle θ of 15 degrees and height h of 5 mm.

1. Introduction

High-pressure technique has been recognized as a powerful tools to investigate the physical properties of solids because pressure can distinctly vary physical parameters due to the contraction of its lattice constants. There is the possibility that other property changes may occur, such as structural phase transition and magnetic transition under high presure[1, 2, 3, 4]. To perform such measurements under high-pressure condition, there is a need to combine a high-pressure cell with other devices that allow modifying parameters such as temperature and magnetic field in addition to pressure in order to explore material properties. However, the size of high-pressure generators is generally limited to combine and use with other instruments. Therefore, it is important to develop space efficient high pressure apparatus to make various experimental techniques accessible to high pressure.

Many trials have been reported using Bridgman anvil cells, diamond anvil cells, and so on[5, 6], where smaller anvil surface areas allow high pressure to be obtained with smaller loads. In general, the pressurizing efficiency of the device is decided based on a variety of factors, such as the compressive strength and tensile strength of the shape of each part. In order to miniaturize the device, the anvil is thinned and the cell thickness is reduced, making this even more difficult. Our aim is to develop a small high pressure cell by optimizing the shapes of the anvil and cells from stress analysis.

2. Basic Construction of the Apparatus

We suppose that the device is being combined with a commercial low-temperature magnetic field generator and set the cell diameter to 20 mmϕ. The basic construction of the anvil was nearly the same as in our previous report[6], as shown in Figure 1. High-pressure generators generally have many large parts, but we attempted to reduce the number of parts in this apparatus as much as possible and simplify the structure of the parts so that pressure can be easily applied. Pressure is generated in a gasket between the anvils made of tungsten carbide (WC). The gasket is made of stainless steel (SUS 304) to maintain pressure. A hole with an internal diameter of 1mm is made and filled with the material as well as a liquid pressure transmitting medium to increase hydrostatic pressure. The cell body is made of hardened beryllium copper (BeCu). Both ends are closed by screws. The load is applied to the top of upper anvil and the force is blocked by tightening the upper screw of the cell. Considering cell thickness, we set the anvil radius to approximately $R = 6$ mmϕ. In order to secure a test material space with a diameter of 1mmϕ, the diameter of the upper surface of the anvil was fixed at 3 mmϕ.

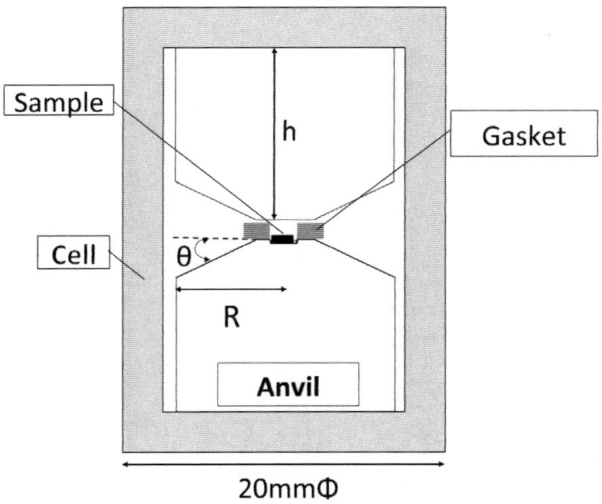

Figure 1. A cross section of the high pressure cell. To simplify the figure, we omitted the screws to fix the anvils after the load is applied.

3. Stress Analysis Thorough The Finite Element Method

A finite element method (FEM) is a powerful tool to investigate irregular and/or complicated systems. In order to examine the conditions that efficiently produced high pressure gives the above restrictions, we demonstrated stress computation of a part of high-pressure cell between the anvils. The model was implemented using COMSOL Multiphysics 3.4, a commercial FEM solver, and the Geoscience Research Laboratory's program Easy-Sigma 2D Lite[7].

In order to simplify the analysis, we made the anvils with top/bottom symmetry and bound the center of the anvil to conduct analysis only on the bottom half. Also, taking account of the anvil with vertical-axis symmetry, we performed two-dimensional analysis on one side of the anvil cross-section. We designed a variety of anvils with different parameters for anvil angle θ and height h, and applied loads F up to 10 ton from the upper anvil in the downward direction. The material properties implemented in our simulations are taken from the literature values for WC and SUS 304. A Young's modulus $E = 580$ GPa and a Poisson's ratio $\nu = 0.22$ were used for the properties of WC, and $E = 193$ GPa and $\nu = 0.3$ for those of SUS 304[8]. Although a liquid has generally a Poisson's ratio $\nu = 0.5$, $\nu = 0.4$ was used for the pressure transmitting medium because of the system limitations of the FEM program. $E = 43$ GPa is estimated from ν and a bulk modulus B. Since B of pressure transmitting medium depends on pressure, the value of a water at 5 GPa[9] was used in the present work.

IOP Conf. Series: Journal of Physics: Conf. Series **950** (2017) 042029 doi:10.1088/1742-6596/950/4/042029

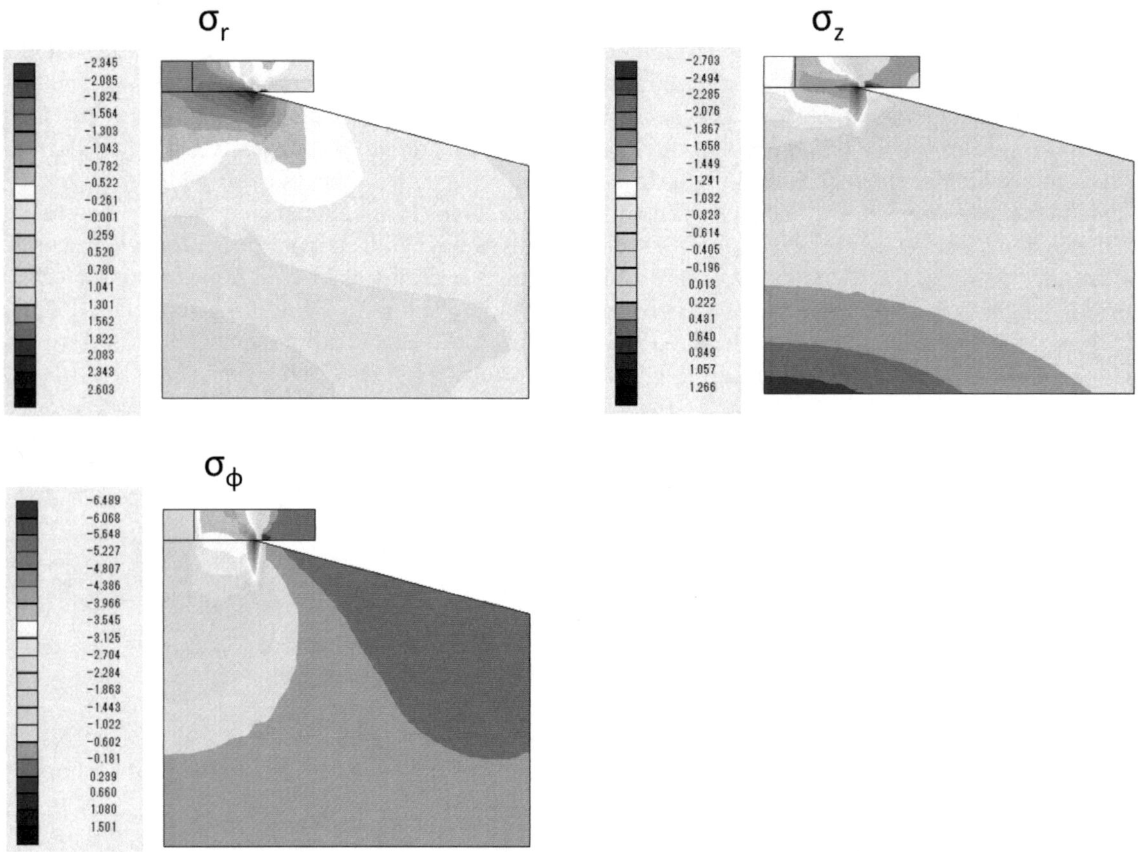

Figure 2. An example of stress distribution of a cross section of a part of high pressure cell. Stresses $\sigma_{\mathrm{r}}, \sigma_{\mathrm{z}}$ and σ_{ϕ} on cylindrical coordinates are calculated through FEM analysis when the load $F = 3$ ton was applied to an anvil of $\theta = 15°$ and $h = 5$ mm. The unit of stress is GPa.

4. Results and discussion

We calculated stresses when a radial direction **r** from the central axis of the high pressure cell, that of central axis **z**, and that perpendicular to **r** and **z**, ϕ as coordinate axes of a cylindrical coordinate system. Figure 2 shows an example of a cross section. Here, the load $F = 3$ ton was applied to an anvil of $\theta = 15°$ and $h = 5$ mm. Each stresses $\sigma_{\mathrm{r}}, \sigma_{\mathrm{z}}$ and σ_{ϕ} is described corresponding to each axes, respectively. It is clear that the magnitudes of every stresses $\sigma_{\mathrm{r}}, \sigma_{\phi}$ and σ_{z} are the highest at the edge on a tip of anvil where an anvil touches a gasket. It is related to the fact that an edge of an anvil is firstly broken in most cases that the load is actually applied.

At the part of a pressure transmitting medium, on the other hand, all values of $\sigma_{\mathrm{r}}, \sigma_{\phi}$ and σ_{z} are almost same within ± 0.1 GPa. It indicates that the sample can be compressed by the pressurization of a hydrostatic pressure in the pressure transmitting medium. Here, we estimated the pressure P in sample space as a deviatoric stress which is an average of $\sigma_{\mathrm{r}}, \sigma_{\phi}$ and σ_{z}. In the case of Figure 2, the magnitude of $P = 1.2$ GPa is much smaller than that in the experimental result, where $P = 4$ GPa is developed by applying the load $F = 3$ ton[6]. It may come from the fact that the parameters used in the FEM analysis are different from the actual value because a 1:1 mix of Fluorinert FC70 and FC77 is used as a pressure transmitting medium in the previous report. Moreover, both E and ν depend on pressure, and a liquid pressure transmitting medium

IOP Conf. Series: Journal of Physics: Conf. Series **950** (2017) 042029 doi:10.1088/1742-6596/950/4/042029

freeze solid and a bulk modulus increases sharply as increasing pressure. To calculate P using exact values of the parameters in the FEM analysis, the measurement of the compression curve of a pressure transmitting medium is planed in future.

Using the stresses found in the analysis, we evaluate the anvil in terms of pressure efficiency and fragility. First, we examine how changes to θ and h affect P in a pressure transmitting medium. Next, to evaluate the fragility, we defined a differential stress Y, which is the difference of the anvil between a maximum main stress σ_z and a minimum main stress σ_r. As Y increases, the anvil becomes more fragile. In all high pressure cells we calculated, the maximum differential stress Y_{max} is the value at the edge on the tip of the anvil. Since it is consistent with the fact that an edge of an anvil is firstly broken in most cases that the load is actually applied, it is reasonable to evaluate the fragility of the anvil by examining how changes to θ and h affect Y_{max}.

Figures 3 and 4 show P and Y_{max} as a function of applied load F, respectively. One condition for breaking the material is when Y_{max} exceeds the compressive strength of the material. For example, when the anvils are made by submicron cemented carbide grades (Sandvik Co. Ltd.) where the compressive strength is about 8 GPa, the pressure cell can be applied the load up to about 10 tons. Both P and Y_{max} increases proportionally as increasing F, while pressure calibration curve is not actually quite simple in a commercial Bridgman anvil cells[6]. Although it comes from the fact that we simplified the structure of a high pressure cell for the FEM analysis, it is possible to evaluate the best pressurization efficiency and lowest fragility at least qualitatively.

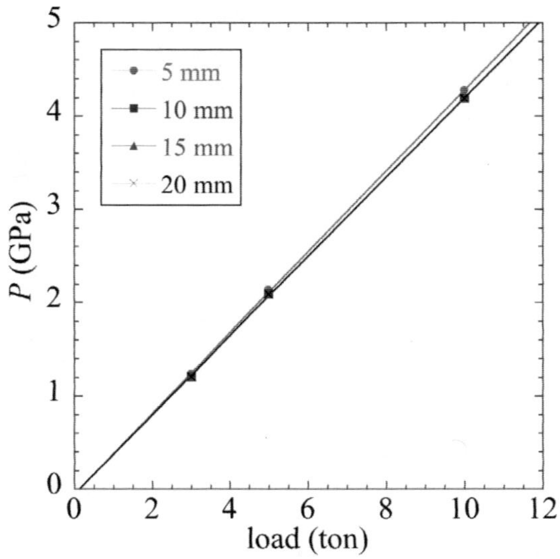

Figure 3. Pressure P of a sample space as a function of the load F (θ=15°).

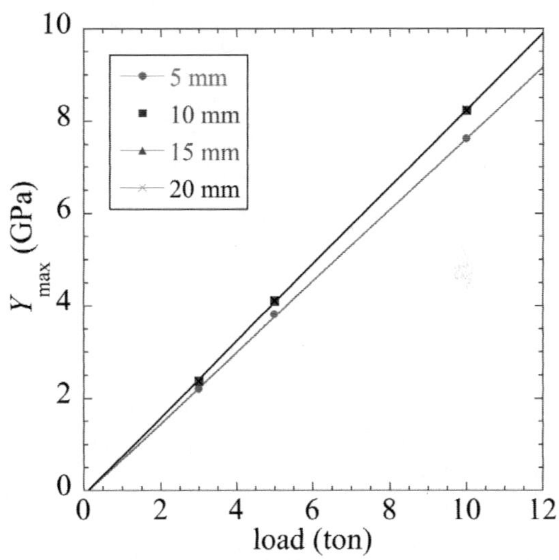

Figure 4. The maximum differential stress Y_{max} as a function of the load F (θ=15°).

Figure 5 shows pressure P of a sample space as a function of the angle θ of the anvils when a 10 ton load is applied. P reaches a maximum when θ is 15 degrees. This is due to the fact that at the angle increases between 5 to 15 degrees, stress is concentrated on the tip, raising P. Above 15 degrees, more stress escapes in the horizontal direction and P decreases. On the other hand, as shown in Figure 6, P tends to be enhanced below h = 10 mm for all angles θ. These results may come from the fact that vertical direction stress σ_z reaches its extreme.

Figure 7 shows the maximum differential stress Y_{max} as a function of angle θ of the anvils when a 10 ton load is applied. Y_{max} decreases monotonically as θ decreases, and tends to saturate

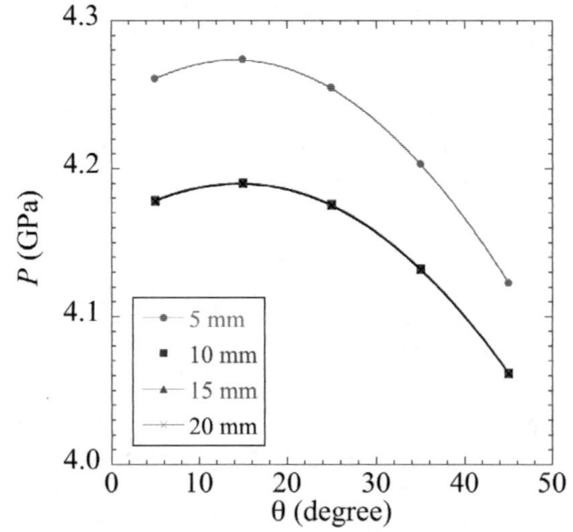

Figure 5. Pressure P of a sample space as a function of angle θ of the anvils when a 10 ton load is applied.

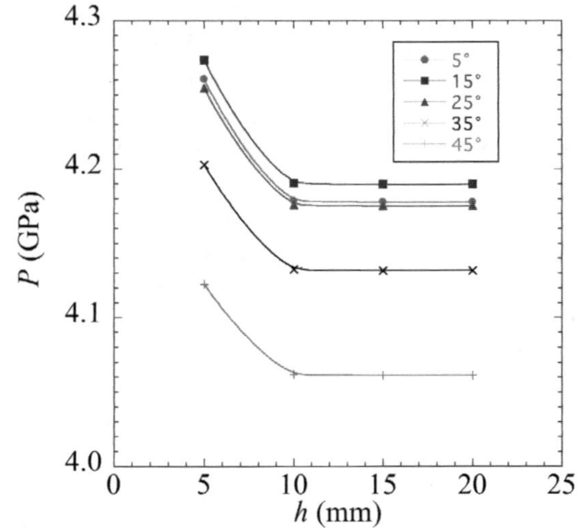

Figure 6. Pressure P of a sample space as a function of height h of the anvils when a 10 ton load is applied.

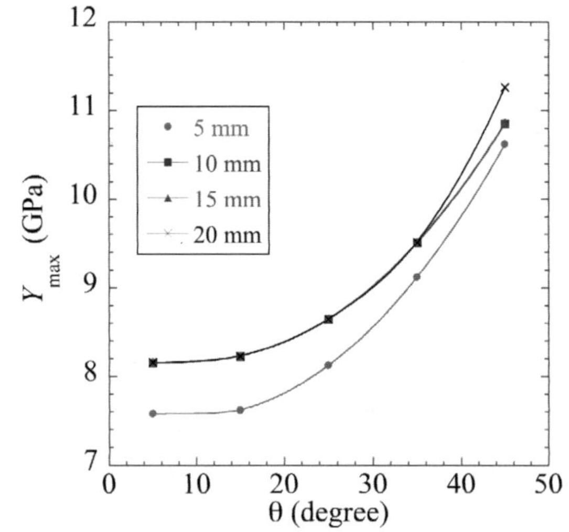

Figure 7. The maximum differential stress Y_{max} as a function of angle θ of the anvils when a 10 ton load is applied.

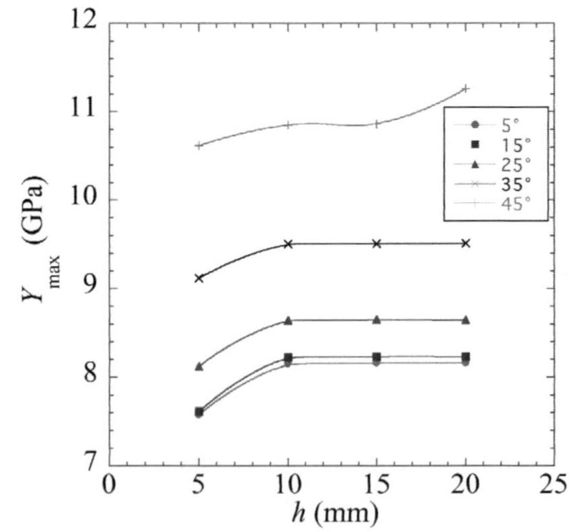

Figure 8. The maximum differential stress Y_{max} as a function of height h of the anvils when a 10 ton load is applied.

below 15 degrees. On the other hand, as shown in Figure 8, Y_{max} tends to be suppressed below $h = 10$ mm for all angles θ. These results indicate that it is hard to break when the low anvils were used in the pressure cell.

From these results about P and Y, we consider the optimal anvil structure using calculation results. First, because P was highest at 15 degrees, a value of 15 degrees is suitable for θ. While smaller values of θ are better for Y_{max}, Figure 7 shows that there is no significant difference in values between 5 and 15 degrees. Thus, we prioritize pressurization efficiency and set the optimal angle at 15 degrees. Next, we found that P was largest and Y_{max} was smallest when h

= 5 mm. This means that pressurization efficiency was suitable and the anvil was less fragile, so we set the optimal value of h at 5 mm.

5. Summary

In the present work, a FEM analysis was performed to optimize the shape of an anvil of opposed anvil cell. We found that the best pressurization efficiency and lowest fragility was achieved with an anvil structure with an angle θ of 15 degrees and height h of 5 mm. In future, we will conduct stress analysis of the overall pressure cell, including the anvil, used to maintain pressure within the sample space. Specifically, we must examine the optimal shape of the overall cell to examine the optimal anvil diameter, cell thickness for maintaining high pressure and the necessary overall cell height.

Acknowledgments

This work was supported in part by Grants-in-Aid from the Ministry of Education, Culture, Sports, Science and Technology of Japan and Suzuki Foundation.

References

[1] Uwatoko Y, Umehara I, Ohashi M, Nakano T and Oomi I 2012 *Handbook on the Physics and Chemistry of Rare Earths* ed. by K. A. Gschneidner Jr. and L. Eyring (North-Holland Amsterdam, 2012) chapter 252 p 1
[2] Ohashi M, Oomi G and Satoh I 2007 *J. Phys. Soc. Jpn.* **76** 114712.
[3] Ohashi M, Tashiro A, Oomi G, Maeda S and Zheng X G 2006 *Phys. Rev. B* **73** 134421.
[4] Ohashi M, Oomi G, Koiwai S, Hedo M and Uwatoko Y 2003 *Phys. Rev. B* **68** 144428.
[5] Eremets M I 1996 *High pressure experimental methods* (Oxford University Press)
[6] Ohashi M and Oomi G 2009 *Jpn. J. Appl. Phys.* vol 48 p 070221
[7] http://www.easy-sigma.jp (in japanese)
[8] National Institutes of Natural Science (NIMS), National Astronomical Observatory of Japan 2015 *Chronological science tables* (Tokyo: Maruzen publishing co., Ltd.)
[9] Bridgman P W 1942 *Proc. Am. Acad., Arts Sci.* vol 74 p399

IOP Conf. Series: Journal of Physics: Conf. Series **950** (2017) 042030 doi:10.1088/1742-6596/950/4/042030

Anomalous compression behavior in a C15 Laves compound $CeAl_2$ under high pressure

Shota Tateno[1], Nobuya Kishii[1], Masashi Ohashi[1,2,3], Hidenori Miyagawa[3], Gendo Oomi[3,4], Isamu Satoh[5], Nobuyoshi Miyajima[6] and Takehiko Yagi[6,7]

[1]Graduate School of Natural Science and Technology, Kanazawa University, Kakuma-machi, Kanazawa, 920-1192, Japan
[2]Institute of Science and Engineering, Kanazawa University, Kakuma-machi, Kanazawa, 920-1192, Japan
[3]Department of Physics, Kyushu University, 6-10-1 Hakozaki, Fukuoka 812-8581, Japan
[4]Kurume Institute of Technology, Kamitsu-machi, Kurume, Fukuoka 830-0052, Japan
[5]Institute for Materials Research, Tohoku University, 2-1-1 Katahira, Aoba-ku, Sendai, 980-8577, Japan
[6]Institute of Solid State Physics, the University of Tokyo, 5-1-5 Kashiwanoha, Kashiwa, 277-8581, Japan
[7]Geochemical Research Center, the University of Tokyo, 7-3-1 Hongo, Tokyo, 113-0033, Japan

E-mail: ohashi@se.kanazawa-u.ac.jp

Abstract. Although there have been quite a number of discrepancies in the compression curves of $CeAl_2$, such results come from the lattice compression at non-hydrostatic condition by solidification of pressure transmitting medium. We have carried out the powder X-ray diffraction measurement of $CeAl_2$ under hydrostatic pressure with helium gas as pressure-transmitting medium which achieves the best hydrostatic conditions. Although no splitting of the peak is observed in the present diffraction pattern, it is found that a full width at half maximum is enhanced above 20 GPa for several Bragg reflections. Such behaviors suggest the existence of a new pressure-induced structural phase transition which is not observed in the previous reports under non-hydrostatic pressure.

1. Introduction

RAl_2 (R : rare earth) intermetallic compounds crystallize in the $MgCu_2$ type structure which belongs to the C15 Laves phase (space group $Fd\bar{3}m$, No. 227), where the rare earth ions have equivalent sites of cubic point symmetry[1, 2, 3]. Considering the Al^{3+} ions as magnetically neutral, the configuration of the magnetic lattice consisting of rare earth ions is a diamond structure type, and the nearest neighbours of rare earth ions are Al atoms. The lattice constant of RAl_2 decreases with increasing f-electron number except for $EuAl_2$ and $YbAl_2$. This decrease is so-called lanthanoid contraction. Although the valence of rare earth ion is trivalent fundamentally, the valence of Eu and Yb ions in $EuAl_2$ and $YbAl_2$ is divalent.

$CeAl_2$ also crystallizes in the $MgCu_2$ type, and has been well known as a prototype of heavy fermion Kondo compound[4, 5, 6]. These characteristic properties are mainly originated from the existence of unstable $4f$ electrons and the hybridization between local f electrons and conduction band electrons. Since the magnitude of hybridization is affected strongly by an application of external pressure and a lot of interesting phase transition is expected to induce, the investigation of these compounds under application of external forces like high pressure has been carried out extensively by many authors[7, 8, 9].

It is well known that some heavy-fermion compounds show a discontinuous volume change due to valence transition in compression curve[10, 11]. As for $CeAl_2$, on the other hand, there have been quite a number of discrepancies in the compression curves. A discontinuous change in the pressure dependence of lattice constant at room temperature was reported around 6 GPa by some authors[12, 13]. But other authors did not find it below 20 GPa[14, 15]. Chandra *et al.* supported the latter group from their calculation about changes in electronic structure, but predicted that it could show transformation to either a $MgNi_2$ or a $MgZn_2$ type structure, which is hexagonal but again belongs to the Laves phase[16].

The origin of these discrepancies seems to be a complicated issue, but it may originate mainly from the lattice compression at non-hydrostatic condition by solidification of pressure transmitting medium. Takemura investigated in Zn that the anomaly on the compression curve is induced by the non-hydrostaticity associated with the solidification of the transmitting medium [17]. In this paper, we report the high-pressure powder X-ray diffraction measurement of $CeAl_2$ with helium pressure-transmitting medium which achieves the best hydrostatic conditions, to discuss about the effect of the hydrostaticity from the compression curve and from the half width of each X-ray diffraction peak.

2. Experimental

A single crystal of $CeAl_2$ was grown by using Czochralski pulling method with a tetra-arc furnace in argon gas atmosphere. The crystal axes were determined for all single crystals by X-ray back Laue method. For the X-ray diffraction measurements, the crystal was crushed and powdered using a mortar and pestle. The sample was found to be of single phase and the lattice parameter of cubic C15 Laves phase (space group $Fd\bar{3}m$, No. 227) was determined to be $a = 8.055 \pm 0.002$ Å at ambient pressure. It is in good agreement with a previous report [4].

Powder X-ray diffraction experiments were carried out under high pressure by means of diamond-anvil cell having a cut face of 0.45 mm in diameter. This diamond anvil type pressure apparatus is an improved Mao-Bell's lever type. SUS 304 is used as the gasket which is the pressure transmitting disk. In order to prevent to blow out the transmitting medium, at first pre-press was carried out up to the half value of target pressure, and then drilled a hole (0.25mmϕ) in the center of the pattern. The powdered sample was put in the hole and a little piece of ruby to monitor the pressure. The helium gas was used as pressure transmitting medium, which has the advantage that hydrostatic pressure nature is very high also under high pressure comparing with liquid pressure transmitting medium. At first, powder sample and pressure marker put into the hole of gasket, and then the gasket was pressed by upper and lower-anvil. X-ray diffraction measurement was made at room temperature using Mo Kα radiation from a 5.4 kW Rigaku rotating anode generator equipped with a 100 μm collimator. An image plate was used as the detector. The d-values were obtained by recording the diffraction.

3. Results

Figure 1 shows the diffraction patterns of $CeAl_2$ at several pressures up to 23.3 GPa. Indices of some of the prominent peaks are also shown. A broad peak is observed at $2\theta = 20 \sim 22°$ the diffraction patterns at every pressures. Such behavior comes from the diffraction of the gasket of SUS 304. It is noted that all of the diffraction peaks are sharp compared with the previous results where a 4:1 methanol: ethanol mixture (ME) is used as the pressure-transmitting medium[18]. It means that good hydrostaticity is retained compared with that by using ME. The diffraction patterns does not change and all the diffraction peaks shifted to higher 2θ as

IOP Conf. Series: Journal of Physics: Conf. Series **950** (2017) 042030 doi:10.1088/1742-6596/950/4/042030

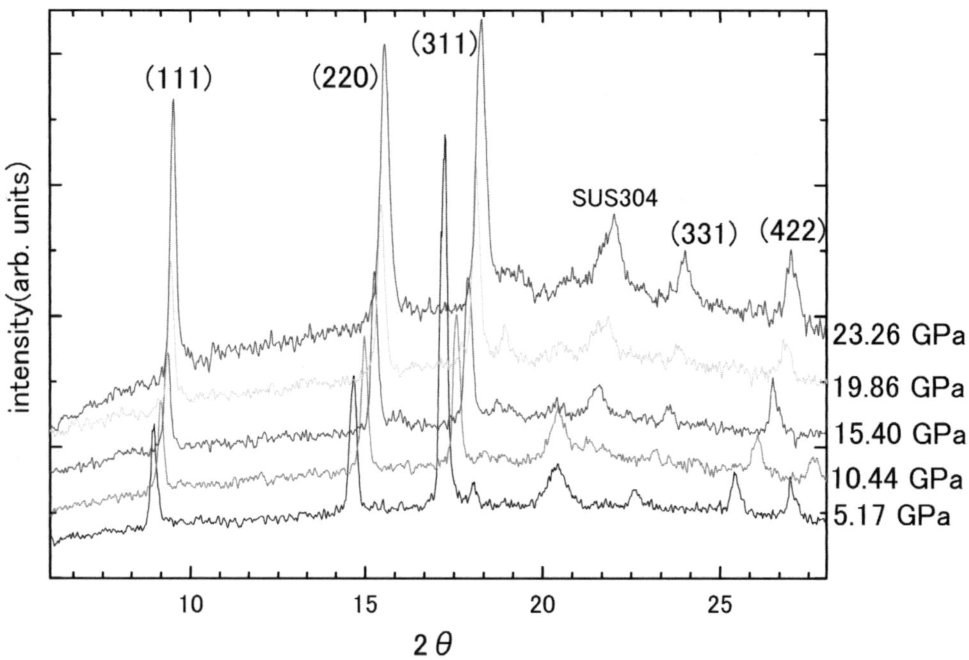

Figure 1. The powder X-ray diffraction patterns of $CeAl_2$ at room temperature under high pressures using helium gas as the pressure-transmitting medium. Indices of some of the prominent peaks are also shown.

increasing pressure. No splitting of the peak is observed within experimental error. The d-values of each Bragg reflections are obtained from the diffraction patterns.

4. Discussion

Assuming that the cubic structure is stable, the lattice constant a is determined mainly from the d-values corresponding to the reflections (111), (200), (311), (331) and (422). Figure 2 shows the volume of $CeAl_2$ as a function of pressure, where the volume is obtained to be $V = a^3$. The figure compares two experimental runs, one with helium gas as the pressure-transmitting medium in the present study and the other with ME in the previous study[18]. It is found that V decreases continuously with increasing pressure below 17 GPa in both compression curves, and that both compression curves are almost coincident with each other. It indicates that the non-hydrostaticity does not cause the uniaxial stress of the sample with keeping cubic symmetry within experimental error. Here, the bulk modulus was estimated by assuming the Murnaghan's equation of state[19], which is described as follows,

$$\frac{V}{V_0} = \left(\left(\frac{B_0'}{B_0}\right) \times P + 1\right)^{\frac{-1}{B_0'}} \tag{1}$$

, where B_0 and B_0' are the bulk modulus and its pressure derivative at ambient pressure, respectively, and V_0 is the volume at high and ambient pressure. By least square fitting of the experimental data in Figure 2 to Eq. (1) below 15 GPa, B_0 and B_0' were obtained to be 62 GPa and 3.0 in the sample with helium gas as the pressure-transmitting medium, respectively. It is comparable to the values $B_0 = 63$ GPa and $B_0' = 3.0$ in another results with ME.

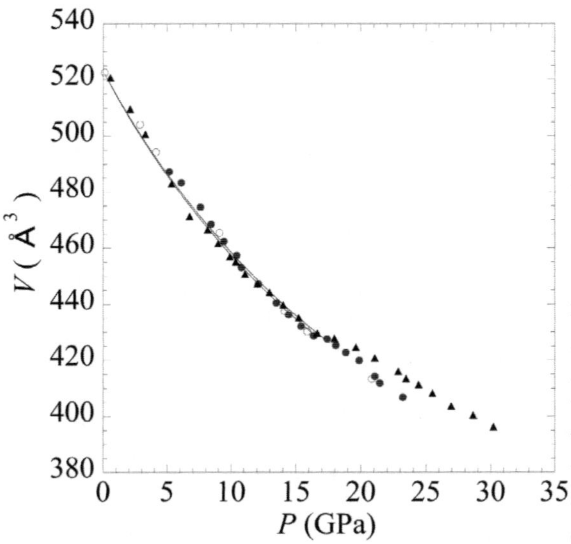

Figure 2. The volume V of $CeAl_2$ as a function of pressure. The solid and open circles show the data on increasing and decreasing pressure, respectively, in which helium gas is used as the pressure-transmitting medium. The solid triangles show the data on increasing pressure, in which ME is used as the pressure-transmitting medium[18]. Solid curves show the results of fitting to equation (1)

On the other hand an anomaly exists in the compression curves under high pressure, where the slope of the compression curve changes above 20 GPa. Such anomaly is obviously observed not only in the compression curve by using ME as the pressure-transmitting medium, but also by using helium gas as the best hydrostatic conditions. It means that the anomaly does not come from the nonhydrostaticity but from physical origins such as the pressure induced phase transition. Furthermore, the compression curve tends to deviate from another one above 20 GPa. Taking account that crystals which has low structural symmetry are generally sensitive to uniaxial stress, such behavior may come from the reduction in the structural symmetry under pressure.

To make it clear, we estimated a full width at half maximum (FWHM) of Bragg peaks of each diffraction pattern of $CeAl_2$. Figure 3 shows the magnitude of FWHM as the function of pressure for several Bragg reflections taken from two experimental runs. For the one with ME as the pressure-transmitting medium shown in Fig. 3(a), FWHM is around 0.2° at low pressure but tend to increase at high pressure above 10 GPa. Such behavior is consistent with the previous report that the nonhydrostatify is induced by the soldification of ME pressure medium above 10 GPa[20].

With helium gas as the pressure-transmitting medium shown in Fig. 3(b), on the other hand, not only the magnitude of FWHM but also the dispersion of FWHM is much smaller than those with ME. Although helium solidifies at 11.5 GPa at room temperature[21], FWHM with helium gas remains nearly constant at least 20 GPa for all reflections in Fig. 3(b). Such behavior indicates that good hydrostaticity suppresses uniaxial strain of $CeAl_2$ crystal in helium gas as the pressure-transmitting medium.

Furthermore, it is found that FWHMs of (220) and (311) reflections are enhanced above 20 GPa in Fig. 3(b). It means that a precursory phenomenon of the splitting of (220) and (311) reflections, while that of (111) reflection is almost constant at all pressure region. Since such results indicate that (hhl) reflections of cubic structure at low pressure is split to (hhl) and (hlh) while (hhh) one is not, it is reasonable to assume that the pressure-induced structural phase transition from cubic to tetragonal structure occurs. However, it is difficult to estimate the lattice parameters at high pressure phase because no splitting of the peak is observed in the X-ray powder diffraction pattern even at the highest pressure 23.3 GPa.

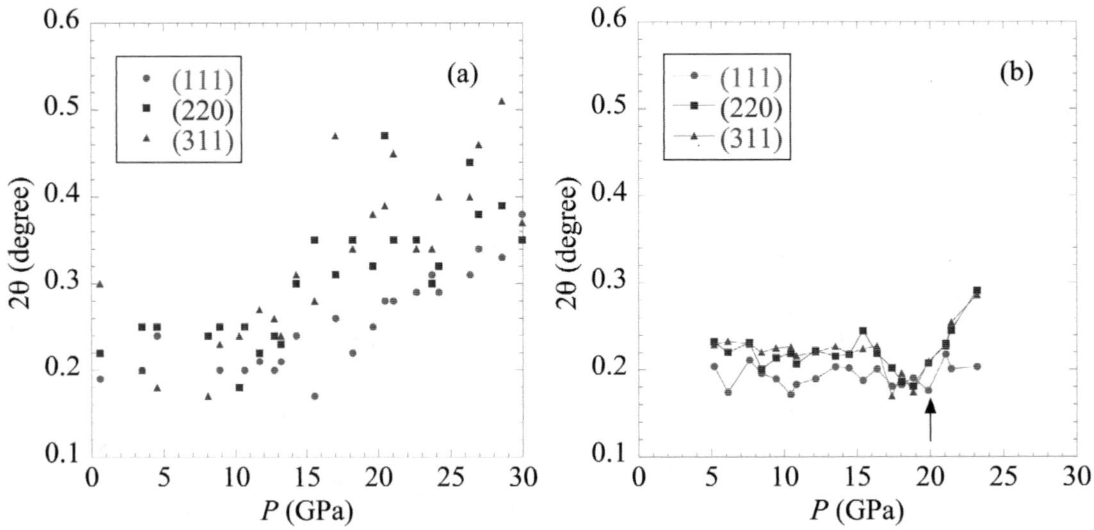

Figure 3. The pressure dependence on FWHM of several Bragg reflections of $CeAl_2$, in which (a) ME is used as the pressure-transmitting medium[18], and (b) helium gas is used as the pressure-transmitting medium. The arrow shows critical pressure where pressure-induced structural transition is suggested.

5. Conclusion

The compression behavior of C15 Laves compound $CeAl_2$ is interpreted through the powder X-ray diffraction measurement under hydrostatic pressure with helium gas pressure-transmitting medium. We support that no discontinuous change in the compression curve below 20 GPa as discussed by the latter group[14, 15]. However, we have suggested that the pressure-induced structural phase transition from cubic to tetragonal structure occurs at around 20 GPa. More precise experiments are needed to determine the space group and all other structural parameters accurately.

Acknowledgments

This work was performed under the Inter-University Cooperative Research Program of the Institute for Materials Research, Tohoku University, and the Visiting Researcher's Program of the Institute for Solid State Physics, the University of Tokyo. This work was supported in part by Grants-in-Aid from the Ministry of Education, Culture, Sports, Science and Technology of Japan, JGC-S Scholarship Foundation and Suzuki Foundation.

References

[1] Purwins H G and Leson A 1990 *Adv. Phys.* **39** 309.
[2] Oishi T, Ohashi M, Suzuki and H, Satoh I 2010 *Journal of Physics: Conference Series* **200** 082022.
[3] Ohashi M, Kishii and N, Tateno S 2016 *accepted to Jpn. J. Appl. Phys.*
[4] Onuki Y, Furukawa Y and Komatsubara T 1984 *J. Phys. Soc. Jpn.* **53** 2734.
[5] Miyagawa H, Oomi G, Ohashi M, Satoh I, Komatsubara T, Hedo and M, Uwatoko Y 2008 *Phys. Rev. B* **78** 064403.
[6] Ohashi M, Miyagawa H, Nakano T, Oomi G, Sechovsky V, Satoh I and Komatsubara T, 2014 *J. Phys. Soc. Jpn.* **83** 024701.

IOP Conf. Series: Journal of Physics: Conf. Series **950** (2017) 042030 doi:10.1088/1742-6596/950/4/042030

[7] Ohashi M, Oomi G, Koiwai S, Hedo M and Uwatoko Y 2003 *Phys. Rev. B* **68** 144428.

[8] Ohashi M, Oomi G and Satoh I 2007 *J. Phys. Soc. Jpn.* **76** 114712.

[9] Ohashi M and Oomi G 2009 *Jpn. J. Appl. Phys.* **48** 070221.

[10] Uwatoko Y, Suenaga K and Oomi G 1992 *J. Magn. Magn. Mater.* **104-107** 645.

[11] Miyagawa H, Oomi G, Ohashi M, Maezawa K, and Kagayama T 2006 *J. Alloys and Compounds* **408-412** 230.

[12] Croft M and Jayaraman A 1979 *Solid State Commun.* **29** 9.

[13] Bartholin H, Waintal A, Parisot G, Kervella F, Senateur J P 1980 *Phys. Status Solidi A Appl. Res.* **61** K87.

[14] Barbara B, Beille J, Cheiato B, Laurant J M, Rossignol M F, Waintal A, and Zemirli S 1986 *Physics Letters* **113** 381.

[15] Vedel I, Redon A M, Léger J M, Mignot J M, Flouquet J 1986 *J. Magn. Magn. Mater.* **54-57** 361.

[16] Chandra Shekar V N V, Sahu P C, Yousuf M, Rajan K G and Rajagopalan M 1999 *Solid State Commun.* **111** 529.

[17] Takemura K 1999 *Phys. Rev. B* **60** 6171.

[18] Miyagawa H, Ohashi M, Oomi G, Satoh I, Komatsubara T, Miyajima and N, Yagi T, 2006 *Topology in Ordered Phases* 203.

[19] Murnaghan F D 1951 *Finite Deformation of an elastic solid* (Dover Publication Inc., New York)

[20] Fujishiro I, Piermarini G J, Block S, and Munro R G 1982 in *High Pressure in Research and Industry, Proceedings of the 8th AIRAPT Conference Uppsala, edited by C. M. Backman, T. Johannisson, and L. Tegner* (ISBN, Sweden), Vol. II, p. 608.

[21] Besson J M and Pinceaux J P 1979 *Science* **206** 1073.

IOP Conf. Series: Journal of Physics: Conf. Series **950** (2017) 042031 doi:10.1088/1742-6596/950/4/042031

High-pressure electrical resistivity studies for $Ba_{1-x}Cs_xFe_2Se_3$

C Kawashima[1], H Soeda[1], H Takahashi[1], T Hawai[2], Y Nambu[3], T J Sato[3], Y Hirata[4] and K Ohgushi[4, 5]

[1] College of Humanities and Sciences, Nihon University, Sakurajosui, Setagaya-ku Tokyo 156-8550, Japan
[2] Neutron Science Laboratory, Institute for Solid State Physics, University of Tokyo, Tokai, Ibaraki 319-1106, Japan
[3] Institute of Multidisciplinary Research for Advanced Materials, Tohoku University, Sendai Miyagi 980-8577, Japan
[4] Institute for Solid State Physics, University of Tokyo, Kashiwa, Chiba 277-8581, Japan
[5] Department of Physics, Tohoku University, Sendai, Miyagi 980-8578, Japan

E-mail: kawashima@phys.chs.nihon-u.ac.jp

Abstract. High-pressure electrical resistance measurements were performed for iron-based ladder material $Ba_{1-x}Cs_xFe_2Se_3$ (x = 0.25 and 0.65) using a diamond anvil cell (DAC). Recent high-pressure study revealed that iron-based ladder material $BaFe_2S_3$ exhibits an insulator-metal transition and superconductivity, and this discovery would provide important insight for understanding the mechanism of iron-based superconductors. Therefore, it is intriguing to investigate the high-pressure properties for the iron-based ladder material $Ba_{1-x}Cs_xFe_2Se_3$ system. The parent compounds $BaFe_2Se_3$ and $CsFe_2Se_3$ show insulating and magnetic ordering features. For $Ba_{1-x}Cs_xFe_2Se_3$ system, no magnetic ordering is observed for x = 0.25 and minimum charge gap was estimated for x = 0.65. The insulator-metal transitions are observed in both materials.

1. Introduction

The discovery of iron-based superconductors had a considerable impact on condensed matter physics, triggering rapid efforts in studying the interplay between crystal structure, magnetism, and superconductivity [1]. Several kinds of iron-based superconductors have been discovered at this moment. These materials have a two-dimensional (2D) iron lattice as a common feature, despite of having different crystal structures. These compounds exhibit characteristic magnetic phases adjacent to the superconducting phase based on the 2D iron lattice. While a striped magnetic order is observed in the magnetic phases for 1111(e.g. LaFeAsO), 122(e.g. $BaFe_2As_2$), 111(e.g. LiFeAs) and 11(e.g. FeSe) type iron-based superconductors, a block type magnetic order was revealed in 245(e.g. KFe_4Se_5) type iron-based superconductor. Such a variety has substantial importance to understand the interplay between magnetism and superconductivity [2].

Recently, an iron-based quasi-one-dimensional compound AFe_2X_3 (A= Ba, K, Cs, and X = S, Se) having the spin-ladder arrangement of Fe atoms has attracted much attention. $BaFe_2S_3$ is an insulator having the stripe magnetic order, in which the magnetic moments are arranged to form ferromagnetic units along the rung direction, stacking antiferromagnetically along the ladder direction. This stripe-type magnetic structure is reminiscent of the 1111 and 122 systems found in the iron-based superconductors. Recently, $BaFe_2S_3$ was revealed to exhibit pressure-induced superconductivity having T_c of 14 K at 11 GPa [3]. Motivated by this discovery, extensive studies have been carried out for other iron-based ladder materials. In copper-based materials, the superconducting phase emerges not only in square lattice structures but also in ladder structures. The two-leg ladder material $Sr_{14-x}Ca_xCu_{24}O_{41}$ exhibits superconductivity under high pressures above 4 GPa with T_c of 12 K, which provided great insight to understanding the high-T_c superconductivity [4].

The magnetic structure of $BaFe_2Se_3$ is block-type, in which the magnetic moments form Fe_4 ferromagnetic units which stack antiferromagnetically along the ladder direction in a one-dimensional analogue of the block magnetism observed in the 245 system [5, 6]. In contrast, the magnetic structure of the $CsFe_2Se_3$ is stripe-type, in which the magnetic moments couple ferromagnetically along the rung and antiferromagnetically along the leg direction. This magnetic structure represents a one-dimensional analogue of the stripe magnetism observed in 1111 and 122 systems [7]. Since these ladder materials exhibit magnetic structures similar to those of the iron-based superconductors, they can be considered as potential superconductors.

In this study, high-pressure electrical resistivity measurements are performed for $Ba_{1-x}Cs_xFe_2Se_3$ (x = 0.25 and 0.65). The crystal is composed of a quasi-one-dimensional structure formed by edge-shared $FeSe_4$ tetrahedra with channels occupied by (Ba, Cs) atoms and these materials for x = 0.25 and 0.65 have the *Cmcm* space group, as shown in figure 1, while $BaFe_2Se_3$ has the *Pnma* space group. The crystal structure is depicted by the VESTA software [8]. The block-type magnetism is suppressed by substitution of Ba into Cs, and no magnetic reflection was observed in the powder neutron diffraction measurements for x = 0.25 [9]. It is interesting to study electrical properties under high pressure for x = 0.25, because the antiferromagnetic ordering of $BaFe_2S_3$ is suppressed with applying pressure and superconductivity appears after suppression of magnetic ordering state [10]. On the other hand, the formal valence of Fe is Fe^{2+} and $Fe^{+2.5}$ for $BaFe_2Se_3$ and $CsFe_2Se_3$, respectively. The material for x = 0.65 is chosen for the second material for high-pressure experiment, since the material for x = 0.65 shows the minimum charge gap state [9].

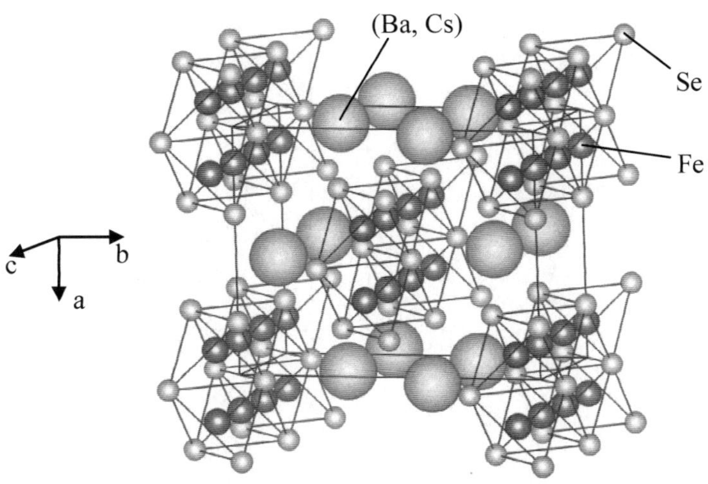

Figure 1. The Crystal structure of $Ba_{1-x}Cs_xFe_2Se_3$, consisting of edge-shared $FeSe_4$ tetrahedra extending along the c axis and of channels occupied by Ba or Cs atoms, in which Fe atoms form a ladder structure.

2. Experimental

High-pressure electric resistance measurements for single crystalline $Ba_{1-x}Cs_xFe_2Se_3$ were performed by a standard d.c. four-probe method using a diamond anvil cell (DAC). Detail of the sample preparation is described elsewhere [9]. The electric currents of 1.0, 0.1 and 0.05mA were made to flow to samples. The sample chamber of DAC was filled with powdered NaCl at the pressure-transmitting medium, using a rhenium gasket, and thin (10μm thick) platinum ribbons were used as leads for the standard d.c. four-probe analysis. The anvils having culet size of 0.5mm were used. The samples are rectangular in shape of about 0.1mm×0.1mm×0.3mm. A thin BN layer acted as electric insulation between the leads and the rhenium gasket and finely ground ruby powder scattered in the sample chamber was used to determine the pressure by the standard ruby fluorescence method.

3. Results and Discussion

3.1. $Ba_{0.75}Cs_{0.25}Fe_2Se_3$ (x = 0.25)

Figure 2(a) and (b) show the temperature dependence of the electrical resistance of $Ba_{0.75}Cs_{0.25}Fe_2Se_3$ under high pressure up to 12 GPa on a logarithmic scale and 13GPa to 18 GPa on a linear scale, respectively. This material has the *Cmcm* space group and no magnetic reflection was observed in the powder neutron diffraction measurements for x = 0.25. The electric resistance decreases with applying pressure. It shows metallic temperature dependence above 12 GPa. However, the upturn of resistance is observed up to the maximum pressure below 50 K. The double-digit reduction of room-temperature resistance is observed from ambient pressure to 18 GPa. Since about triple-digit reduction of room-temperature resistance was reported for $BaFe_2S_3$ at 11 GPa where the metal insulator transition takes place and superconductivity appears [3], the pressure of 18 GPa might be not enough pressure to induce superconductivity for $Ba_{0.75}Cs_{0.25}Fe_2Se_3$.

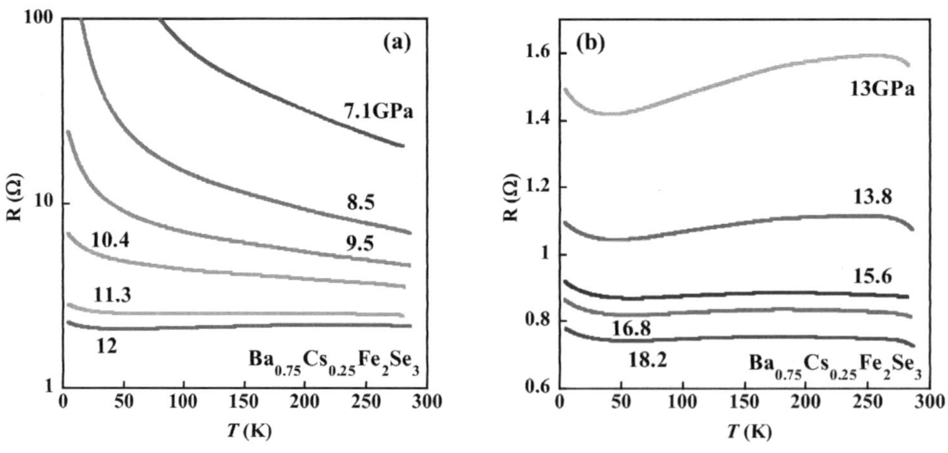

Figure 2. (a)Temperature dependence of the electrical resistance of $Ba_{0.75}Cs_{0.25}Fe_2Se_3$ up to 12GPa on a logarithmic scale. (b)Electrical resistance up to 18.2GPa on a linear scale.

3.2. $Ba_{0.35}Cs_{0.65}Fe_2Se_3$ (x = 0.65)

Figure 3(a) and (b) show the temperature dependence of the electrical resistance of $Ba_{0.35}Cs_{0.65}Fe_2Se_3$ under high pressure up to 19 GPa on a logarithmic scale and 20 GPa to 30 GPa on a linear scale, respectively. This material has the *Cmcm* space group and shows the minimum charge gap state. The electric resistance decreases with applying pressure. It shows metallic temperature dependence above 14 GPa. No upturn of resistance is observed up to the maximum pressure in the low temperature range. About 1.5 digit reduction of room-temperature resistance is observed from ambient pressure to 30 GPa. Though this material shows metallic behaviour, the pressure of 30 GPa might be still not enough to induce superconductivity for $Ba_{0.35}Cs_{0.65}Fe_2Se_3$. It is necessary to carry out the electrical resistance measurement at much higher pressure.

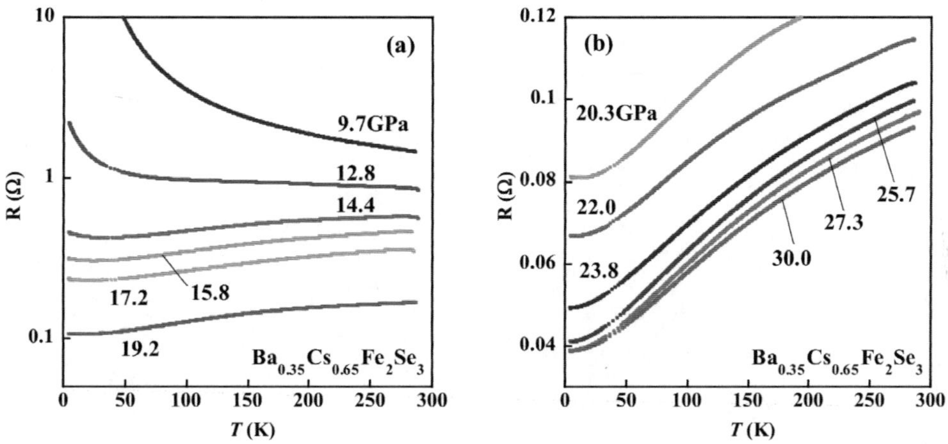

Figure 3. (a)Temperature dependence of the electrical resistance of $Ba_{0.25}Cs_{0.65}Fe_2Se_3$ up to 19.2 GPa on a logarithmic scale. (b)Electrical resistance up to 30.0GPa on a linear scale.

4. Summary

High-pressure electrical resistance measurements were performed for iron-based ladder material $Ba_{1-x}Cs_xFe_2Se_3$ (x = 0.25 and 0.65) using a diamond anvil cell (DAC). These samples exhibit the insulator-metal transition at about 12GPa and 14 GPa for x = 0.25 and x = 0.65, respectively. Superconductivity could not be observed in these experimental pressures. It has to be needed to perform higher pressure experiment for these materials.

Acknowledgements

This work was partly supported by JSPS Grants-in-Aid for Scientific Research (B) (24340088 and 26287073) and the Strategic Research Base Development Program for Private Universities (2009, S0901022) of MEXT.

References

[1] Kamihara Y, Waranabe T, Hirano M and Hosono H 2008 *J. Am. Chem. Soc.* **130** 3296
[2] Hosono H and Kuroki K 2015 *Physica C* **514** 399
[3] Takahashi H *et al* 2015 *Nat. Mater.* **14** 1008

[4] Uehara M, Nagata T, Akimitsu J, Takahashi H, Mori N and Kinoshita K 1996 *J. Phys. Soc. Jpn.* **65** 2764
[5] Nambu Y *et al* 2012 *Phys. Rev. B* **85** 064413
[6] Caron J M, Neilson J R, Miller D C, Llobet A and McQueen T M 2011 *Phys. Rev. B* **84** 180409
[7] Du F, Ohgushi K, Nambu Y, Kawakami T, Avdeev M, Hirata Y, Watanabe Y, Sato T J and Ueda Y 2012 *Phys. Rev. B* **85** 214436
[8] Momma K and Izumi F 2011 *J. Appl. Crystallogr.* **44** 1272
[9] Hawai T *et al* 2015 *Phys. Rev. B* **91** 184416
[10] Yamauchi T *et al* to be published in 2015 *Phys. Rev. Lett.*

IOP Conf. Series: Journal of Physics: Conf. Series **950** (2017) 042032 doi:10.1088/1742-6596/950/4/042032

High-pressure Synthesis and Physical Properties of New Filled Skutterudite Compound $BaOs_4P_{12}$

S Deminami, Y Kawamura, Y Q Chen, M Kanazawa, J Hayashi, T Kuzuya, K Takeda, M Matsuda and C Sekine

Muroran Institute of Technology, 27-1 Mizumoto-cho, Muroran 050-8585, Japan

E-mail: sekine@mmm.muroran-it.ac.jp

Abstract. We have succeeded in synthesizing samples of new filled skutterudite compound $BaOs_4P_{12}$ using the high-pressure synthesis technique. The physical properties of the compound are reported for the first time. The electrical resistivity decreases with decreasing temperature, and drop sharply around 1.8K. This indicates that $BaOs_4P_{12}$ is a new superconductor. Before the high-pressure synthesis, using synchrotron radiation x-ray, we tried to observe synthesizing processes of $BaOs_4P_{12}$ in-situ at high temperature and high pressure to obtain the optimum condition for synthesis.

1. Introduction

Filled skutterudite compounds AT_4X_{12} (A = alkali metal, alkaline earth metal, lanthanide (Ln), and actinide; T = Fe, Ru, Os, and Pt; X = P, As, Sb, and Ge) crystallize in a body-centered cubic structure with a space group $Im\overline{3}$ (T^5_h, No. 204) [1]. Filled skutterudite compounds, especially compounds including Ln, have attracted much attention owing to their wide variety of strongly correlated electron behaviors, such as anomalous metal-insulator transition [2], multipole ordering [3] and unconventional superconductivity [4]. The various attractive features in these systems are believed to be mainly due to the large hybridization between f electrons of Ln and p electrons of X. The d electrons of Fe, Ru and Os atoms should also play important roles in the physical properties. However, the roles have not been well understood yet. Matsuoka *et al.* reported that filled skutterudite compounds including alkaline earth metal AFe_4Sb_{12} (A = Ca, Sr, and Ba) have revealed strong ferromagnetic spin fluctuations of the Fe $3d$ electrons and the properties of nearly ferromagnetic metals [5]. Furthermore, AOs_4Sb_{12} (A=Sr, Ba) show the character of enhanced Pauli paramagnets where Os $5d$ electrons play important roles [6]. The filled skutterudite compounds including alkaline earth metal could be a favorable system to systematically investigate the effect of d electrons in skutterudite systems because alkaline earth ions carry no magnetic moments. However, only preliminary studies have been conducted on the properties of P and As-based compounds including alkaline earth metal because the compounds are quite difficult to prepare. In this study, we focused on P-based filled skutterudite compounds including alkaline earth metal as part of our search for new materials using a high-pressure synthesis technique.

2. Experimental

In-situ x-ray diffraction patterns were taken by an energy-dispersive method using synchrotron radiation and a solid-state detector. High pressure was applied using a multi-anvil assembly 6-6 with a cubic-anvil high-pressure apparatus, the MAX80 system, installed at the beam line AR-NE5C, at Photon

Factory (PF) in High Energy Accelerator Research Organization (KEK) (Tsukuba, Japan). The sample container made of pyrophyllite is formed a cube. The starting materials, which are mixture of Ba, Os and P, are put into a BN crucible. The crucible with a graphite heater is inserted in a cube-shaped pyrophyllite solid pressure medium. Pressure was determined by the lattice constant of NaCl internal pressure marker. The details of the in-situ observation method were described in previous reports [7, 8].

$BaOs_4P_{12}$ samples for measurements were prepared at high temperatures and high pressures using a double-stage multi-anvil high-pressure apparatus. The sample cell assembly is similar to that used for the high-pressure synthesis of $CeRu_4As_{12}$ [9]. The prepared samples were characterized by powder x-ray diffraction using $CoK_{\alpha 1}$ radiation and silicon as a standard. Resistivity was measured by a standard dc four-probe method in the temperature region from 0.4K to 300K.

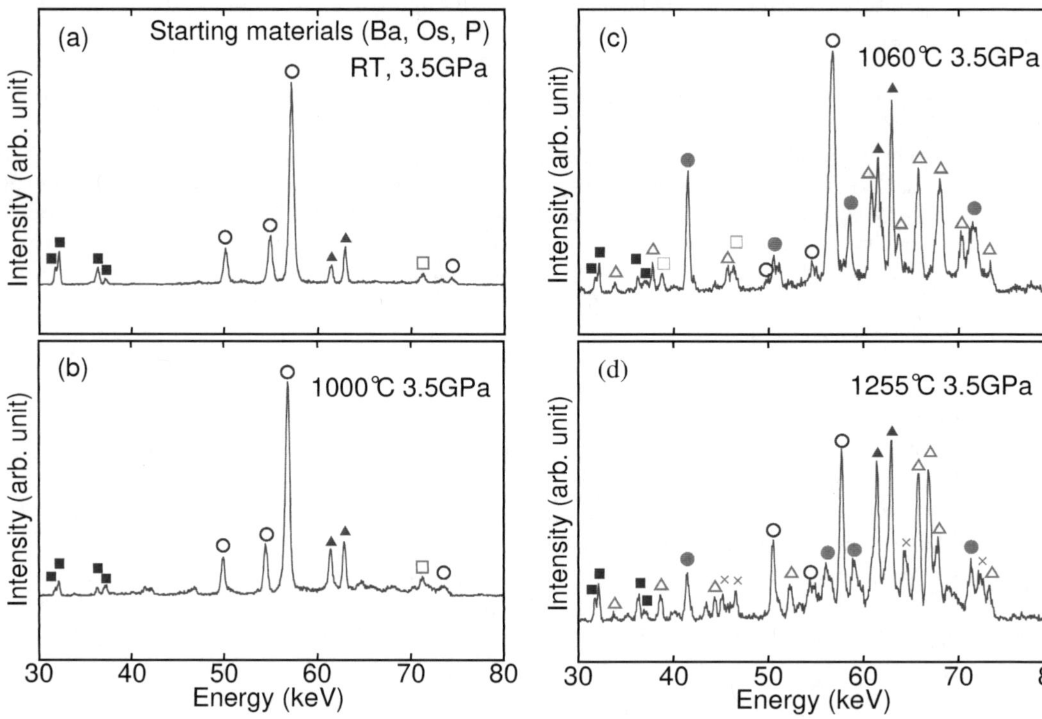

Figure 1. X-ray diffraction patterns of synthesizing process of $BaOs_4P_{12}$ at 3.5 GPa. Open blue circles designate the Bragg peaks of Os. Solid blue squares and solid green triangles indicate the characteristic x-ray of Ba and Os. Open purple squares indicate the peaks of BN used for crucible. Solid red circles designate the Bragg peaks of the skutterudite structure. Open brown triangles, open green squares and blue crosses indicate the peaks of an impurity phase of OsP_4, P_4O_8 and OsP_2, respectively. (a) The starting materials at room temperature, (b)1000℃, (c) 1060℃ and (d) 1255℃.

3. Results and discussion

The results of in-situ x-ray diffraction experiments are shown in figure 1. Figure 1(a) shows x-ray diffraction pattern of the starting materials, which are mixture of each metal and phosphorus powder in the atomic ratio of Ba : Os : P = 1 : 4 : 12, at room temperature and 3.5 GPa. Open blue circles designate the Bragg peaks of Os. Solid blue squares and solid green triangles indicate the characteristic x-ray for Ba and Os, respectively. Open purple squares indicate the Bragg peaks of BN used for crucible. The diffraction peaks show no change up to 1000°C (figure 1(b)). With increasing temperature above 1000°C, the peaks of the skutterudite structure were observed above 1060°C (closed red circles in figure 1(c)). The peaks for impurity phases of OsP_4 and P_4O_8, which are indicated by open brown triangles and open green squares, respectively, also appeared at this temperature. Elevating temperature up to 1255°C, the peaks of the skutterudite structure weakened and another impurity phase of OsP_2 was observed (blue crosses in figure 1(d)). In series of the experiments, we found an appropriate temperature range of 1050-1200°C for synthesizing $BaOs_4P_{12}$ at 3.5 GPa.

Based on the synthesis condition determined by the in-situ experiments, we have actually synthesized $BaOs_4P_{12}$ at 4GPa and 1050°C using a Kawai-type two-stage anvil apparatus. Figure 2 shows a powder x-ray diffraction pattern of $BaOs_4P_{12}$ prepared under high pressure. Most of the observed diffraction peaks were indexable using the skutterudite structure. The lattice constant determined by a least-squares fit to the data was 8.124Å. Figure 3 plots the lattice constant of $BaOs_4P_{12}$ along with values reported earlier for filled skutterudite phosphides $LnOs_4P_{12}$ [10]. Starting with La, one observes the typical reduction of the lattice constant due to the contraction of the trivalent ionic radii of the lanthanides with increasing atomic number except for Ce and Eu compounds. The small lattice constant of Ce compound is due to the strong hybridization between 4*f* electrons and conduction electrons, while the large lattice constants of Eu compound is caused by the Eu ions being in the divalent or mixed-valence state. The large lattice constant of $BaOs_4P_{12}$ is considered to be due to the divalent Ba ions.

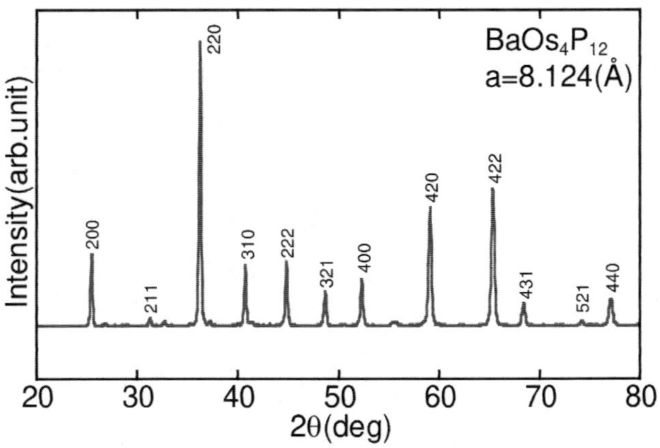

Figure 2. Powder x-ray diffraction pattern of $BaOs_4P_{12}$ prepared under high pressure.

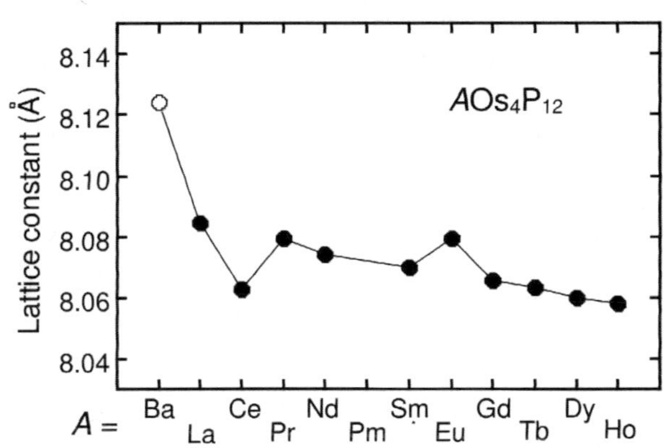

Figure 3. Lattice constants of filled skutterudite phosphides AOs_4P_{12}.

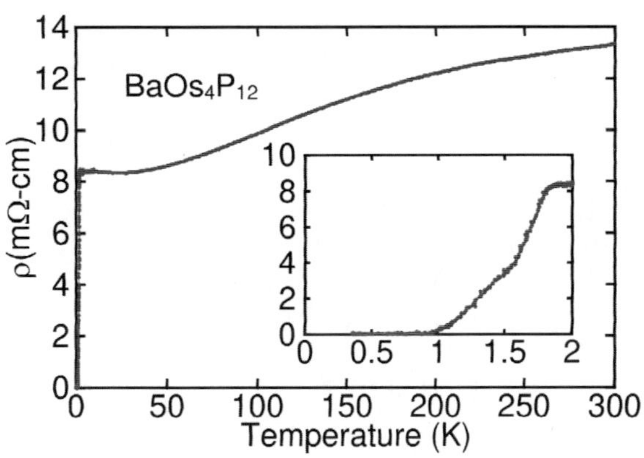

Figure 4. Temperature dependence of electrical resistivity of $BaOs_4P_{12}$. The inset shows the low temperature data on an expanded scale.

IOP Conf. Series: Journal of Physics: Conf. Series **950** (2017) 042032 doi:10.1088/1742-6596/950/4/042032

Figure 4 shows the temperature dependence of electrical resistivity $\rho(T)$ for $BaOs_4P_{12}$. The inset shows the low temperature data on an expanded scale. The $\rho(T)$ exhibits a positive temperature dependence like typical metals below room temperature and shows a sharp drop below 1.8K. This compound is a new superconductor. The application of a magnetic field suppresses superconductivity at fields above 1T. Furthermore, a broad shoulder around 170K in $\rho(T)$ was observed. The strong curvature of the compound may be due to strong electron-phonon interactions, which was discussed for AOs_4Sb_{12} (A = Sr, Ba). Matsuoka *et al.* proposed that the strong electron-phonon interaction in the presence of tunneling states of A^{2+} ions in the Os_4Sb_{12} cage is the origin of a similar shoulder observed in the resistivity of AOs_4Sb_{12} (A = Sr, Ba) [6]. The electron-phonon interaction seems to be responsible for the formation of Cooper pairs in this system.

4. Summary

Using the high-pressure synthesis technique, we have succeeded in synthesizing a new skutterudite compound $BaOs_4P_{12}$ with lattice constant of 8.124Å. $BaOs_4P_{12}$ was found to be a new superconductors around 1.8K. We could obtained the optimum condition for synthesizing $BaOs_4P_{12}$ under high pressure by the in-situ x-ray experiments. This is an excellent method for determining the condition for obtaining only target material without impurity phases for solid-phase reaction synthesis under high pressure.

Acknowledgments
This work was performed under the approval of the Photon Factory Program Advisory Committee (Proposal No. 2013G124, 2015G031). This work was supported by JSPS KAKENHI Grant Number 23340092.

References
[1] Jeitschko W and Braun D 1977 *Acta Crystallogr.,* Sect. B **33** 3401
[2] Sekine C, Uchiumi T, Shirotani I and Yagi T 1997 *Phys. Rev. Lett.* **79** 3218
[3] Yoshizawa M, Nakanishi Y, Oikawa M, Sekine C, Shirotani I, Saha S R, Sugawara H and Sato H 2005 *J. Phys. Soc. Jpn.* **74** 2141
[4] Aoki Y, Tayama T, Sakakibara T, Kuwahara K, Iwasa K, Kohgi M, Higemoto W, MacLaughlin D E, Sugawara H and Sato H 2007 *J. Phys. Soc. Jpn.* **76** 051006
[5] Matsuoka E, Hayashi K, Ikeda A, Yanaka K, Takabatake T and Matsumura M 2005 *J. Phys. Soc. Jpn.* **74** 1382
[6] Matsuoka E, Narazu S, Hayashi K, Umeo K, and Takabatake T 2006 *J. Phys. Soc. Jpn.* **75** 014602
[7] Sekine C, Kachi T, Yoshida T, Abe R, Namiki T, Akahira K and Ito K 2010 *J. Phys.: Conf. Ser.* **215** 012141
[8] Sekine C, Kato H, Kanazawa M, Kawamura Y, Takeda K, Matsuda M, Kihou K, Lee C H and Gotou H 2014 *J. Phys.: Conf. Ser.* **502** 012017
[9] Sekine C, Kawata T, Kawamura Y and Yagi T 2013 *J. Korean Phys. Soc.* **63** 359
[10] Sekine C, Ando H, Sugiuchi Y, Shirotani I, Matsuhira K and Wakeshima M 2008 *J. Phys. Soc. Jpn.* **77** Suppl. A, 135

IOP Conf. Series: Journal of Physics: Conf. Series **950** (2017) 042033 doi:10.1088/1742-6596/950/4/042033

T_c enhancement under pressure in the La1111 pnictide $LaFeAsO_{1-x}H_x$ (x=0.2) studied by NMR

N. Fujiwara*, N. Kawaguchi, S. Iimura[1], S. Matsuishi[1], and H. Hosono[1]

[1]Graduate School of Human & Environmental Studies, Kyoto University, Kyoto, Kyoto 606-8501, Japan
[2]Material and Structures Laboratory (MSL) & Frontier Research Center (FRC), Tokyo Institute of Technology, 4259 Nagatsuda, Midori-ku, Yokohama 226-8503, Japan

*naoki@fujiwara.h.kyoto-u.ac.jp

Abstract. Nuclear magnetic resonance (NMR) measurements under a pressure of 3.0 GPa were performed for a prototypical iron-based high-T_c superconductor $LaFeAsO_{1-x}H_x$ (x=0.2). At ambient pressure, this compound has superconducting (SC) double domes with the T_c minimum of 26 K at x=0.2 in the electronic phase diagram. Pressure application merges the double domes into a single dome, and the T_c minimum at x=0.2 goes up to 45 K, the T_c maximum, at 3.0 GPa. Despite that T_c enhancement under pressure was confirmed from the ^{75}As relaxation rate divided by temperature (T), $1/T_1T$ around T_c, $1/T_1T$ at high temperatures was almost the same with that at ambient pressure and hardly showed Cure-Weiss behaviour. These features are also seen for 14% F-doped $LaFeAsO_{1-x}F_x$, suggesting that spin fluctuations are not essential to achieve high T_c over 45 K.

1. Introduction

The R1111 series such as $RFeAsO_{1-x}F_x$ or $RFeAsO_{1-x}H_x$ (R=La, Ce, Ne, Pr, Sm) is very unique among iron-based high-T_c superconductors in the points that high T_c over 50 K has been marked only in this series and only this series can accept high electron doping due to H substitution. Owing to the capability of electron doping, the H-doped La1111 series undergoes novel phenomena in the electronic phase diagram: (i) H doping leads to superconducting (SC) phase with double domes following a stripe-type antiferromagnetic (AF) phase [1], and (ii) further H doping leads to spatially inhomogeneous AF ordering following double SC domes [2-4]. Pressure application merges the SC double domes to a single dome: at 20% H-doping level, T_c marks the minimum of 26 K at ambient pressure, and intriguingly the minimum turns to the maximum of 45 K at 3.0 GPa [1, 5]. This specific situation allows one to investigate the high T_c mechanism in iron-based superconductors.

To investigate what causes the T_c maximum on a microscopic level, we performed ^{75}As nuclear magnetic resonance (NMR) measurements on the 20% H-doped La1111 samples at 3.0GPa and compared the results with those obtained at ambient pressure.

2. Experimental conditions

2.1. Pressure cell

Figure 1 shows an illustration of a piston-cylinder type NiCrAl pressure cell [6]. The length of the NiCrAl cylinder is 30.0 mm, and it's inner and outer diameters are 6.0-, and 16.0-mm, respectively. A

IOP Conf. Series: Journal of Physics: Conf. Series **950** (2017) 042033 doi:10.1088/1742-6596/950/4/042033

CuBe sleeve with an outer diameter of 40.0 mm covers the NiCrAl cylinder, as shown in Fig. 1. A Teflon capsule of 6.0 mm outer diameter and 15.0 mm length and a CuBe sealing ring were attached on a CuBe plug, forming a liquid container. Powder samples packed in a sample holder was inserted in the liquid container. Ruby powders were attached on the top of an optical fiber, and they were also inserted in the liquid container. Pressure inside the sample cell was monitored by Ruby fluorescence measurements. The initial volume of the sample space was 4.4 mm diameter × 11.0 mm length. A mixture of fluorinate liquids, Fluorinert FC-70 and FC-77, was used as a pressure-mediation liquid. The pressure cell potentially reaches 4.6 GPa by applying a load of 16 ton, despite that the length of the sample space becomes extremely small.

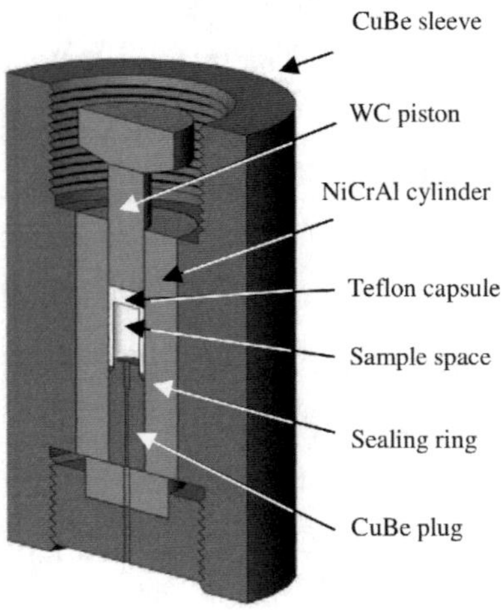

Figure 1. Illustration of a piston-cylinder-type NiCrAl pressure cell.

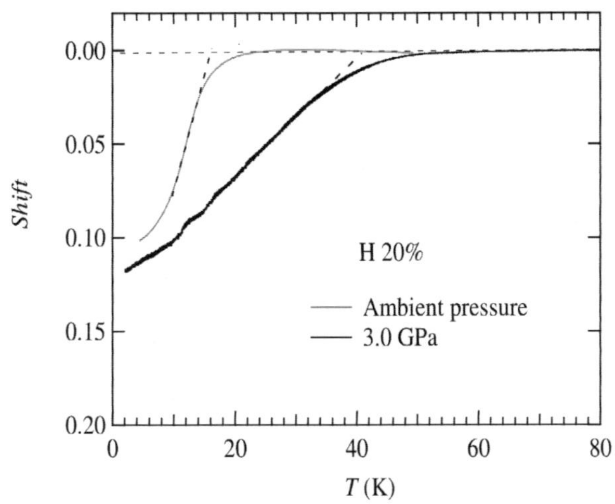

Figure 2. Shift of the resonance frequency of the tank circuit, which corresponds to the AC susceptibility measurements.

2.2. AC susceptibility and NMR measurements

NMR measurements were performed using a conventional coherent-pulsed NMR spectrometer. A NMR probe equipped with two variable capacitors was used for AC susceptibility measurements as well as NMR measurements. For the measurements under pressure, a single coil wounded around the sample holder and capacitors outside the pressure cell form a tank circuit. The AC susceptibility was measured via the resonance frequency of the tank circuit. The resonance frequency was monitored by using a network analyzer.

The resonance frequency shifts to high frequency region when the samples undergo a superconducting state owing to the large demagnetization. On the other hand, T_c can be also determined from the relaxation rate divided by temperature (T), $1/T_1T$: a superconducting gap opens below T_c and thus $1/T_1T$ decreases more rapidly than that in the normal state. The measurements of T_1 were performed for the signals of the central transition (I=−1/2 ⇔ 1/2) of ^{75}As (I=3/2). The NMR spectra broaden owing to the electric quadrupole interaction, and T_1 was measured for the powder samples with FeAs planes parallel to the applied field, as described below.

Despite that one can estimate T_c from both measurements, the T_c values may be not always coincide with each other: all powder samples are involved in the resonance frequency measurements, whereas only powder samples with FeAs planes parallel to the applied field are involved in the $1/T_1T$ measurements.

3. Experimental results

3.1. Resonance frequency

Figure 2 shows the shift of the resonant frequency. We define *Shift* = f/f_0-1, where f is the resonance frequency and f_0 is the estimation extrapolated from the normal state. At ambient pressure, T_c determined from the onset is about 22 K, and that determined from the extrapolation is about 18 K, as shown by a dashed line in Fig. 2. At 3.0 GPa, T_c determined from the onset is about 48 K, and that determined from the extrapolation is about 40 K. The discrepancy of T_c at 3.0 GPa is about 8 K and is larger than that estimated at ambient pressure. In fact, the resonance frequency at 3.0 GPa changes gradually with decreasing temperature. This could be caused since applied pressure becomes less hydrostatic at low temperature than at room temperatures.

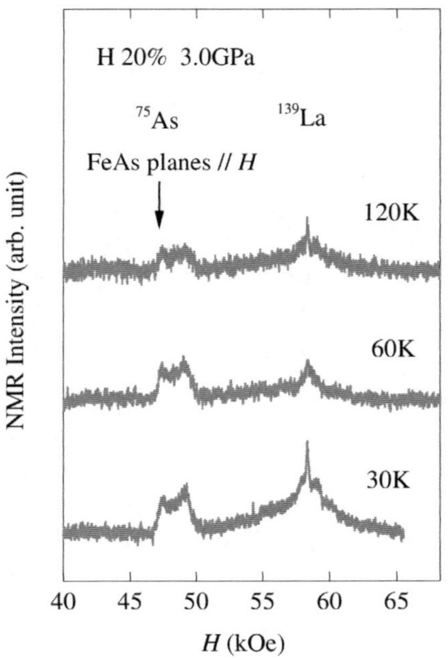

Figure 3. NMR spectra of ^{75}As (I=1/2) and ^{139}La (I=7/2) measured at 3.0 GPa at 35.1MHz.

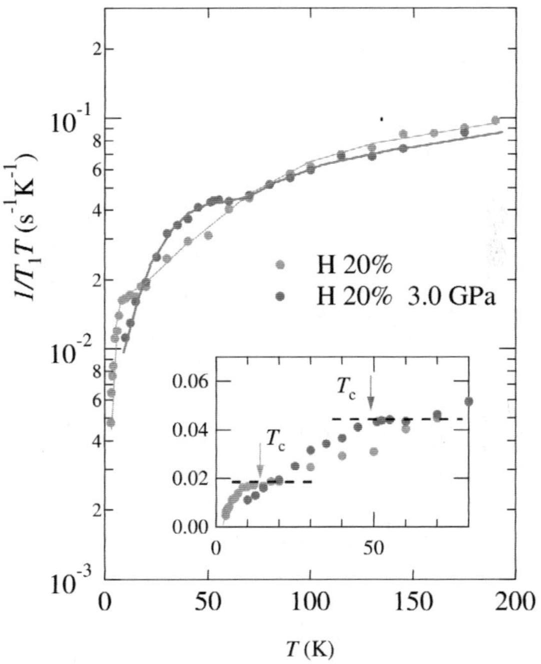

Figure 4. Temperature dependence of $1/T_1T$ for ^{75}As. The inset shows the expansion around T_c.

3.2. NMR spectra

Figure 3 shows the NMR spectra of ^{75}As (I=1/2) and ^{139}La (I=7/2) measured at 3.0 GPa at 35.1MHz. The ^{75}As signals of the central transition (I=-1/2 ⇔1/2) appear at 47-50 KOe as a double peak pattern due to large electric field gradient (EFG). All ^{139}La signals appear around 58 KOe due to small EFG. The central peak and six satellite edges of ^{139}La should be observable in pure samples, however, broad signals were observed instead of six edges due to the distribution of EFG at La sites. The distribution mainly originates from H doping. The NMR spectra were almost unchanged as shown in Fig. 3, implying that the EFG in FeAs planes is almost T independent. The relaxation rate was measured at the lower-field peak as shown by an arrow in Fig. 3. Only the powders with FeAs planes parallel to the applied field are involved in this peak.

3.3. Relaxation rate

Figure 4 shows the T dependence of $1/T_1T$, which gives a measure of spin fluctuations. At both ambient pressure and 3.0 GPa, $1/T_1T$ exhibits a plateau just above T_c and increases monotonously with increasing temperature. One can determine T_c as deviation from the plateau. By applying pressure, T_c goes from 15 K at ambient pressure up to 48-50 K. These values are consistent with T_c determined from the resonance frequency. The increase in $1/T_1T$ at high temperatures is not attributable to spin fluctuations but to the increase in the density of states $N(\varepsilon_F)$ involved in Fermi surface at high temperatures. In conventional metals where strong spin fluctuations are absent, $1/T_1T$ is described by Korringa relation,

$$\frac{1}{T_1T} \propto A_{hf}^2 N(\varepsilon_F)^2 K(\alpha) \quad (1)$$

where A_{hf} is the hyperfine coupling, and $K(\alpha)$ gives a measure of electron interactions. At high temperatures, $1/T_1T$ at 3.0 GPa is almost the same as that at ambient pressure, implying that A_{hf} and $K(\alpha)$ are unchanged by pressure application up to 3.0 GPa. In the present case, the T dependence of $1/T_1T$ is attributable to that of $N(\varepsilon_F)$; $N(\varepsilon_F)$ is a constant just above T_c and increases monotonously with increasing temperature. The similar features have also been observed for 14% F-doped $LaFeAsO_{1-x}F_x$, the maximally F-doped samples [7, 8]. In this case, T_c rises from 20 K to 40 K by applying a pressure of 3.0 GPa. These doping levels are located away from the AF orderings in the phase diagram, which may be a key factor for high T_c over 40 K as well as the phase segregation between AF and SC states.

4. Conclusion

We performed NMR measurements on $LaFeAsO_{1-x}H_x$ (x=0.2) at 3.0 GPa. Pressure application raises T_c from 15 K at ambient pressure to 48 K without observable enhancement of AF fluctuations. The effect of high T_c is observable in $1/T_1T$ as enhancement of $N(\varepsilon_F)$. These facts give advantage for the superconducting mechanism via orbital fluctuations rather than the mechanism via spin fluctuations.

References

[1] Iimura S, Matsuishi S, Sato H, Hanna T, Muraba Y, Kim SW, Kim JE, Takata M, and Hosono H, 2012 *Nat. Commun.* **3** 943
[2] Fujiwara N, Tsutsumi S, Iimura S, Matsuishi S, Hosono H, Yamakawa Y, and Kontani H, 2013 *Phys. Rev. Lett.* **111** 097002
[3] Sakurai R, Fujiwara N, Kawaguchi N, Y. Yamakawa, Kontani H, Iimura S, Matsuishi S, Hosono H, 2015 *Phys. Rev. B* **91** 064509
[4] Yamakawa Y, Onari S, Kontani H, Fujiwara N, Iimura S, Hosono H, 2013 *Phys. Rev. B* **88** 041106(R)
[5] Takhahashi H, Soeda H, Nukii M, Kawashima C, Nakanishi T, Limura S, Muraba Y, Matsuishi S, and Hosono H, 2015 *Sci. Rpt.* **5** 7829
[6] Fujiwara N, Matsumoto T, Koyama-Nakazawa K, Hisada A and Uwatoko Y, 2007 *Rev. Sci. Instrum.* **78** 073905
[7] Nakano T, Fujiwara N, Tatsumi K, Okada H, Takahashi H, Kamihara Y, Hirano M, and Hosono H, 2010 *Phys. Rev. B* **81** 100510(R)
[8] Tatsumi K, Fujiwara N, Okada H, Takahashi H, Kamihara Y, Hirano M, and Hosono H, 2009 *J. Phys. Soc. Jpn.* **78** 023709

IOP Conf. Series: Journal of Physics: Conf. Series **950** (2017) 042034 doi:10.1088/1742-6596/950/4/042034

High-pressure electrical resistivity studies for $Ba_{1-x}Cs_xFe_2Se_3$

C Kawashima[1], H Soeda[1], H Takahashi[1], T Hawai[2], Y Nambu[3], T J Sato[3], Y Hirata[4] and K Ohgushi[4, 5]

[1] College of Humanities and Sciences, Nihon University, Sakurajosui, Setagaya-ku Tokyo 156-8550, Japan
[2] Neutron Science Laboratory, Institute for Solid State Physics, University of Tokyo, Tokai, Ibaraki 319-1106, Japan
[3] Institute of Multidisciplinary Research for Advanced Materials, Tohoku University, Sendai Miyagi 980-8577, Japan
[4] Institute for Solid State Physics, University of Tokyo, Kashiwa, Chiba 277-8581, Japan
[5] Department of Physics, Tohoku University, Sendai, Miyagi 980-8578, Japan

E-mail: kawashima@phys.chs.nihon-u.ac.jp

Abstract. High-pressure electrical resistance measurements were performed for iron-based ladder material $Ba_{1-x}Cs_xFe_2Se_3$ (x = 0.25 and 0.65) using a diamond anvil cell (DAC). Recent high-pressure study revealed that iron-based ladder material $BaFe_2S_3$ exhibits an insulator-metal transition and superconductivity, and this discovery would provide important insight for understanding the mechanism of iron-based superconductors. Therefore, it is intriguing to investigate the high-pressure properties for the iron-based ladder material $Ba_{1-x}Cs_xFe_2Se_3$ system. The parent compounds $BaFe_2Se_3$ and $CsFe_2Se_3$ show insulating and magnetic ordering features. For $Ba_{1-x}Cs_xFe_2Se_3$ system, no magnetic ordering is observed for x = 0.25 and minimum charge gap was estimated for x = 0.65. The insulator-metal transitions are observed in both materials.

1. Introduction

The discovery of iron-based superconductors had a considerable impact on condensed matter physics, triggering rapid efforts in studying the interplay between crystal structure, magnetism, and superconductivity [1]. Several kinds of iron-based superconductors have been discovered at this moment. These materials have a two-dimensional (2D) iron lattice as a common feature, despite of having different crystal structures. These compounds exhibit characteristic magnetic phases adjacent to the superconducting phase based on the 2D iron lattice. While a striped magnetic order is observed in the magnetic phases for 1111(e.g. LaFeAsO), 122(e.g. $BaFe_2As_2$), 111(e.g. LiFeAs) and 11(e.g. FeSe) type iron-based superconductors, a block type magnetic order was revealed in 245(e.g. KFe_4Se_5) type iron-based superconductor. Such a variety has substantial importance to understand the interplay between magnetism and superconductivity [2].

Recently, an iron-based quasi-one-dimensional compound AFe_2X_3 (A= Ba, K, Cs, and X = S, Se) having the spin-ladder arrangement of Fe atoms has attracted much attention. $BaFe_2S_3$ is an insulator having the stripe magnetic order, in which the magnetic moments are arranged to form ferromagnetic units along the rung direction, stacking antiferromagnetically along the ladder direction. This stripe-type magnetic structure is reminiscent of the 1111 and 122 systems found in the iron-based superconductors. Recently, $BaFe_2S_3$ was revealed to exhibit pressure-induced superconductivity having T_c of 14 K at 11 GPa [3]. Motivated by this discovery, extensive studies have been carried out for other iron-based ladder materials. In copper-based materials, the superconducting phase emerges not only in square lattice structures but also in ladder structures. The two-leg ladder material $Sr_{14-x}Ca_xCu_{24}O_{41}$ exhibits superconductivity under high pressures above 4 GPa with T_c of 12 K, which provided great insight to understanding the high-T_c superconductivity [4].

The magnetic structure of $BaFe_2Se_3$ is block-type, in which the magnetic moments form Fe_4 ferromagnetic units which stack antiferromagnetically along the ladder direction in a one-dimensional analogue of the block magnetism observed in the 245 system [5, 6]. In contrast, the magnetic structure of the $CsFe_2Se_3$ is stripe-type, in which the magnetic moments couple ferromagnetically along the rung and antiferromagnetically along the leg direction. This magnetic structure represents a one-dimensional analogue of the stripe magnetism observed in 1111 and 122 systems [7]. Since these ladder materials exhibit magnetic structures similar to those of the iron-based superconductors, they can be considered as potential superconductors.

In this study, high-pressure electrical resistivity measurements are performed for $Ba_{1-x}Cs_xFe_2Se_3$ (x = 0.25 and 0.65). The crystal is composed of a quasi-one-dimensional structure formed by edge-shared $FeSe_4$ tetrahedra with channels occupied by (Ba, Cs) atoms and these materials for x = 0.25 and 0.65 have the *Cmcm* space group, as shown in figure 1, while $BaFe_2Se_3$ has the *Pnma* space group. The crystal structure is depicted by the VESTA software [8]. The block-type magnetism is suppressed by substitution of Ba into Cs, and no magnetic reflection was observed in the powder neutron diffraction measurements for x = 0.25 [9]. It is interesting to study electrical properties under high pressure for x = 0.25, because the antiferromagnetic ordering of $BaFe_2S_3$ is suppressed with applying pressure and superconductivity appears after suppression of magnetic ordering state [10]. On the other hand, the formal valence of Fe is Fe^{2+} and $Fe^{+2.5}$ for $BaFe_2Se_3$ and $CsFe_2Se_3$, respectively. The material for x = 0.65 is chosen for the second material for high-pressure experiment, since the material for x = 0.65 shows the minimum charge gap state [9].

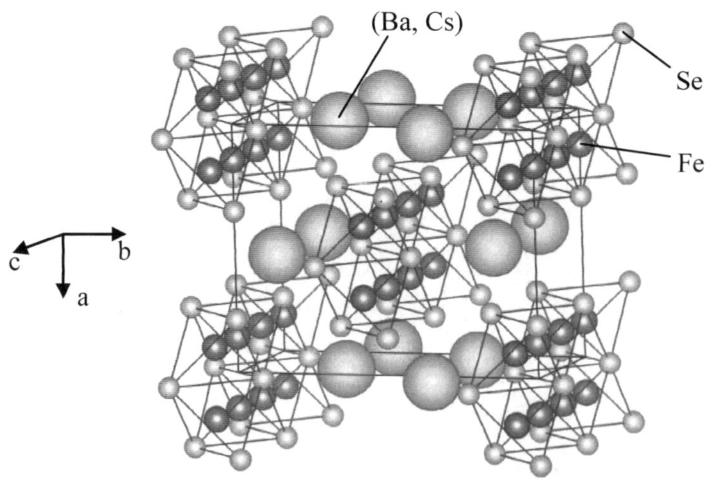

Figure 1. The Crystal structure of $Ba_{1-x}Cs_xFe_2Se_3$, consisting of edge-shared $FeSe_4$ tetrahedra extending along the c axis and of channels occupied by Ba or Cs atoms, in which Fe atoms form a ladder structure.

2. Experimental

High-pressure electric resistance measurements for single crystalline $Ba_{1-x}Cs_xFe_2Se_3$ were performed by a standard d.c. four-probe method using a diamond anvil cell (DAC). Detail of the sample preparation is described elsewhere [9]. The electric currents of 1.0, 0.1 and 0.05mA were made to flow to samples. The sample chamber of DAC was filled with powdered NaCl at the pressure-transmitting medium, using a rhenium gasket, and thin (10μm thick) platinum ribbons were used as leads for the standard d.c. four-probe analysis. The anvils having culet size of 0.5mm were used. The samples are rectangular in shape of about 0.1mm×0.1mm×0.3mm. A thin BN layer acted as electric insulation between the leads and the rhenium gasket and finely ground ruby powder scattered in the sample chamber was used to determine the pressure by the standard ruby fluorescence method.

3. Results and Discussion

3.1. $Ba_{0.75}Cs_{0.25}Fe_2Se_3$ (x = 0.25)

Figure 2(a) and (b) show the temperature dependence of the electrical resistance of $Ba_{0.75}Cs_{0.25}Fe_2Se_3$ under high pressure up to 12 GPa on a logarithmic scale and 13GPa to 18 GPa on a linear scale, respectively. This material has the *Cmcm* space group and no magnetic reflection was observed in the powder neutron diffraction measurements for x = 0.25. The electric resistance decreases with applying pressure. It shows metallic temperature dependence above 12 GPa. However, the upturn of resistance is observed up to the maximum pressure below 50 K. The double-digit reduction of room-temperature resistance is observed from ambient pressure to 18 GPa. Since about triple-digit reduction of room-temperature resistance was reported for $BaFe_2S_3$ at 11 GPa where the metal insulator transition takes place and superconductivity appears [3], the pressure of 18 GPa might be not enough pressure to induce superconductivity for $Ba_{0.75}Cs_{0.25}Fe_2Se_3$.

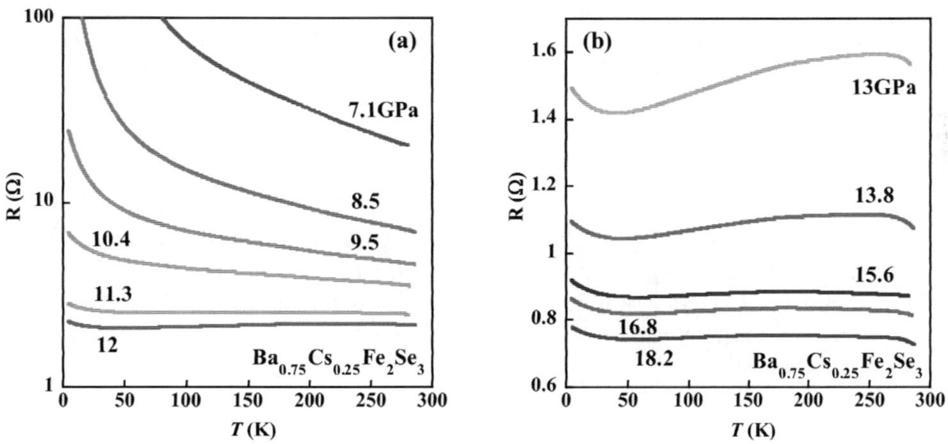

Figure 2. (a)Temperature dependence of the electrical resistance of $Ba_{0.75}Cs_{0.25}Fe_2Se_3$ up to 12GPa on a logarithmic scale. (b)Electrical resistance up to 18.2GPa on a linear scale.

3.2. $Ba_{0.35}Cs_{0.65}Fe_2Se_3$ (x = 0.65)

Figure 3(a) and (b) show the temperature dependence of the electrical resistance of $Ba_{0.35}Cs_{0.65}Fe_2Se_3$ under high pressure up to 19 GPa on a logarithmic scale and 20 GPa to 30 GPa on a linear scale, respectively. This material has the *Cmcm* space group and shows the minimum charge gap state. The electric resistance decreases with applying pressure. It shows metallic temperature dependence above 14 GPa. No upturn of resistance is observed up to the maximum pressure in the low temperature range. About 1.5 digit reduction of room-temperature resistance is observed from ambient pressure to 30 GPa. Though this material shows metallic behaviour, the pressure of 30 GPa might be still not enough to induce superconductivity for $Ba_{0.35}Cs_{0.65}Fe_2Se_3$. It is necessary to carry out the electrical resistance measurement at much higher pressure.

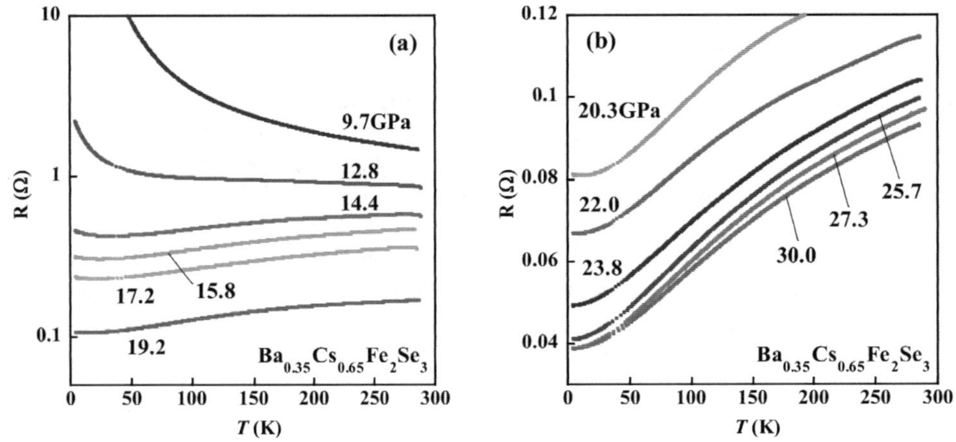

Figure 3. (a)Temperature dependence of the electrical resistance of $Ba_{0.25}Cs_{0.65}Fe_2Se_3$ up to 19.2 GPa on a logarithmic scale. (b)Electrical resistance up to 30.0GPa on a linear scale.

4. Summary

High-pressure electrical resistance measurements were performed for iron-based ladder material $Ba_{1-x}Cs_xFe_2Se_3$ (x = 0.25 and 0.65) using a diamond anvil cell (DAC). These samples exhibit the insulator-metal transition at about 12GPa and 14 GPa for x = 0.25 and x = 0.65, respectively. Superconductivity could not be observed in these experimental pressures. It has to be needed to perform higher pressure experiment for these materials.

Acknowledgements

This work was partly supported by JSPS Grants-in-Aid for Scientific Research (B) (24340088 and 26287073) and the Strategic Research Base Development Program for Private Universities (2009, S0901022) of MEXT.

References

[1] Kamihara Y, Waranabe T, Hirano M and Hosono H 2008 *J. Am. Chem. Soc*. **130** 3296
[2] Hosono H and Kuroki K 2015 *Physica C* **514** 399
[3] Takahashi H *et al* 2015 *Nat. Mater.* **14** 1008

[4] Uehara M, Nagata T, Akimitsu J, Takahashi H, Mori N and Kinoshita K 1996 *J. Phys. Soc. Jpn.* **65** 2764
[5] Nambu Y *et al* 2012 *Phys. Rev. B* **85** 064413
[6] Caron J M, Neilson J R, Miller D C, Llobet A and McQueen T M 2011 *Phys. Rev. B* **84** 180409
[7] Du F, Ohgushi K, Nambu Y, Kawakami T, Avdeev M, Hirata Y, Watanabe Y, Sato T J and Ueda Y 2012 *Phys. Rev. B* **85** 214436
[8] Momma K and Izumi F 2011 *J. Appl. Crystallogr.* **44** 1272
[9] Hawai T *et al* 2015 *Phys. Rev. B* **91** 184416
[10] Yamauchi T *et al* to be published in 2015 *Phys. Rev. Lett.*

IOP Conf. Series: Journal of Physics: Conf. Series **950** (2017) 042035 doi:10.1088/1742-6596/950/4/042035

Charge ordering in Sm_4Bi_3 under pressure

H. Kobayashi[1], S. Yamaoka[1], S. Ikeda[1], N. Hirao[2], Y. Ohishi[2], Y. Yoda[2], and A. Ochiai[3]

[1] Graduate School of Material Science, University of Hyogo, Hyogo 678-1297, JAPAN
[2] Japan Synchrotron Radiation Institute, Hyogo 679-5198, JAPAN
[3] Graduate School of Physics, Tohoku University, Sendai, 980-8578, JAPAN

E-mail: kobayash@sci.u-hyogo.ac.jp

Abstract. We have investigated the charge ordering in Sm_4Bi_3 under pressure by magnetic susceptibility and x-ray diffraction measurements. We observed discontinuous anomalies in the temperature dependences of magnetic susceptibility and volume at the charge ordering transition owing to the first-order transition and then the estimated charge ordering transition temperature decreases linearly with increasing pressure. The volume ratio between the $SmBi_6$ octahedra at two nonequivalent Sm sites indicates the partial charge ordering of Sm ions in the charge ordered phase. The evidence of the partial charge ordering in Sm_4Bi_3 is consistent with the smaller electronic component in the entropy change at the charge ordering transition.

1. Introduction

Sm_4Bi_3 and Yb_4As_3 are metals that crystallize into a cubic anti-Th_3P_4 structure with the $I\bar{4}3d$ space group at room temperature. Assuming a charge of -3 for the Bi and As ions, this indicates that the Sm and Yb ions have an averaged non-integral valence of +2.25. Thus, Sm_4Bi_3 and Yb_4As_3 are mixed valence systems and show charge ordering at $T_o = 260$ and 290 K, respectively, accompanied by a structural change from the cubic to a trigonal structure with the $R3c$ space group [1, 2]. In Yb_4As_3 with the trigonal structure, neutron and x-ray diffraction measurements [3, 4, 5] revealed that pure trivalent Yb ions order along one of the four chains along the $\langle 111 \rangle$ direction in the cubic structure.

Although Eu_4As_3 has the trigonal structure at room temperature, it was confirmed using single-crystal samples that Eu_4As_3 is the mixed valence compound and charge ordering occurs at T_o = 343 K [6]. It was demonstrated by high-pressure x-ray diffraction using synchrotron radiation that the local volume around the Eu site and the valence state of the Eu ion are closely correlated and pure trivalent and divalent Eu ions occupy two nonequivalent Eu sites in the charge-ordered state [7]. The charge ordering in mixed valence systems is one of important subjects since the mechanism of charge ordering is directly related to the feature of valence fluctuating state. However, the mechanism of these charge orderings in the rare-earth compounds with the cubic anti-Th_3P_4 structurer remains open question despite considerable efforts.

2. Experimental

Single-crystalline samples of Sm_4Bi_3 were prepared by a self-flux method. The samples obtained were analyzed by x-ray diffraction and confirmed to be a single phase. Charge ordering occurs at $T_o = 261.8(3)$ K, which is in very good agreement with previous results [1, 6, 8].

IOP Conf. Series: Journal of Physics: Conf. Series **950** (2017) 042035 doi:10.1088/1742-6596/950/4/042035

The temperature dependences of the magnetic susceptibility under pressure were measured using a piston cylinder cell by a commercial SQUID magnetometer. Hydrostatic pressure was maintained using a Teflon cell filled with Fluorinert as a pressure transmitting medium. The pressure was calibrated by a measurement of the superconducting transition temperature of Sn metal. The magnetization of the sample under pressure was evaluated from the best fit of the obtained SQUID response curve of the sample in the pressure cell by using the idealized function of the SQUID response curve as described in the literature [9].

The powder x-ray diffraction data were collected at low temperatures under pressure on BL10XU at SPring-8 with an image-plate detector. The incident x-ray wavelength was 0.4114 Å, and finely ground samples were loaded into a diamond anvil cell with ruby chips and helium as a pressure-transmitting medium. Pressure was calibrated at each experimental condition by measuring the wavelength shift of the R_1 luminescence line of the ruby chips.

3. Experimental Results and Discussion

Figure 1 shows the temperature dependences of the inverse magnetic susceptibilities, $1/\chi$, under pressure. The $1/\chi$ values under pressure represent the Curie-Weiss behavior above 150 K with a discontinuous anomaly around 250 K. The anomaly corresponds to the charge ordering of Sm ions in Sm_4Bi_3. The obtained T_o values decrease linearly with increasing pressure as shown in Fig. 2 and the pressure coefficient is found to be -12.2(3) K/GPa which agrees qualitatively with the previous result [10].

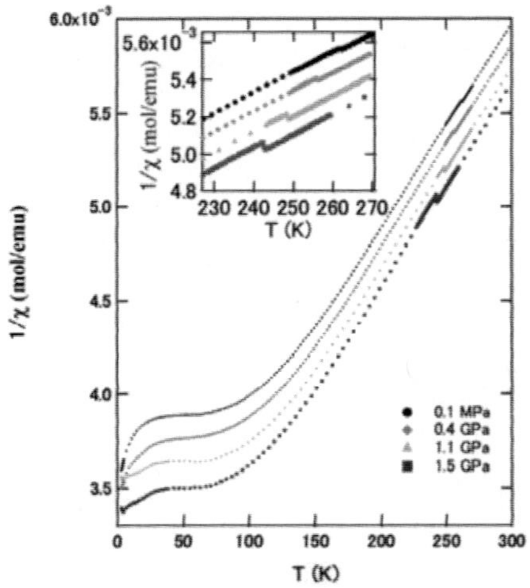

Figure 1. Temperature dependences of the inverse magnetic susceptibilities, $1/\chi$, of Sm_4Bi_3 under pressure. The inset shows the anomalies in $1/\chi$ around the charge ordering transition temperature, T_o.

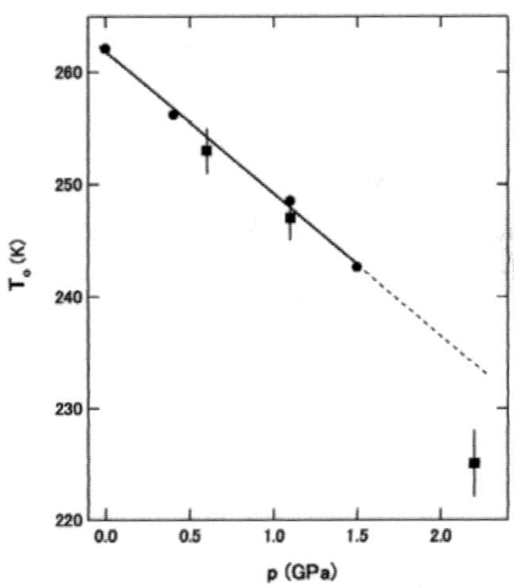

Figure 2. Pressure dependence of the charge ordering transition temperature, T_o, of Sm_4Bi_3. The solid circles and squares with error bars represent T_o estimated using magnetization and x-ray diffraction measurements, respectively. The solid line indicates the results of fitting.

Figure 3 shows selected integrated x-ray diffraction patterns of Sm_4Bi_3 at 1.1 GPa. All diffraction lines in the x-ray diffraction patterns above $\sim$ 255 K are labeled with the indices of the cubic $I\bar{4}3d$ structure and the splitting of the fundamental diffraction lines appears below $\sim$ 240 K. The diffraction patterns below $\sim$ 240 K are very similar to those of Eu_4As_3 and Yb_4As_3 observed below T_o at ambient pressure [5, 6, 7, 11]. The observed x-ray diffraction patterns were analyzed with the Rietveld refinement program RIETAN-2000 [12] using the cubic $I\bar{4}3d$ and trigonal $R3c$ structures above and below T_o, respectively. All observed x-ray diffraction lines gave good fits at low temperatures up to 2.2 GPa. Figure 4(a) shows the temperature dependence of volume, V, evaluated at 1.1 GPa. The change in V at T_o, ΔV, decreases linearly with increasing pressure as shown in Fig. 4(b). The pressure dependence of T_o determined by x-ray diffraction measurements is shown in Fig. 2. The temperature of the structural change below 1.1 GPa corresponds to the anomaly in the temperature dependence of $1/\chi$ within experimental accuracy. These results indicate that the structural change from the cubic $I\bar{4}3d$ to the trigonal $R3c$ occurs simultaneously with the charge ordering of the Sm ions in Sm_4Bi_3. The negative value of ΔV is consistent with the decrease in T_o with increasing pressure. Thus, the valence fluctuating phase is stabilized in Sm_4Bi_3 by applied pressure. This result is similar to that of Yb_4As_3, where T_o decreases with increasing pressure and the charge-ordered phase disappears above 9 GPa [13].

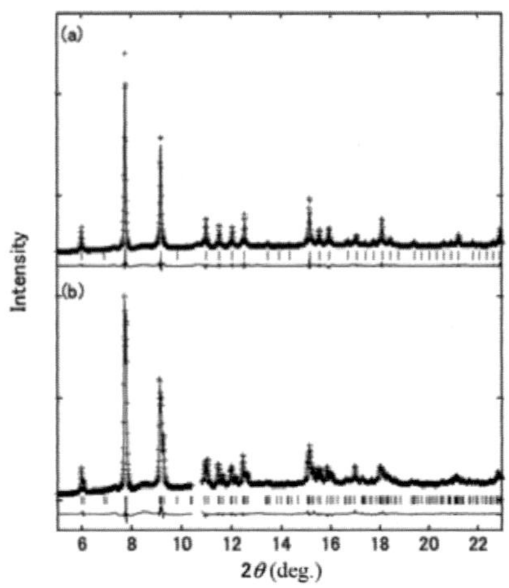

Figure 3. Selected integrated x-ray diffraction patterns of Sm_4Bi_3 at 1.1 GPa: (a) 255 and (b) 240 K, where the crosses show the observed diffraction intensities. The solid lines represent the results of Rietveld refinement fitting. The tick marks show the positions of all reflections allowed by the cubic $I\bar{4}3d$ and trigonal $R3c$ structures. The differences between the observed and calculated intensities are shown below the tick marks.

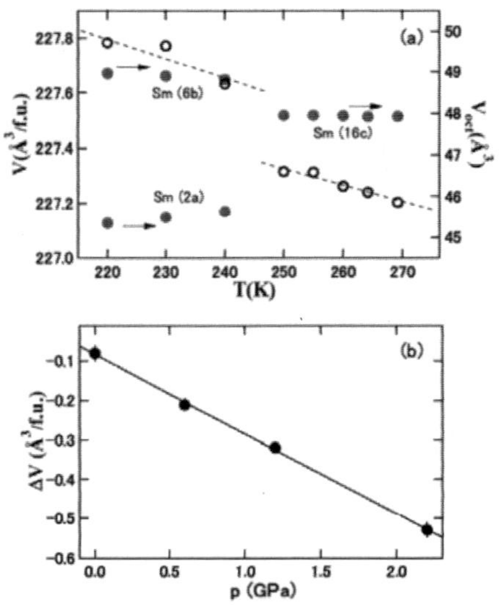

Figure 4. Temperature dependences of the volume of Sm_4Bi_3, V, and the volume of the distorted $SmBi_6$ octahedron, V_{oct}, at 1.1 GPa, (a) and the pressure dependence of volume change at the charge ordering transition temperature, ΔV, (b). In (a), the open and blue circles represent V and V_{oct}, respectively. The black solid line in (b) represents the result of fitting.

IOP Conf. Series: Journal of Physics: Conf. Series **950** (2017) 042035 doi:10.1088/1742-6596/950/4/042035

In the charge-ordered phase with the trigonal $R3c$ structure, Sm ions are in two nonequivalent 2a and 6b sites, which are coordinated by six Bi ions forming distorted $SmBi_6$ octahedra and expected to be occupied by different valence Sm ions. Since there is one Bi site in both the structures, the local volume around the Sm site is correlated with the valence state of the Sm ion, as shown in the case of Eu_4As_3 [7]. Thus, the volume of a distorted $SmBi_6$ octahedron, V_{oct}, was calculated by the refined lattice and atomic coordination parameters. The temperature dependence of V_{oct} at 1.1 GPa is shown in Fig. 4(a). The V_{oct}(6b) value is larger than V_{oct}(2a) below T_o and the weighted mean volume of V_{oct}(6b) and V_{oct}(2a) is the same as V_{oct}(16c) in the cubic $I\bar{4}3d$ structure within our experimental accuracy. The $V_{oct}(6b)/V_{oct}(2a)$ ratio was found to be 1.07(1) just below T_o at 1.1 GPa and increases gradually with decreasing temperature, reaching 1.083(3) at 16 K.

In Sm^{3+}Bi with the rock-salt structure, Sm^{3+} ions are coordinated octahedrally by six Bi ions. The expected $V_{oct}(Sm^{3+})$ was evaluated to be 42.8 ^{3} which is smaller than V_{oct}(2a) at 1.1 GPa. Since some Sm ions in Sm monochalcogenides with the rock-salt structure are divalent, the expected $V_{oct}(Sm^{2+})/V_{oct}(Sm^{3+})$ ratio was evaluated to be 1.18 in each Sm monochalcogenide from the series of the rare earth monochalcogenides [14]. Thus, the evaluated $V_{oct}(6b)/V_{oct}(2a)$ ratios are smaller than the expected $V_{oct}(Sm^{2+})/V_{oct}(Sm^{3+})$ ratio. Since the weighted mean volume of V_{oct}(6b) and V_{oct}(2a) is the same as V_{oct}(16c), the weighted mean valence of Sm ions remains below T_o. The averaged valences of Sm ions at the 2a and 6b sites were evaluated to be around +2.6 and +2.1, respectively. Accordingly, these results reveal a roughly 50% partial charge ordering in the charge-ordered phase of Sm_4Bi_3. This partial charge ordering is inconsistent with the results for Eu_4As_3 and Yb_4As_3, where the almost pure trivalent and divalent Eu and Yb ions occupy the 2a and 6b sites, respectively, in the trigonal $R3c$ structure just below T_o [4, 5, 7, 15].

The Clausius-Clapeyron relation, $dp/dT = \Delta S/\Delta V$, is applicable at the charge ordering transition of Sm_4Bi_3 because the transition is of first order. The entropy change, ΔS, was estimated to be 3.93(1) J/molK at ambient pressure from dT_o/dp = -12.2(3) K/GPa and ΔV = -0.080(7) Å^3/f.u. The estimated ΔS of Sm_4Bi_3 at ambient pressure is about ten times smaller than that (40.6 J/mol K) of Eu_4As_3 [11, 16]. If Sm_4Bi_3 is a thermal valence fluctuating state in the cubic $I\bar{4}3d$ structure, a random arrangement of Sm^{2+} and Sm^{3+} ions would be expected. The entropy of this randomness is simply calculated to be 18.7 J/mol K [8]. Because of the partial charge ordering in Sm_4Bi_3, the entropy change is expected to be approximately 10 J/molK, assuming the simple thermal valence fluctuation model in the cubic $I\bar{4}3d$ structure. This entropy change is 2.5 larger than ΔS estimated using the Clausius-Clapeyron relation.

The total ΔS includes electronic and vibrational components. On the basis of the specific heat and magnetization measurements, the antiferromagnetic magnetic ordering in Sm_4Bi_3 may occur at 2.7 K [17], which is much lower than T_o. Thus, the contribution of magnetic ordering to ΔS is negligible at T_o. In the case of Eu_4As_3, the change of the vibrational entropy of Eu_4As_3 was evaluated to be 13.4 J/molK at T_o= 343 K by using ^{151}Eu nuclear resonant inelastic scattering (NRIS) [16]. We have roughly estimated the change of the vibrational entropy of Sm_4Bi_3 to be -8.0 J/molK by using ^{149}Sm NRIS. Consequently, the electronic contribution in the total ΔS is approximately 12 J/molK at ambient pressure, which is comparable to the configuration entropy in the partial charge ordering.

In summary, we observed discontinuous anomalies in the temperature dependences of $1/\chi$ and V at T_o owing to the first-order transition and the estimated T_o decreases linearly with increasing pressure. Since the $V_{oct}(6b)/V_{oct}(2a)$ ratio was about 90% of the expected $V_{oct}(Sm^{2+})/V_{oct}(Sm^{3+})$ ratio, it is concluded that partial charge ordering occurs in the charge-ordered phase. The evidence for partial charge ordering in Sm_4Bi_3 is consistent with the smaller electronic component in total ΔS.

IOP Conf. Series: Journal of Physics: Conf. Series **950** (2017) 042035 doi:10.1088/1742-6596/950/4/042035

Acknowledgments
The experiments using synchrotron radiation were performed at SPring-8 with approval of Japan Synchrotron Radiation Research Institute (JASRI) (Proposal Nos. 2004A0154, 2004B0726, 2008B1398, 2010B1517). H.K. acknowledges the partial support of the Japanese Ministry of Education, Culture, Sports, Science and Technology, and Grants-in-Aid for Scientific Research (B) 17340098.

References

[1] Ochiai A, Suzuki T and Kasuya T 1985 *J. Magn. Magn. Mater.* **52** 13
[2] Ochiai A, Suzuki T and Kasuya T 1990 *J. Phys. Soc. Japan* **59** 4129
[3] Kohgi M, *et al* 1997 *Physica* B **230-232** 638
[4] Staub U, *et al* 2002 *Physica* B **318** 284
[5] Staub U, *et al* 2005 *Phys. Rev.* B **71** 075115
[6] Ochiai A, Shima Y and Shirakawa M 2003 *J. Phys. Soc. Japan* **72** 3174
[7] Kobayashi H, *et al* 2008 *J. Phys.: Condens. Matter* **20** 415217
[8] Ochiai A, Li D X, Haga Y, Nakamura O and Suzuki T 1993 *Physica* B **186-188** 437
[9] Sakaguchi Y, Ikeda S, Kuse T and Kobayashi H 2014 *J. Phys.: Condens. Matter* **26** 295702
[10] Jayarama A, Maines R G and Bucher E 1978 *Solid State Commun.* **27** 709
[11] Kobayashi H, Kazekami Y, Ohishi Y, Shirakawa M and Ochiai A 2005 *J. Phys.: Condens. Matter* **17** S895
[12] Izumi F and Ikeda T 2000 *Mater. Sci. Forum* **321-324** 198
[13] Môri N, *et al* 1997 *Physica* B **230-232** 630
[14] Jayaraman A *Handbook on the Physics and Chemistry of Rare Earths* Gschneidner K A Jr. and Eyring L (eds) Vol 2 (North-Holland, Amsterdam) p 575 and references therein.
[15] Iwasa K, *et al* 1998 *J. Magn. Magn. Mater.* **177-181** 393
[16] Kobayashi H, Yoda Y, Shirakawa M, and Ochiai A 2006 *J. Phys. Soc. Japan* **75** 034602
[17] Bonville P, Ochiai A, Suzuki T and Vincent E 1994 *J . Phys.* I France **4** 593

Investigation of topological phase transition in BiTeBr under high pressure

A Ohmura [1], Y Higuchi [2], T Ochiai [3], M Kanou [4], S Nakano [5], F Ishikawa [3], A Nakayama [1], Yuh Yamada [3], T Sasagawa [4]

[1] Center for Transdisciplinary Research, Niigata University, 8050, Ikarashi 2-no-cho, Nishi-ku, Niigata, 950-2181, Japan
[2] Graduate School of Science and Technology, Niigata University, 8050, Ikarashi 2-no-cho, Nishi-ku, Niigata, 950-2181, Japan
[3] Faculty of Science, Niigata University, 8050, Ikarashi 2-no-cho, Nishi-ku, Niigata, 950-2181, Japan
[4] Materials and Structures Laboratory, Tokyo Institute of Technology, 4259 Nagatsuta Midori-ku, Yokohama-city, Kanagawa 226-8503, Japan
[5] Advanced Key Technologies Division, National Institute for Materials Science (NIMS), 1-1 Namiki, Tsukuba, Ibaraki, 305-0044, Japan

E-mail: ohmura@phys.sc.niigata-u.ac.jp

Abstract. The polar semiconductor BiTeBr has the $P3m1$ symmetry, similar to that of BiTeI, in which a pressure-induced topological phase transition from a trivial semiconductor to a topological insulator has been theoretically predicted. To investigate the topological phase transition in BiTeBr, we performed X-ray diffraction and electrical resistivity measurement under high pressure, and the band calculation using experimental structural parameters. The $P3m1$ structure is stable up to the pressure of ~7 GPa; the ratio of lattice constants (c/a) reaches a minimum at around 3 GPa. Furthermore, the following phenomena are observed at around this pressure; 1) the band-gap energy closes and reopens with pressure, 2) the temperature dependence of resistivity changes from metallic to semiconducting. From theoretical prediction in BiTeI, the topological phase transition is accompanied by the band gap closing/reopening. Therefore, the results obtained in BiTeBr suggest that the topological phase transition occurs at around 3 GPa.

1. Introduction

Polar-semiconductors BiTeX (X = Cl, Br, I) are well known as materials exhibiting the giant bulk Rashba spin-split, which is caused by the combination of non-centrosymmetric structure and strong spin-orbit interaction (SOI) [1-4]. Recently, Bahramy *et al.* theoretically predicted the occurrence of pressure-induced topological phase transition in BiTeI [5]. Their report concludes that the bulk electronic states of BiTeI change from a trivial semiconductor to a topological insulator at a few gigapascals under hydrostatic condition. The valence and conduction bands near Fermi energy are mainly composed of Bi-6p and Te, I-5p orbitals, respectively. The transition is accompanied by the band-gap closing and re-opening, and the band inversion between the bands mentioned above occurs at higher pressures. Since the publication of Bahramy *et al.*, the X-ray diffraction, the spectroscopy, and the band calculation were performed on BiTeI to investigate the topological phase transition [6-8].

IOP Conf. Series: Journal of Physics: Conf. Series **950** (2017) 042036 doi:10.1088/1742-6596/950/4/042036

These experiments suggested that a minimum ratio of lattice constants, *c*/*a*, makes a significant contribution to the topological transition; the free-carrier spectral weight shows a maximum and the band-gap closes when the *c*/*a* has a minimum.

BiTeBr has a non-centrosymmetric layered structure with space group *P*3*m*1 (space group No.156), which is the same structure as BiTeI. The Bi-Te and Bi-Br bonds in each layer have covalent and ionic properties, respectively, and a unit of Te-Bi-Br three layers is weakly bonded by van der Waals forces along the *c*-axis. Since the crystal structure and the electronic state of BiTeBr are similar to those of BiTeI, the topological phase transition is also expected to occur in BiTeBr.

Therefore, we have performed synchrotron radiation X-ray diffraction and electrical resistivity measurement of BiTeBr under high pressure, and then the first-principles calculations in the framework of density functional theory using the experimental structural parameters to obtain the pressure variation of bulk electronic structures.

2. Experiments

Single crystals of BiTeBr were grown utilizing the vapor transport technique [3]. We used the modified Bridgman anvil cell for electrical resistivity measurement under high pressure. A fragment of BiTeBr with a size of $0.30\times0.75\times0.07$ mm^3 was placed into a Teflon capsule filled with a mixture of Fluorinert (FC70 : FC77 = 1:1) as the pressure-transmitting medium for quasi-hydrostatic compression. The pressure scale in the sample chamber was calibrated in advance by transition pressures of structural phase transition in bismuth. The high-pressure cell was cooled down to the temperature of ~3 K by the Gifford-McMahon 4K cryogenic refrigerator developed by Iwatani Industrial Gases Corporation.

The powder sample for X-ray diffraction was prepared by grinding fragments of single crystals in an alumina mortar that is cooled by liquid nitrogen in a glove-bag. We used a diamond anvil cell and loaded the sample into a hole drilled in the SUS301 gasket. For hydrostatic compression, the sample hole was filled with fluid helium (He), which was compressed to ~180 MPa by the gas-loading system [8]. Synchrotron radiation X-ray diffraction at room temperature was carried out at the beamline of AR-NE1 of Photon Factory in High-Energy Accelerator Research Organization (KEK) in Japan. The incident beam was tuned to the energy of 29.723 keV (λ = 0.41718(9) Å). The imaging plate was used as a detector. The experimental pressure was determined using the ruby fluorescence method [9].

First-principles calculations in the framework of density functional theory (DFT) were performed within the generalized gradient approximation (GGA), using the pseudopotential plane-wave method as implemented in the ABINIT code [10]. Electronic structures were calculated using structural parameters, which were experimentally obtained at each pressure in this study. Within the experimental unit cell, atomic coordinates were optimized based on the minimum of force on each atom. The effect of SOI on the band structure was examined by tuning the SOI parameters for the pseudopotential.

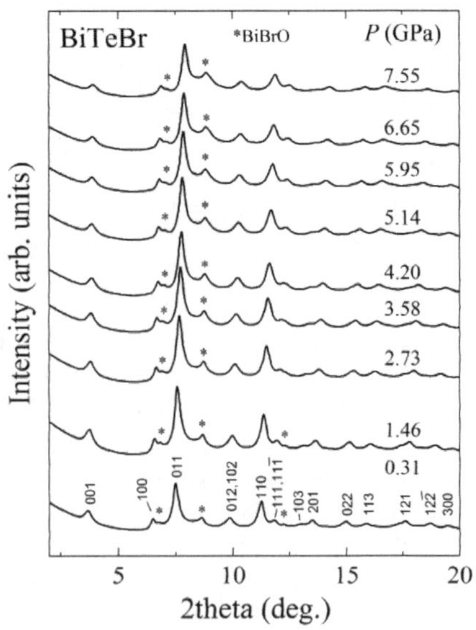

Figure 1. X-ray diffraction patterns of BiTeBr up to a pressure of 7.55 GPa at room temperature. Asterisks indicate the reflections from BiBrO.

3. Results and Discussion

Figure 1 shows the X-ray diffraction patterns with increasing pressure at room temperature. The DAC, after being filled with He, was sealed at the pressure of 0.31 GPa. The *P*3*m*1 symmetry of BiTeBr indexes most reflections at this pressure. The remaining peaks indicated with an asterisk are reflections from BiBrO (bismuth oxybromide). We think that

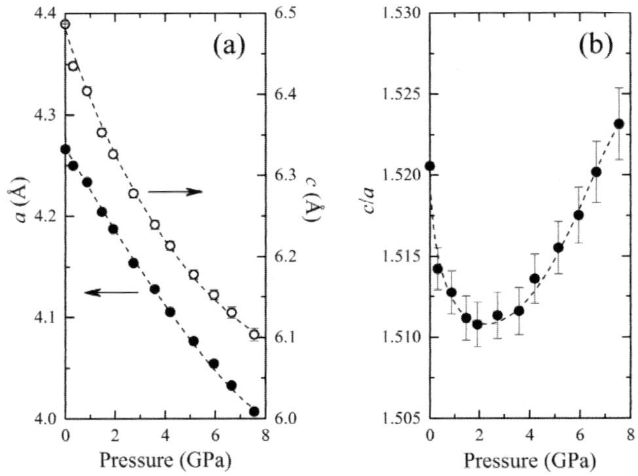

Figure 2. Pressure variation of (a) lattice constants a (closed circle), c (open circle) and (b) their ratio, c/a. Dashed lines are the guides for the eye.

BiBrO results from the reaction of BiTeBr with atmospheric moisture during the handling of fine powder. The $P3m1$ structure is stable at the pressure of 7.55 GPa since the pattern's characteristics do not change with pressure.

We applied the Rietveld analysis on the coexistent pattern of BiTeBr and BiBrO using the program of RIETAN-FP [11], and refined only lattice parameters of the $P3m1$ structure, a and c, due to the quality of diffraction patterns. Figure 2 (a) and (b) show pressure variations of a, c, and the ratio of them, c/a, respectively. Lattice constants monotonically decrease as a function of pressure. The c-axis largely decreases at lower pressures, reflecting the layered structure, while the a-axis linearly decreases. The linear compressibility of the c-axis is about twice as large as that of the a-axis. The c/a reaches a minimum between 2 and 3 GPa similar to the behavior of BiTeI.

Pressure variation of the band-gap energy, E_g, obtained from the DFT calculations in this study, is shown as solid circles in Fig. 3(a). As pressure increases, the E_g first decreases and then increases again beyond ~3 GPa. To estimate a critical pressure, P_c, at which the band-gap closes, we inverted the values along the zero-gap point, creating a mirror-like effect. Both variations, namely the actual data (positive) and the mirrored data (negative), smoothly relate to each other. Consequently, we can estimate the P_c of 3 GPa from the position of an intersection point on the zero-gap line.

This result suggests the topological phase transition in BiTeBr occurs at the pressure of 3 GPa, since the band-gap closing is necessary for the phase transition from a trivial state to a topological one. Hence, we examined the effect of SOI on the bulk electronic states at P = 1 atm and 7.55 GPa to judge whether or not the quantum states at lower and higher pressures of P_c are topologically different. The results are shown in Fig. 3 (b) and (c), respectively. The electronic states at 1 atm vary monotonically

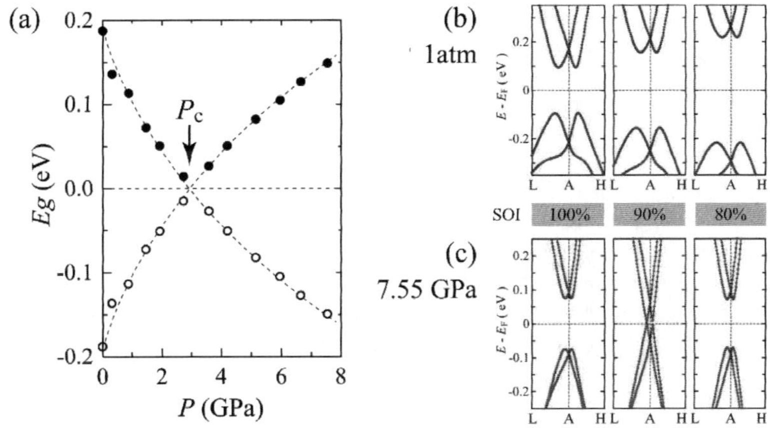

Figure 3. (a) E_g calculated at each pressure (closed circles) and its mirror image (open circles) and the effect of spin-orbit interaction on bulk band structures at (b) 1 atm and (c) 7.55 GPa.

with the decreasing effect of SOI; the Rashba-type spin-split becomes smaller, and the magnitude of E_g increases. In contrast, those at 7.55 GPa show the band-gap closing at the SOI-strength of ~ 90% and reopening at ~80 %, indicating that the quantum state returns from topological to trivial.

Finally, we show the temperature dependences of electrical resistivity of BiTeBr below 3 GPa in Fig. 4. The resistivity has the value of ~2.4 mΩ·cm at ambient conditions and shows metallic-like behavior up to pressures of 2 GPa. Applying pressure causes a change in the temperature derivative of resistivity from positive to negative. Considering the E_g closes/reopens at around 3 GPa, the semiconducting behavior suggests that the bulk carriers localizes with the opening of E_g. Furthermore, the temperature dependence of resistivity at 3 GPa has a plateau region between 50 and 150 K. We consider that this is proof of the topological phase because the plateau region suggests the dominance of surface conduction at low temperatures, and it is observed in the topological insulators [12].

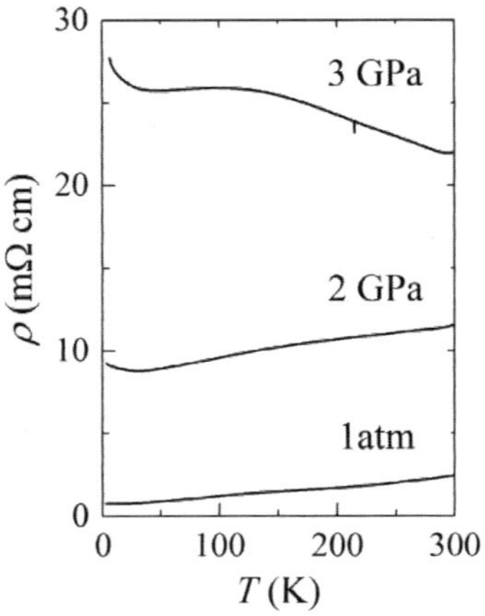

Figure 4. Temperature dependence of electrical resistivity with an increase in pressure.

4. Summary

We investigated pressure-induced topological phase transition in BiTeBr by combining theoretical and experimental studies. When the ratio of lattice constants in the *P*3*m*1 structure reaches a minimum at around 3 GPa, the closing/reopening of band-gap occurs and the transport property changes from metallic to semiconducting behavior. Furthermore, the effect of SOI on the electronic structure is clearly different above and below P_c. At 7.55 GPa above P_c, a decrease in the SOI-strength causes the reverse transition from a topological to a trivial state. From these results, we concluded that the topological phase transition in BiTeBr occurs at 3 GPa.

Acknowledgement

The present work was performed under Proposal No. 2012G162 and 2014G107 of Photon Factory, KEK. This work was supported by Grant-in-Aid for Young Scientists (B) (No. 24740229) and for Scientific Research (B) (No. 24340078), and the "Topological Quantum Phenomena" (No. 25103710) Grant-in-Aid for Scientific Research on Innovative Areas from MEXT of Japan, and Collaborative Research Projects 2013-2015 in Materials and Structures Laboratory, Tokyo Institute of Technology.

References

[1] Bahramy M S, Arita R, Nagaosa N 2011 *Phys. Rev.* B **84** 041202(R)
[2] Ishizaka K, Bahramy M S, Murakawa H, Sakano M *et al.* 2011 *Nat. Mater.* **10** 521
[3] Kanou M, Sasagawa T 2013 *J. Phys.: Condens. Matter* **25** 135801
[4] Chen Y L, Kanou M, Liu Z K, Zhang H J, Sobota J A *et al.* 2013 *Nat. Phys.* **9** 704
[5] Bahramy M S, Yang B -J, Arita R, and Nagaosa N 2012 *Nat. Commun.* **3** 679
[6] Xi X, Ma C, Liu Z, Chen Z, Ku W *et al.* 2013 *Phys. Rev. Lett.* **111** 155701
[7] Tran M K, Levallois J, Lerch P, Teyssier J *et al.* 2014 *Phys. Rev. Lett.* **112** 047402
[8] Ponosov Yu S, Kuznetsova T V, Tereshchenko O E *et al.* 2013 *JETP Letters* **98** 557
[9] Takemura K, Sahu P Ch, Kunii Y, and Toma Y 2001 *Rev. Sci. Instrum.* **72** 3873
[10] Zha C -S, Mao H -K, Hemley R J 2000 *Proc. Nat. Acad. Sci. USA* **97** 13494
[11] Gonze X *et al.* 2009 Computer Phys. Commun. **180** 2582
[12] Izumi F, Momma K 2007 *Solid State Phenom.* **130** 15
[13] Taskin A A, Ren Z, Sasaki S, Segawa K, Ando Y 2011 *Phys. Rev. Lett.* **107** 016801

IOP Conf. Series: Journal of Physics: Conf. Series **950** (2017) 042037 doi:10.1088/1742-6596/950/4/042037

Implementation of Hydrodynamic Simulation Code in Shock Experiment Design for Alkali Metals

A L Coleman[1], R Briggs[1], M G Gorman[1], S Ali[2], A Lazicki[2], D C Swift[2], P G Stubley[3], E E McBride[4], G Collins[2], J S Wark[3] and M I McMahon[1]

[1] SUPA, School of Physics & Astronomy, and Centre for Science at Extreme Conditions, The University of Edinburgh, Edinburgh EH9 3FD, UK

[2] Lawrence Livermore National Laboratory, 6000 East Avenue, Livermore, California 94500, USA

[3] Department of Physics, Clarendon Laboratory, Parks Road, University of Oxford, Oxford OX1 3PU, UK

[4] DESY Photon Science, Notkestr. 85, D-22607 Hamburg, Germany

E-mail: s0936764@sms.ed.ac.uk

Abstract. Shock compression techniques enable the investigation of extreme P-T states. In order to probe off-Hugoniot regions of P-T space, target makeup and laser pulse parameters must be carefully designed. HYADES is a hydrodynamic simulation code which has been successfully utilised to simulate shock compression events and refine the experimental parameters required in order to explore new P-T states in alkali metals. Here we describe simulations and experiments on potassium, along with the techniques required to access off-Hugoniot states.

1. Introduction

The alkali metals are commonly considered to be "simple" at ambient conditions, and are well explained by the nearly-free electron model. At higher pressures these metals have been shown to depart from this simple behaviour and exhibit more complex, low-symmetry structures [1, 2, 3, 4]. Due to limitations in resistive and laser heating in DACs, the investigation of such exotic behaviour in alkali metals has been limited to moderate temperatures. There have, however, been several computational studies which predict transitions at higher temperatures than are inaccessible using static compression techniques [5, 6]. The use of dynamic compression techniques allows the study of these transitions in previously unexplored P-T regimes.

Shock compression generates states on a Hugoniot: the loci of P-T states that can be achieved with a single shock. Off-Hugoniot states can be accessed using ramp compression, were the ideal pathway is an isentrope [7], and the region between the Hugoniot and the isentrope can be explored with multiple shocks or ramps [8]. The Hugoniot and isentrope of potassium are shown in Figure 1, with the region accessible through multi-shock compression indicated by the shaded area. When combining multiple shocks in a multi-layered target, the complexity of the wave interactions is such that the final P-T states achieved must be determined with the aid of hydrodynamic simulations [9]. Such simulations are essential both in the choice of experimental parameters and also in aiding the interpretation of experimental data in real time.

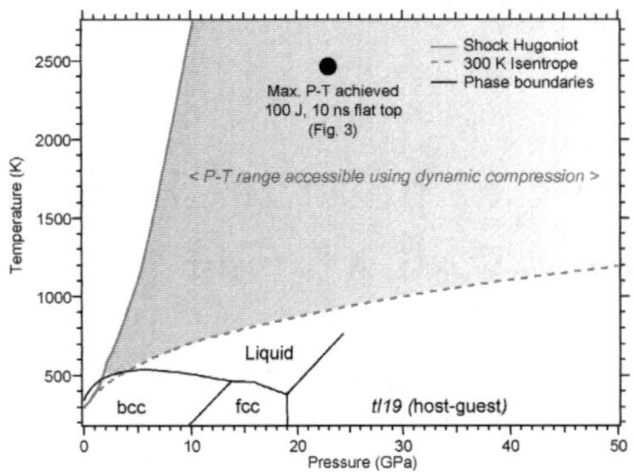

Figure 1. The potassium phase diagram [10, 11, 12] showing the P-T conditions accessible using laser compression techniques between the potassium principal Hugoniot and isentrope. The maximum P-T state achieved using a 100 J, 10 ns shock pulse is also shown.

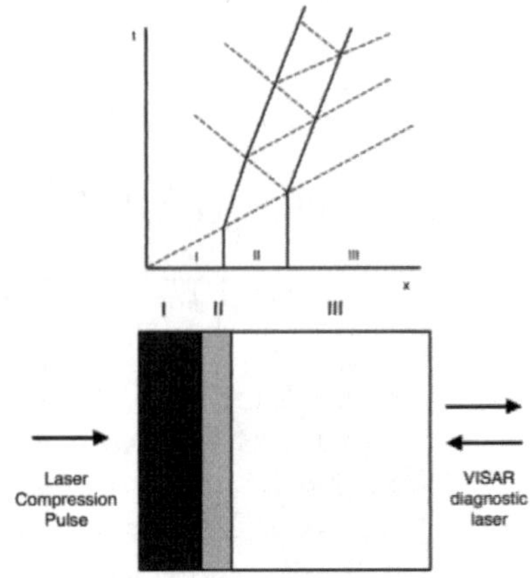

Figure 2. The lower image shows a typical target schematic while the upper image illustrates the interactions of a single shock entering the target [13].

The work presented here focuses on developing shock/multi-shock HYADES simulations in order to investigate unexplored regions of the P-T phase diagram of potassium. Potassium was chosen due to its structural similarities with Li and Na at high pressures, but with the phase transitions occurring at much lower pressures which are readily accessible at large-scale laser facilities. Preliminary laser compression data have been compared to the simulations presented here.

2. Methods

Simulations using the HYADES radiation hydrocode have been used to determine the laser compression parameters (pulse length, intensity, and double-shock pulse timing) and target parameters (ablator and sample thickness) required to compress potassium into previously unexplored P-T conditions. HYADES simulations utilise a Lagrangian mesh that moves with the target material [9]. Each zone in the mesh has a fixed mass and it is the movement of the mesh which determines the volume and the density of each zone through conservation of mass, energy, and momentum [14]. HYADES allows meaningful calculations to be executed rapidly with minimal computing expense which is advantageous in the planning of high shot-rate experimental campaigns and the interpretation of the results *in situ*.

The design of the target package is a crucial aspect of the experimental setup which must be optimised in order to access the required P-T conditions. Careful consideration of each target material's shock impedance [13] must be taken into account. The impedance matching technique allows a low-impedance material of interest to be layered between two higher-impedance materials such that a series of reverberating compression waves 'ring up' between the two higher-impedance materials until pressure uniformity in the target is achieved. These interactions are illustrated in figure 2 [13]. The lower image shows the target design consisting of the high impedance ablator (I), the lower impedance material of interest (II) and the high-impedance window material (III), while the upper image illustrates the ringing up of a shock wave through a series of reflections in the material of interest. For the potassium targets we used Al as the ablator material, with windows of LiF, as these two materials are well impedance matched, thus ensuring that there are no significant pressure gradients within the sample at peak compression. The SESAME database was used for equation of state and opacity data for

IOP Conf. Series: Journal of Physics: Conf. Series **950** (2017) 042037 doi:10.1088/1742-6596/950/4/042037

for Al, K and LiF [15].

The experimental work on potassium was conducted using the Janus laser at the Jupiter Laser Facility, LLNL, USA. This facility offers two Nd:glass (527 nm) laser beams which can provide flat-top pulses with energies up to 500 J and pulse lengths between 0.35 and 20 ns. A $1mm^2$ phase plate was selected for use with pulses between 6 and 10 ns in length using varying energies.

3. Results and Discussion

Initially, single-shock simulations were used to understand the effects of a shock propagating through the multi-layer target package discussed above. This allowed the optimisation of the target parameters, laser pulse length and energy within the physical constraints of the Jupiter Laser Facility and the availability of materials. The effect of the 'ringing up' of the shock wave was determined, along with the length of pulse required in order to sustain compression through the ablator to the rear surface of the potassium. For a flat-top pulse length of 10 ns, a series of laser energies between 25 J and 175 J were simulated for a target consisting of 25 μm Al, 10 μm K and 500 μm LiF. These simulations showed initial pressures between 1 and 10 GPa in the K layer, with a subsequent jump to pressures between 2 and 17 GPa following the immediate re-shock from the K/LiF interface. After the ringing-up of the re-shocks between the high-impedance Al and LiF layers, the peak pressures in the K layer ranged from 4.5 to 32 GPa.

These simulations agreed well with the experimental data, as illustrated in Figure 3, which compares the experimentally-determined particle velocity in the K layer resulting from a 100 J, 10 ns flat-top pulse to the output from a HYADES simulation using the same parameters. The good agreement between the simulated and experimental velocity data means that the P and T obtained from the simulation (23 GPa and 2400 K) may be used to infer a location on a P-T diagram for the shock event (as shown on Figure 1).

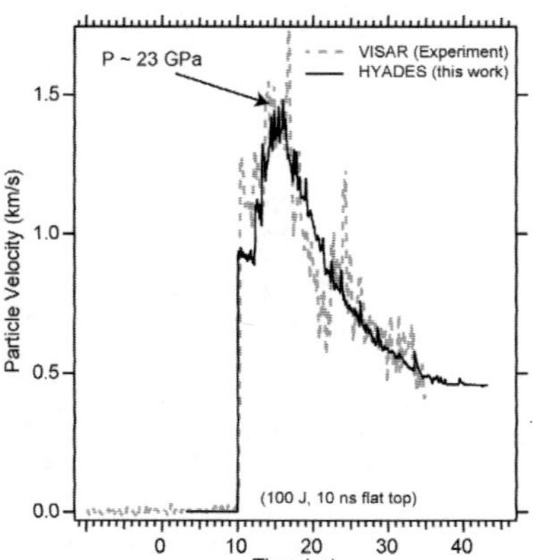

Figure 3. The experimental VISAR trace from a single 100 J, 10 ns flat-top pulse shows good agreement with the particle velocity data produced by the corresponding HYADES simulation.

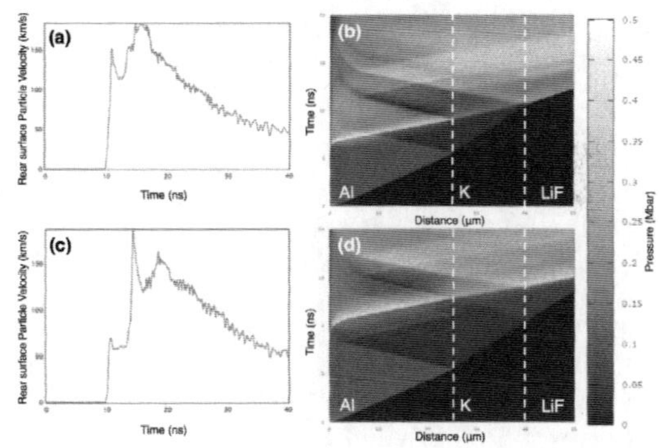

Figure 4. The upper VISAR trace and pressure plot show the results of a simulated double-shock event (50 J, 10 ns pulse and 150 J 10 ns pulse) with a delay of 6 ns between pulses, while the lower plots show the results of the same setup but with a delay of 9 ns between pulses.

More complex, double-shock experiments were then designed to further extend the range of off-Hugoniot P-T conditions that could be explored. This was achieved by first varying the initial shock in the system to take the potassium to an on-Hugoniot P-T state, with subsequent secondary shocks and further 'ring-ups' taking the compression pathway away from the Hugoniot towards an isentrope. One important aspect of this design process is the relative timing of the two laser pulses, as illustrated in figure 4, which shows a double-shock simulation in which an initial 50 J, 10 ns flat-top pulse is followed by a 150 J, 10 ns flat-top pulse. Figure 4(a) shows a simulated VISAR trace and 4(b) shows the pressure distribution in the distance-time plane, for a 6 ns delay between the beginning of the first and second laser pulses. Figures 4(c) and 4(d) show the same information for a 9 ns delay between the laser pulses. In the case of the 6 ns delay, the first compression wave is overtaken by that produced by the second pulse before the former has fully propagated through the K layer and reached the K/LiF interface. As a result, the VISAR diagnostic used during the experiment to obtain the particle velocity at the K/LiF interface will be unable to identify the particle velocity resulting from just the first shock, and instead observes only the jump to the higher velocity resulting from the combination of the two compression waves. The loss of information about the on-Hugoniot state achieved by the initial shock means that the final P-T state is not obtainable from the VISAR data. However, for the 9 ns delay, the initial shock is able to propagate fully through the K layer before the second wave catches up, allowing the on-Hugoniot state achieved by the first pulse to be determined. The final P-T state produced by the second shock is then more easily determined.

4. Conclusions

HYADES has been used to plan a successful experimental campaign on shock-compressed potassium, by optimising target design and laser pulse parameters. It has proven to be an invaluable tool in the interpretation of the complex wave interactions, and the subsequent P-T states induced, in a laser-compressed multi-layer target. The results are directly applicable to our on-going campaigns on Li and Na.

5. Acknowledgements

We thank the Jupiter Laser Facility at Lawrence Livermore National Laboratory for provision of the laser facility time and support. This work was supported by research grant EP/J017256/1 from the UK Engineering and Physical Sciences Research Council.

References

[1] McMahon M I and Nelmes R J 2006 *Chem. Soc. Revs.* **35**, 943
[2] Guillaume C L, Gregoryanz E, Degtyareva O *et al* 2011 *Nat. Phys.* **7**, 211
[3] Gregoryanz E, Lundegaard L F, McMahon M I, Guillaume C L *et al* 2008 *Science* **320**, 1054
[4] Lundegaard L F, Stinton G W, Zelazny M, Guillaume C L, Proctor J E *et al* 2013 *Phys. Rev. B* **88**, 054106
[5] Ma Y, Oganov A R and Xie Y 2008 *Phys. Rev. B* **78**, 014102
[6] Kim H, Su J T and Goddard III W A *et al.* 2011 *PNAS* **108**, 15101
[7] Smith R F, Eggert J H, Jeanloz R, Duffy T S, Braun D G, Patterson J R *et al* 2014 *Nature* **511**, 330
[8] Boehly T R, Vianello E, Miller J E, Craxton R S, Collins T J B *et al* 2006 *Phys. Plasmas* **13**, 056303
[9] Larsen J T and Lane S M 1994 *J. Quant. Spectrosc. Radiat. Transfer* **51**, 179-186
[10] McBride E E, Munro K A, Stinton G W, Husband R J, Briggs R *et al* 2015 *Phys. Rev. B* **91**, 144111
[11] McBride E E, Narygina O, Stinton G W *et al* 2012 *Journal of Physics: Conference Series* **377**, 012040
[12] Narygina O, McBride E E, Stinton G W and McMahon M I 2011 *Phys. Rev. B* **84**, 054111
[13] Forbes J W 2012 *Shockwave compression of condensed matter* (Springer, New York)
[14] Swift D C and Kraus R G 2008 *Phys. Rev. E* **77**, 066402
[15] Lyon S P and Johnson J D *Private Communications on SESAME database*

IOP Conf. Series: Journal of Physics: Conf. Series **950** (2017) 042038 doi:10.1088/1742-6596/950/4/042038

Phase transitions in shock compressed bismuth identified using single photon energy dispersive X-ray diffraction (SPEDX)

R Briggs[1], MJ Suggit[2], MG Gorman[1], A Coleman[1], R Heathcote[3], A Higginbotham[4], S Patel[2], JS Wark[2] and MI McMahon[1]

[1]SUPA, School of Physics and Astronomy, and Centre for Science at Extreme Conditions, The University of Edinburgh, Edinburgh, EH9 3FD, UK
[2]Department of Physics, Clarendon Laboratory, University of Oxford, Parks Road, Oxford, OX1 3PU, UK
[3]Department of Physics, University of York, Heslington, York, YO10 5DD, UK
[4]Science and Technology Facilities Council, Rutherford Appleton Laboratory, Harwell, Didcot, OX11 0QX, UK

E-mail: m.i.mcmahon@ed.ac.uk

Abstract. We present evidence for phase transitions in shock-compressed bismuth using the SPEDX x-ray diffraction technique. Experiments were performed on the Vulcan laser at the Central Laser Facility, RAL, Didcot, UK. We observed diffraction from the (110) bcc peak of Bi-V, and from its calculated lattice parameter the pressure was determined to be approximately 17 GPa. Upon further compression (higher laser intensities), no further diffraction from solid phases was observed. Shock melting of bismuth is thought to occur between 18 and 27 GPa. Diffraction results at lower pressures as a function of delay time are also presented.

1. Introduction

Dynamic compression of materials using high-power lasers allows access to extreme P-T states that lie well beyond the current limits of diamond anvil cell techniques. Laser facilities such as the National Ignition Facility are capable of compressing samples to 10s of megabars (> 1 TPa), whilst ensuring that the sample remains sufficiently cool to investigate solid-solid phase transitions [1]. However, collecting X-ray diffraction data from such samples is difficult, as the very high laser intensities used to produce the nanosecond-duration plasma X-ray source create a hostile environment in which the X-ray background from the drive lasers can eclipse any diffraction signal from the sample [2]. Obtaining X-ray diffraction from dynamically-compressed samples is therefore challenging at even modest pressures. Single photon energy dispersive X-ray diffraction (SPEDX) is a relatively new diffraction technique that uses CCD cameras in single photon counting mode to directly record the energy of X-rays diffracted from a laser-compressed sample [3]. Two X-ray CCDs record the diffraction at different Bragg angles, with each camera covering a wide range of k-space, allowing for recording of a significant number of Bragg peaks arising from the sample's crystal structure.

Recent shock experiments on bismuth, combined with *in situ* synchrotron X-ray diffraction, have revealed complex structural dynamics and phase transitions upon shock release [4]. After peak compression to the Bi-V phase (body centered cubic; bcc) at $\sim$ 10 GPa, the pressure in the sample then released to ambient pressure via three successive phase transitions – Bi-V (bcc) $\rightarrow$Bi-III (host-guest) $\rightarrow$ Bi-II $\rightarrow$ Bi-I – within 30 ns. More recently, we showed that on shock release from pressures above 11 GPa Bi-V melts within 3 ns [5].

Here we present evidence for phase transitions in shock-compressed bismuth using the SPEDX diffraction technique. Experiments were performed on the Vulcan laser at the Central Laser Facility, RAL, Didcot, UK. We observe diffraction from the (110) bcc peak of Bi-V, and from the measured lattice parameter the sample's pressure was determined to be $\sim$ 17 GPa. Upon further compression (higher laser intensities) no diffraction from crystalline Bi was observed, consistent with the incipient melting thought to occur between 18 and 27 GPa [6]. Also presented and discussed are lower-pressure diffraction data on compressed Bi to < 10 GPa.

2. Experimental methods

The experiment was carried out at the TAW, Vulcan facility (Central Laser Facility, Rutherford Appleton Laboratory, Didcot, UK). Two laser beam pathways were used to either drive the sample target package (with 1 or 2 drive beams) or to generate a broadband X-ray spectrum from a backlighter target foil (Fig. 1). The drive lasers delivered up to 200 J of 527 nm light onto target using a flat top laser pulse with pulse length of 6 ns, which shock-compressed samples to pressures of $\sim$ 20 GPa and below. The drive spot was 3.2 mm in diameter, generating intensities of $\sim 10^{11}$-10^{12} W/cm^2. Similar laser intensities were shown to shock compress Bi to peak pressures of $\sim$ 14 GPa using the same ablating material [5]. Hydrocode simulations in Gorman *et al.* revealed a steady shock in 15 μm of Bi with a 20 ns pulse length. The flat top laser pulse of 6 ns used in these experiments ensures a steady shock within the diffracting layers of the sample (since X-ray diffraction here is collected in reflection geometry).

Six beams were used to drive the x-ray backlighter, delivering up to 600 J of 1053 nm light in 1 ns with a spot size of < 1.0 mm and intensities of $\sim 10^{14}$ W/cm^2.

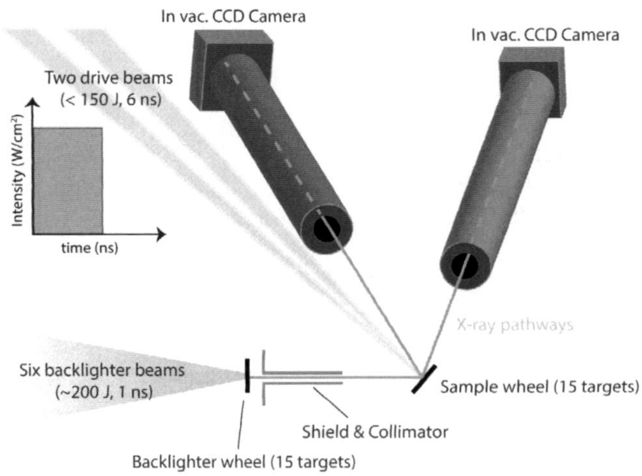

Figure 1. Experimental setup within the TAW target chamber.

Samples consisted of 50 μm Bi metal foil coated with 20 μm of CH-N polyimide ablator with a 100 nm Al flash coating. Backlighter targets (used to generate a quasi-white light source between 3-9 keV) were mixed-metal coatings, produced by the CLF target fabrication group.

Both sample and backlighter targets were mounted on a multi-target wheel that was remotely controlled to rotate and align new targets after each shot, a technique developed to increase the shot rate during each experimental shift.

Two in-vacuum X-ray CCD cameras were used to collect scattered X-rays at two different Bragg angles. The experimental setup is shown in Fig. 1. The two cameras were placed approximately 50 cm from the sample wheel and were shielded with thick Al tubes to reduce the background noise. Plastic filtering was also added to the end of the Al tubes to preferentially reduce the number of background photons near 3 keV that were generated from the mid-Z backlighter samples. Both cameras were water-cooled *via* an external chiller to minimize the number of dark counts recorded by the CCDs. The X-rays were collimated using a Mo collimator with Pb shielding, thereby ensuring that only an X-ray spot size of $\sim$ 0.5 mm^2 was incident upon the driven area of the sample.

The two drive laser beams were timed with respect to each other such that the samples could be studied before, during and after the shock wave reached the bismuth sample. This allowed us to collect ambient diffraction from undriven targets at early times, diffraction from driven samples at the peak compression, and observe possible phases on shock release from the peak state, such as liquid-Bi or Bi-III / Bi-II, as observed by Hu *et al.* in their synchrotron-based study [4].

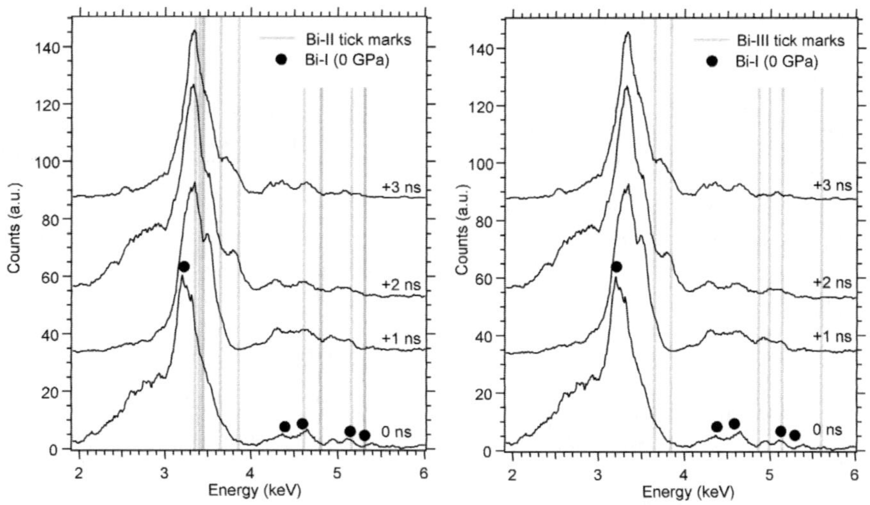

Figure 2. SPEDX diffraction data collected at low pressure P < 5 GPa with tick marks (grey shaded lines) of the Bi-II (left) and Bi-III (right) crystal structures. Filled circles identify the ambient Bi-I diffraction peaks.

3. Results

We first investigated shock-compressed Bi at low pressures (P < 10 GPa) by using laser intensities of $\sim 10^{11}$ W/cm^2. From a previous experiment on Bi we were able to generate a pressure versus laser-intensity calibration for the same target configuration (pressure-intensity calibrated to P < 20 GPa) [5]. Figure 2 shows diffraction data obtained just before the shock entered the Bi (0 ns profile) and then three further diffraction profiles obtained at increasing time delays. These low-pressure data reveal new diffraction peaks that cannot be fitted with either the compressed Bi-I or Bi-V crystal structures. It is possible that the peaks arise from Bi-II, only stable between 2.55 and 2.7 GPa, and the vertical grey lines shown in (Fig. 2 left) show the expected Bi-II peak positions at $\sim$ 2.7 GPa [7]. Whilst there is a reasonable agreement with the observed peaks at +1 ns (Fig. 2 left), at +2 ns a new peak appears at $\sim$ 3.75 keV that does not fit very well to the Bi-II structure. Fits to the Bi-III crystal structure are shown in Fig. 2 right. Both the Bi-II and Bi-III crystal structures have some agreement to the position of this peak (Bi-II peak within 0.6 keV, Bi-III within 0.3 keV), but it should be noted that the Bi-II peak (-201) has a calculated intensity that is significantly weaker than the strong Bi-III (310) peak at this energy.

When the laser intensity was increased to beyond $\sim 10^{12}$ W/cm^2 heavy filtering of the CCD cameras was required in order to reduce the large drive noise at 3-4 keV (Fig. 3). For data collected using such drive laser intensities, the total number of counts in the signal was significantly reduced and some weak fluorescence peaks were also observed. The position of these fluorescence peaks on the CCD cameras are independent of the Bragg angle (the peaks are present at the same energy on both cameras) and we can therefore easily distinguish them from the diffracted Bi photons. Figure 3 shows an example of data collected from a driven target that shows evidence of the high-pressure Bi-V phase. The fluorescence peaks arising from the backlighter foils are identified in Fig. 3 with filled circles above them. By knowing the Bragg angle of each detector (obtained using ambient diffraction from Ta/Bi and calibrating the detector angles), the peak position of the Bi-V (110) reflection ($\sim$ 4.03 keV at 35°) can be used to determine the lattice parameter for the driven sample, which was found to be 3.68 Å. From the known equation of state of Bi [8], this gives an estimated sample pressure of $\sim$ 17 GPa. All of the remaining peaks can be fitted to the ambient Bi-I phase arising from uncompressed material ahead of the shock wave (rectangular tick marks in main figure).

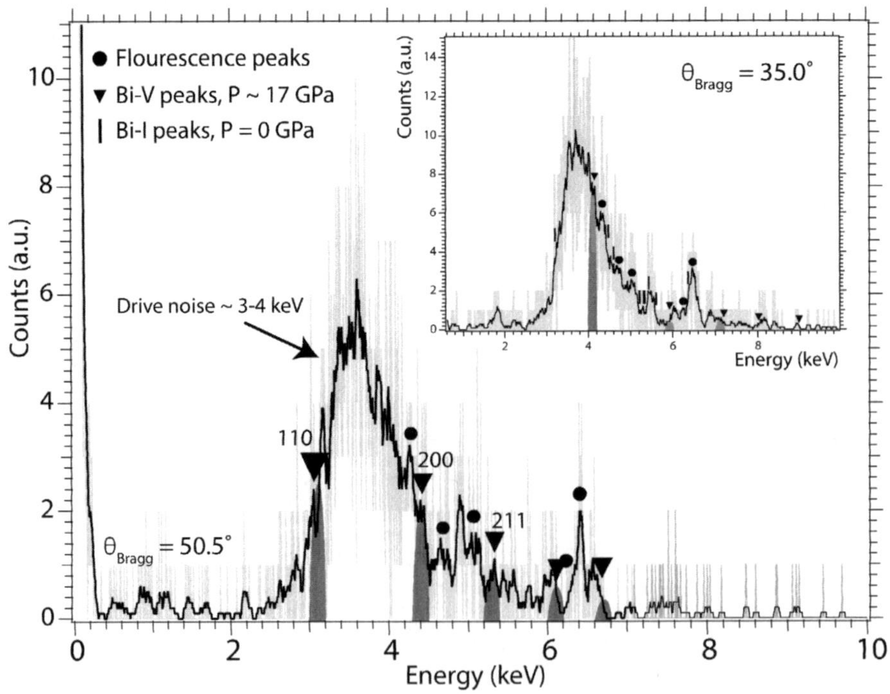

Figure 3. Diffraction data collected from the two CCD cameras at different Bragg angles. Triangles indicate the location of Bi-V peaks at a pressure of 17 GPa. Filled circles identify peaks that are present at the same energy on both cameras, and which are therefore fluorescence peaks from the backlighter foil.

Recent experiments carried out at the LCLS x-ray free electron laser (XFEL) indicate that melting of the bismuth should also be observed on shock release from pressures above 11 GPa [5]. Unfortunately, the weak diffraction signal from the liquid is difficult to extract in the current experiment as it is overwhelmed by diffraction from several high-pressures phases. In some later shots in the campaign, we used higher laser energies than those where the Bi-V phase was observed and found that we were no longer able to observe any Bragg peaks from the Bi-V phase. This could be an indication that shock melting had occurred (previously reported to occur between 18 and 27 GPa [6]) as the Hugoniot crosses the melting curve. However, the drive noise was significantly higher for these data collections and the expected liquid diffraction peak is overlapped by the drive noise. In several cases the detectors were also unable to operate in single photon counting mode, since excessive numbers of noisy pixels from the drive saturated the CCD camera, overwhelming the diffraction signal from the liquid.

4. Conclusions

X-ray diffraction studies of bismuth under shock compression have been carried out and reveal phase transitions at low pressure ($P < 10$ GPa) and observe the high-pressure Bi-V phase at $P \sim 17$ GPa, the highest pressure that Bi-V has been observed on the shock Hugoniot. It is not yet clear as to what structure/structures the samples transform to at low pressures, and further work is planned to investigate the lower-pressure phase transitions using different detector angles. This would both improve our ability to distinguish Bi-II and Bi-III peaks, and move key Bragg peaks away from the region of drive noise for the highest energy drives. We will plan to investigate using the SPEDX technique in a transmission geometry, using thinner sample targets, thereby completely removing the drive noise signal present on the CCD camera, and enabling the weak liquid diffraction signal to be observed.

Acknowledgments

We acknowledge the expertise of the staff at the Central Laser Facility, Rutherford Appleton Laboratories for their help during the experiment. M.I.M. and J.S.W. would like to acknowledge support from EPSRC under Grant No. EP/J017256/1 and EP/J017051/1.

References

[1] R Smith *et al.* 2014 *Nature*, **511**, 330-333
[2] A Lazicki *et al.* 2015 *Phys. Rev. Lett.*, **115**, 075502
[3] A Higginbotham *et al.* 2014 *Rev. Sci. Instrum.*, **85**, 033906
[4] J Hu *et al.* 2013 *Appl. Phys. Lett.*, **103**, 161904
[5] MG Gorman *et al.* 2015 *Phys. Rev. Lett.*, **115**, 095701
[6] Y Tan, Y Yu, C Dai, K Jin, Q Wang, J Hu, and H Tan 2013 *J. Appl. Phys.*, **113**, 093509
[7] O Degtyareva, MI McMahon, and RJ Nelmes 2004 *High Pressure Res.*, **24**, 319-356
[8] Y Akahama, H Kawamura, and AK Singh 2002 *J. Appl. Phys*, **92**, 5892-5897

IOP Conf. Series: Journal of Physics: Conf. Series **950** (2017) 042039 doi:10.1088/1742-6596/950/4/042039

Jetting formation of the explosively loaded powders

Kun Xue, Qiqi Yu

State Key Laboratory of Explosive Science and Technology, Beijing Institute of Technology, Beijing 100081, China

xuekun@bit.edu.cn

Abstract. The formation of jet-like structures is widely reported in the explosive dispersal of powders surrounding high explosive charges. The jetting of powder beds initiates upon the shock wave reaches the outer edge of the charge. Opposed to the interface instability theory, a hollow sphere based bulk fragmentation model is established to account for the jetting of powders. A two-phase process, namely the nucleation and free expansion of hollow spheres, corresponds to the unloading process of the powder compact caused by the rarefaction waves which governs the fragmentation of the powders. The separation between adjacent hollow spheres dictates the size of the particle clusters, which would evolve into particle jets in later times. The predicted breakup time and the size of particle jets agree well with the experimental results. The increased moisture content in powders results in an increased number of particle jets. This moisture effect can be understood in light of the varied energy distribution due to the incompressibility of the interstitial liquids trapped inside the inter-grain pores. The portion of shock energy which is not consumed in the shock compaction of the wet powders would be dissipated through the viscous shear flows during the unloading of the wet powder compact. The excessive viscous energy requires to activate more localized shear flows, accordingly leading to an increased number of particle jets. Introduction

1. Introduction

When a layer of solid particles is dispersed by a blast or shock wave, the expanding particle cloud typically forms a non-uniform structure comprised of coherent jets [1-7]. A growing body of photographic and radiographic evidence suggests that the jets form early during the dispersal process, on the timescale of the propagation of the shock wave through the particle layer [2, 3]. The observed instability timescale is much shorter than that predicted by the hydrodynamic instability theories, such Rayleigh-Tailor, and Richtmyer-Meshkov (RM) instability [2, 3]. Besides, the length scale of the instabilities assumed by the RM instability model is on the order of the particle scale, which obviously contradicts with the macroscropic jets consisting of a large number of constituent grains [1-8]. Thus investigators increasingly focus their attentions to the dynamic fragmentation of the shock compressed particle bed which is regarded as the dominant mechanism pertaining to the particle jetting.

In this work, we establish a bulk fragmentation model based on the expansion of hollow spheres to describe the decomposition process of the shocked powder rings caused by the rarefaction waves. The matrix of the hollow sphere represents the dense particle flows whose behavior satisfies the Bingham constitutive relations. Via keeping the thermodynamic variables derived from the expansion of the microscopic hollow spheres consistent with the those obtained from the macroscopic continuum deformation, we can determine the separation of the neighboring hollow spheres which corresponds to the jet size. The consistency between the theoretical jet size and the experimental results validates our model.

Both experiments and our theoretical model have found that increasing moisture content leads to an increased number of jets. This moisture effect is found to be correlated with the viscous energy dissipated through the localized shear flows. Wet powders endure much higher overpressure during the shock compaction due to the reduced compressibility. The ensuing viscous energy dissipation which depends on the overpressure becomes much intensified in the wet powders. Accordingly more particle clusters, equally particle jets, as the carriers of shear flows emerge in wet powders

2. Theoretical model based on the expansion of the hollow sphere

Our previous hydrodynamic simulations of the shock interaction with the dry and saturated sand shells revealed that the decomposition of the shocked powder compact coincides with the propagation of the rarefaction waves [9]. What is worth noting is that the overpressure in the saturated sand is higher than that in the dry sand by one order of magnitude.

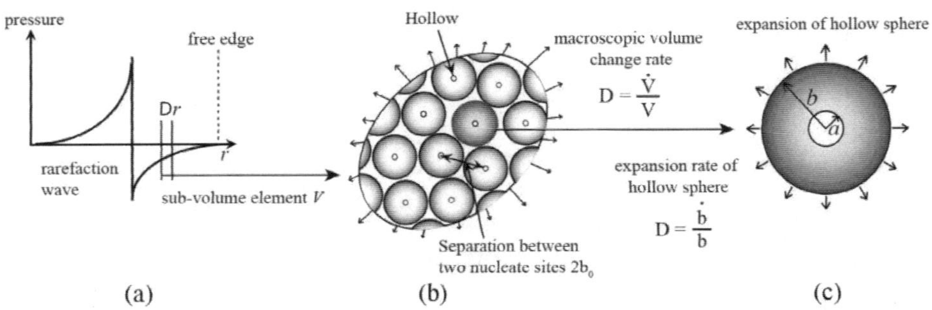

Figure 1. Illustration of the hollow sphere model applied on the spallation. (a) the unloaded sub-volume element in the wake of the rarefaction wave. (b) sub-volume element consisting of a large number of hollow spheres; (c) the expansion of the hollow sphere.

The illustration of the hollow sphere model is shown in figure 1. This model assumes that the sub-volume element V in the relaxed body is in the thermodynamic uniform state, and the growth of the voids inside can be approximated by the expansion of the hollow spheres. The inner and outer radii of the hollow sphere are $a(t)$ and $b(t)$, respectively, where the $b(t)$ represents the separation between two neighboring nucleate sites of voids. The porosity can be calculated by $f = a^3/b^3$. In the frame of the hollow sphere model, the matrix outside the hollow is treated as the continuum consisting of the dense particle flows. The hollow relates to the porosity rather than the actual voids.

The expansion of the hollow sphere consists of two sequent phases:1) $0 < t < \tau$, the hollow sphere expands while being subjected to the rarefaction, the expansion rate kept identical with the total volume change rate; 3) $t > \tau$, the hollow sphere undergoes free expansion through the momentum obtained in the first phase. The loading conditions applied on the hollow spheres is prescribed by two parameters, the expansion rate D and the duration of rarefaction waves τ derived from the hydrodynamic simulations. Table 1 presents the parameters pertaining the rarefaction waves in the dry and saturated sands.

Table 1. rarefaction wave loading parameters

materials	Expansion rate D (s^{-1})	Duration τ (μs)
Dry sand	$7\text{-}8.4\times10^3$	20
Saturated sand	$1.0\text{-}1.46\times10^3$	2

The strain rate tensor of the radially expanding hollow sphere in the spherical coordinates can be expressed as equation (1)

$$d_{rr} = -\frac{2\dot{\omega}}{3x_r^3}, d_{\theta\theta} = d_{\varphi\varphi} = \frac{\dot{\omega}}{3x_r^3} , \qquad (1)$$

where x_r is the radial coordinate of the particle, $\omega(t)$ represents the volume expansion of the hollow, $\omega(t) = x_a^3 - a^3$. The behavior of the matrix of the hollow sphere is described by the constitutive relation of the dense particle flows [10]. The stress tensor is composed by the hydrostatic pressure P and the shear stress τ_{ij} :

$$\sigma_{ij} = -P\delta_{ij} + \tau_{ij} \qquad (2)$$

$$\tau_{ij} = \eta_{eff} d_{ij}, \quad \eta_{eff} = \frac{\mu(I,\beta)P}{|d|} \qquad (3)$$

where $|d|$ is the second order invariant of the shear rate tensor, $|d| = \sqrt{1/2 d_{ij} d_{ij}}$, η_{eff} is the effective viscosity. The effective fictional coefficient μ is the function of the inertial number I and the cohesion number β as shown in equation (4)

$$\begin{gathered} \mu(I,\beta)=\mu_{\min}(\beta)+b(\beta)I, \\ I = |d|\sqrt{\frac{m}{P}}, \quad \beta = \frac{N^c}{Pd_p^2} \end{gathered} \qquad (4)$$

where m is the mass of the particle, d_p is the diameter of particle; N^c is the largest cohesion between particles, for dry sand, N^c($\sim10^{-4}$ N·m) is the van der Waals force, for saturated sand, N^c($\sim10^{-4}$ N·m) is the cohesion induced by the surface tension of the liquid. For $P\sim10^7\text{-}10^8$Pa, $d_p\sim10^{-3}$m, $m\sim7\times10^{-7}$ kg, shear rate $|d|\sim10^4 s^{-1}$, we can obtain

$$I \sim 10^{-3}, \beta \sim 10^{-3}, \quad \mu \sim \mu_{\min}(0)\sim0.25 \qquad (5)$$

Substituting equations (3-5) into equation (2), we have

$$[\sigma] = -P\begin{pmatrix}1 & 0 & 0\\ 0 & 1 & 0\\ 0 & 0 & 1\end{pmatrix} + \frac{\sqrt{3}\mu(I)P}{3}\begin{pmatrix}-2 & 0 & 0\\ 0 & 1 & 0\\ 0 & 0 & 1\end{pmatrix} \tag{6}$$

The Euler form of the momentum equation under the spherical symmetric condition is

$$\frac{\partial\sigma_{rr}}{\partial x_r} + \frac{1}{x_r}\left(2\sigma_{rr} - 2\sigma_{\theta\theta}\right) = \rho\gamma_r\,. \tag{7}$$

For dry sand, there is no traction on the inner surface of the hollow. For the saturated sand,

$$\sigma_{rr}\left(r = a, t\right) = 2\gamma/a\,. \tag{8}$$

Integrating equation (7) from a to b, for dry sand it becomes

$$\left(1 + \frac{2\sqrt{3}}{3}\mu(I)\right)P + 2\sqrt{3}\mu(I)P\ln\frac{b}{a} = \rho\left[\frac{\ddot{\omega}}{3}\left(\frac{1}{b} - \frac{1}{a}\right) + \frac{\dot{\omega}^2}{18}\left(\frac{1}{a^4} - \frac{1}{b^4}\right)\right] \tag{9}$$

Substituting (8) into (7), for the saturated sand, integrating the momentum equation leads to

$$\left(1 + \frac{2\sqrt{3}}{3}\mu(I)\right)P + \frac{2\gamma}{a} + 2\sqrt{3}\mu(I)P\ln\frac{b}{a} = \rho\left[\frac{\ddot{\omega}}{3}\left(\frac{1}{b} - \frac{1}{a}\right) + \frac{\dot{\omega}^2}{18}\left(\frac{1}{a^4} - \frac{1}{b^4}\right)\right] \tag{10}$$

Keeping the expansion rate of the hollow sphere identical with the macroscale volume change rate:

$$D = 3\frac{\dot{b}}{b} = \frac{\dot{\omega}}{b_0^3 + \omega} \tag{11}$$

The initial condition is $\omega(0) = 0$, $\dot{\omega}(0) = Db_0^3$.

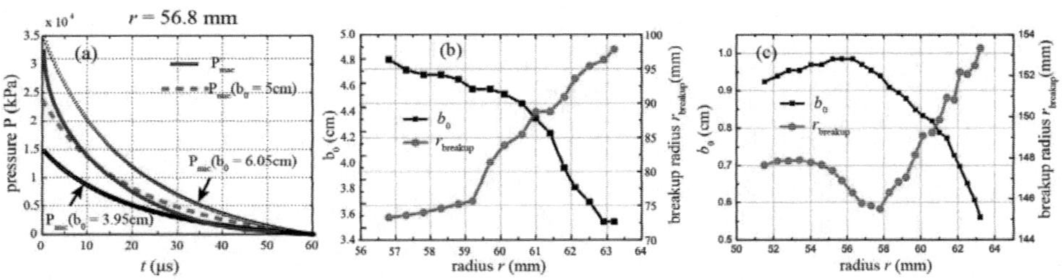

Figure 2. (a) Temporal variations in the microscopic pressure with different b_0 and the microscopic pressure in the layer with r = 56.8 mm. The theoretically estimated breakup radius and the fragment size of layers with increasing radius in dry (b) and saturated sand (c).

By solving the integration equation (9) or (10), the microscopic pressure $P_{\mathrm{mic}} = -\sigma_{rr}(b,t)$ and the porosity $f = a^3(a_0,t)/b^3(b_0,t)$ can be derived from the $\omega(t)$. Figure 2(a) presents the comparison of P_{mic} which is a function of b_0 and the macroscopic pressure, P_{mac}, obtained by the hydrodynamic simulations. The b_0 is chosen in such way that the deviation between P_{mic} and P_{mac} is the smallest. The predicted fragment size of the saturated sand, $b_{0,\mathrm{sat}} \sim$ 0.6-1cm is one order smaller than the predicted one for the dry sand, $b_{0,\mathrm{dry}} \sim$ 3-5 cm (see figure 2 (b) and (c)), indicating the refinement effect of the moisture content on particle jets observed in experiments. The dependence of the fragment size on the radius of particle layers can be attributed to the reduced strain rate rendered by the declined rarefaction waves. The number of jets around the outer perimeter of the charge is around 100 reported by experiments, consistent with the theoretical estimation on the order of magnitude [2-3].

Applying the fragmentation criteria, $f_c \sim 0.8$, the disintegration of the outer particle layers initiates at t_c ~170-200 μs. t_c is very close to the inception time of the particle jetting recorded by the X-ray radiographic photos. The breakup of the saturated particle layers occurs in the later times, suggesting that the particle jets are formed in the wet particles at delayed times and form more uniform structure [5].

3. Discussion

Grady fragmentation theory argues that the material fragments when the input energy is larger than the energy required by the creation of new surfaces combined by that dissipated during this process [11]. The surface tension is trivial in both dry and saturated particles subjected to the high strain rate loadings. Thus surface energy is negligible during the fragmentation of the particle systems. By contrast, due to the proportionate dependence of viscosity on pressure, the viscous energy dissipation in the shocked particles becomes significant due to the high overpressure.

The energy dissipation rate caused by the inelastic shear flows is

$$\varphi_v = \int_a^b \tau_{ij} d_{ij} 4\pi x_r^2 dx_r = 4\pi\mu P \int_a^b \frac{d_{kk}^2 x_r^2}{|d|} dx_r \quad . \tag{12}$$

Throughout the unloading process of the sand compact, the cumulative viscous energy in the unit volume, Φ_v is

$$\Phi_v = \int_0^\tau \frac{\varphi_v}{\frac{1}{6}\pi b^3} dt = \frac{48}{\sqrt{3}} \mu \int_0^\tau \frac{P(t)\dot{\omega}(t)}{b^3(t)} \ln\frac{b(t)}{a(t)} dt \tag{13}$$

where τ is the duration of the rarefaction waves.

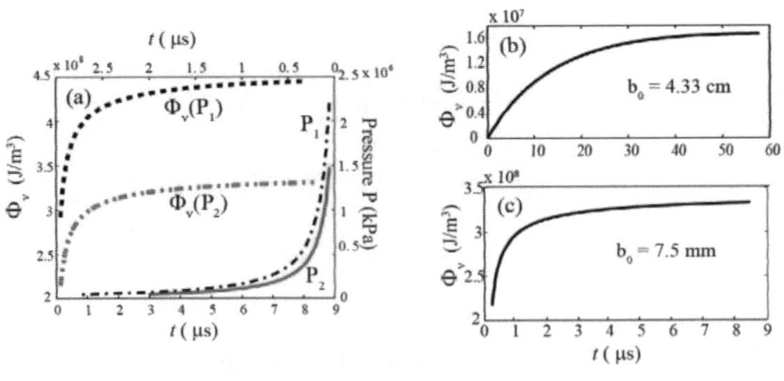

Figure 3 (a) the cumulative viscous energy dissipated in the saturated sand under different pressures, where b_0 = 7.5 mm. the cumulative viscous energy dissipated in the dry (b) and saturated sand (c)

Φ_v increases with the pressure and b_0. Figure 3 demonstrates the pressure dependence of Φ_v. A trivial increase of pressure suffices to result in the significant increase of Φ_v dissipated in the unloading process of the saturated sand. Although the b_0 is smaller in the saturated sand, the pressure effect dominates Φ_v which is one order of magnitude higher in the saturated sand than that in the dry sand. Viscous energy dissipation occurs wherever the shear flows are present. Instead of being prevalent throughout the particle system, the shear flows are localized and intensified within a number of particle clusters which would involve into particle jets in later times. The intensity of shear flows as a function of velocity gradient doesn't change much regardless of the composition of the particle system. Consequently increasing the number of particle clusters where the shear flows occur is the most effective way to dissipate the excessive viscous energy. Increased viscous energy dissipated in the saturated sand demands an increased number of particle clusters, equally more localized shear flows be activated, resulting in an increased number of particle jets.

4. Conclusion

The particle jetting is initiated upon the reflection of the shock wave on the outer edge of the particle ring, and evolves during the unloading process of the compact particles caused by the rarefaction waves. A two-phase hollow sphere model is developed to account for the unloading and decomposition of the particle compact. The predicted breakup time, the breakup radius of the expanding shell as well as the jet size are all consistent with the experimental results as least on the order of magnitude. Increased viscous energy is dissipated in the saturated sand due to its incompressibility, resulting in the intensified shear flows. Accordingly the fragmentation of the saturated sand involves an increased number of particle clusters as the carriers of shear flows, which would evolve into particle jets in later times.

References

[1] D L Frost, C Ornthanalai, Z Zarei, V Tanguay and F Zhang 2007 *J. App. Phys.* **101** (11), 113529.

[2] A M Milne, E Floyd, A W Longbottom and P Taylor 2014 *Shock Waves* **24** (5), 501-513.

[3] A M Milne, C Parrish and I Worland 2009 *Shock Waves* **20** (1), 41-51.

[4] V Rodriguez, R Saurel, G Jourdan and L Houas 2013 *Phy. Rev. E* **88** (6), 063011.

[5] K Xue, F. Li, C Bai 2013 *Eur. Phys. J. E* **36**, 95.

[6] F Zhang, D L Frost, P A Thibault and S B Murray 2001 *Shock Waves* **10** (6), 431-443.

[7] F Zhang, R C Ripley, A Yoshinaka, C R Findlay, J. Anderson and B. von Rosen 2014 *Shock Waves* **25** (3), 239-254.

[8] D L. Frost, Y. Grégoire, O. Petel, S. Goroshin and F. Zhang 2012 *Phy. Fluids* **24** (9), 091109.

[9] K. Xue, Q. Yu, C. Bai, 2014 *Eur. Phys. J. E* **37**,88.

[10] P. G. Rognon, J. Roux, M. Naaïm, F. Chevoir 2008 *J. Fluid Mech.* **21** 47.

[11] D.E. Grady, Dynamic fragmentation of solids, Shock wave science and technology, Springer Berlin Heidelberg, 3:1-108.

Experimental Modelling of Hydrocarbon Migration Processes

E D Mukhina[1,2], A Yu Kolesnikov[1], A Yu Serovaiskii[1,2] and V G Kutcherov[1,2]
[1] Gubkin Russian State University of Oil and Gas, Moscow 119991, Russia
[2] KTH Royal Institute of Technology, Stockholm SE-100 44, Sweden

E-mail: mukhina.e@gubkin.ru

Abstract. One of the most important questions in the frame of the concept of deep abiogenic origin of hydrocarbons is how hydrocarbons generated under the upper mantle conditions could migrate upward to the Earth's crust to form hydrocarbon deposits. Two different ways of fluid migration were proposed and simulated – slow migration during hundreds of years and fast migration-eruption. Influence of the fluid's migration speed on the final hydrocarbon mixture composition was studied. The received results show that the relative chemical composition of the hydrocarbon mixtures probably does not depend on the cooling conditions (the speed of the fluid migration).

1. Introduction

Our knowledge about the behavior of hydrocarbons under extreme thermobaric conditions is very limited. This knowledge is important for a wide range of topics that include the Earth's global carbon cycle, the origin of life and global size, location and distribution of hydrocarbon reserves.

According to the concept of deep abiogenic origin of hydrocarbons, complex hydrocarbon mixtures can be generated from inorganic substances under thermobaric conditions corresponding to the conditions of the deep Earth's layers. Several experimental works were devoted to this process [1]–[7]. The experiments confirming the possibility of the formation of hydrocarbon mixtures with composition similar to natural gas was published in [8].

This work continues series of authors' research devoted to the origin and phase behavior of hydrocarbons under extreme thermobaric conditions.

One of the most important questions in the frame of the above-mentioned research is how hydrocarbons generated under the upper mantle conditions could migrate upward to the Earth's crust to form hydrocarbon deposits.

It is proposed, that the process of hydrocarbon fluid's migration from the upper mantle may occur in two different ways. The way of slow ascent lasts for thousands of years; during eruption it may proceed in hours (figure 1). Different speed of the fluid depends on geological setting between the zone of hydrocarbon generation and reservoir rock in the crust – crystalline, volcanic, sedimentary – where hydrocarbons can be accumulated. We assume that rocks overlying generation zone occasionally can carry large deep faults. These deep faults can be caused by meteorite impact or seismic activity, such as shift of tectonic plates. In case of their presence upward migration should be fast, along with eruption processes. The author of [9] gives examples of magma ascent rates up to 0,04 m/s, that is relatively fast in geological timescale. Considering hydrocarbon fluid as more pervasive than melts, the speed should be even higher.

In some geological areas such as the middle of thick tectonic plates, it is possible to expect absence of the deep faults. In this case rocks surrounding the generation zone that bear fractures, fissures or pores will have to transmit the substances. Hydrocarbon fluid is presumably able to seep through any kind of rock defect. The fluid's speed in this case will be low and much more time will be necessary to migrate through tight fractures compared to deep faults.

During the process of migration variety of factors are changing around the fluid. These factors are pressure, temperature, chemical composition of surroundings. All these factors can affect the final hydrocarbon mixture, causing its phase and chemical transformations.

We attempted to simulate the migration process by changing the most important parameters: the duration of thermobaric exposition and the cooling rate.

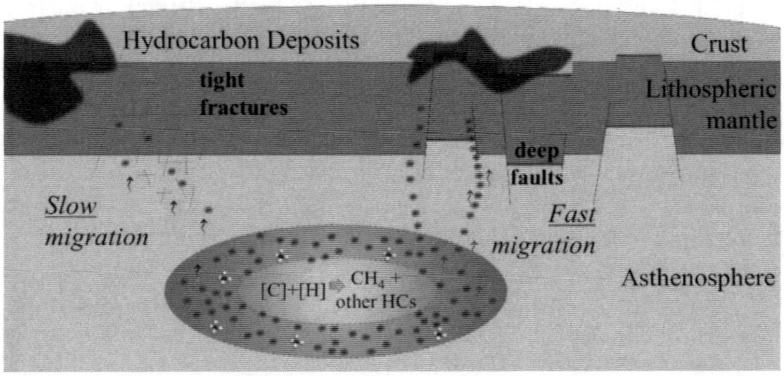

Figure 1. Two possible ways of the of hydrocarbon fluid migration.

2. Equipment and Methods

2.1. Method of high pressure and high temperature experiment with large volume

The large volume high pressure apparatus "URS-2" with "Toroid" type chamber and resistive heating was chosen to perform simulation of both generation of hydrocarbons and migration processes. The working volume of used metallic capsules (figure 2) is ~0,8 cm^3. Such volume was enough to observe the difference in gas composition between experiments recognizable by means of gas chromatography. The other important advantage of large volume technique is control the amount of initial components.

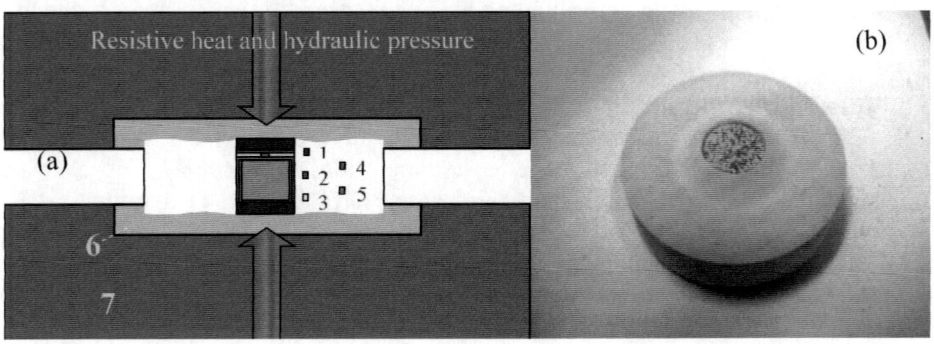

Figure 2. (a) Scheme of filled capsule inside Toroid chamber of "URS-2": 1 - graphite heaters, 2 - foil, 3 - thick paper, 4 - standard substance, 5 - filled capsule, 6 - hard alloy base, 7 - matrices-anvils, (b) photo of the Toroid chamber with a capsule inside covered by a graphite heater.

Initial inorganic substances – calcium carbonate, water and iron oxide (II) – were weighted and mixed using a mortar and placed into the metallic capsule in the exactly same relative amount for every experiment. The closed capsule filled with initial mixture was placed inside the Toroid chamber and covered by graphite heaters from both sides.

In calibration experiments the reference substance (Bi, PbTe, PbSe for pressure; Ni, Al, Sn, Pb for temperature) was placed between one heater and the capsule, covered by aluminum foil. Reference substance changed its structure during increasing pressure or temperature and corresponded resistivity jump was detected by "URS-2" device control system.

After certain exposure at chosen pressure and temperature sample was cooled and hydrocarbon gases were analyzed by gas chromatography.

2.2. Gas chromatography analysis

The gas chromatograph injecting system was modified by adding the specially constructed device where an experimental capsule was opened in the helium atmosphere and gas products from the capsule were transferred directly to the gas chromatograph.

The gas extracting cell (figure 3) consists of the body (1) with a recess and connection to the gas chromatograph, and the cap with a screw-needle (2) inside, that can move up and down, penetrating the capsule (3). After the capsule was inside the cell, the cell was closed and the air from inside of the cell was blown away with the flow of carrier gas (He). Then, the needle penetrated into the capsule and the gas products were immediately released (4) from the capsule. The sealing rubber ring (5) between the body of the cell and the cap didn't allow the gas to escape from the cell. After some time the gas spread throughout the cell and was taken to the gas chromatograph (7) with the flow of carrier gas (He).

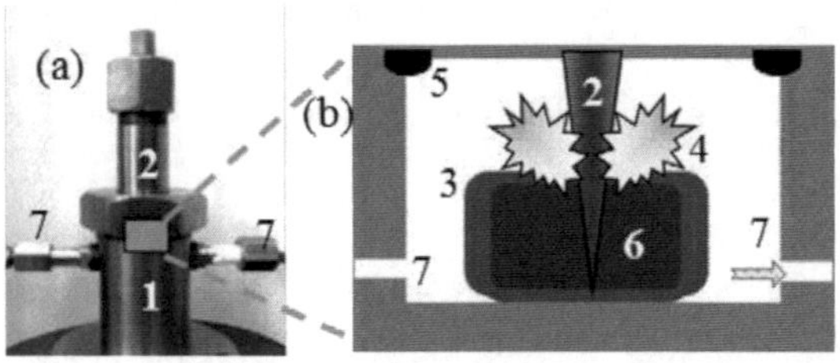

Figure 3. The gas-extracting cell outside (a) and inside (b) 1 – body of the cell, 2 – moving up and down screw-needle, 3 – capsule, 4 – hydrocarbon gas spread inside the cell, 5 – sealing rubber ring, 6 – solid product, 7 – contact with a chromatograph.

The described way of gas extracting decreases the possibility of losing some of gas compounds, and gives an opportunity to make the most reliable analysis.

3. Experiments

The main idea of the experiments was to simulate the hydrocarbon formation process and to change the experimental conditions in the one of three ways. Quenching gave us information about a "fresh-generated" hydrocarbon system. A long exposure experiment ending with quenching simulated the slow migration processes where thermobaric parameters are almost not changing in the scale of hours. The experiments with the slow cooling to the ambient temperature after certain exposure at chosen PT-parameters fulfilled general picture and modelled eruption migration process. The large volume high pressure device "URS-2" with "Toroid" chambers can perform almost any duration of PT exposition and any rate of cooling.

Five experiments were conducted, each of them was repeated at least three times. In every experiment pressure was 60 kbar and temperature was 900°C, which corresponds to the depth of ~100-150 km according to the cold geotherm [10].

Initial inorganic substances were calcium carbonate as the donor of carbon, water as the donor of hydrogen and iron oxide (II) as the simulator of oxygen fugacity near the wustite-magnetite fO_2 buffer. Assumed reaction is taken from [8]:

$$\boldsymbol{n\mathrm{CaCO_3} + (9n+3)\mathrm{FeO} + (2n+1)\mathrm{H_2O} \rightarrow n\mathrm{Ca(OH)_2} + (3n+1)\mathrm{Fe_3O_4} + \mathrm{C}_n\mathrm{H}_{2n+2}} \qquad (1)$$

Durations of exposure and cooling are presented in the table 1.

Table 1. Conditions of experiments.

	Initial substances	P, kbar	T, °C	Depth, km	Duration of exposure, s	Duration of cooling, s	Calculated migration rate, m/s
1		60	900	100-150	300	21600	~ 5,5
2		60	900	100-150	300	14400	~ 8,3
3	$CaCO_3$ + H_2O + FeO	60	900	100-150	300	2400	~ 50
4		60	900	100-150	300	1	-
5		60	900	100-150	2400	1	-

Experiments no.1 and 2 represent formation of hydrocarbon fluid during 5 minutes and slow migration from generation zone to the surface. Experiment no.3 represents fast fluid migration after 5 minutes formation. Experiments no.4 and 5 represent conditions of the formed fluid after 5 and 40 minutes of exposure respectively, or, in other words, very slow migration of the order of meter per years.

4. Results

Figure 4 shows chromatogram of the gas product from one of iteration of experiment no.2

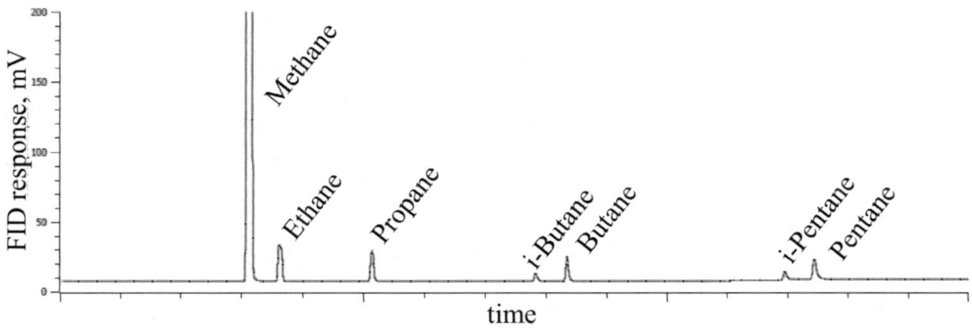

Figure 4. Chromatogram of gas product from the experiment no.5 (exposure 40 min, cooling 1 s)

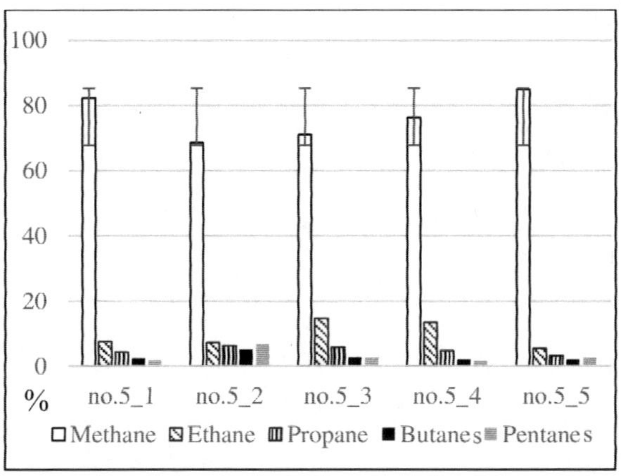

Figure 5. Standard deviation of methane in experiment no.5. Exposure = 40 min, cooling = 1 s

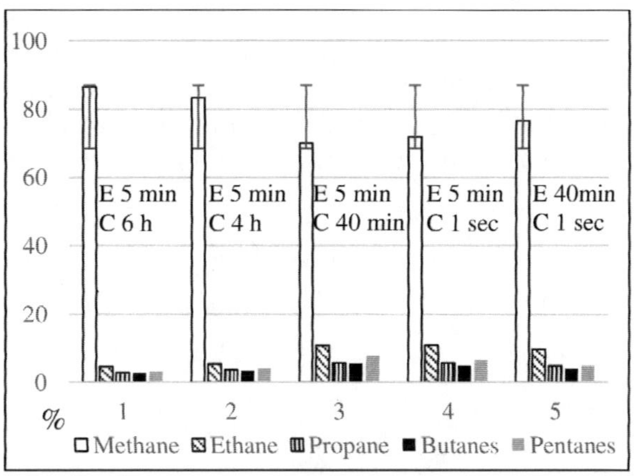

Figure 6. Relative distribution of gas product components (average values) in comparison with experiment no.5 and its standard deviation. E – exposure, C – cooling.

Figure 6 shows relative distribution of methane in the gas component in the experiments no.1-5. For better perception only first seven members of gas products are presented – methane (CH_4), ethane (C_2H_6), propane (C_3H_8), n-butane and i-butane together (C_4H_{10}), n-pentane and i-pentane together (C_5H_{12}).

The bars on the figure 6 are the standard deviation of methane in five repeated experiments no.5 (figure 5).

5. Conclusions

Hydrocarbon-forming reactions from inorganic substances under the conditions corresponding to the conditions of the Earth's asthenosphere were investigated. Composition of synthesized complex hydrocarbon mixture was similar to natural gas. To model different types of upward fluid migration we proposed several high pressure experimental procedures. Cooling during 40 minutes represented fast fluid migration, cooling during 4 and 6 hours represented slower migration and quenching represented no migration – fluid as it was formed.

The experimental results showed that the relative chemical composition of the generated hydrocarbon mixture did not depend on the cooling conditions. According to the suggested model these results mean that mantle generated natural gas does not change its composition during its upward migration to the Earth's crust to form hydrocarbon deposits.

References

[1] Chekaluk E B, Boyko G E, and Bakul V N 1969 *Probl. Geol. i sverhglubokogo burenya na Neft. i gaz v USSR (in Russ.)* 63–70
[2] Kenney J F, Kutcherov V A, Bendeliani N A, and Alekseev V A 2002 *PNAS* **99**(17) 10–15
[3] Kutcherov V G and Krayushkin V A 2010 *Rev. Geophys.* **48** (RG1001) 1–30
[4] Kolesnikov A Yu, Kutcherov V G and Goncharov A F 2009 *Nat. Geosci.* **2** 566–70
[5] Chen J Y, Jin L J, Dong J P, Zheng H F and Liu G Y 2008 *Chinese Chem. Lett.* **19** 475–8
[6] Scott H P, Hemley R J, Mao H, Herschbach D R, Fried L E and Howard W M 2004 *PNAS* **101**(39) 14023–6
[7] Mukhina E D, Kolesnikov A Yu and Serovaiskii A Yu 2014 Experimental Investigation of Natural Gas Synthesis at High Pressure and Temperature: Lower Thermobaric Limit *Deep Genesis of Oil and Gas All-Russian Conference "3rd Kudryavtsevskie Chteniya" (Moscow, Russia, 20-23 October 2014) (Proceedings of 3rd Kudryavtsevskie Chteniya)* pp. 1–3
[8] Kucherov V G, Kolesnikov A Yu, Dyuzheva T I, Kulikova L F, Nikolaev N N, Sazanova O A and Brazhkin V V 2010 *Dokl. Phys. Chem.* **433**(3) 132–35
[9] Rutherford M J and Nicholis M G 2004 *Geol. Soc. Am.* **6** 489–92
[10] Karpov I K and Zubkov V S 1998 *Geol. I Geofiz.* **11**(39) 1518–28

IOP Conf. Series: Journal of Physics: Conf. Series **950** (2017) 042041 doi:10.1088/1742-6596/950/4/042041

High Pressure Serpentinization Catalysed by Awaruite in Planetary Bodies

J Neto-Lima, M Fernández-Sampedro, O Prieto-Ballesteros

Centro de Astrobiología, INTA-CSIC, Carretera de Torrejón a Ajalvir, Km4, 28850 Torrejón de Ardoz - Madrid Spain

E-mail: jlima@cab.inta-csic.es

Abstract. Recent discoveries from planetary missions show that serpentinization process may act significantly on the geological evolution and potential habitability of the icy bodies of the Solar System, like Enceladus or Europa. Here we review the available experimental data so far about methane formation occurring during serpentinization, which is potentially relevant to icy moons, and present our results using awaruite as a catalyst of this process. The efficiency of awaruite and high pressure in the Fischer-Tropsch and Sabatier Type reactions are evaluated here when olivine is incubated.

1. Introduction

Olivine is ubiquitous throughout the Solar System, its presence is detected in a wide range of planetary surfaces, from planets to asteroids, meteorites and interstellar dust. Therefore, it is a good mineral to study the evolution of silicate-rich materials throughout our Solar system's history.

Serpentinization consists in the aqueous alteration of olivine to form minerals of the serpentinite group. In the Earth, they are part of rocks from the ultramafic upper mantle and mafic oceanic crust. Serpentinization is proposed to be an important geological process that affects the cycle of bio-essential elements not only in the Earth but in other planetary bodies of the Solar system. Oxidation of the ferrous iron of olivine to magnetite results in the release of dihydrogen (H_2) and precipitation of secondary minerals. When this serpentinization-derived fluids contact with CO_2 rich water, under certain constraints it leads to the formation of methane and other phases, as shown in the equation:

$$(Mg,Fe)_2SiO_4 + H_2O + CO_2 + Ni_3Fe \rightarrow Mg_3Si_2O_5(OH)_4 + Mg(OH)_2 + Fe_3O_4 + H_2 + CH_4$$

The most popular environments for serpentinization on Earth are hydrothermal systems such as those found in Lost City, the alkaline hydrothermal field discovered in 2000, at the mid-Atlantic ocean.

Serpentinization occurs under a wide range of conditions of pressure and temperature (from 40ºC to 500ºC), pH values have to be between 9-11, otherwise serpentinization is inhibited. Temperature is rate controlling, meaning higher the temperature faster the olivine hydrolysis; and pressure assists the diffusion rates of water through olivine and serpentinized rocks, not enough water at the serpentinite-olivine interface limits the rate of the reaction [1].

Under the temperature constraints expected for the icy moons seafloor, and without the presence of a catalyst, abiotic CH_4 formation would be extremely slow. Such catalyst has to have a reliable and

IOP Conf. Series: Journal of Physics: Conf. Series **950** (2017) 042041 doi:10.1088/1742-6596/950/4/042041

constant source, and to remain stable under these conditions. Two naturally occurring, serpentinization products that can act as methane catalysts are, iron- oxides (such as magnetite (Fe_3O_4)) and awaruite. Magnetite can further react with dihydrogen, producing ferrous iron that can produce a nickel-iron alloy, like awaruite, if nickel is present in the olivine that was serpentinized.

Awaruite is a nickel-iron alloy with a composition varying from Ni_2Fe to Ni_3Fe. It was first discovered embedded in black grains of sand collected from the West Coast of the south island of New Zealand associated to a subduction context and the occurrence of serpentinites [2]. Since then, awaruite is usually connected to low temperature serpentinization processes of ultramafic and mafic rocks. During this alteration process, awaruite acts as a surface catalyst for the series of reactions that lead to the formation of methane, from inorganic carbon [3].

The discovery of serpentine-hosted vent systems on Earth's seafloor coupled with fossil records to support the sustainability of high-biomass communities by them, indicates the possibility that such systems may have played important roles in the emergence of life on Earth's primitive oceans. Hydrothermal vents are proposed to exist in the rocky layer in contact with the global ocean of Europa [4] [5], which was indirectly detected by the Galileo spacecraft. Some of the structures observed on the surface strongly point towards the idea that a liquid subsurface ocean exists and periodically, through different processes exchanges materials to and from Europa's surface (e.g. cracks on the icy shell or local melting or subduction episodes). These endogenous materials could be measured by future missions, including methane vented from plumes, if they occur. Understanding serpentinization is paramount to grasp how methane behaves beneath the icy moons ice shell and how it can originate redox gradients that can be used by chemotrophic communities.

2. Materials and Methods

The experiments we report are carried under relatively low pressure so the influence of the presence of awaruite could be observed and quantified. To simulate the hydrothermal environment in which methane can be formed, a constant temperature of 120ºC is maintained.

Although iron is the main driver for methane formation, forsterite is selected as the olivine phase in these experiments because of Galileo's Near Infrared Mapping Spectrometer (NIMS) detection of Mg-hydrated materials on Europa's icy shell [6]. In addition to that, forsterite is detected in many meteorites, thus is suggested to be a primary condensate of the solar nebula [7].

Natural forsterite is used, which XRD patterns are compared to reference samples (Joint Committee on Powder Diffraction Standards file nº. 34-189). Composition is determined to be almost pure forsterite (Mg_2SiO_4) and enstatite ($MgSiO_3$).

Our experiments are run in order to understand the best olivine:awaruite:water ratio for CH_4 catalysis reactions.

3. Results

3.1. Awaruite Synthesis

The awaruite crystals used throughout our experimental assemblages are synthesized via hydrazine hydrate reduction in an ethanol solution (2:3 ratio) [8].

$$3Ni^{2+} + Fe^{2+} + 2N_2H_4 + 8OH^- \rightarrow Ni_3Fe + 2N_2 + 8H_2O$$

Our main goal being the production of enough amount of this alloy for the serpentinization simulation experiments, the perfect crystallization became secondary, therefore the amount of the iron and nickel salts used in the reaction is increased tenfold. Collection of the highly magnetic particles is carried using a magnet, thus limiting impurities. The collected particles are first washed with a solution of ultrapure water and absolute alcohol (4:1 ratio), and then placed in an oven to dry. Characterization of the awaruite is obtained through Scanning Electron Microscope (SEM) coupled with EDS microanalysis system and X-ray powder diffraction (Fig. 1).

IOP Conf. Series: Journal of Physics: Conf. Series **950** (2017) 042041 doi:10.1088/1742-6596/950/4/042041

Element	Weight%	Atomic%
O K	3.78	12.47
Na K	-0.32	-0.74
S K	0.41	0.67
Cl K	0.23	0.34
Fe K	23.95	22.62
Ni K	71.96	64.65
Totals	100.00	

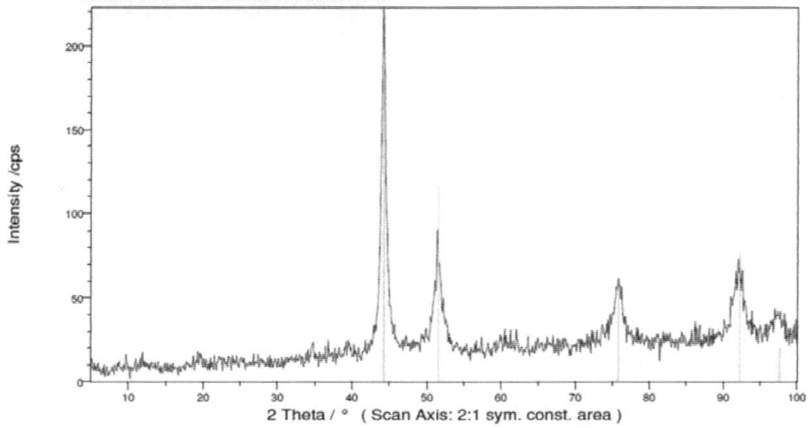

Figure 1. Awaruite SEM image with EDS microanalysis results. The impurities detected result from the initial metallic salts, and NaOH used during synthesis. X-Ray diffraction pattern of resultant powder synthesized by hydrazine reduction, in an ethanol solution (2:3 ratio). XRD patterns obtained are compared to natural awaruite from Awarua Bay, New Zealand (Joint Committee on Powder Diffraction Standards.

3.2 Serpentinization Runs

The synthesized awaruite together with natural pulverized forsterite into a fine grain powder, are used in this serpentinization simulation experiment under relatively low pressures. The experiment is assembled according to Table 1:

Table 1. Experimental setup for the low pressure experiment

Reactant mass (gr)	Run					
	0	**1**	**2**	**3**	**4**	**5**
Olivine	-	1	1	1	2	-
Awaruite	-	-	0,005	0,010	0,005	0,005
CO_2	CO_2 pellet [dry ice]					
Water	5ml [ion chromatography purity]					

Glass vials hermetically sealed with a vulcanized rubber cork, Teflon and an aluminium cover, are heated up to 120ºC.

The pressure inside each one, which is slightly higher than 1bar, is the result of the release of CO_2 by the dry ice pellets and the increasing water vapor as the temperature rose.

After 34 days at 120ºC, the solid fractions are analyzed with XRD to verify the alterations suffered by awaruite and olivine.

A mixture of forsterite olivine and awaruite, respecting the ratios used on the low pressure experiments is loaded into twin high pressure chambers. In it, a constant temperature is maintained using external heating and 300bar of pressure are kept using a RUSKA hydraulic system. The carbon dioxide for this serpentinization run is added from a pressurized bottle, 2mol of CO_2 were injected into the twin chambers before pressurizing them at 300bar.

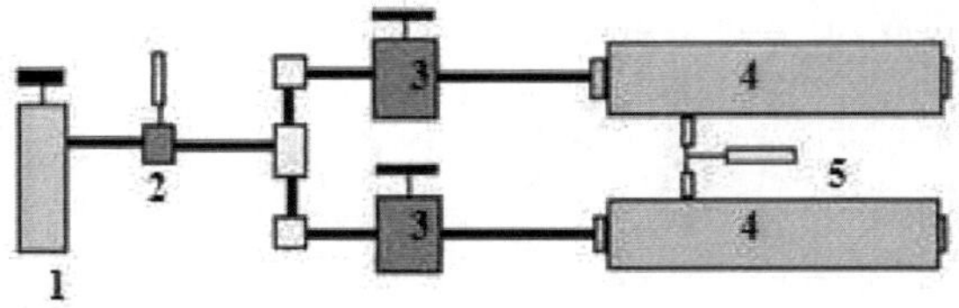

Figure 2. High Pressure Experimental Assembly (1.CO2 bottle; 2.Pressure Sensor; 3.Valves; 4.High volume high pressure chambers; 5.Temperature sensors)

4. Results

The use of CO_2 pellets in our experiments allows a faster dissolution of the gas in the water and consequently the acidification of the aqueous system. Serpentinization can only occur under alkaline conditions, with pH values between 9-11, therefore new experiments are planned for the near future in which a redox/carbon-source buffer will be used, in this case a sodium carbonate. Sodium is chosen because it is detected in Europa's atmosphere by Galileo's Near Infrared Mapping Spectrometer (NIMS)[9] and is proposed to originate from the global liquid ocean bellow Europa's ice-shell.

Subtle changes in the crystallinity of the solid samples are evident in the diffractograms of the products.

At low pressures, hydrogen formation is favoured, and it is more thermodynamically favourable than methane, as pressure increases methane concentration in the system becomes comparable to hydrogen. Higher the pressure, higher the concentration of methane compared to hydrogen. According to experimental data, the high pressure conditions observed on Earth's upper mantle favour the formation of methane. [10] Combining high pressure, with the presence of a nickel-iron alloy such as awaruite, the rate of abiogenic formation of methane raises considerably [3].

5. Summary and Conclusions

The understanding of the action of methane catalysts such as the nickel-iron alloy awaruite, during the serpentinization process of the magnesium end-member of the olivine family, forsterite, becomes a priority in order to grasp the complex processes involved in the carbon cycling in Europa.

References

[1] Macdonald AH and Fyfe W S 1985 *Tectonophysics* **116** 123-35
[2] Ulrich FGS 1890 *QJGS* **66** 619-32
[3] Horita J and Berndt M E 1999 *Science* **285** 1055-57
[4] McCollom TM 1999 *J. Geophys. Res.* **104-E12** 30.729-42
[5] Zolotov MY and Shock E L 2003 *J. Geophys. Res.* **108(E4)** 5022
[6] McCord T B, Hansen G B, Combe J-P, Hayne P 2010 *Icarus* **209** 639-50
[7] Weinbruch S, Palme H, Spettel B 2000 *Meteorit. Planet. Sci.* 35 161-71
[8] Abellán G, Carrasco J A, Coronado E and Prima-García H 2014 *J Sol-Gel Sci Technol* **70** 292-99
[9] Johnson, R E 2000 *Icarus* **143**, 429-33
[10] Scott H P, Hemley R J, Mao H-K, Herschbach D R, Fried L E, Howard W M and Bastea S 2004 *PNAS* **101** 39 14023-26

IOP Conf. Series: Journal of Physics: Conf. Series **950** (2017) 042042 doi:10.1088/1742-6596/950/4/042042

Salting-out phenomenon induced by the clathrate hydrates formation at high-pressure

A S J Méndez[1], V Muñoz-Iglesias[2], F Izquierdo-Ruíz[1, 3], O Prieto-Ballesteros[1]

[1]Centro de Astrobiología (INTA-CSIC). Ctra. Ajalvir km. 4. 28850 Madrid. Spain.
[2]Jet Propulsion Laboratory, California Institute of Technology, 4800 Oak Grove Drive, Pasadena, CA 91109, United States.
[3]Dept. Química Física y Analítica, U. Oviedo, Julián Clavería, 8, 33006 Oviedo, Spain.

E-mail: asanjose@cab.inta-csic.es

Abstract. Here we show the salting out phenomenon assessment occurring on $MgSO_4$-CO_2-H_2O system due to the presence of clathrate hydrates under temperature and pressure conditions of the ocean of Jupiter's icy moon Europa. In order to form clathrate hydrates, water molecules are removed from the aqueous solution. This fact causes great impact on solutes concentration, including available nutrients in solution so habitability is directly affected. Both, thermodynamical history recorded in the experiments and visual features following of the system help to understand the truly effect of salting-out on geology and its expression on Europa's surface in application to space exploration.

1. Introduction

Salting-out is the chemical response of a system in which electrolyte and non-electrolyte compounds compete with each other for their dissolution in an aqueous environment. Commonly, when any salt and gas are dissolved in water, gas solubility goes down if the ionic strength increases, which is directly proportional to salt concentration. However, under certain conditions, a similar phenomenon takes places in the opposite way.

One example of geological feature from salting out phenomenon in terrestrial environments is the called brinicle [1]. They are tubular formations, which form in polar regions. Beneath the icy crust of these regions, two-dimensional crystals of water ice known as platelets crystallize from seawater. This process implies salt exclusion, and therefore the formation of brines in interstitial spaces between platelets. As water freezes in the brine inclusions, it generates a volume expansion and the crystal lattice undergoes an increment of internal pressure. This fact may provoke cracks in the ice pack causing release of trapped brine. This dense, cold brine drains downwards from sea ice freezing the water that is in direct contact with and forming a tube of ice.

$MgSO_4$-CO_2-H_2O system is proposed as an approach of Europa's ocean composition [2]. In this particular system, CO_2 may form clathrate hydrates under the properly conditions. Clathrate hydrates are ice-like crystalline structures, which trap individual gas molecules in cages. This fact makes these minerals to have a starring role in planetary science due to their ability to act as efficient reservoirs of carbon-bearing species like CH_4 and CO_2 [3]. In order to form clathrate hydrates, water molecules are removed from the liquid phase. As a consequence, salt concentration increases and whether this situation continues long enough, hydrated salts may precipitate [4].

Several simulation experiments are carried out at high pressure followed by Raman spectroscopy analysis, which characterize involved chemical species in formation-dissociation of clathrate hydrates from salty water. Here we evaluate how Europa's geochemistry and even habitability could be conditioned by this phenomenon.

2. Materials and methods

In order to characterize the salting out process, $MgSO_4$-CO_2-H_2O system is studied under pressure. Reagents used to perform the experiments are $MgSO_4 \cdot 7H_2O$ ($\geq$ 99.5% purity, Sigma–Aldrich, USA), CO_2 gas (99.998% of purity, Praxair, USA) and Milli-Q water with total organic carbon (TOC) lower than 5 ppb. To run the salting-out experiment it is used a thermostated high pressure chamber. The chamber is equipped with a sapphire window, which allows both optical examination and Raman spectroscopy analysis to follow all changes involved in the evolution of the system. Circulation Thermostat Unistat 815 w (Huber, Germany) is used to control sample temperature, while the cell is connected to a tank of CO_2 (g) by high pressure tubing. A Nd:YAG solid state laser with a wavelength of 532 nm non polarized is used for Raman analysis. After focusing onto a monochromator (Horiba JobinYvon HRi550, 550 mm optical length), with a diffraction grating of 1800 grooves/mm, the scattered light is detected with a Charge Coupled Device (CCD) with 1024x256 pixels cooled to 203 K for thermal-noise reduction. Pixel resolution of the equipment is 2.16 cm^{-1}/pixel (binning factor=1). Spectral resolution, with a width slit of 50 μm, results better than 10 cm^{-1}.

Thermodynamical history of the ternary system is recorded to identify all involved transformations. For this purpose, pressure and temperature data are monitored and saved using a personalized Labview program.

Salting-out induced by clathrate hydrate formation is proposed to occur in the aqueous ocean of Europa [5]. Although its distribution is uncertain yet, it is suggested that lays between 20-100 km (24-120 MPa, considering that gravity is 1.314 $m \cdot s^{-2}$ and crust composition is dominated by water ice Ih). On this basis, the pressure at Europa's ocean-icy crust interphase would correspond to CO_2 (l) stability region thus P-T conditions of the experiment should respect this constrain.

For convenience, the initial composition of the aqueous solution is 17 wt% $MgSO_4$, which is the composition of $MgSO_4$-H_2O system eutectic point. Employed volume of sample is around 30 mL, which is the half of the total volume of the chamber in order to differentiate gas-liquid interphase at half height of the sapphire window. The system is saturated with CO_2 at about 283 K and 3.3 MPa during few days and then it is brought to 273 K in order to form clathrates. The pressure is maintained in the CO_2 (l) stability region of H_2O-CO_2 phase diagram. The system is periodically refilled with CO_2 for maintaining the formation rate nearly constant. Liquid phase is immediately homogenised by shaking the chamber after every gas replenishment in order to avoid the presence of gradients.

2.1. Chemical system characterization by Raman spectroscopy

All phases and chemical changes in the ternary system are studied by Raman spectroscopy.

In solution, sulphate ion has a Raman band at 982 cm^{-1}, which is quite strong and narrow (figure 1). Raman shift increases as hydration does for the hydrated salts (Table 1): hexahydrate ($MgSO_4 \cdot 6H_2O$), epsomite ($MgSO_4 \cdot 7H_2O$) and meridianiite ($MgSO_4 \cdot 12H_2O$) respectively. CO_2 occurs as a diad in the spectrum due to Fermi resonance. Raman shift and band shape of the diad is modified for different CO_2 states. Water Raman band may show different shapes regarding alterations such as concentration or dilution of salt in solution, thus it allows to discriminate between clathrates and water ice as well.

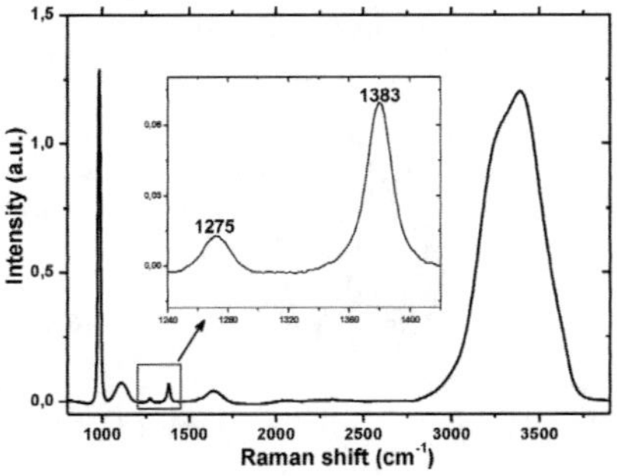

Figure 1. Raman spectrum of liquid phase in the chamber.

Vibration mode	Raman shift (cm^{-1})
ν_1-SO_4^{2-}	981.9 (aqueous) 983 ($MgSO_4{\cdot}6H_2O$) 986 ($MgSO_4{\cdot}7H_2O$) 990 ($MgSO_4{\cdot}12H_2O$)
Fermi doublet-CO_2	1285/1388 (gas) 1280/1286 (liquid) 1275/1383 (aqueous) 1276/1381 (clathrate)
OH-stretching	2900-3900

Table 1. Frequencies of key bands.

2.2 Calibration.

In order to quantify salt concentration, the spectrometer is calibrated with 5 aqueous solutions 5, 10, 20, 25 and 30 wt% $MgSO_4$. The ratio between sulphate and water molecules vibration band areas, i.e. $a = \mathrm{A}(SO_4^{2-})\cdot[\mathrm{A}(H_2O)]^{-1}$, are related to salt concentration by a second order polynomial (1), after spectrum normalization using Raman isosbestic point from liquid water [6].

$$\mathrm{wt\%}=(-10{,}67\pm0{,}45)+(6{,}7\mathrm{E}2\pm1{,}7\mathrm{E}1)\cdot a+(2\mathrm{E}3\pm1\mathrm{E}2)\cdot a^2 \quad (1)$$

3. Results and discussion

Thermodynamical evolution of the system during the formation/dissociation of clathrate hydrates is recorded continuously and plotted on a P-T-time diagram, while spectral changes are followed at discrete points (figure 2). The most relevant points are labeled from 1 to 6 and they are also pointed on H_2O-CO_2 phase diagram in figure 3. Involved thermal transformations as eutectic and peritectic points of $MgSO_4$-H_2O binary system are labeled as well.

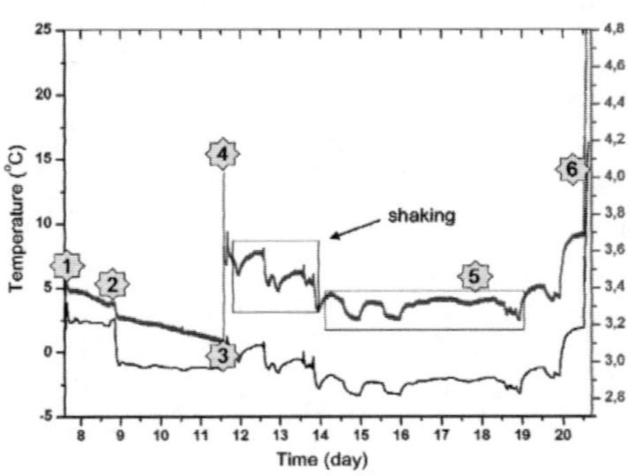

Figure 2. P (upper line)-T (bottom line)-time diagram registered along the salting-out experiment.

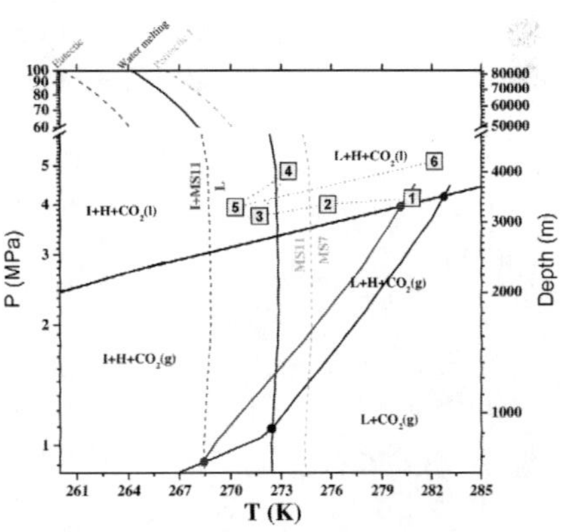

Figure 3. Phase diagram of H_2O-CO_2 system. Numbers correspond to explored points along one run experiment (as in figure 2).

The onset of clathrate hydrate formation occurs at the turbidity time [7], which denotes the first appearance of stable clathrate crystals (see figure 4). The crystals growth upwards from the liquid-gas interphase, therefore clathrate formation requires starting mainly from aqueous CO_2. This stage

IOP Conf. Series: Journal of Physics: Conf. Series **950** (2017) 042042 doi:10.1088/1742-6596/950/4/042042

corresponds to point 1 on the P-T-time diagram. When clathrate hydrates start to grow, several drops in pressure at constant temperature are observed (points 1-2-3 on the P-T-time diagram).

Figure 4. Pictures showing the liquid-gas interphase through the sapphire window. **(a)** Progress of a turbid front in the liquid phase at the turbidity time. **(b)** Clathrate hydrate layer over liquid-gas interphase.

When clathrate hydrates layer form, Raman spectrum of the Fermi diad of CO_2 (g) is observed (figure 5). In spite of maintaining the pressure into the stability region of liquid CO_2, the explored points along the experiment are very close to the limit between the stability regions of both liquid CO_2 and gas, thus it is explained that there is CO_2 (g) in the upper part of the chamber. Entrapment of CO_2 molecules in cages of clathrates modifies the vibration frequency of molecular bonds and therefore Fermi doublet is shifted to lower frequencies (figure 5). Shape of hydroxyl stretching band evolves as well, showing a new broad peak around 3200 cm^{-1}.

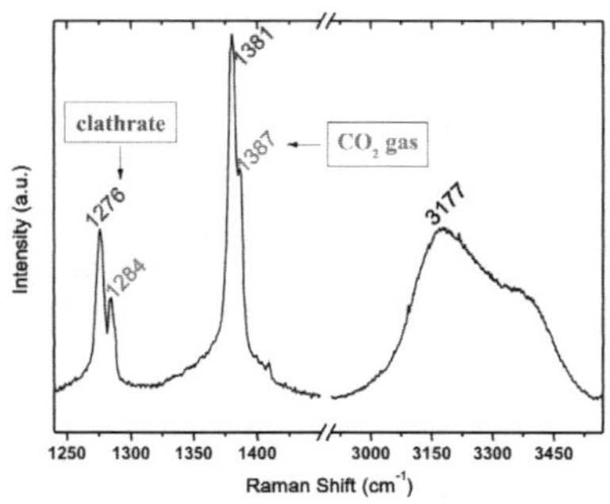

Figure 5. Raman spectrum of CO_2 (g) and CO_2-clathrates. A clear displacement of CO_2-clathrates Fermi diad is observed upon clathrate formation.

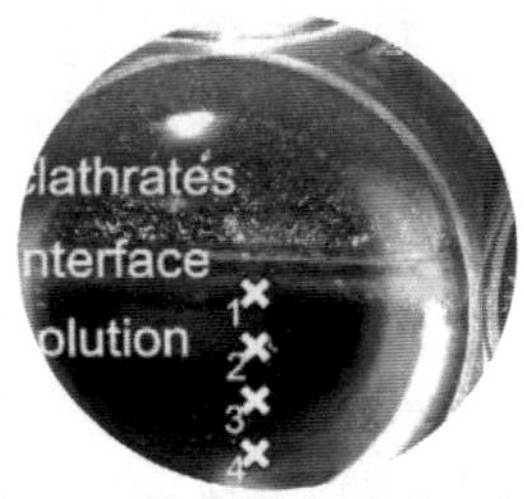

Figure 6. Measured points as a function of depth.

X-points	1	2	3	4
$A(SO_4^{2-})\cdot[A(H_2O)]^{-1}$	0.047	0.049	0.057	0.060
$[SO_4^{2-}]$ (% w/w)	15.1	16.2	19.5	20.5

Table 2. Salt concentration at different depth X-points of the liquid phase.

Point 4 on P-T-time diagram corresponds to a refilling step that helps CO_2 to diffuse throughout the clathrate layer and replenish the quantity of CO_2 in solution and maintain the clathrate growth almost constant. After that, chamber is shaking assisting clathrates to nucleate. As more clathrates form, more water molecules are removed from the solution and salt concentration raises near the interphase. Then, the proximity becomes denser and tends to diffuse to the bottom, appearing a gradient of salt along the liquid phase (Table 2). This gradient varies from 15 wt% to 20 wt% $MgSO_4$. If salt concentration goes on to increase, aqueous solution can be supersaturated until hydrated sulphate precipitate and it implies registering wide variations in pressure due to the volume expansion caused mostly by meridianiite ($MgSO_4 \cdot 11H_2O$) precipitation than that of epsomite ($MgSO_4 \cdot 7H_2O$). That step corresponds to the P-T-time diagram region labeled as point 5. Large amounts of solids at the bottom of chamber are visually detectable but also by Raman spectroscopy (figure 7) which shows a mixture of both salts. Finally, point 6 corresponds to a heating step when clathrate dissociation occurs. This is an endothermic process which is recorded at around 282 K.

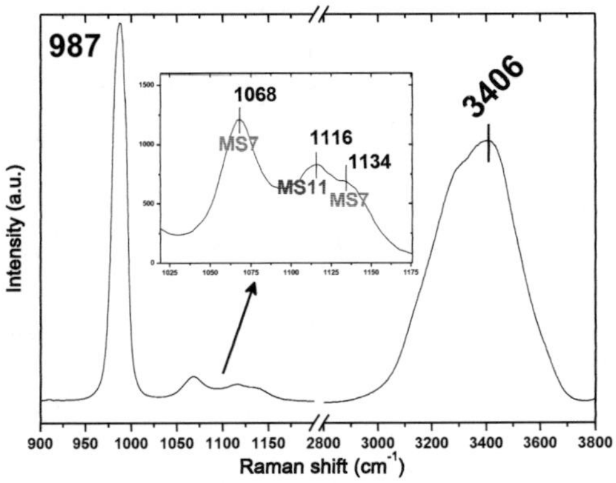

Figure 7. Raman spectrum of the solids at 271 K and 3.3 MPa shows a mixture of epsomite and meridianiite, as it is expected attending to the phase diagram.

In order to follow the effects of clathrates formation in salty water, it is feasible to attend to salt concentration evolution along the experiments (figure 8). For this assessment, measures are taken with the Raman laser focused on the middle of the liquid phase. It is clear that first, salt concentration raises due to clathrate formation, then it decreases due to salt precipitation and after clathrate dissociation salt concentration recovers its initial value.

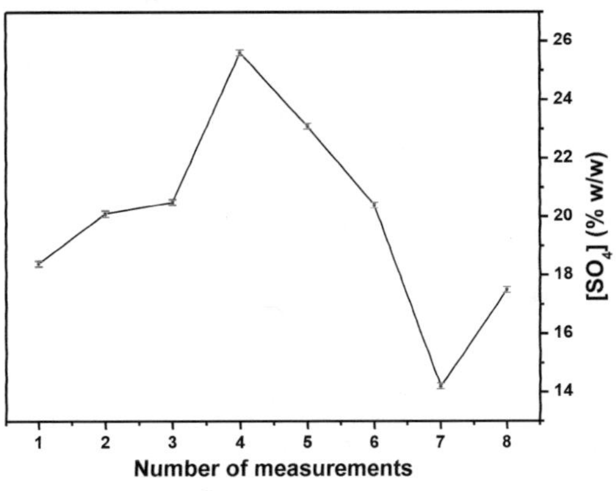

Figure 8. Variation in salt concentration according to the number of the most representative measurements along the salting-out experiment. It is in a great agreement with experimental error.

4. Conclusion

Raman spectroscopy analysis and thermodynamical information are used here to follow the evolution of salt-gas-rich fluids during cooling. It allows detecting clathrate hydrate formation/dissociation processes and precipitation of minerals involved in $MgSO_4$-CO_2-H_2O system.

Salting-out takes places in $MgSO_4$-CO_2-H_2O system after clathrate hydrates formed appearing a gradient of mineral density which drives fluid convection. Besides, gradients in salt concentration may derive to transient stratification before fluid convects. Salting-out associated to the presence of clathrates may have a great influence on Europa's geological activity due to changes in several physical properties of the system (e.g. total volume, density, or thermal properties).

It is assumed that Ganymedes' fluids have a similar chemistry to Europa, but they are distributed in deeper layers up to 1.3 GPa. Fluids are sandwiched between different high pressure phases of water ice instead of being in direct contact with rock, like occurs in Europa [8]. We predict that formation of CO_2 high pressure clathrate hydrates may have different consequences on salting-out and therefore on remnant brines density. HP-clathrates require lower amount of water dictated by their stoichiometry so salting –out impact would be less pronounced at the bottom of the ocean.

Salting-out effects may be expressed on surfaces of the icy moons as geological structures, which will be explored by future space missions to the icy moons.

References

[1] Cartwright J H E, Escribano B, González D L, Sainz-Díaz C I and Tuval I 2013 *Langmuir* **29** 7655-60
[2] Muñoz-Iglesias V, Bonales L J, Santamaría-Pérez D, Taravillo M, Baonza V G and Prieto-Ballesteros O 2012 *Spectrosc. Lett.* **45** 407-12
[3] Lombardi S, Altunina L K and Beaubien S E 2006 *Advances in the Geological Storage of Carbon Dioxide* (Netherlands: Springer) p. 111
[4] Muñoz-Iglesias V, Prieto-Ballesteros O and Bonales L J 2014 *Geochim. Cosmochim. Acta* **125** 466-75
[5] Dalton J B, Prieto-Ballesteros O, Kargel J S, Jamieson C S, Jolivet J and Quinn R 2005 *Icarus* **177** 472-90
[6] Walrafen G E, Hokmabadi M S and Yang W H 1986 *J. Chem. Phys.* **85** 6964-69
[7] Servio P and Englezos P 2001 *Fluid Phase Equilib.* **190** 127-34
[8] Vance S, Bouffard M, Choukroun M, Sotin C 2014 *Planet. Space Sci.* **96** 62-70

IOP Conf. Series: Journal of Physics: Conf. Series **950** (2017) 042043 doi:10.1088/1742-6596/950/4/042043

Experimental Simulation of Methane Hydrate Extraction at High Pressure Conditions: Influence of the Sediment Bed

J R Agudo[1], J Park[1], G Luzi[1], M Williams[2], C Rauh[3], A Wierschem[4] and A Delgado[4]

[1] Institute of Fluid Mechanics, FAU Busan, Busan, Republic of Korea
[2] Chair of Chemical Reaction Engineering, University of Erlangen-Nuremberg, Erlangen, Germany
[3] Department of Food Biotechnology and Food Process Engineering, Technical University of Berlin, Berlin, Germany
[4] Institute of Fluid Mechanics, University of Erlangen-Nuremberg, Erlangen, Germany

E-mail: jose.a.rodriguez@fau.de

Abstract. Being a clean alternative to other fossil fuels, Methane Hydrate (MH) is currently considered as one of the most important potential sources for hydrocarbon fuels [1]. In addition, the high energy density of MH and its stability at higher temperatures as compared to LNG (Liquefied Natural Gas) makes MH a potential greener method for energy transportation. At the same time, the low thermodynamic stability of MH strongly questions the future exploitation of gas hydrate deposits, turning its extraction into a possible geohazard [2]. Fluctuations in pressure, temperature, salinity, degree of saturation or sediment bed properties may cause methane gas release from the water lattice. We experimentally study the influence of the sediment bed geometry during formation-dissociation of MH. For this purpose, MH is synthesized within regular substrates in a 93 cm^3 high pressure vessel. The regular substrates are triangular and quadratic arrangements of identical glass spheres with a diameter of 2 and 5 mm, respectively. MH formation within regular substrate reduces the possibility of spontaneous nucleation to a unique geometrical configuration. This fact permits us to characterize the kinetics of MH formation-dissociation as a function of the sediment bed geometry. Preliminary experimental results reveal a strong dependence of MH formation on the geometry of the regular substrate. For instance, under the same pressure and temperature, the kinetics of MH production is found to change by a factor 3 solely depending on the substrate symmetry, i.e. triangular or quadratic.

1. Introduction

Methane Hydrate (MH) are formed spontaneously in appropriate physical conditions, i.e. low temperature, high pressure, appropriate amount of gas former and water [3]. Other phenomena, however, enhance hydrate formation and dissociation. It is well known that the nucleation sites, given by the geometrical properties of the sediment bed within which MH is formed, strongly affect the kinetics of MH formation-dissociation [4]. The formation and decomposition of MH within the sediment bed is thus of vital importance for realistic reproduction of the process occurring in Nature either in permafrost or in deep-sea sediments [5]. The structure of the sediment bed arrangement influences the kinetics of the process [5], the equilibrium conditions for MH formation [6], the internal fluid flow migration processes during formation-dissociation process [7] and the stability of the hydrate [8].

The literature is lacking in an experimental study for the kinetics of MH formation-dissociation within regular geometries. In irregularly arranged structures, as considered until now, the spontaneous MH nucleation is favoured in optimal local volumes inside the sediment bed. This results in a stochastic phenomenon with limited predictability for the onset of hydrate formation, which is strongly dependent on the laboratory equipment [9]. Studying the MH nucleation in regularly arranged sediment beds according to triangular or quadratic symmetries reduces the possibility of spontaneous MH nucleation to a unique geometrical configuration. This fact, therefore, permits us to characterize the kinetics of MH formation-dissociation as a function of the bed geometry.

2. Experimental Set-up

Methane Hydrate is produced in regular substrates placed in a stainless steel vessel with a total volume of 93.4 cm^3. The regular substrates consist of a monolayer of uniform size glass beads of 2 ± 0.02 mm and 5 ± 0.02 mm in diameter, arranged and fixed in triangular and quadratic symmetries, respectively. To build the substrates, we follow a similar procedure as described in [10-12]. Figure 1(a)-(b) shows pictures of the different regular bed geometries. The vessel is filled with 6 mL of deionized water and is well distributed on the sediment bed surface. After flushing the system, high purity methane gas (99.90%) is injected into the vessel to achieve an initial pressure of 100 bar. The vessel is submerged into a water tank. A thermostat is used to control the vessel temperature, which is set at 273.5 K, for optimal conditions for MH nucleation inside the cavities of the sediment bed. The pressure on the system is continuously measured using a pressure transducer while the temperature is monitored using a thermocouple placed at the center of the vessel. Figure 1(c) depicts the experimental set-up.

We indirectly measure the methane consumption with time during MH formation via the pressure drop in the vessel in a similar experimental procedure as described in [5, 13]. The kinetics of MH formation are measured experimentally for each of the regular substrate geometries.

3. Experimental results and discussion

In isothermal conditions, the pressure gradually decreases as the gas phase is enclosed inside the MH cages. The increase of temperature due to MH formation is rapidly dissipated in the vessel by the temperature control tank.

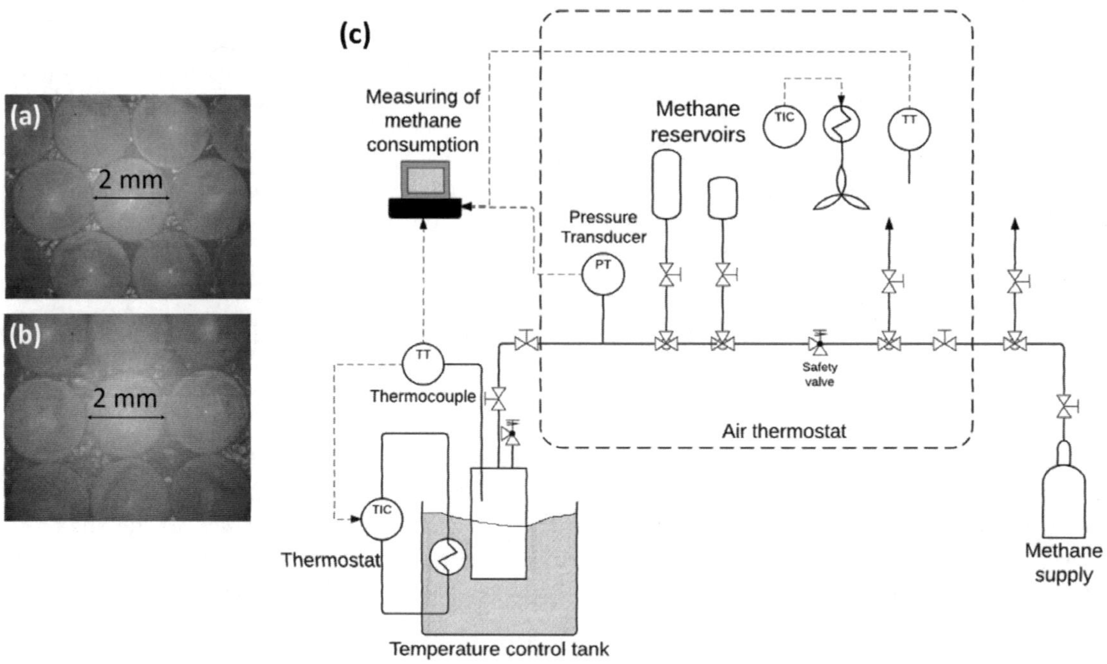

Figure 1. Substrates of identical and regularly arranged spherical glass bead. Triangular (a) and quadratic (b) configuration. Sketch of the experimental setup (c)

We record pressure and temperature every 2 minutes in order to study the progress of MH formation. To study the kinetics of MH formation for different regular geometries, the methane consumption was evaluated using ideal gas and Peng-Robinson equations of state as recommended by [13]. At the pressure and temperature considered, the choice of the equation of state hardly affected the mole of consumed gaseous methane. Figure 2 plots the methane consumption in moles as function of time for a triangular (a) and quadratic (b) substrate made of 5 mm diameter beads. Error bars in the figure represents the range of uncertainty of the measurements. The data for methane consumption are depicted every hour to simplify the diagram. (Red) dashed lines are drawn to guide the eyes. The experiments were repeated three times for the triangular substrate and twice for the quadratic one. Different symbol colours on the diagram represent different experiments performed under the same initial condition, i.e. temperature and initial pressure. As shown in Figure 2, the experiments roughly coincide on the diagram showing a good repeatability of the results. We observe a similar rate of MH production within the pores of the same size if other thermodynamic parameters remain constant. This fact corroborates that our experimental results remain independent of boundary effects.

Figure 3 shows the methane consumption as a function of time for each of the substrate geometries. The data are depicted again each hour and the error bars represent the range of uncertainty of the measurements. For the experiments performed using the substrate made of 5 mm, the average value of the repeated experiments is shown. Figure 3 reveals a strong influence of the substrate geometry on the MH formation. For same initial conditions, the total amount of methane consumed for MH formation was increased by a factor of two for the substrate of 5 mm beads, compared to the substrates of 2 mm beads (compare solid and open symbols in Figure 3). This increase in methane consumption represents a clear enhancement of the hydrate saturation level depending on the size of the sediment pores.

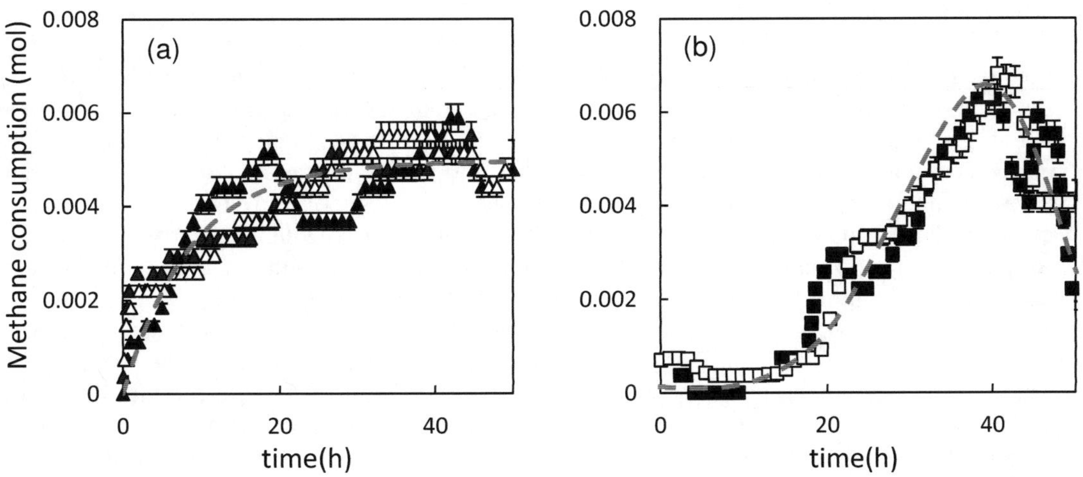

Figure 2. Kinetics of MH formation within the triangular (a) and quadratic (b) substrates made of 5 mm diameter beads. Different symbol colours represent different experiments performed under the same condition. The initial pressure was set at 100 bar and the temperature was fixed at 0.5°C. The error bars represent the range of uncertainty of the measurements. Dashed lines are drawn to guide the eyes.

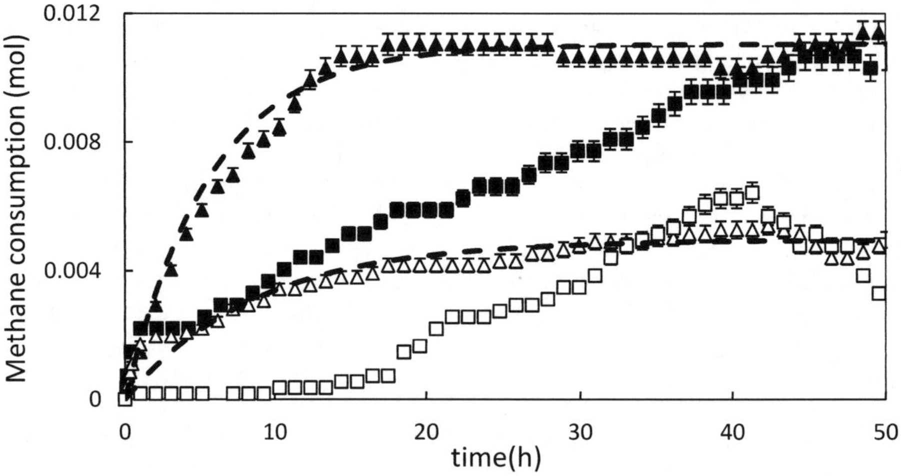

Figure 3. Kinetics of MH formation within different regular substrates. Solid and open triangles: triangular substrates formed by beads of 2 and 5 mm in diameter, respectively. Solid and open squares: quadratic substrates formed by beads of 2 and 5 mm in diameter, respectively. For experiments performed using the substrate made of 5 mm diameter beads, the average value of repeated experiments is depicted in the figure. The initial pressure was set at 100 bar and the temperature was fixed at 0.5°C. The error bars represent the range of uncertainty of the measurements. Dashed lines represent a fit to the experimental data according to Eq. 1 for n^* equal to one as proposed in [5].

On the other hand, for substrates made of beads with the same size, the kinetics of MH formation strongly depends on the geometrical symmetry (compare triangles and squares in Figure 3). For the experiments performed with the triangular substrate made of 2 mm beads (solid triangles), the maximum methane consumption was achieved after around 15 hours. For the quadratic one, however, the maximum methane consumption achieved a plateau at around the same level but after 45 h. This fact corresponds to an enhanced kinetics of MH formation within the triangular geometry, by a factor of about 3, as compared to the quadratic configuration. Note that comparing triangular and quadratic substrates made of 5 mm beads, we also observe different kinetics of MH formation.

Moreover, MH formed in quadratic substrates made of 5 mm beads (open squares) showed a dissociation phase after around 40 h. This dissociation phase is not observed in triangular substrates with 5 mm beads (open triangles), pointing to a higher stability of the MH sample produced within the triangularly arranged substrate as compared to the quadratic one. This dissociation phase may be due to the decrease of pressure as methane changes from gas to solid phase. For the experiments performed with beads of 5 mm diameter, the final pressure drop achieved typical values of about 3 bar in the system. This decrease of pressure apparently affected the sample formed within the quadratic substrate, but did not affect the hydrate generated within the triangularly arranged one.

Kono *et al.* studied the dependency of kinetics of MH formation and dissociation on the sediment bed properties [5]. For that purpose, they formed MH in various custom-designed porous sediments with glass beads and cylindrical particles according to irregular arrangements. The glass bead diameter in their experiments varied from 0.1 to 5 mm. In order to characterize the different kinetics of MH formation within irregular beds, they defined an overall formation rate constant defined in terms of reaction engineering by the following equation [5]:

$$\frac{dn_{CH4}}{dt} = -k_f n_{CH4}^{n^*} \quad (1)$$

where n_{CH4} is the number of mol of consumed methane, k_f is an overall formation rate constant that includes the effects of heat transfer, mass transfer and reaction and n^* is the overall order of MH formation process. While the overall order of the formation process n^* was found to be equal to one independent of the sediment bed, the overall formation rate was observed to depend on the irregular bed configuration, which was varied on the experiments by the sediment shape and size but not by the arrangement [5]. Dashed lines in Figure 3 represent a fit of the overall kinetic model to our experimental data according to Eq. 1 for n^* equal to one. As shown in Figure 3, the overall kinetics of first order agrees within the range of uncertainty of our experimental data for MH formed in triangular substrates. Under the same pressure and temperature conditions, the overall formation rate constants are 2.9x10^{-3} and 1.9x10^{-3} min^{-1} for the triangular substrates with 2 mm and 5 mm beads, respectively. Despite variation on the experimental set-up, these values are in accordance with the data obtained by Kono *et al.* for irregularly arranged sediment bed made of uniformly sized spherical beads of 0.1 mm in diameter, $k_f = 2.6\text{x}10^{-3}$ min^{-1} [5]. We remark that operating pressure and temperature used during our experiments are similar to those performed in [5]. We hypothesize that the good agreement between our data for a triangular geometry and the overall first order kinetic model proposed by Kono *et al.* is due to the fact that a naturally arranged sediment may represent a close-packed sediment bed with a packing coefficient close to the maximum as in our triangular geometries. Note that the experimental data obtained for the quadratic substrate strongly differ from the overall first order reaction (fit is not shown in Figure 3).

4. Conclusions and perspectives

We experimentally study the kinetics of MH formation as a function of the sediment bed geometry. For that purpose, MH is synthesized within the regular substrates, triangular or quadratic monolayers of glass beads of 2 and 5 mm in diameter, respectively. After placing the substrate into a vessel of 93.4 cm^3, we added 6 mL of distilled water and injected high purity methane in order to achieve an initial pressure of 100 bar. The temperature was fixed at 0.5°C. Proper conditions for methane hydrate formation were, therefore, achieved inside the pressure vessel.

Experiments performed using the same regular substrate and under the same initial conditions showed a good repeatability of results. On the other hand, preliminary results pointed to a strong influence of the substrate geometry on the MH formation and the MH stability within the sediment bed. Under the same initial conditions, we measured a different total amount of consumed methane, by a factor of 2, depending only on the glass bead diameter. Furthermore, depending only on the substrate symmetry, we observed a favoured kinetics of MH formation, by a factor up to 3, for the triangular substrate as compared to the quadratic one. Finally, we observed dissociation phases in quadratic geometries but not in triangular substrates under the same initial conditions.

Further experiments for MH formation and dissociation must be performed in three dimensional regular structures with different width of the bed as suggested by [14]. In addition to the importance to quantify the role of the sediment bed geometry on the MH formation-dissociation, optimal configurations for MH stability within the sediment bed can be developed. An optimum bed configuration may permit to enclose a larger number of gas molecules per unit of volume within the MH (i.e. higher gas saturation), increasing the energetic density for transport purposes.

Acknowledgments
The authors are thankful to Mr. J. Lee for collaborating in setting up the experiment.

References

[1] Demirbas A 2010 *Mehane gas hydrate*, (London, Dordrecht, Heidelberg, New York: Springer) p 133
[2] Kennett J P, Cannariato K G, Hendy I L and Behl R J 2000 *Science*, **288** 128
[3] Max N D, Johnson A H and Dillon W P 2006 (Dordrecht: Springer)
[4] Carroll J J 2009 *Natural Gas Hydrates, a guide for Engineers* (Oxford: Elsevier)
[5] Kono H O, Sridhar N, Song F and Smith D 2000 *Powder Technol.* **122** 239-246
[6] Yang M, Song Y, Ruan X, Liu L, Zhao J and Li Q 2012 *Energies* **5** 925-937
[7] Kneafsey T, Tomutsa L, Moridis G J, Seol Y, Freifeld B M, Taylor C E and Gupta A 2007 *J. Petrol. Sci. Eng.* **56** 108-126
[8] Waite F W, Santamarina J C, Cortes D D, Dugan B, Espinoza D N, Germaine J, Jang J, Jung J W, Kneafsey T J, Shin H, Soga K, Winters W J and Yun T S 2009 *Rev. Geophys.* **47** RG4003

[9] Sloan E D and Koh A C 2008 *Clathrate Hydrates of Natural Gases* (CRC Press: New York)
[10] Agudo J R and Wierschem A 2012 *Phys. Fluids.* **093302** 24
[11] Agudo J R, Dasilva S and Wierschem A 2014 *Phys. Fluids.* **053303** 26
[12] Agudo J R and Wierschem A 2015 *Chemie Ingenieur Technik* **87** 8 1107
[13] Guanli J, Qingbai W and Jing Z 2010 *Fluid Phase Equilibr.* **298** 225-230
[14] Linga P, Haligva C, Sung C N, Ripmeester J A and Englezos P 2009 *Energ. Fuel* **23** 5508-5516

IOP Conf. Series: Journal of Physics: Conf. Series **950** (2017) 042044 doi:10.1088/1742-6596/950/4/042044

Transition mechanism of sH to filled-ice Ih structure of methane hydrate under fixed pressure condition

H Kadobayashi[1], H Hirai[1], H Ohfuji[1], Y Kojima[1], Y Ohishi[2], N Hirao[2], M Ohtake[3] and Y Yamamoto[3]

[1]Geodynamics Research Center, Ehime University, Matsuyama, Ehime 790-8577, Japan
[2]Japan Synchrotron Radiation Research Institute, Hyogo 679-5198, Japan
[3]The National Institute of Advanced Industrial Science and Technology, Tsukuba, Ibaraki 305-8568, Japan

kadobayashi@sci.ehime-u.ac.jp

Abstract. The phase transition mechanism of methane hydrate from sH to filled-ice Ih structure was examined using a combination of time-resolved X-ray diffractometry (XRD) and Raman spectroscopy in conjunction with charge-coupled device (CCD) camera observation under fixed pressure conditions. Prior to time-resolved Raman experiments, the typical C-H vibration modes and their pressure dependence of three methane hydrate structures, fluid methane and solid methane were measured using Raman spectroscopy to distinguish the phase transitions of methane hydrates from decomposition to solid methane and ice VI or VII. Experimental results by XRD, Raman spectroscopy and CCD camera observation revealed that the structural transition of sH to filled-ice Ih occurs through a collapse of the sH framework followed by the release of fluid methane that is then gradually incorporated into the filled-ice Ih to reconstruct its structure. These observations suggest that the phase transition of sH to filled-ice Ih takes place by a typical reconstructive mechanism.

1. Introduction

Methane hydrate, called "fiery ice," is expected to be a clean energy resource, although methane is, on the other hand, a greenhouse gas even more potent than carbon dioxide. It is also thought to be an important constituent of icy planets and their satellites such as Neptune and Titan. Gas hydrates are clathrate compounds composed of hydrogen-bonded water cages (host) with molecules or atoms (guests) included in the cages [1]. Many guest species can be incorporated into clathrate structures, which have three types of varieties, sI, sII, and sH structures depending on the guest sizes at ambient to relatively low pressures [1]. At higher pressures, more varieties of structures are formed [2-6]. With increasing pressure, a cage structure is compressed and forced to change to one that can accommodate more guests per mole of water [7, 8]. At further higher pressures, the cage structures are no longer retained, and consequently well-designed structures, called filled-ice structures, are formed [9]. The filled-ice structures consist of an ice or icy framework and guests filling the voids between them [9-11]. Almost all filled-ice structures decompose at pressures below 6–7 GPa [12-14]. Among them, methane and hydrogen hydrates exhibit notable stability up to 86 GPa at room temperature and low temperatures according to previous experimental and theoretical studies [15, 16]. Methane hydrate (hereafter denoted as MH) has an sI structure composed of two 12-hedra and six 14-hedra components in a unit cell at low (< 0.8 GPa) pressures and room temperature, but transforms to an sH cage structure formed by three 12-hedra, two modified 12-hedra, and a 20-hedra at approximately 0.8–1.0 GPa. A further transformation to a filled-ice Ih structure (hereafter denoted as FIIhS), which consists of a framework of ice (similar to ice Ih) and methane molecules that fill the channels running along the a- and c-axes [14]. Water is released through each of the transitions and subsequently crystallizes as either ice VI or VII depending on the pressure. Thus, MH increasingly becomes methane-rich as it transitions from one structure to another. The sequence of these phase transitions of MH with pressure have been studied [7, 13, 17], and the structural details including the precise positions of the host oxygen atoms, multi-occupancies, and site disorders of guests have been determined by neutron diffraction [14, 18]. In contrast, there are only a few studies about the essential mechanism of the phase transitions [19, 20].

A recent work by Hirai *et al.* [19] studied the mechanism of sI to sH phase transition using time-resolved in-situ X-ray diffraction and Raman spectroscopy at a constant pressure. They revealed that although the 12-hedra of sI are retained, the 14-hedra are replaced sequentially by additional (newly formed) 12-hedra, modified 12-hedra, and 20-hedra cages of sH, which suggests that the cage-recombination is responsible for the sI–sH transition mechanism [19]. On the other hand, in-situ observation under a dynamic loading condition [20] and that at a fixed pressure [19] both showed that the sH-FIIhS transition occurs by a reconstructive mechanism, but their signatures are quite different [19, 20]. In the dynamic loading experiment, the sH-FIIhS transition was detected by the rapid pressure decreased and low-frequency sift of the C-H vibration mode [20], while in the static pressure experiment such low-frequency sift of the C-H mode was interpreted as a result of the decomposition to solid methane and ice [21]. It is also noted that the phase transition kinetics is greatly influenced by the rate of compression [19, 20]. Therefore, to understand the essential mechanism of the phase transition an experiment where the pressure is carefully and precisely controlled from beginning to end of the transition is required.

The C–H stretching modes of methane molecules, ν_1 and ν_3, are very sensitive to the surrounding conditions and the vibrational frequency of these modes vary in response to changes in the cage framework [1, 2, 21]. In this study prior to time-resolved measurements of the MH transition, the C–H vibration modes of the each MH phase (sI, sH and FIIhS) as well as solid and fluid methane, which are produced through decomposition of the hydrate structure, were collected as a function of pressure, in which the each structure was simultaneously confirmed by XRD. We then use those data as reference for the assignment of Raman peaks and identified the cage structure of MH and phase state of the decomposition product.

This study employs in-situ, time-resolved observations using synchrotron XRD and Raman spectroscopy to understand the essential mechanism of sH-FIIhS phase transition.

2. Experimental procedure

A clamp-type and a lever-spring-type of diamond anvil cells (DAC) were used to generate high pressures. A pair of anvils with a culet size of 600 μm and a stainless steel gasket were used for each run. Pressure (0.2-6.0 GPa) was measured using the ruby fluorescence method and carefully controlled so as to have a constant pressure distribution across the sample chamber. All the experiments were performed at room temperature (297 K). The initial methane hydrate sI sample was synthesized by a conventional ice–gas interface method at the National Institute of Advanced Industrial Science and Technology (AIST). The powdered sample (2–3 μm in average grain size) was loaded together with a few ruby balls for pressure measurement in a sample hole in a gasket in a cryogenic vessel cooled by liquid nitrogen. The initial pressure was carefully adjusted to below 0.2 GPa at low temperature to seal the MH sample in the cell without being transformed to the sH structure. In-situ XRD observations were performed by using synchrotron radiations at BL-10XU beamline of SPring-8 and at BL-18C beamline of the Photon Factory of the High Energy Accelerator Research Organization (KEK). Monochromatic X-rays with wavelengths of 0.04134 and 0.06198 nm with a beam size of 100 μm were used in SPring-8 and KEK, respectively. Raman spectroscopy was conducted by using a highly confocal Raman system (PhotonDesign, RSM 800) equipped with a diode laser (λ = 473 nm). For time-resolved measurements, diffraction patterns and Raman spectra were collected every 4.3 and 8 seconds using an X-ray CCD detector and the Raman system, respectively.

Prior to time-resolved XRD measurements, we collected a series of XRD patterns using an imaging plate detector during pressure increase by 0.1 GPa step in order to precisely determine the phase transition pressure. As the result, the transition pressure of the sH–FIIhS transition was determined to be 1.8 GPa, which is consistent with previous studies [2, 4]. The sample pressure was finely controlled by an automatic gas flow system by adjusting the flow rate and pressure of helium gas flowing into a membrane in the DAC. The helium gas pressure was increased very slowly at a rate of 1.0 bar/min until the sample pressure reached to just below the transition pressure of the sample, and the pressure was kept constant for approximately 1 h to make sure no phase transition occurred before starting time-resolved measurements. Helium gas was then re-introduced to the membrane at the same flow rate to provide a driving force for the phase transition, although the absolute pressure would not change in this process (i.e. the pressure in the sample chamber was kept constant throughout the time-resolved measurements). The time-delay of the phase transition was 40 to 160 s after the gas introduction in the present runs. The pressure was also kept constant during the time-resolved Raman spectroscopy measurements in which a driving force for the transition was introduced by slightly rotating the clamp screw by less than 1 deg.

3. Results and discussion

Figures 1a and 1b show typical Raman spectra of C-H vibration modes for sI, sH and FIIhS and variation of Raman shift as a function of pressure from 0.2 to 6.0 GPa, respectively. The open circles show the variation of Raman shift of sI below 0.9 GPa, sH between 0.9 and 1.8 GPa and FIIhS above 1.8 GPa. The crosses indicate the data points in which the structure was also identified as FIIhS by XRD. The solid triangles and solid squares are liquid and solid methane, respectively, which were obtained from a separate run where MH was decomposed to water and fluid methane at 0.7 GPa and 423 K. The C-H vibration modes of both liquid and solid methane incresed almost linearly with increasing pressure, and the observed Raman shift values are in good agreement with those of the previous study [21, 22]. Two C-H vibration modes observed below 1.8 GPa are assigned to the small and large cages of sI and sH, respectively and linearly increase with pressure. The abrupt changes at about 1.0 GPa are attributed to the phase transition from sI to sH. Note that the Raman shift values obtained from FIIhS are about 10 cm^{-1} higher than those of solid methane. The Raman shifts observed in this study are well comparable to those reported by Hirai *et al.* (2001) [2] and Bezacier *et al.* (2014) [21], but are not consistent with those by Kumazaki *et al.* (2004) [23]. The result of the latter study [23] seems to be explained more resonably by solid methane.

Figures 2a and 2b show the time-resolved Raman spectra and the typical Raman spectra of the C-H vibration modes observed during the transition from sH to FIIhS. Although the C–H modes from the three types of cages were almost superimposed just after the formation of sH at 0.9 GPa, that from the 20-hedra appeared as a shoulder peak at the higher frequency side of the main peak just before the transition to FIIhS at 1.8 GPa (Figs. 2a, 2b bottom). In the sH structure, the cage occupancy of the 20-hedra increases with increasing pressure to maintain such large cages [13, 18]. This leads to changes in the vibration mode of the 20-hedra, as observed in the present and previous studies [2, 23]. Thus, the spectrum collected after 8 s (Figs. 2a and 2b) shows the typical C–H modes of sH at 1.8 GPa. After approximately 100 s, the peaks of sH cages suddenly broadened and a fluid methane peak appeared together with the peak characteristic of FIIhS. The relative intensity of fluid methane then gradually decreased, whereas that of FIIhS became intense and sharp (Figs. 2a and 2b). At the beginning of the FIIhS formation (between 100 and 160 s), its Raman shift was slightly lower by 5-6 cm^{-1} than that of typical FIIhS (Fig. 2b), but became comparable (~2945 cm^{-1}) after 160 s (Fig. 2b).

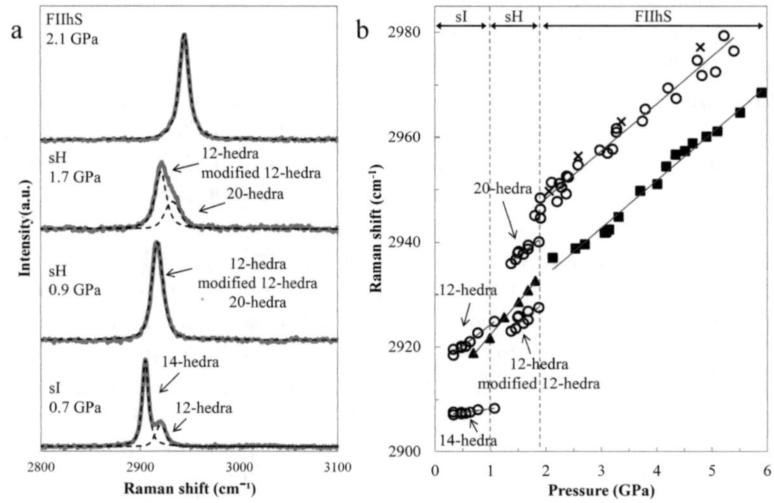

Figure 1. (a) Typical Raman spectra of C-H vibration modes for methane molecules included in sI, sH, and FIIhS. (b) Variations of Raman shift of C-H vibration modes as a function of pressure. The open circles indicate Raman shift of MH. The crosses in the FIIhS region show data of which structure is confirmed to be FIIhS by XRD. The solid triangles and solid squares are fluid and solid methane, respectively.

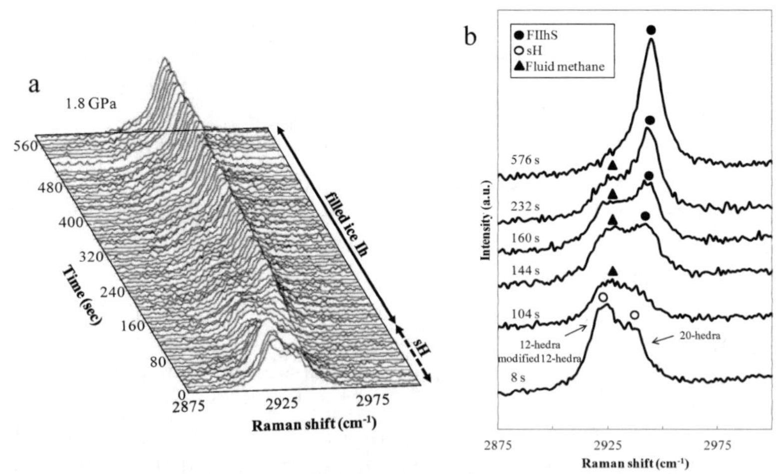

Figure 2. (a) Time-resolved Raman spectra and (b) typical Raman spectra obtained during sH–FIIhS transition at a fixed pressure of 1.8 GPa. The black circles, black triangles and open circles indicate the C-H vibration modes of FIIhS, fluid methane and sH, respectively.

Figures 3a and 3b show the time-resolved XRD patterns and typical XRD patterns during the transition from sH to FIIhS at 1.8 GPa. Up to ~160 s, the observed diffraction peaks were indexed with sH and ice VI. After 160 s, the sH peaks suddenly disappeared and instead, peaks of FIIhS appeared. This means that the construction of the FIIhS framework requires a relatively short time period of 50–60 s. In addition to FIIhS peaks, small peaks of solid methane also appeared (Fig. 3b). The Raman observation clearly shows that fluid methane was released during the transition to FIIhS. It is most likely that a part of such formed methane crystalized during the XRD measurement, since the pressure condition was very close to its solidification pressure (1.7 GPa at room temperature [22]).

IOP Conf. Series: Journal of Physics: Conf. Series **950** (2017) 042044 doi:10.1088/1742-6596/950/4/042044

Figure 4 shows the CCD camera images that were taken simultaneously with the XRD measurements. Drastic changes in color and transparency were observed during the sH–FIIhS transition: the transparent sH suddenly turned to brown to dark gray, which probably indicated the release of fluid or solid methane, according to a previous study reporting the decomposition of MH [19]. The release of solid and fluid methane was also detected by the Raman and XRD observations, as described above. After the drastic color change (i.e., collapse of the sH framework), the sample became gradually transparent again, suggesting that methane was being consumed for the construction of FIIhS.

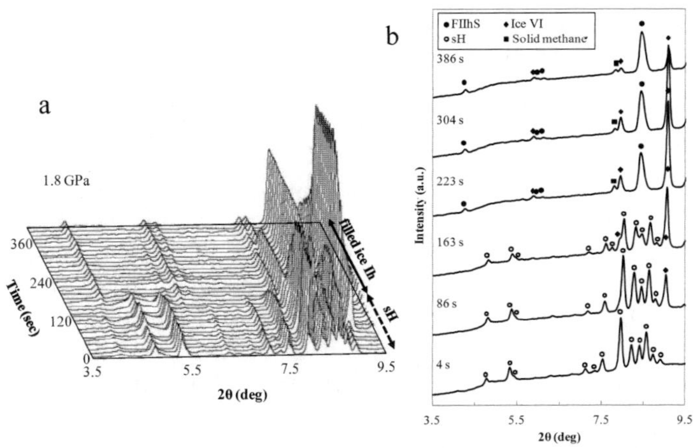

Figure 3. (a) Time-resolved XRD patterns and (b) typical XRD patterns obtained during sH–FIIhS transition at a fixed pressure of 1.8 GPa. Black circles, black diamonds, black squares and open circles indicate the diffraction lines of FIIhS (Imcm, a=4.7960, b=8.1159, c=7.9691 Å), Ice VI ($P4_2/nmc$, a=6.0959, c=5.6177 Å) and sH (P6/mmm, a=11.8135, c=9.9958 Å), respectively.

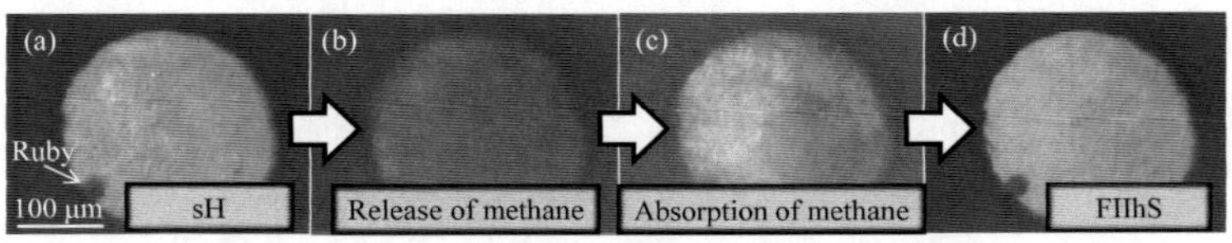

Figure 4. Optical changes in the sample chamber observed during the sH-FIIhS transformation by CCD camera.

The present in-situ observations using XRD, Raman spectroscopy, and CCD imaging demonstrates that the sH-FIIhS transition occurs though a sudden collapse of sH structure followed by the release of fluid or solid methane that are then gradually incorporated into the FIIhS to reconstruct its structure. This clearly shows that the phase transition takes place by a typical reconstructive mechanism, which seems to be reasonable taking into account the large difference in crystal structure and chemical composition between the two phases.

References

[1] Sloan E D and Koh K A 2008 *Clathrate Hydrates of Natural Gases* 3rd ed (Taylor & Francis, London, New York) p. 45, 257, 320, and 537.
[2] Hirai H, Uchihara Y, Fujihisa H, Sakashita M, Katoh E, Aoki K, Nagashima K, Yamamoto Y, and Yagi T 2001 *J. Chem. Phys.* 115, 7066
[3] Mao W L, Mao H K, Goncharov A F, Struzhkin V V, Guo Q, Hu J, Shu J, Hemley R J,

Somayazulu M and Zhao Y 2002 *Science* 29, 72247
[4] Loveday J S, Nelmes R J, Guthrle M, Belmonte S A, Allan D R, Klug D D, Tse J S, and Handa Y P 2001 *Nature* 410, 661
[5] Kurnosov A V, Manakov A Y, Komarov V Y, Voronin V I, Teplykh A E and Dyadin Y A 2001 *Dokl. Phys. Chem.* 381, 303
[6] Ripmeester J A, Tse J S, Ratcliffe C I and Powell B M 1987 *Nature* 325, 135
[7] Hirai H, Tanaka T, Kawamura T, Yamamoto Y and Yagi T 2004 *J. Phys. Chem. Solid* 65, 1555
[8] Mao W L, Koh C A and Sloan E D 2007 *Physics Today* 60, 42
[9] Loveday J S, Nelmes R J, Guthrie M, Klug D D and Tse J S 2001 *Phys. Rev Lett.* 87, 215501
[10] Vos W L, Finger L W, Hemley R J and Mao H K 1993 *Phys. Rev. Lett.* 71, 3150
[11] Vos W L, Finger L W, Hemley R J and Mao H K 1996 *Chem. Phys. Lett.* 257, 524
[12] Hirai H, Tanaka T, Kawamura T, Yamamoto Y and Yagi T 2003 *Phys. Rev.* B 68, 172102
[13] Loveday J S, Nelmes R J, Klug D D, Tse J S and Desgreniers S 2003 *Canadian J. Phys.* 81, 539
[14] Loveday J S and Nelmes R J 2008 *Phys. Chem. Chem. Phys.* 10, 913
[15] Machida S, Hirai H, Kawamura T, Yamamoto Y and Yagi T 2006 *Phys. Earth Planet. Interiors* 155, 170
[16] Iitaka T and Ebisuzaki T 2003 *Phys. Rev.* B 68, 172105
[17] Shimizu H, Kumazaki T, Kume T and Sasaki S 2002 *J. Phys. Chem.* 106, 30
[18] Tulk C A, Klug D D, dos Santos A M, Karotis G, Guthrie M, Molaison J J and Pradhan N 2012 *J. Chem. Phys.* 136, 054502
[19] Hirai H, Kadobayashi H, Hirao N, Ohishi Y, Ohtake M, Yamamoto Y and Nakano S 2015 *J. Chem. Phys.* 142, 024707
[20] Chen J Y and Yoo C S 2012 *J. Chem. Phys.* **136**, 114513
[21] Bezacier L, Menn E Le, Grasset O, Bollengier O, Oancea A, Mezouar M, Tobie G 2014 *Phys. Earth Planet. Interiors* **229**, 144-152
[22] Hebert P, Polian A, Loubeyre P and Toullec R Le 1987 *Phys. Rev.* B **36** (17), 9196-9201
[23] Kumazaki T, Kito Y, Sasaki S, Kume T, Shimizu H 2004 *Chem. Phys. Lett.* **388**, 18

IOP Conf. Series: Journal of Physics: Conf. Series **950** (2017) 042045 doi:10.1088/1742-6596/950/4/042045

On origin and evolution of diamond-forming lower-mantle systems: physicochemical studies in experiments at 24 and 26 GPa

YuA Litvin[1], AV Spivak[1], DA Simonova[1] and LS Dubrovinsky[2]

[1] Institute of Experimental Mineralogy, Chernogolovka, Moscow Region, RF
[2] Bayerishes Geoinstitut, University of Bayreuth, Bayreuth, DE

E-mail: litvin@iem.ac.ru

Abstract. The MgO – FeO – SiO_2 – CaO system is representative for the dominant lower-mantle ultramafic assemblage ferropericlase $(MgO{\cdot}FeO)_{ss}$ + bridgmanite $(Mg,Fe)SiO_3$ + Ca-perovskite $CaSiO_3$. Melting relations of the system were studied over its polythermal section $(MgO)_{70}(FeO)_{30}$ – $(SiO_2)_{70}(FeO)_{30}$ at 24 GPa. A peritectic reaction of bridgmanite with melt was revealed to form the mafic assemblage stishovite SiO_2 + periclase-wustite solid solution phases $(MgO{\cdot}FeO)_{ss}$ (effect of "stishovite paradox"). Experimental phase relations demonstrate the possibility of ultramafic-mafic fractional evolution of the lower-mantle magma and *in situ* formation of stishovite. Moreover, origin of ultradeep diamonds and associated phases in melts of the MgO – FeO – SiO_2 – CaO – (Mg-Fe-Na-K-carbonate) – carbon system have been examined at 26 GPa. A key role of the peritectic reaction of bridgmanite in genesis of mafic inclusions in lower-mantle diamonds is verified.

1. Introduction

No lower-mantle rocks have turned up among the mantle derived xenoliths. Experiments on subsolidus conversion of pyrolite composition at 25 – 30 GPa [1] indicate a dominant role of the ultramafic lower-mantle assemblage ferropericlase $(MgO{\cdot}FeO)_{ss}$ + bridgmanite $(Mg,Fe)SiO_3$ + Ca-perovskite $CaSiO_3$. Similar minerals are known as primary inclusions in ultradeep diamonds [2, 3]. But, the inclusions were trapped by lower-mantle diamonds as fragments of heterogeneous oxide-silicate-carbonate-carbon parental medium [4]. Accumulation of diamond-parental melts creates their chambers inside the enclosing lower-mantle rocks. Hence, the ways of origin of rock-forming and diamond-hosted minerals is different in spite of their similarity. The lower-mantle rocks are the in situ products of earlier magmatic events. The diamond-parental melts were produced by dissolution of rock-forming ferropericlase, bridgmanite and Ca-perovskite minerals in primary carbonate melts (most probably, of metasomatic origin). Inclusions of these minerals together with carbonate minerals aragonite $CaCO_3$, dolomite $CaMg(CO_3)_2$, nyerereite $Na_2Ca(CO_3)_2$, nahcolite $NaHCO_3$ are symptomatic for compositions of the multicomponent completely miscible oxide-silicate-carbonate-carbon diamond-parental melts. Inclusions in ultradeep diamonds are the crystallization products from the lower-mantle diamond-parental melts. In essence, the inclusions are products of rock-forming minerals re-crystallized in the parental melts.

The main discrepancy between the experimentally estimated lower-mantle ultramafic mineralogy and minerals included in "super-deep" diamonds lies in the fact that stishovite SiO_2 is quite representative among the primary inclusions. Moreover, diamond-hosted

IOP Conf. Series: Journal of Physics: Conf. Series **950** (2017) 042045 doi:10.1088/1742-6596/950/4/042045

stishovite is closely associated with periclase-wustite solid solution phases (MgO·FeO)$_{ss}$, even forming closely related paragenetic intergrowths. The association is paradoxial because

(MgO·FeO)$_{ss}$ + SiO_2 paragenesis is forbidden for quartz and coesite due to formation of enstatite and ferrosilite under the upper mantle and transition zone conditions. Thus, not only ultramafic but stishovite-bearing mafic material is drawn into the in situ lower-mantle diamond-forming process.

In this work, melting relations of lower-mantle rock-forming oxide-silicate MgO – FeO – SiO_2 – CaO and diamond-parental oxide-silicate-carbonate-carbon MgO – FeO – SiO_2 – CaO – (Mg-Fe-Ca-Na-carbonate) – C systems are studied in experiments at 24 and 26 GPa. The main goals are following: (1) to reveal the physico-chemical mechanism of ultrabasic-basic magmatic evolution of the lower-mantle MgO – FeO – SiO_2 – CaO system at 24 GPa; (2) to estimate the possibilities for ultrabasic-basic evolution of partially molten MgO – FeO – SiO2 – CaO – (Mg-Fe-Ca-Na-carbonate) – C system parental for ultradeep diamonds and their primary inclusions at 26 GPa.

2. Experimental and analytical methods

Melting relations of the oxide-silicate system at 24 GPa and ultradeep diamonds + paragenetic minerals formation in the oxide-silicate-carbonate-carbon system at 26 GPa were studied in multianvil apparatus designed and calibrated by pressure and temperature at the Bayerishes Geoinstitut in Bayreuth [5]. The starting materials were powdered mixtures of oxides MgO, FeO, CaO, SiO_2, carbonates $MgCO_3$, $FeCO_3$, $CaCO_3$, Na_2CO_3 and graphite. Diamond and paragenetic minerals formation and melting relations were studied with the use of a $LaCrO_3$ heater for generation of high temperature inside the pressure chamber. The starting materials have been introduced in a rhenium capsules. Accuracies in pressure determinations were about 1 GPa. Uncertainties in the temperature measurements using the W-Re thermocouple were estimated as ±50°C. The recovered samples were examined using a CamScan M2300 SEM (VEGA TS 5130M. The compositions of the phases were analyzed using a Link INCA energy-dispersive microprobe at the Institute of Experimental Mineralogy in Chernogolovka, Moscow Region.

3. Experimental oxide-silicate and oxide-silicate-carbonate-carbon lower-mantle systems

The MgO – FeO– SiO_2 – CaO system is representative for the dominant lower-mantle ultramafic assemblage ferropericlase (MgO·FeO)$_{ss}$ + ferrobridgmanite (Mg,Fe)SiO_3 + Ca-perovskite $CaSiO_3$. Subsolidus assemblages of the MgO– FeO– SiO_2 join at 24 GPa (Fig. 1) demonstrate that periclase MgO and wustite FeO form unlimited solid solutions; ferrobridgmanite solid solutions (Mg,Fe)SiO_3 are limited, and $FeSiO_3$ is not stable and decomposes to FeO and SiO_2. Melting relations over the polythermal section $(MgO)_{70}(FeO)_{30}$ – $(SiO_2)_{70}(FeO)_{30}$ at 24 GPa (Figs. 2 and 3) reveal a bridgmanite and melt peritectic reaction resulted in formation of mafic assemblage stishovite SiO_2 + periclase-wustite solid solution phases (MgO·FeO)$_{ss}$. The reaction acts also in the system MgO– FeO– SiO_2 – $CaSiO_3$ (Fig. 4) providing the physicochemical means for ultramafic-mafic magmatic evolution over the "ultramafic" cotectic curve FPer + FBrd +CaPrv +L, peritectic point FPer + FBrd + Sti + CaPrv +L and "mafic" cotectic curve Sti + MWus + CaPrv +L.

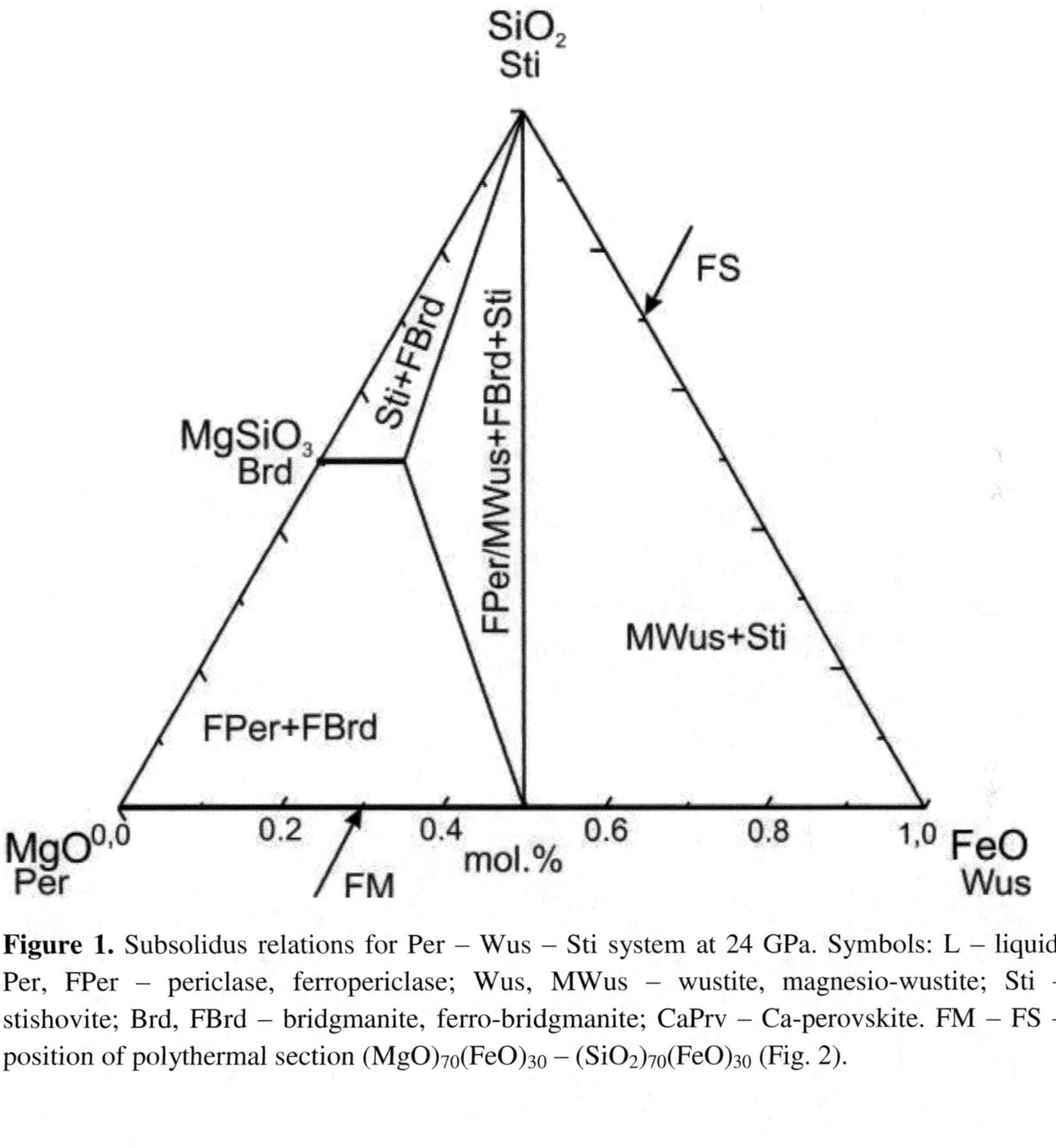

Figure 1. Subsolidus relations for Per – Wus – Sti system at 24 GPa. Symbols: L – liquid, Per, FPer – periclase, ferropericlase; Wus, MWus – wustite, magnesio-wustite; Sti – stishovite; Brd, FBrd – bridgmanite, ferro-bridgmanite; CaPrv – Ca-perovskite. FM – FS – position of polythermal section $(MgO)_{70}(FeO)_{30}$ – $(SiO_2)_{70}(FeO)_{30}$ (Fig. 2).

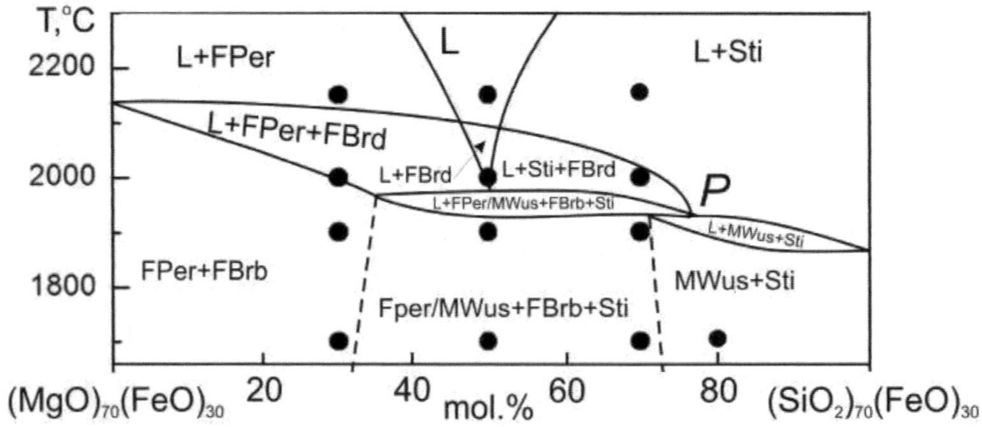

Figure 2. Melting relations for $(MgO)_{70}(FeO)_{30}$ – $(SiO_2)_{70}(FeO)_{30}$ polythermal section at 24 GPa. Symbols as for Fig. 1.

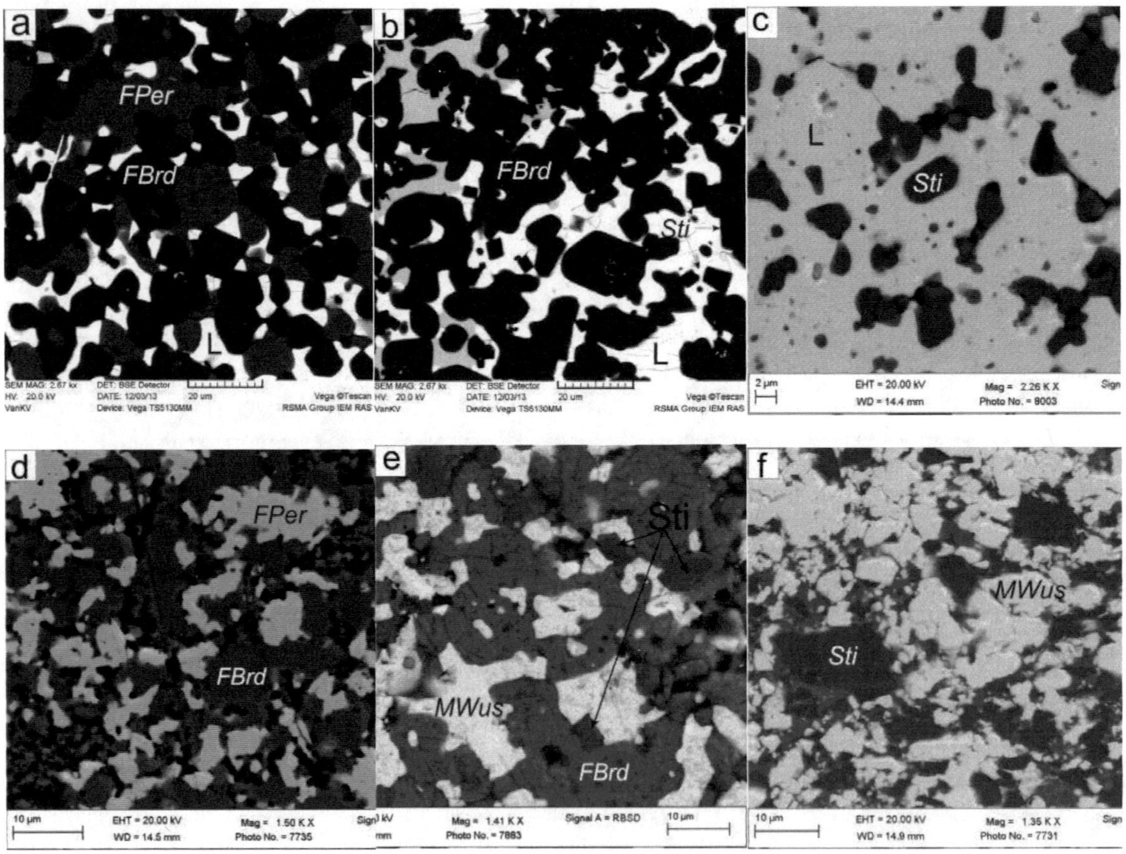

Figure 3. SEM images of experimental samples over polythermal section $(MgO)_{70}(FeO)_{30}$ – $(SiO_2)_{70}(FeO)_{30}$ at 24 GPa after quenching. Starting compositions and temperatures: (a) $(MgO)_{49}(FeO)_{30}(SiO_2)_{21}$, 2000°C; (b) $(MgO)_{35}(FeO)_{30}(SiO_2)_{35}$, 2000°C; (c) $(MgO)_{21}(FeO)_{30}(SiO_2)_{49}$, 2150°C; (d) $(MgO)_{49}(FeO)_{30}(SiO_2)_{21}$, 1700°C; (e) $(MgO)_{35}(FeO)_{30}(SiO_2)_{35}$, 1700°C; (f) $(MgO)_{14}(FeO)_{30}(SiO_2)_{56}$, 1700°C.

Crystallization of ultradeep diamonds and coprecipitated paragenetic minerals proceeds in melts of ultramafic and mafic compositions of the oxide-silicate-carbonate-carbon diamond-parental system at 26 GPa (Fig. 5).

4. Discussion and concluding remarks

Experimental physicochemical data reveal the effect of "stishovite paradox" [6] responsible for origin of the "forbidden" assemblage SiO2 + (MgO·FeO)ss with stishovite polymorph due to peritectic reaction of bridgmanite and lower-mantle ultrabasic magma. The reaction is also important for joint in situ genesis of ultradeep diamonds and primary mafic inclusions.

The results are compatible with the mantle-carbonatite concept of diamond genesis [4] and reveal the physicochemical mechanism for ultramafic-mafic fractional evolution of lower-mantle magmas and diamond –parental melts of the MgO – FeO – SiO2 – CaO – (Mg-Fe-Na-K-carbonate) – C system and in situ formation of stishovite and other mafic inclusions in ultradeep diamonds.

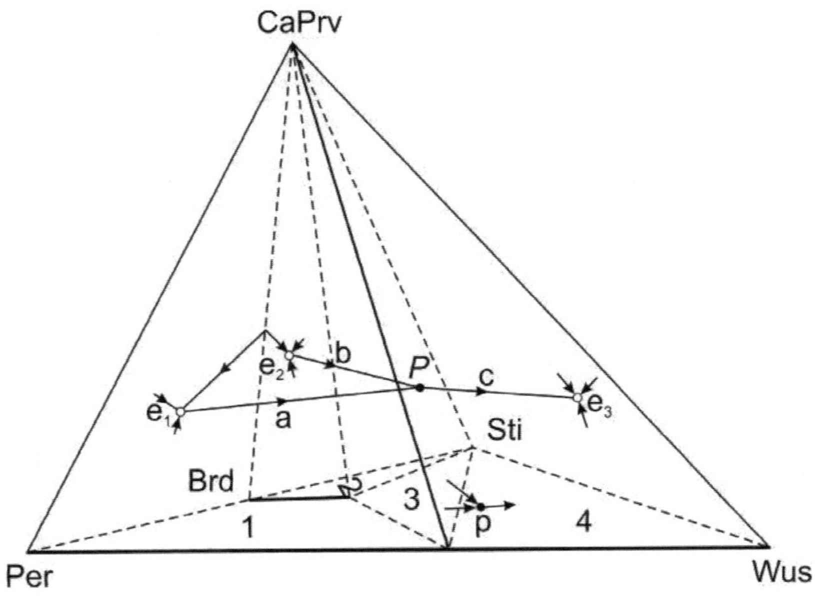

Figure 4. Melting relations for Per (MgO) – Wus (FeO) – Sti (SiO2) – CaPrv (CaSiO3) system at 24 GPa: 1 – 4 – subsolidus phase volumes (symbols in Fig. 1). Symbols: e1 – e3 eutectic points; p – boundary peritectic point (Per·Wus)ss + FBrd + Sti + L; P – principal peritectic point (Per·Wus)ss + FBrd + Sti + CaPer + L. Cotectic curves: (a) ultramafic FPer + FBrd +CaPrv + L, mafic (b) Sti + FBrd +CaPrv + L, mafic (c) Sti + MWus +CaPrv + L.

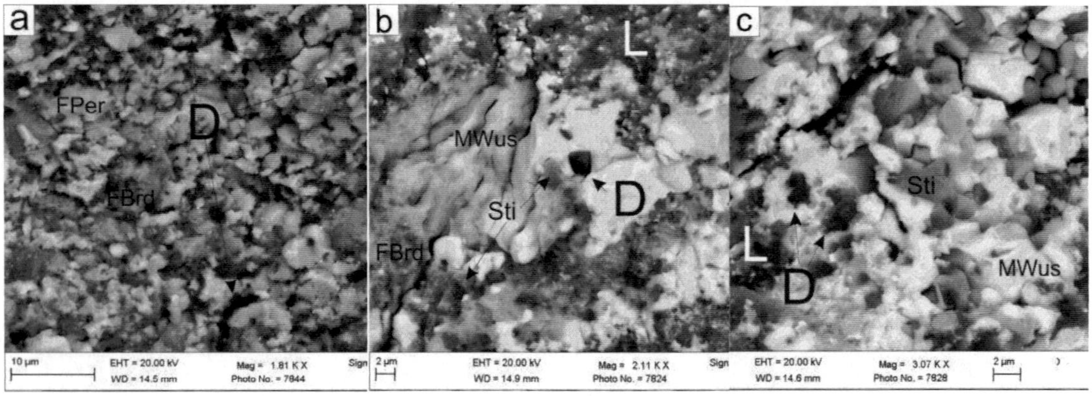

Figure 5. SEM images of experimental samples with co-precipitated diamond and paragenetic minerals at 26 GPa in growing melts of the system [(MgO)x(FeO)y(SiO2)z]18Carb42graphite40 with variation of indexes x, y, z: (a) (MgO)49(FeO)30(SiO2)21; (б) (MgO)35(FeO)35(SiO2)30; (c) (MgO)21(FeO)30(SiO2)49 (after quenching).

Acknowledgments. This work was funded by Russian Academy of Sciences (program P34) and Russian Foundation for Basic Research (grants 13-05-00835 and 14-05-00537).

References

[1] Akaogi M 2007 *High-Pressure Mineralogy* (Special Paper) ed E Ohtani. (Geological Society of America) 421 pp 1-13
[2] Harte B Harris J W 1994 *Miner. Mag.* **58A** 384
[3] Kaminsky F 2012 *Earth. Sci. Rev.* **110** 127
[4] Litvin Yu, Spivak A, Solopova N, Dubrovinsky L 2014 *Phys. Earth. Planet. Inter.* **228** 176
[5] Frost D J, Poe B T, Trønnes R G, Liebske C, Duba A and Rubie D C 2004 *Phys. Earth. Planet. Inter.* **143** 507
[6] Litvin Yu 2014 *Doklady Earth Sciences* **455 1** 274

IOP Conf. Series: Journal of Physics: Conf. Series **950** (2017) 042046 doi:10.1088/1742-6596/950/4/042046

Magnetic fields of Uranus and Neptune: Metallic fluid hydrogen

W J Nellis

Department of Physics, Harvard University, Cambridge, MA 02138, USA

nellis@physics.harvard.edu

Abstract. The unusual magnetic fields of Uranus and Neptune (U/N) have been a major scientific issue since their discovery in the 1980s. Based on an extensive database of planetary fluids measured under dynamic compression at pressures and temperatures up to 200 GPa and a few 1000 K, the complex magnetic fields of U/N are made primarily by dynamo convection of metallic fluid hydrogen (MFH) near crossovers from their H-He envelopes to "Ice" cores at ~100 GPa pressures at ~90% of radii of U/N. Because those fields are made close to outer surfaces, non-dipolar magnetic fields can be expected. "Ice" cores are heterogeneous fluid mixtures of nebular Ice and Rock that accreted, sank below the H-He envelopes into the cores, decomposed at high pressures and temperatures and formed new chemical species. Such complex mixtures are expected to have many nucleation sites for convection. Rotational and effective dipolar axes probably have large relative tilts because rotational motions of U/N are weakly coupled to convective motions that make their magnetic fields. "Polar wander" is probably a better descriptor for variations of magnetic field over time than "polar reversal" as for Earth. There is probably little "Ice" in the Ice Giants.

1. Introduction

Dynamic compression achieves extreme pressures P and temperatures T in fluids in the Giant Planets. In the 1970s the Voyager 2 spacecraft departed Earth to visit the Giant Gas Planets Jupiter, Saturn and then on for flybys of the Giant Ice Planets Uranus and Neptune (U/N) [1]. U/N are expected to contain substantial amounts of "ices" H_2O, NH_3, CH_4 etc. Dynamic compression experiments on planetary fluids have been performed intermittently for almost 30 years at the two-stage gun (2SG) at Lawrence Livermore National Laboratory (LLNL). Those experimental measurements are an extensive database for models of interiors of Giant Planets. Metallic fluid H (MFH) is of particular importance because dominant contributions to the unusual fields of U/N are probably made near boundary regions between their Gas envelopes and "Ice" cores at ~100 GPa pressures.

NASAs Voyager 2 spacecraft discovered the non-dipolar and non-axisymmetric magnetic fields of Uranus and Neptune (U/N) in the 1980's [1]. Those fields are made primarily by degenerate, metallic, fluid H (MFH) [2,3] and are ~2×10^{-5} Tesla [4], comparable to Earth's. If the measured fields of U/N are force-fit to effective dipolar fields, the effective magnetic axes are tilted 59^0 and 47^0 from their respective rotational axes and the effective dipole centers are offset by ~0.33 R_U and ~0.55 R_N, respectively, from the physical centers of U/N, where R_U and R_N are the radii of U and N. Those fields are illustrated in figure 1. The magnetic fields of Earth, Saturn, Jupiter and its moon Ganymede are dipolar with their magnetic axes aligned within ~10 degrees of their respective rotational axes.

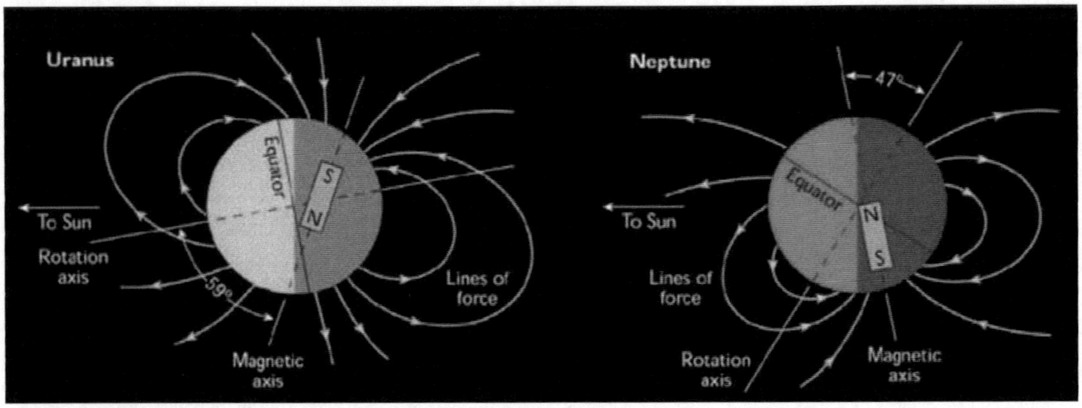

Figure 1. Non-dipolar non-axisymmetric fields of Uranus and Neptune measured by Voyager 2 [1].

The cause of the non-dipolar non-axisymmetric magnetic fields of U/N has been a major scientific question for decades. Planetary fields are made by dynamos: convection of electrically conducting fluids across lines of magnetic flux [5]. Pressures and temperatures in U/N range up to ~700 GPa and 7000 K near their centers [6,7]. Dynamic compression with a 2SG achieves pressures up to ~300 GPa and several 1000 K, conditions at which the external magnetic fields of U/N are made.

2. Chemical compositions and properties

U/N have similar sizes, densities, magnetic and gravitational fields and probably chemical compositions, although exact compositions are unknown. U/N are composed of [8-10]: (i) Gas, an H-rich mixture of H and He, (ii) Ice, a mixture primarily of H_2O, CH_4, and NH_3, and (iii) Rock, a mix of silicates and Fe/Ni. H and He have solar abundances of about 92 at.% and 7 at.%, respectively [11]. In space "icy" compounds are "nebular". U/N have masses 14.5 and 17.2 M_E, respectively, where M_E is mass of Earth. Total mass of the Giant Planets Jupiter, Saturn, Uranus and Neptune is 445 M_E. Total mass of the Rocky Planets Earth, Venus, Mars and Mercury is 2.0 M_E. Most mass of the solar system is MFH. Experiments on planetary fluids began when Voyager 2 departed Saturn for U/N [12,13].

3. Dynamic compression experiments

The vast majority of measured properties of fluids at high planetary P/Ts have been made under dynamic compression. MFH has been made by impact of a plate onto a cryogenic sample holder containing liquid H_2 or D_2 at 20 K. The impactor was accelerated to velocities as high as 7 km/s with a 2SG and to higher velocities in four experiments at Sandia National Laboratories Albuquerque (SNLA). H_2 samples in the 2SG experiments were 25 mm in diameter and 0.5 mm or more thick.

Historically dynamic compression has been achieved by a single step-increase in pressure that rises in ~ps and ~nm. The sharp rise is called the front of a shock wave, which travels at supersonic velocity in the medium ahead of the front. Supersonic velocity causes the front to "snow-plow" material ahead of it, which compresses rapidly and adiabatically generating dissipation as T and entropy S. Recently, higher density ρ and lower T have been achieved by quasi-isentropic (QI) compression by tailoring rise time of the wave front to be longer than that of a shock [14]. An isentrope is the limit of a large number of weak shocks [15].

Pressure tuning is illustrated in figure 2. A Hugoniot is the locus of points along the pressure-density curve in right panel, each point of which is achieved by a single jump from given initial pressure to a higher pressure. Multiple shocks lengthen time to achieve maximum pressure, as illustrated in the left panel. A ramp wave (not shown) increases pressure continuously from zero or a relatively small initial shock pressure. The affect of a ramp is similar to that of multiple-shocks.

Multiple-shock (MS) compression of liquid H_2 was achieved in the holder in figure 3 to achieve MFH at 140 GPa. Liquid H_2 was contained between two disks of sapphire (single-crystal Al_2O_3). This assembly is a cylindrically symmetric Al cryostat (liquid-H_2 coolant chambers not shown). Two current and two voltage probes are inserted from the right side of the sample holder. Photographs of

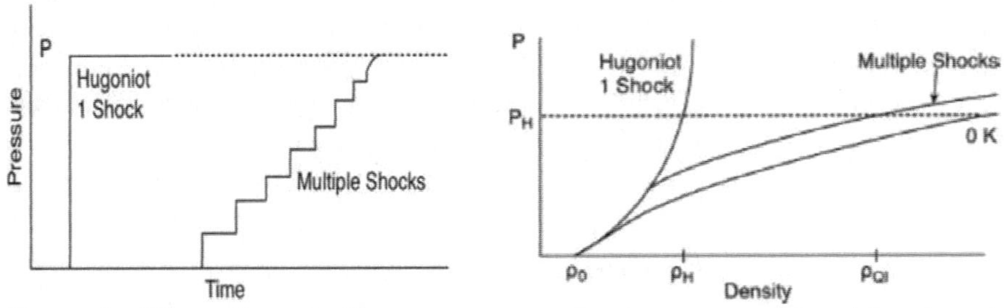

Figure 2. Illustration of affect on pressure-density states achieved by sharp jump to Hugoniot pressure versus multiple shocks to same final pressure. Dotted horizontal lines are same Hugoniot pressure in both figures.

the facility and sample holder are published [16]. Right panel in figure 3 indicates more than 8 shocks reverberated in liquid H_2 to make MFH. Model calculations with ideal-gas EOS [17] show 9 shocks produce an isentrope starting from the first shock state on the Hugoniot up to maximum pressure. That is, the calculated MS P(t) curve in figure 3 produces the corresponding MS P(ρ) curve in figure 2.

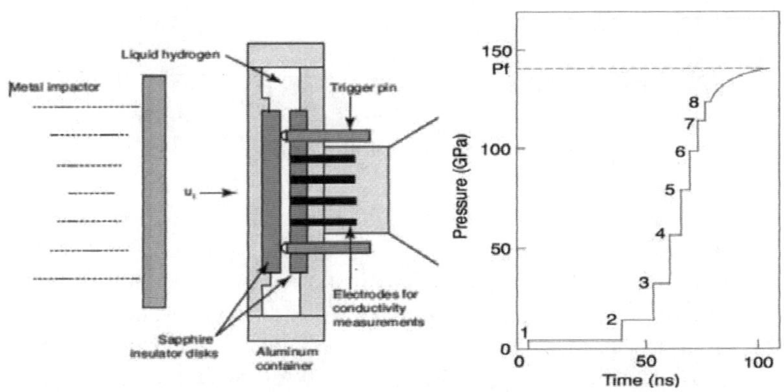

Figure 3. Experiment to measure conductivity of MFH at pressures P_f. Sample is liquid H_2 at 20 K. Impactor launched by 2SG is about to impact sample holder. Right: calculated pressure history of shock reverberations at midpoint of liquid H_2 layer made in place by condensing H_2 gas. Copyright 1999 by the American Physical Society.

4. Hydrogen: Two-stage light-gas gun

The advantage of measuring the pressure dependence of electrical conductivities of fluid H is their values and radial dependences in U/N are known to estimate the radial positions in U/N at which their external magnetic fields are made. MFH is degenerate condensed matter with $T/T_F<<1$, where T_F is Fermi temperature. At final temperatures of ~2000 K quantum differences between fluid H and D are negligible. However, different ρ and T are achieved because quantum differences at 20 K between H_2 and D_2 cause different initial number densities. Hydrogen samples "rang-up" to maximum pressure P_f in Al_2O_3 in each of ten experiments in ~100 ns [18-21]. P_f in each experiment was determined experimentally by measuring impactor velocity and determining P_f with shock-impedance matching [20]. Measured electrical resistivities versus P_f are plotted in figure 4. Conductivity reaches that of MFH (~2000/(Ω-cm)) at 140 GPa, 0.64 mol H/cm^3 (9-fold compressed H density in liquid H_2), ~2600

K and T_F =220,000 K. This density of MFH is essentially metallization density predicted by WH, 0.62 mol H/cm^3 [22]. Similar hydrogen conductivities were measured by Fortov et al [23].

Measured resistivities of fluid H in figure 4 indicate a semiconductor from 93 GPa to 140 GPa, above which conductivity is ~2000/(Ω-cm), typical of strong-scattering metals with minimum metallic conductivity (MMC). Metallization density ρ_{met} was chosen simply as the density at which mobility gap $E_{gap}(\rho) \rightarrow k_BT$, where k_B is Boltzmann's constant. Mobility gap $E_{gap}(\rho)$ was obtained by fitting measured conductivities to the standard expression for a thermally activated semiconductor. $E_{gap}(\rho)$ was assumed linear between 90 GPa and 124 GPa, a range away from the nonmetal-metal transition at 140 GPa in which values of $E_{gap}(\rho)$ are positive. MFH is probably monatomic [14].

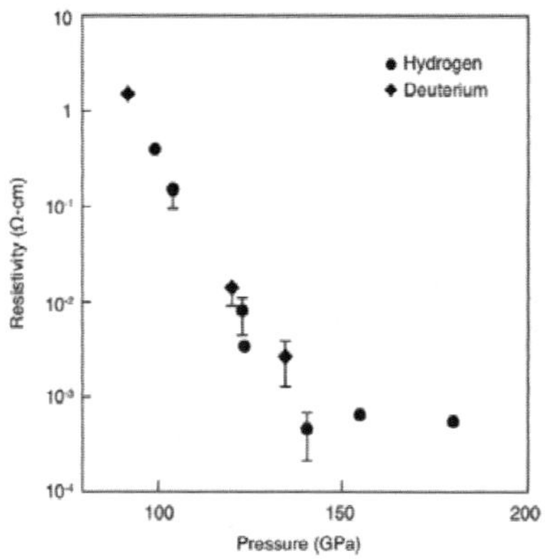

Figure 4. Measured electrical resistivities of fluid D and H versus multiple-shock pressure P_f determined experimentally by shock-impedance matching in each of 10 experiments. Copyright 1999 by the American Physical Society.

Because neither ρ nor T could be measured, they were calculated with an EOS of D_2/H_2 developed with a substantial body of *previously measured experimental data*. The initial shock in each reverberation is on the Hugoniot. The EOS was derived with: (i) measured Hugoniot data [24], (ii) measured single (Hugoniot)- and double-shock temperatures [25] and (iii) peak pressure in hydrogen obtained experimentally by shock-impedance matching each conductivity experiment [20]. 40% of each calculated temperature was caused by measured Hugoniot temperatures [25]. To determine systematic sensitivity of ρ_{met} to hydrogen EOS, a second EOS developed by Kerley was also used to calculate ρ, T. Two values of ρ_{met} using both sets of (ρ,T) values were determined from $E_{gap}(\rho_{met}) \approx k_BT$. Although ten values of $E_{gap}(\rho)$ differ by ~20% between the two EOS models, the two ρ_{met} values differ by 2%, sufficient to determine ρ_{met} = 0.64 mol H/cm^3 [20].

5. Wigner and Huntington's prediction

In 1935 Wigner and Huntington (WH) predicted the density of a first-order dissociative transition from insulating H_2 to metallic H at density ρ_{met}=0.62 mole H/cm^3 [22] at which static pressure is 73 GPa at 300 K [26]. A metallic solid phase of H_2 has yet to be observed at static pressures approaching the 400 GPa range. In contrast dynamic compression deposits sufficient T and S to dissociate H_2 to H in a cross over that completes at WH's predicted metallic density by overlap of $1s^1$ wave functions on adjacent atoms at temperatures cold enough to make degenerate condensed matter ($T \ll T_F$) [27].

6. Deuterium: Z Accelerator

Liquid D_2 has been compressed quasi-isentropically in two stages by current-generated magnetic pressure in 4 experiments at the Z Accelerator [28]. The first stage generated three shocks by reverberation in liquid-D_2. The second stage then drove a ramp wave that compressed deuterium isentropically up to ~300 GPa at temperatures in the range 800 K to 1700 K. Diagnostics were optical VISAR signals to measure velocity histories v(t), which were used to derive P(t) in deuterium that generated v(t), using a hydrodynamic computer code. ρ(t) and T(t) were calculated theoretically. Calculated Ts were nearly constant between 150 GPa and 300 GPa, a surprising and unusual observation that warrants temperature measurements. Despite this unusual reported behavior of T(P), no experimental or systematic uncertainties in P and T are given. Reflectivities were measured up to ~300 GPa at which reflectivity of fluid D increases abruptly to ~45%, which suggests an electrical conductivity of a few 10^5 S/m, as measured for MFH with a 2SG. This Z transition at 300 GPa is claimed to be *first-order* at substantial finite temperatures. Other transitions of fluid hydrogen achieved under dynamic compression are observed as crossovers at finite temperatures. The nature of this transition at 300 GPa needs to be determined experimentally. For example, could this observation of transparency be caused by dynamics of the longitudinal density distribution of fluid D, its corresponding longitudinal distribution of optical depth and the bright illumination laser used in the reflectivity measurement? D reflectivity decreases rapidly by orders of magnitude below the transition.

H densities are greater than 1.0 mol H/cm^3 at 300 GPa, factors of ~2 and ~60% greater than P and ρ on completion of the crossover at 140 GPa at the 2SG. These substantial differences indicate the transitions in the Z and 2SG experiments are different transitions. Agreement in metallic H conductivities in both experiments implies both measured strong-scattering MMCs. The P/T states of the Z experiments are sufficiently below estimates of T(P) adiabats of Giant Planets that the transition observed at Z probably makes negligible contribution to external magnetic fields of the Giant Planets.

7. Ices: Two-stage light-gas gun

Experiments have been performed on many small-molecular liquids under single- and multiple-shock compression [16]. Multiple-shock compression achieves relatively high pressures, densities and low temperatures with respect to single-shock compression [14]. *Trends in those measurements are used as a basis for speculation on the nature of the interiors of U/N, which might produce their unusual magnetic fields*. Liquids investigated under multiple-shock compression include H_2/D_2 [19-21], O_2 [29,30], N_2 [30], H_2O [31,32], and Synthetic Uranus (SU) [7]. SU is a single-phase liquid mixture of polar molecules H_2O, NH_3, and C_3H_8O (isopropanol). Liquids investigated under single-shock compression include He [33], H_2/D_2 [24,25,34-37], N_2 [38-41], O_2 [38,42], CO [43], CO_2 [44], air [44] CH_4 [43,45-47], NH_3 [13,46], H_2O [13,48,49], C_6H_6 [47,50], C_4H_8 [50,51], CH_2 [47, 50] and SU [52].

Measured properties indicate small-molecular fluids decompose below ~100 GPa and a few 1000 K. Measured electrical conductivities of liquid H_2, N_2, O_2 and H_2O under QI multiple-shock compression are plotted in Fig. 5. Conductivity of SU is within a factor of 2 of that of H_2O [7]. Conductivity curves of fluid H, N and O are similar because they are dominated by disorder scattering, which depends weakly on element. Figure 5 implies MFH, N, and O have similar electronic structures and are mutually soluble. Chemical ionization of H_2O causes positive protons to dominate conductivity below ~30 GPa. At higher pressures it appears electron carriers delocalize from OH^- ions with increasing pressure, which causes electrical conductivity to increase slowly with increasing pressure. Thus, once accreted, nebular molecules sink to depths in the Cores and decomposed nebular molecules re-react to form new chemical species. Also at depths in U/N at pressures above ~300 GPa most species might be monatomic because ρ and T might be too high for chemical bonds to exist.

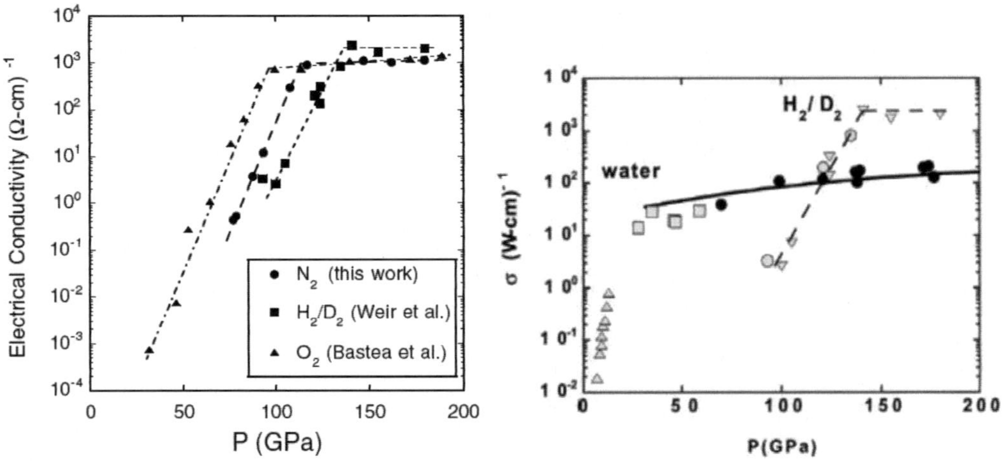

Figure 5. Measured electrical conductivities achieved by QI multiple-shock compression. Copyright 2003 by the American Physical Society (Left); copyright 2001 by the APS (Right).

8. Planetary fluids: Static compression

Using a pulsed-heating technique Deemyad and Silvera measured melting temperatures of hydrogen in the pressure range 50 to 80 GPa with a maximum of 1050 K at 65 GPa [53]. Dzyabura et al achieved pressures and temperatures in the ranges 1000 K to 1700 K at 119 GPa and 900 K to 1500 K at 125 GPa [54]. Those states are comparable to semiconducting ones under dynamic compression [20]. The melting curve of water ice has been measured in the range 20-90 GPa with melt temperatures between 1000 K and 2400 K [55]. Melting temperatures of water have also been measured at lower pressures as well [56,57]. Chemically ionized H_2O is observed up to 56 GPa and 1500 K [58]. Decomposition and reaction of ices to form new species have been observed with spectroscopic measurements in NH_3 [59] and CH_4 [60] at P/Ts up to 60 GPa and 2500 K and 80 GPa and 2000 K, respectively.

9. Interior picture of Uranus and Neptune

Voyager 2 measured gravitational moments, which lead to radial density distributions [**61**]. This process depends on the EOS of fluids in U/N with large systematic errors. Densities increase from zero at R_U/R_N up to ~0.6 g/cm^3 at radii of ~90% R_U/R_N. Densities then increase to ~4-5 g/cm^3 near planetary centers. Lighter outer regions are Gas envelops; heavier inner regions are Ice/Rock cores. Measured pressure dependence of electrical conductivities of fluid H suggest radii at which fields are made, which suggests a likely explanation for the shapes of those fields. At depths below H-rich envelops, details of P/Ts of whichever nebular molecules accreted and decomposed to form U/N are not needed to conclude that resulting mixtures probably make relatively small contributions to the planetary fields, which are coupled weakly to dominant contributions made by convection of MFH.

Conductivities of fluid H at 93 GPa and ~1500 K reach 2000/(Ω-cm) at 140 GPa, 0.64 g/cm^3 and ~2600 K. 2000/(Ω-cm) is 100 times larger than 20/(Ω-cm) measured for "Ices" and thought to cause magnetic fields of U/N prior to Voyager 2 [**62**]. The crossover in fluid H to MFH probably occurs in P/T intervals as much as ~100 GPa and ~1500 K. Because gravitational moments are delocalized constraints and although 0.6 g/cm^3 [**61**] might be caused by Gas mixed with Rock, it is reasonable to expect that at 0.6 g/cm^3 in H-rich U/N pressure is ~100 GPa near bottoms of envelopes and tops of cores. A similar estimate is based on measurements at static pressures. At 0.6 g/cm^3, pressure of solid H_2 is 73 GPa [26]. At ~2000 K H_2 is dissociated or nearly so, which means more particles to increase thermal pressure caused by temperatures up to ~3000 K. Decomposed "Ice" probably has substantial fractions of MFH, which likely exists in cores out to ~0.9 R_U. Unreacted N and O formed on decomposition of Ices are probably soluble in MFH. Dynamos probably make dominant contributions to the magnetic fields by convection of fluid H from lower envelops down into outer cores.

Voyager 2 implies the fields of U/N are made relatively close to their outer surfaces [63,64]. A dipolar field falls off with distance r as $1/r^3$. Multi-polar fields fall off faster than $1/r^3$ and are more likely to be observed if made relatively close to outer surfaces. Fields of U/N are probably made at radii as large as ~0.9 R_U/R_N. Earth's dipolar field is made in convecting fluid Fe outer core at ~0.5 R_E.

A clue to finding the cause of non-axisymmetric fields of U/N is why Earth's field is axisymmetric. Small deviations from axial symmetry might be caused by fluctuations in convective flows that generates Earth's magnetic field, which is not perfectly dipolar, and its dipole axis wanders over time and periodically magnetic North and South poles reverse in times short compared to times dipole axis points North or South [65]. Because rotational motion (RM) of Earth is strongly coupled into convective dynamo (CD) motions of its fluid-Fe outer core, planetary R-M stabilizes convective motions that generate a dipolar magnetic field. If a convective fluctuation occurs which tends to destabilize a given dipolar axis, then strong RM-CD coupling either drives convective motions that essentially restore the initial orientation or CD fluctuations that drive the initial magnetic axes out of orientational equilibrium are so strong that RM-CD coupling eventually drives the dipolar axis into an alignment anti-parallel to its initial one.

Recent analysis of deep-Earth seismic signals supports this picture. Earth has a strong, solid mantle that rotates with constant angular velocity, except that strong RM-CD coupling causes tiny variations in the length of a day observed for 130 years. Possible interactions mechanisms that cause field reversal include surface roughness of ~0.5 km on inner radius (~3500 km) of Earth's mantle [66].

U/N do not have solid mantles to strongly influence convective fluid motions in their cores and thus magnetic fields. Fluid H-He envelopes have little strength. However, complex fluid mixtures in the cores of U/N are probably heterogeneous [3], implying relatively active convective cores, with many nucleation sites for convection. Thus local convective dynamo motions of fluids that produce the magnetic fields are essentially decoupled from global rotational motions, as is implied by Fig. 1. The dynamos of U/N would then be relatively free to wander as local convective fluctuations dictate. Thus, tilt angles and center-offsets of their fields would vary slowly over the age of the Solar System. "Polar wander" is probably a better term for the magnetic fields of U/N than "polar reversal" for Earth.

10. References

[1] http://www.nasa.gov/voyager
[2] Stevenson D J 1998 *J Phys: Condens Matter* **10** 11227-11234
[3] Nellis W J 2015 *Mod Phys Lett B* **29** 1430018
[4] Stevenson D J 2010 *Space Sci Rev* **152** 651-664
[5] Stevenson D J 1983 *Reports Prog Phys* **46** 555-620
[6] Hubbard W B Fig. 2 in Ref. 7
[7] Chau R, Hamel S and Nellis W J 2011 *Nature Comm* |2:203|DOI:10.1038/ncomms1198 |
[8] Stevenson, D J 1982 *Ann Rev Earth Planet Sci* **10** 257-295
[9] Hubbard, W B 1984 *Planetary Interiors* New York Van Nostrand-Rheinhold
[10] Irwin P G J 2003 *Giant planets of our solar system* New York Springer-Verlag
[11] Arnett D 1960 *Supernovae and Nucleosynthesis* Princeton, Princeton University Press
[12] Hubbard W B 1981 *Science* **214** 145-148
[13] Mitchell A C and Nellis W J 1982 *J Chem Phys* **76** 6273-6281
[14] Nellis W J 2015 *J Phys Chem Sol* **84** 49-56
[15] Zeldovich Ya B and Raizer Yu P 1966 *Physics of Shock Waves and High-Temperature Hydrodynamic Phenomena* Volumes 1 and II New York Academic
[16] Nellis W J 2007 *Rev High Press Sci Tech (Japan)* **17** 328-333
[17] Nellis W J 2006 *Rep Prog Phys* **69** 1479-1580
[18] Weir S T, Mitchell A C and Nellis W J 1996 *Phys Rev Lett* **76** 1860-1863
[19] Nellis W J, Louis A A and Ashcroft N W 1998 *Phil Trans Roy Soc (London)* **356** 119-138
[20] Nellis W J, Weir S T and Mitchell A C 1999 *Phys Rev B* **59** 3434-3439
[21] Nellis W J 2013 *High Press Res* **33** 369-376

IOP Conf. Series: Journal of Physics: Conf. Series **950** (2017) 042046 doi:10.1088/1742-6596/950/4/042046

[22] Wigner E and Huntington H B 1935 *J Chem Phys 3* 764-770
[23] Fortov V E *et al* 2003 *JETP* **97** 259-278
[24] Nellis W J *et al* 1983 *J Chem Phys* **79** 1480-1486
[25] Holmes N C, Nellis W J and M. Ross M (1995) *Phys Rev B* **52** 15,835-15,845
[26] Loubeyre P *et al* 1998 *Nature (London)* **383** 702-704
[27] Mott N F 1936 *The theory of the properties of metals and alloys* New York Dover 176.
[28] Knudson M D *et al* 2015 *Science* **348** 1455-1459
[29] Bastea M, Mitchell A C and Nellis W J 2001 *Phys Rev Lett* **86** 3108-3111
[30] Chau R, Mitchell A C, Minich R W and Nellis W J 2003 *Phys Rev Lett* **90** 245501
[31] Chau R et al 2001 *J Chem Phys* **114** 1361-1365
[32] Yakushev V V 1997 in *High Pressure Research*, edited by R. J. Hemley *et al* Boston Materials Research Society 121-126
[33] Nellis W J *et al* 1984 *Phys Rev Lett* **53** 1248-1251
[34] Holmes N C, Nellis W J and Ross M 1998 in *Shock Compression of Condensed Matter – 1997*, edited by S. C. Schmidt *et al* Woodbury NY American Institute of Physics 61-65
[35] Nellis W J *et al* 1992 *Phys Rev Lett* **68** 2937-2940
[36] Nellis W J 2002 *Phys Rev Lett* **89** 165502
[37] Boriskov G V *et al* 2005 *Phys Rev B* **71** 092104
[38] Nellis W J and Mitchell A C 1980 *J Chem Phys* **73** 6137-6145
[39] Nellis W J *et al* 1984 *Phys Rev Lett* **53** 1661-1664
[40] Radousky H B *et al* 1986 *Phys Rev Lett* **57** 2419-2422
[41] Nellis W J *et al* 1991 *J Chem Phys* **94** 2244-2257
[42] Hamilton *et al* 1988 *J Chem Phys* **88** 5042-5050
[43] Nellis et al 1981 *J Chem Phys* **75** 3055-3063
[44] Nellis W J *et al* 1991 *J Chem Phys* **95** 5268-5272
[45] Nellis *et al* 1980 in *High Pressure Science and Technology Vol. 2*, edited by B. Vodar and Ph. Marteau Oxford Pergamon 1043-1047
[46] Radousky H B *et al* 1990 *J Chem Phys* **93** 8235-8239
[47] Nellis W J *et al* 2001 *J Chem Phys* **115** 1015-1019
[48] Lyzenga G A *et al* 1982 *J Chem Phys* **76** 6282-6286
[49] Holmes N C *et al* 1985 *Phys Rev Lett* **55** 2433-2436
[50] Nellis W J *et al* 1984 *J Chem Phys* **80** 2789-2799
[51] Hamilton D C *et al* 1988 *J Chem Phys* **88** 7706-7708
[52] Nellis W J *et al* 1997 *J Chem Phys* **107**, 9096-9100
[53] Deemyad S and Silvera I F 2008 *Phys Rev Lett* **100** 155701
[54] Dzyabura V *et al* 2013 *Proc Natl Acad Sci USA* **110** 8040-8044
[55] Schwager B *et al* 2004 *J Phys: Condens Matter* **16** S1177-S1179
[56] Lin J-F *et al* 2005 *Geophys Res Lett* **32** L11306
[57] Dubrovinskaia N and Dubrovinsky L 2003 *High Press Res* **23** 307-311
[58] Goncharov A F *et al* 2005 *Phys Rev Lett* **94** 125508
[59] Ojwang J G O *et al* 2012 *J Chem Phys* **137**,064507
[60] S. S. Lobanov S S *et al* 2013 *Nature Comm* **4** 2446-2450
[61] Helled R et al, 2011 *Ap J* **726** 15
[62] Nellis W J *et al* 1988 *Science* **240** 779-781
[63] Podolak M, Hubbard W B and Stevenson, D J 1991 in *Uranus*, edited by J. T. Bergstralh, E.D. Miner and M. S. Matthews Tuscon University of Arizona 29-61
[64] Hubbard W B, Podolak M and Stevenson D J 1995 in *Neptune and Triton*, edited by D. P. Cruikshank Tuscon University of Arizona 109-138
[65] Sagnotti L et al 2014 *Geophys J Int* **199** 1110-1124; doi: 10.1093/gji/ggu287
[66] Hide R et al 2013 in *Relating Geophysical Structures and Processes: The Jeffreys Volume*, Geophysical Monograph 76, edited by K. Aki and R. Dmowska, IUGG Volume 16

IOP Conf. Series: Journal of Physics: Conf. Series **950** (2017) 042047 doi:10.1088/1742-6596/950/4/042047

Polarization mode switching in p-AlGaAs/GaAsP/n-AlGaAs diodes in presence of compressive stress

E V Bogdanov[1], K I Kolokolov[1], N V Melnikova[2], N Ya Minina[1] and G V Tikhomirova[3]

[1]Low Temperature Physics Department, Moscow State University, 119991 Moscow, Russia
[2]Low Temperature Physics Department, Ural Federal University, 620002 Ekaterinburg, Russia
[3]Physics Department, Ural Federal University, 620002 Ekaterinburg, Russia

E-mail: min@mig.phys.msu.ru

Abstract. Numerical calculations and experimental results show that, for the broad range of tensile strained *p*-$Al_xGa_{1-x}As/GaAs_{1-y}P_y$/*n*-$Al_xGa_{1-x}As$ heterostructures widely used in commercial laser diodes emitting at 766 – 808 nm, polarization of emitted light may be extremely sensitive to external uniaxial stress due to the change of wave functions symmetry and possible optical transitions in the quantum well levels system. In some heterostructures with quantum well width of 10 nm and phosphorus content below 0.08, TM/TE polarization mode relation showcases a several times decrease and even dominant polarization mode switching under moderate compression of about 5 - 6 kbar in [100] and [110] directions. At the same time switching from TE to TM mode is possible under compression in [001] direction.

1. Introduction

In comparison with uniform compression, uniaxial stress magnitude that can be applied to a material is not very high and limited by sample fracture. However, being kept below this limit it often breaks the crystal symmetry and cause strong anisotropic effects in electron energy spectrum, transport and optical properties [1-4] that can not be observed under hydrostatic pressure. It was reported previously about electroluminescence spectra blue shift up to 27 meV in *p*-$Al_xGa_{1-x}As/GaAs_{1-y}P_y$/*n*-$Al_xGa_{1-x}As$ ($y = 0.16$) laser diode heterostructure under uniaxial compression up to $P \approx 5$ kbar along [110] direction [5]. According to the numerical calculations, this shift is connected with about 2% increase of effective optical gap.

In this paper we show both theoretically and experimentally that, on a level with this relatively moderate effect, dramatic changes could take place under uniaxial compression in correlation between transverse electric (TE) and transverse magnetic (TM) polarization modes of emitting light due to the change of wave functions symmetry. Though the influence of uniaxial stress on semiconductor laser polarization was investigated previously [6, 7], in the present paper we for the first time demonstrate the possibility of the efficient TE/TM polarization mode tuning and switching by uniaxial compression in different crystallography directions for a wide range of *p*-$Al_xGa_{1-x}As/GaAs_{1-y}P_y$/*n*-$Al_xGa_{1-x}As$ heterostructures with tensile strained quantum wells (QWs) of different thickness and composition. The numerical calculations are compared with experimental results.

IOP Conf. Series: Journal of Physics: Conf. Series **950** (2017) 042047 doi:10.1088/1742-6596/950/4/042047

2. Samples and experimental details

In the present experiment, the structures under investigation were grown on Si doped (001)-oriented GaAs substrates by metalorganic vapor phase epitaxy. The tensile strained $GaAs_{0.84}P_{0.16}$ QW with a width of 14 nm was not intentionally doped. It was embedded in 1000 nm thick $Al_{0.45}Ga_{0.55}As$ waveguide layers and 800 nm $Al_{0.70}Ga_{0.30}As$ cladding layers. The first 100 nm $Al_{0.45}Ga_{0.55}As$ around the QW were not intentionally doped. Then, the *p*- and *n*-doping started with 1×10^{17} cm^{-3} and increased up to 2×10^{18} cm^{-3} at the top of the cladding layers. Because of the strong built-in tensile strain in the $GaAs_{0.84}P_{0.16}$ QW of 0.58% arising during the epitaxial growth, the LH1 subband becomes the ground state of the valence band leading to predominant TM emission at laser operation.

Not being a conventional technique, the method of uniaxial compression of a sample firmly fastened in the elastic steel ring has been already mentioned in our previous papers [2, 4] and has been described in a special issue [8]. This approach is advantageous because of the axial distribution of the stress and a rigid fastening of the ends that protect the sample from premature destruction. The compressive stress may be applied in liquid nitrogen or helium with the help of a special device described in [8]. The magnitude of uniaxial strain (and stress) in the sample is determined using X-ray diffraction measurements [8].

The electroluminescence (EL) spectra were investigated at 77 K under uniaxial compression up to $P = 5$ kbar along the [110] direction in an optical cryostat with the uniaxial stress device inside. They were recorded using a PC controlled MDR-12 monochromator with a spectral resolution of up to 0.1 nm. Phase sensitive detection provided a dynamic range of operation for the system of up to five orders of magnitude. Noise and interferences limited the intensity region for a particular spectrum to the level of 10^{-2} - 10^{-3} of the maximum signal intensity. A Glan prism and thin-film polarizer were used for polarization analysis of the emitted light.

3. Experimental results

The intensity of light *I* in the plane normal to the heterostructure, taken at liquid nitrogen temperature, is represented in dependence on polarizer turn angle at $P = 0$ and 5.1 kbar in figure 1(a). Figure 1(a) demonstrates not only the strong increase of intensity under compression in [110] direction, but also evident decrease of relative TM/TE intensity anisotropy: from 1.8 at $P = 0$ to 1.2 at $P = 5.1$ kbar (that is a factor of 1.5). It has to be mentioned, that the spectral shift and the increase of the EL intensity with applied stress as well as the ratio between TM and TE modes (polarized perpendicular and parallel to the plane of heterostructure, respectively) were absolutely reversible in the investigated uniaxial stress interval under loading and unloading in different cycles of the experiment [figure 1(b)].

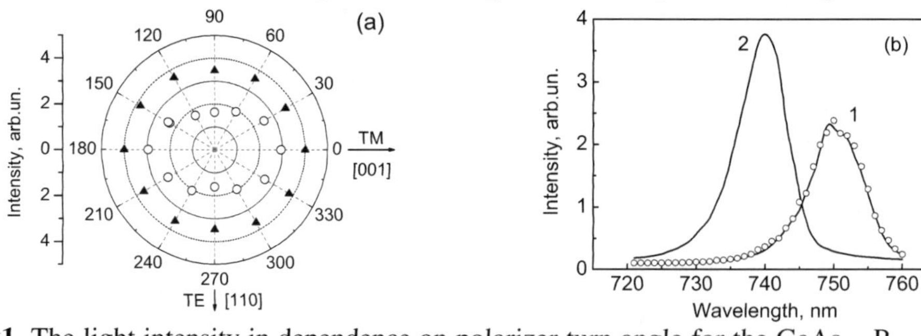

Figure 1. The light intensity in dependence on polarizer turn angle for the $GaAs_{0.84}P_{0.16}$ QW of 14 nm width under uniaxial compression along [110] at $P = 0$ (circles) and 5.1 kbar (triangles) (a) and EL spectra measured at polarizer turn angle 300° under $P = 0$ (before the loading) (curve 1), $P = 5.1$ kbar (curve 2) and $P = 0$ (after the loading cycle) (circles) (b).

4. Numerical calculations and discussion

Numerical calculations of uniaxial stress influence on the energy spectrum, wave functions, matrix elements of electron-photon interaction as well as the optical gain of TE and TM polarization modes in strained p-$Al_xGa_{1-x}As/GaAs_{1-y}P_y/n$-$Al_xGa_{1-x}As$ laser diode structures have been performed for a number of samples with different combinations of the phosphorus contents $y = 0 - 0.20$ (with 0.02 intervals) and the QW widths $d = 4$, 10, 14 and 20 nm. The calculations were carried out under uniaxial stress up to 10 kbar along [110], [100] and [001] directions at a temperature of 77 K. The Luttinger-Kohn Hamiltonian [9] with strain terms [10, 11] was self-consistently solved together with Poisson's equation for the electrostatic potential using the finite difference **k•p** method. Numerical calculations were performed using the program Heterostructure Design Studio 2.1 in the vicinity of the zone center at the Γ point. We used the model developed in [2]. In the calculations, internal strain as well as external strain effects were considered. The internal biaxial tension at $P = 0$ is considered to be determined by the lattice mismatch between the $GaAs_{1-y}P_y$ QW and the $Al_{0.45}Ga_{0.55}As$ barriers (not by the GaAs substrate).

For illustration of the uniaxial compression influence, figure 2 shows the strong change of the energy spectrum of light hole (LH) and heavy hole (HH) quantized levels h1, h2,... h7 in the QW (d = 4 nm, y = 0.16) [figure 2(a)] together with the analysis of LH and HH states mixing under the applied stress [figure 2(b)]. The results of calculations depicted in figure 2(b) demonstrate that the uppermost level h1 is a pure LH state at $P = 0$. After h1-h2 anticrossing at $P \approx 4$ kbar [figure 2(a)], the share of HH states dominates at level h1 [figure 2(b)], i.e. the uppermost level h1 in the QW starts to become almost of HH nature under uniaxial compression. At the same time h2 level, which is a pure HH state at $P = 0$, starts to become almost LH nature after h1-h2 anticrossing [figure 2(c)]. As a result, the change of the ground state symmetry in the QW removes the prohibition on the interband transitions only with the TM mode illumination. Therefore, we conclude that an essential change in the TE/TM mode intensity ratio occurs under uniaxial compression: the stronger the HH states admixture to LH states at the upper level, the larger the advantage of the TE mode.

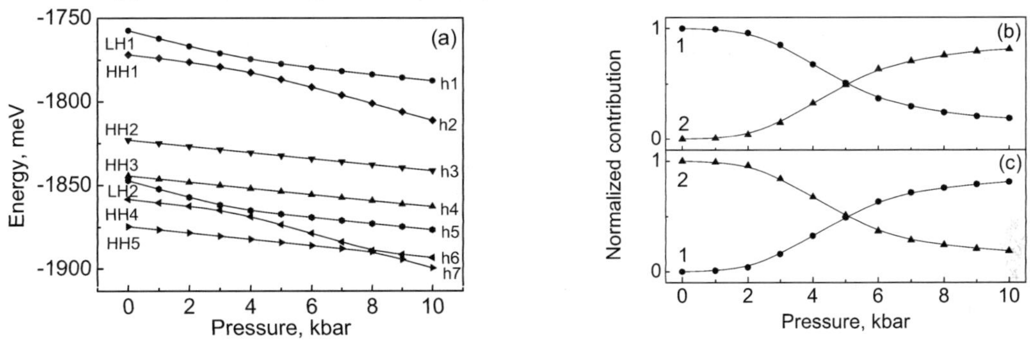

Figure 2. Calculated energy shifts of seven hole levels at the zone center under uniaxial compression along [110] direction (a) and pressure dependence of normalized contribution of basic functions with different angular momentum projection $m_J = \pm 1/2$ (curve 1) and $m_J = \pm 3/2$ (curve 2) into the wave function of h1 (b) and h2 (c) hole levels for the $GaAs_{0.84}P_{0.16}$ QW of 4 nm width.

After the calculation of the energy spectrum and wave functions, the program Heterostructure Design Studio 2.1 admits to introduce the scheme of possible optical transitions in the level system and calculates the matrix elements of the electron-photon interaction operator between two different states. The optical gain $g(\hbar\omega)$, which is in fact a negative absorption coefficient, is determined at the photon energy $E_{ph} = \hbar\omega$ according to the well-known method [3, 12]. The nonequilibrium carrier concentration $n = p = 10^{12} cm^{-2}$ in the QW was evaluated according to the average value of the injected current, ~ 10 mA [4].

In figure 3, we present the results of calculations for the experimentally investigated structure. Our analysis demonstrates that in the QW (d = 14 nm, y = 0.16) the optical gain g_{TM} of the TM mode at $P = 0$ is 8 times larger compared to the gain g_{TE} of TE mode. This fact is quite natural for a structure designed for TM emitting laser diodes. Under compression, the g_{TM}/g_{TE} ratio strongly drops (except for the case $P \parallel [001]$ direction). At the conditions of the present experiment (P = 5.1 kbar in [110] direction) $g_{TM}/g_{TE} \approx 5$ reveals a decrease by a factor of 1.6 that is in very good agreement with the experimental data (factor of 1.5) shown in figure 1(a). It should be noted that the TM/TE intensity ratio in a real laser operation is strongly determined by cavity parameters, and there is no direct proportionality between I_{TM}/I_{TE} and g_{TM}/g_{TE} ratios. Nevertheless, taking into account the fact that the lasing cavity practically is not changed under compression, we have very good agreement between calculations and experimental data.

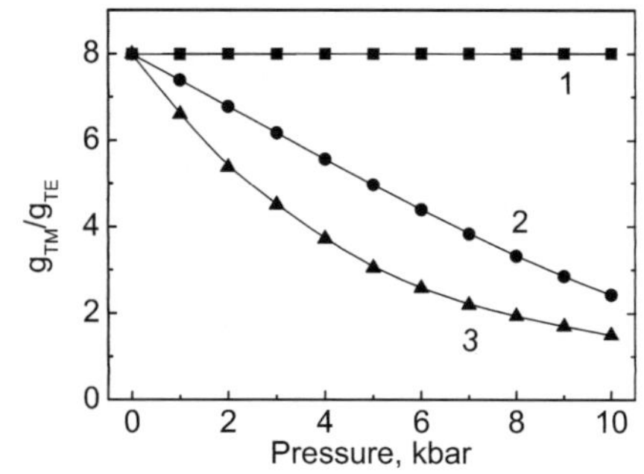

Figure 3. Pressure dependence of the g_{TM}/g_{TE} ratio for the $GaAs_{0.84}P_{0.16}$ QW of 14 nm width under uniaxial compression along [001] (1), [110] (2) and [100] (3) directions.

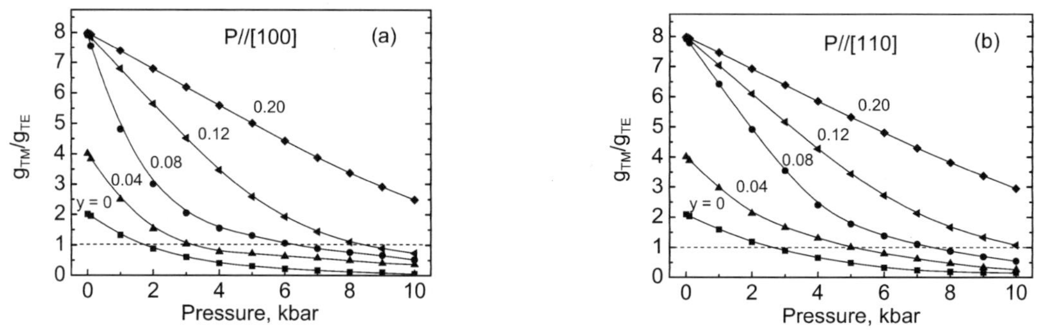

Figure 4. Pressure dependence of g_{TM}/g_{TE} ratio under uniaxial compression along [100] (a) and [110] (b) directions for structures with QWs of 10 nm width and phosphorus content ranging from $y = 0$ to $y = 0.2$.

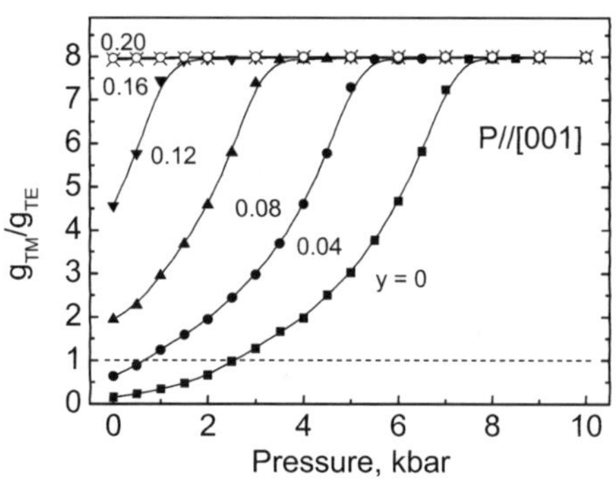

Figure 5. Pressure dependence of the g_{TM}/g_{TE} ratio in $GaAs_{1-y}P_y$ QWs of 4 nm width and phosphorus content ranging from $y = 0$ to $y = 0.2$ under uniaxial compression along [001] direction.

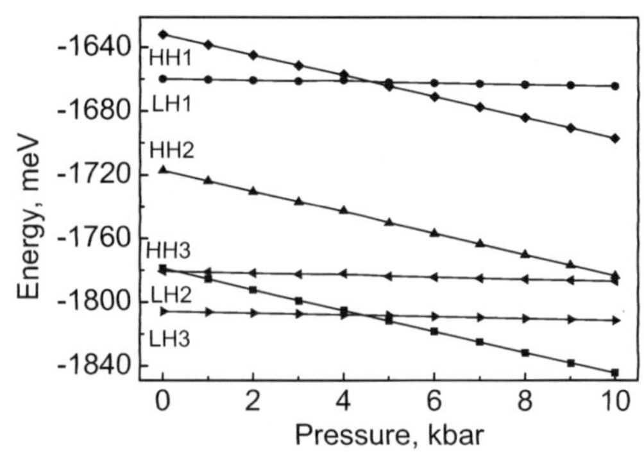

Figure 6. Calculated energy shifts of six hole levels at the zone center under uniaxial compression along [001] direction for the GaAs QW of 4 nm width.

Figures 4 and 5 demonstrate the results of $g_{TM}/g_{TE}(P)$ calculations in the structures with two different QWs: (i) $d = 10$ nm, $P \parallel [100]$ and $P \parallel [110]$ (figure 4) and (ii) $d = 4$ nm, $P \parallel [001]$ (figure 5), for a phosphorus content range of $0 \leq y \leq 0.2$. The results indicate a strong change of the g_{TM}/g_{TE} ratio under compression in the plane of the heterostructure both for $P \parallel [110]$ and $P \parallel [100]$. This effect is connected with the variation of the energy spectrum and symmetry of the size quantized levels in a QW, similar to the effects shown in figure 2. The most interesting result of our calculations is the indication that in the structures with rather small energy separation between the upper LH1 and HH1 states at $P = 0$, the TM $\rightarrow$ TE polarization mode switching is possible under relatively low pressure (figure 4).

In the case of compression along [001] direction that is perpendicular to the plane of the structures under investigation, wave functions do not change their symmetry, and light hole – heavy hole mixing in the Γ point does not take place for so called “light hole up” configuration of hole levels in a QW like represented in figure 2. If LH1 – HH1 energy difference Δ is large (for example, $\Delta = 43$ meV in the case depicted in figure 3, curve 1) g_{TM}/g_{TE} ratio does not depend on stress. At the same time, if HH1 is the ground level in a QW at $P = 0$ (see figure 6) and TE mode is dominant in the emitting light, compression along [001] direction can switch the output light from TE mode to TM mode emission (figure 5).

5. Conclusion

In summary, we have correlated experimental studies and numerical calculations to show that in laser diode structures based on $GaAs_{1-y}P_y$ QWs an effective TM and TE polarization modes tuning and switching under external uniaxial stress is possible. The best candidates for TM to TE switching are tensile strained heterostructures with low phosphorus content and narrow QW widths compressed in-plane along [110] and [100] directions (see figure 4). Switching from TE to TM mode is possible only under compression normal to the structure layers. The represented calculations are useful to anticipate and evaluate mounting and temperature induced strain effects on the diode laser performance [13,14]. A continuous tuning of TM/TE mode relation can be applicable in polarization spectroscopy of size quantized level states in QW structures. The demonstrated numerical analysis can be applied to many other layered QW structures of III-V semiconductor materials.

Acknowledgements

We are grateful to our colleagues Prof. A.E. Yunovich and Dr. H. Kissel for cooperation and valuable discussions.

References

[1] Cardona M 1996 *Phys. Status Solidi (b)* **198** 5
[2] Kolokolov K I, Savin A M, Beneslavski S D, Minina N Ya and Hansen O P 1999 *Phys. Rev.* B **59** 7537
[3] Kolokolov K I, Beneslavski S D, Minina N Ya and Savin A M 2001 *Phys. Rev.* B **63** 195308
[4] Bogdanov E V, Minina N Ya, Tomm J W and Kissel H 2012 *J. Appl. Phys.* **112** 093113
[5] Berman I V, Bogdanov E V, Kissel H, Minina N Ya, Shirokov S S and Yunovich A E 2009 *Phys. Status Solidi (b)* **246** 522
[6] Dutta N K and Craft D C 1984 *J. Appl. Phys.* **56** 65
[7] Adams C S and Cassidy D T 1988 *J. Appl. Phys.* **84** 6631
[8] Minina N Ya, Brandt N B, Savin A M and Bogdanov E V 2008 *J. Phys. Conf. Ser.* **121** 032002
[9] Luttinger J M and Kohn W 1955 *Phys. Rev.* **97** 869
[10] Bir G L and Pikus G E 1974 *Symmetry and Strain-Induced Effects in Semiconductors* (New York: Wiley)
[11] Lee J and Vassell M O 1988 *Phys. Rev.* B **37** 8855
[12] Chuang S L 1995 *Physics of Optoelectronic Devices* (New York: Wiley)
[13] Piechel B, Tomm J W, Bercha A, Trzeciakowski W, Reufer M and Gomez-Iglesias A 2009 *Appl. Phys.* A **97** 179
[14] Gerhardt A, Weik F, Tran T Q, Tomm J W, Elsaesser T, Biesenbach J, Muntz H, Seibold G and Biermann M L 2004 *Appl. Phys. Lett.* **84** 3525

Hydrothermal synthesis and Er^{3+} ion exchange of sodium GTS-type titanosilicate

Keiko Fujiwara, Kae Shinohara and Akihiko Nakatsuka

Graduate School of Science and Engineering, Yamaguchi University, Ube 755-8611, Japan

E-mail: keiko@yamaguchi-u.ac.jp

Abstract. Powder samples of Na-GTS have been prepared using a hydrothermal method. Er^{3+}-exchanged Na-GTS [$Na_{4(1-x)}Er_{(4/3)x}Ti_4Si_3O_{16}\cdot nH_2O$] with the compositions in the range of $0.28 \leq x \leq 0.64$ was obtained by shaking the single-phase sample of Na-GTS in the $ErCl_3$ aqueous solutions (25 mL, 0.01~0.1 M) at 25 °C for 6 hours. The chemical analyses by atomic absorption spectrometry (AAS) show that the maximum allowance of Er^{3+} incorporated into Na-GTS is x = 0.64. TG-DTA measurements show that the incorporation of Er^{3+} slightly decreases the water contents and the dehydration temperatures. The powder X-ray diffraction (XRD) patterns indicate that the diffraction-peak intensities vary systematically with increasing the Er^{3+} composition (x). The simulation of XRD patterns suggests the cation-distribution model that Er^{3+} and Na^+ ions preferentially occupy the 4*e* and 6*g* sites, respectively.

1. Introduction

Grace titanosilicates (GTS), represented by pharmacosiderite, have three dimensional tunnel-type structures. Na-GTS ($Na_4Ti_4Si_3O_{16}\cdot 6H_2O$) crystallizes as a rhombohedral phase (space group $R3m$; a = 7.812 Å, α = 88.79° [1]) close to cubic system. Other forms such as K-GTS ($K_3HTi_4Si_3O_{16}\cdot 4H_2O$) show the cubic $P\bar{4}3m$ symmetry ($a \approx 7.8$ Å). In GTS-type structures (Fig. 1), four TiO_6 octahedra linked by edge-sharing form a Ti_4O_4 cubic cluster; the clusters are linked through SiO_4 tetrahedra to form a three-dimensional framework with an interconnected pore system of 8-ring channels, occupied by the alkali-metal ions and adsorbed water molecules. The ion-exchange properties of GTS for Cs^+ or divalent cations (Co^{2+}, Ca^{2+}, Sr^{2+}, Ba^{2+}, *etc.*) have been investigated for the application to the radioactive waste treatments [2–4]. However, the ion-exchange properties of GTS for trivalent cations have not yet been reported, although the optical and catalytic properties of trivalent-cation-exchanged forms are interesting. In the present study, Er^{3+}-exchanged forms of Na-GTS are investigated by TG-DTA and powder XRD to examine the dehydration behavior and the occupied positions of Er^{3+}.

2. Experimental

2.1. Sample preparation

Sodium GTS-type titanosilicates (Na-GTS: $Na_4Ti_4Si_3O_{16}\cdot 6H_2O$) were prepared by a hydrothermal method according to the procedure reported by the previous studies [5, 6]. The starting materials were NaOH, amorphous SiO_2 fine powder, $TiCl_4$ aqueous solution and HCl. The mixture with the compositional ratios of TiO_2/SiO_2 = 0.32 and Na_2O/TiO_2 = 5.62 was heat-treated at 100 °C for 24 hours in a closed pressure-resistant vessel. The obtained sample was filtered with distilled water and dried at 40 °C for 48 hours to gain a single phase of Na-GTS. The Er^{3+}-exchanged forms were obtained by shaking 0.5 g of Na-GTS in the $ErCl_3$ aqueous solutions (25 mL) at 25 °C for 24 hours. The concentrations of Er^{3+} (C_{Er}) in the aqueous solutions were varied between 0.01 and 0.1 M, where the 25 mL of 0.036 M $ErCl_3$ aqueous solution is necessary for complete exchange of Na^+ in Na-GTS for Er^{3+}. The obtained samples were filtrated, washed and dried at 80 °C.

2.2. Characterization
The Er^{3+} concentration in the supernatant solutions was analyzed with atomic absorption spectrometry (AAS) to evaluate the ion-exchange amount [x in $Na_{4(1-x)}Er_{(4/3)x}Ti_4Si_3O_{16}\cdot nH_2O$]. The phase identification was performed using a Rigaku RINT2200 powder X-ray diffractometer. The dehydration process was examined by TG-DTA. The heating rate of 10 °C/min and the cooling rate of 20 °C/min were applied in the TG-DTA measurements.

3. Results and discussion
3.1. Ion-exchange amounts for Er^{3+}
Figure 2 shows the variation of the Er^{3+} composition (x) in the Er^{3+}-exchanged Na-GTS [$Na_{4(1-x)}Er_{(4/3)x}Ti_4Si_3O_{16}\cdot nH_2O$] as a function of the concentration of $ErCl_3$ aqueous solution (C_{Er}), where the x values were evaluated from AAS for Na^+ in the filtrate after the ion-exchange experiments. The figure shows that the Er^{3+} composition (x) increases monotonously with increasing C_{Er} up to 0.05 M and then is kept constant in $C_{Er} \geq 0.05$ M. Thus, the x value of 0.64 obtained in $C_{Er} \geq 0.05$ is the maximum allowance of the Er^{3+} exchange.

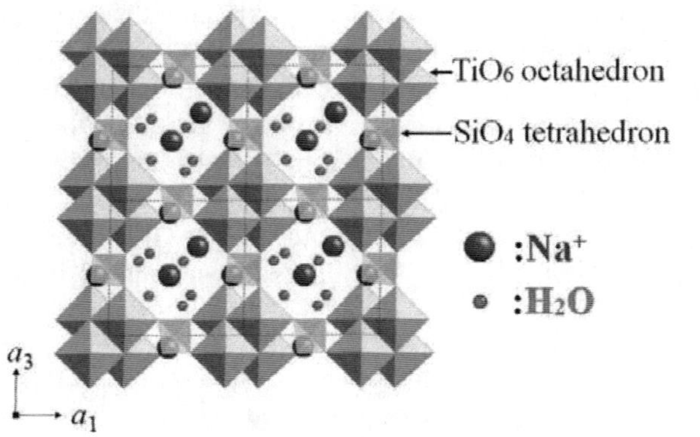

Figure 1. Crystal structure of Na-GTS [$Na_4Ti_4Si_3O_{16}\cdot 6H_2O$] projected on (010) [1].

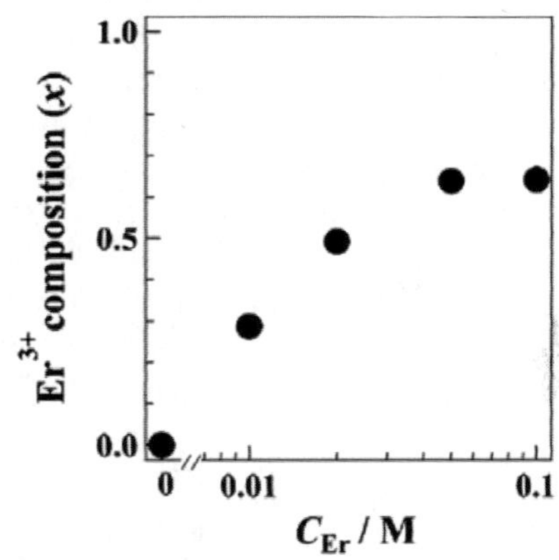

Figure 2. Variation of Er^{3+} composition (x) as a function of C_{Er}. The horizontal axis is represented in a logarithmic scale.

3.2. TG-DTA
Figures 3a and 3b show TG and DTA curves of the Er^{3+}-exchanged samples, respectively. The TG curves show that the weights of the samples decrease steeply up to about 280 °C and then their weight losses are completed by 800 °C. The weight-loss ratios of each sample vary between 18 and 20% with heating up to 800 °C. Because the DTA curves of each sample show the endothermic peaks between 230 and 250 °C, these weight-losses can be due to dehydration from the samples. The water contents (Fig. 4a) and the DTA peak temperatures (Fig. 4b) decrease slightly with increasing C_{Er} [hence, with increasing the Er^{3+} composition (x)].

IOP Conf. Series: Journal of Physics: Conf. Series **950** (2017) 042048 doi:10.1088/1742-6596/950/4/042048

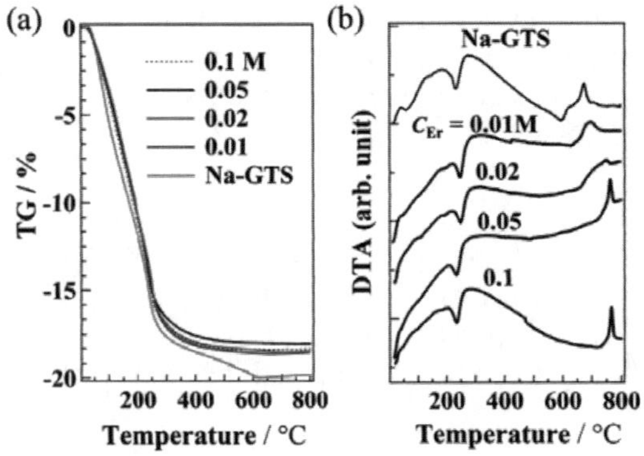

Figure 3. (a) TG and (b) DTA curves of Er^{3+}-exchanged Na-GTS operated at a heating rate of 10 °C/min.

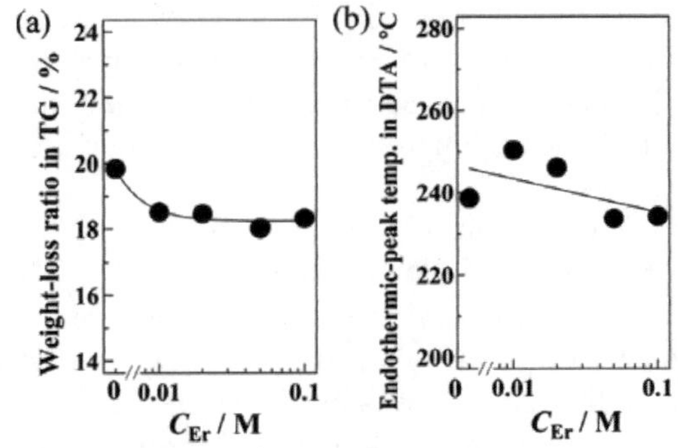

Figure 4. Variations of (a) the weight-loss ratios in TG and (b) the endothermic-peak temperatures in DTA as a function of C_{Er}. The horizontal axes in (a) and (b) are represented in a logarithmic scale.

3.3. XRD patterns

Figure 5a shows the variation of observed XRD patterns as a function of C_{Er}. We here evaluate the lattice parameters and peak intensities by a profile fitting method assuming a rhombohedral unit cell. The C_{Er} dependence of the rhombohedral lattice parameters is shown in Figure 5b. The increase in C_{Er} decreases the lattice parameter *a* and increases the lattice parameter α to decrease the unit-cell volume *V*. The α angles range between 89.08° and 89.31°, and are very close to 90°, corresponding to cubic system, over the investgated C_{Er} range. The diffraction peaks in Figure 5a are therefore indexed as pseudo-cubic system for the sake of convenience. The C_{Er} dependence of the integrated intensities *I*(*hkl*) of pseudo-cubic *hkl* peaks normalized by the integrated intensity *I*(100) of pseudo-cubic 100 peak is shown in Figure 6. The intensity ratios *I*(*hkl*)/*I*(100) vary systematically with increasing C_{Er}. With increasing C_{Er}, the intensity ratios *I*(110)/*I*(100), *I*(111)/*I*(100), *I*(221+300)/*I*(100), *I*(310)/*I*(100) and *I*(400)/*I*(100) increase, but *I*(200)/*I*(100), *I*(211)/*I*(100), *I*(220)/*I*(100) and *I*(311)/*I*(100) decrease, where *I*(221+300) is the intensity of overlapped peak of 221 and 300. These observations can be attributed to the exchange of Na^+ in Na-GTS for Er^{3+}.

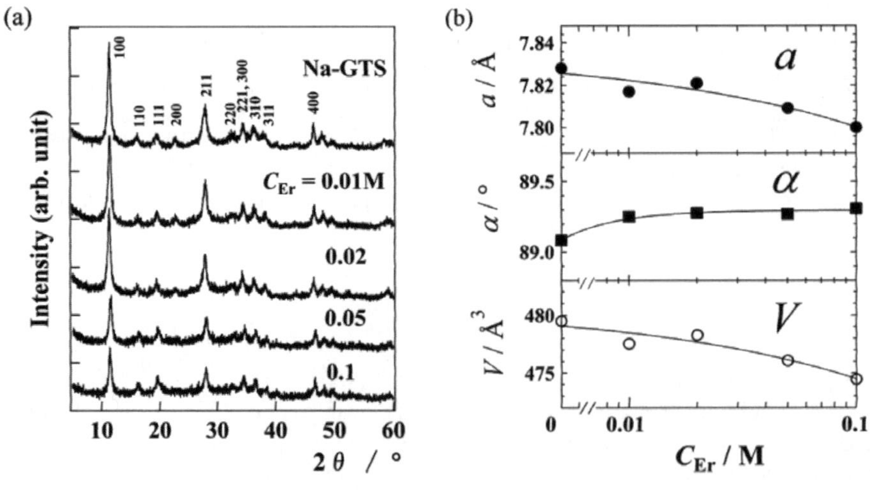

Figure 5. Variations of (a) observed XRD patterns of Er^{3+}-exchanged Na-GTS and (b) rhombohedral lattice parameters as a function of C_{Er}. The diffraction indices in (a) are labeled as those in pseudo-cubic system. The horizontal axis in (b) is represented in a logarithmic scale.

IOP Conf. Series: Journal of Physics: Conf. Series **950** (2017) 042048 doi:10.1088/1742-6596/950/4/042048

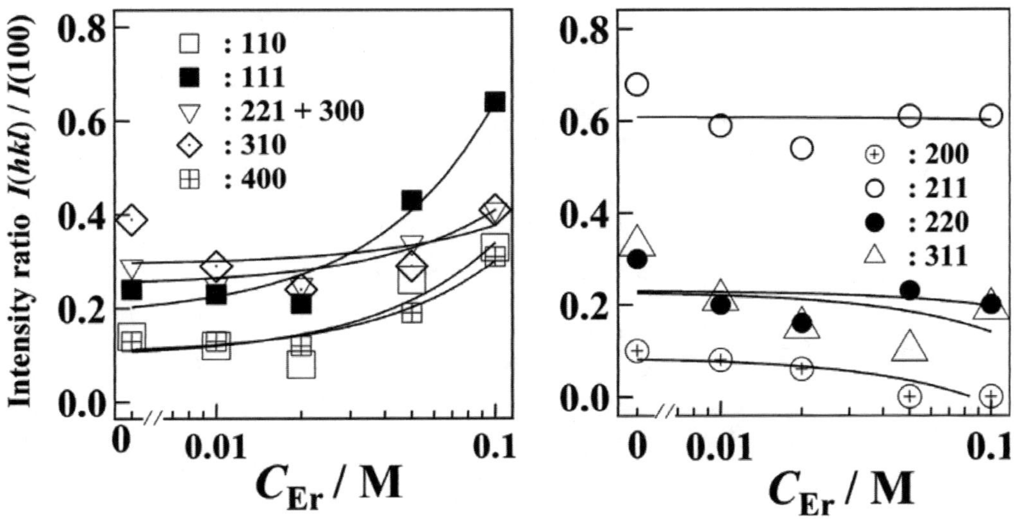

Figure 6. Variation of integrated-intensity ratios $I(hkl)/I(100)$ of observed XRD peaks as a function of C_{Er}. The diffraction indices are labeled as those in pseudo-cubic system. The horizontal axes are represented in a logarithmic scale.

3.4. Possible occupied site of Er^{3+}

The previous structural analysis of a natural sample of Sr^{2+}-exchanged GTS [7] showed that Sr^{2+} ions occupy both 4*e*-site at (0.63, 0.63, 0.63) and 6*g*-site at (0.96, 0.5 0.5) in the GTS cavity. On the assumption that Er^{3+} in Er^{3+}-exchanged Na-GTS is also distributed on the same sites as the case of the Sr^{2+}-exchanged GTS, we here examine several cation-distribution models of Er^{3+} in the present Er^{3+}-exchanged Na-GTS as shown in Table 1. Figure 7 shows the observed XRD pattern of the Er^{3+}-exchanged Na-GTS with $x = 0.64$, together with the calculated ones from each Er^{3+}-distribution model. The peak intensities of the observed XRD pattern are in good agreement with the calculated one from the model (A). This suggests that Er^{3+} and Na^{+} ions preferentially occupy the 4*e* and 6*g* sites, respectively.

Table 1. Cation-distribution models of Er^{3+}-exchanged Na-GTS based on the structural analysis of Sr^{2+}-exchanged GTS [7].

Models	Site occupancy factors			
	Er^{3+}		Na^{+}	
	4*e*-site	6*g*-site	4*e*-site	6*g*-sie
A	0.213	0	0	0.240
B	0.213	0	0.360	0
C	0	0.142	0	0.240
D	0	0.142	0.360	0
E	0.107	0.071	0	0.240
F	0.107	0.071	0.360	0

IOP Conf. Series: Journal of Physics: Conf. Series **950** (2017) 042048 doi:10.1088/1742-6596/950/4/042048

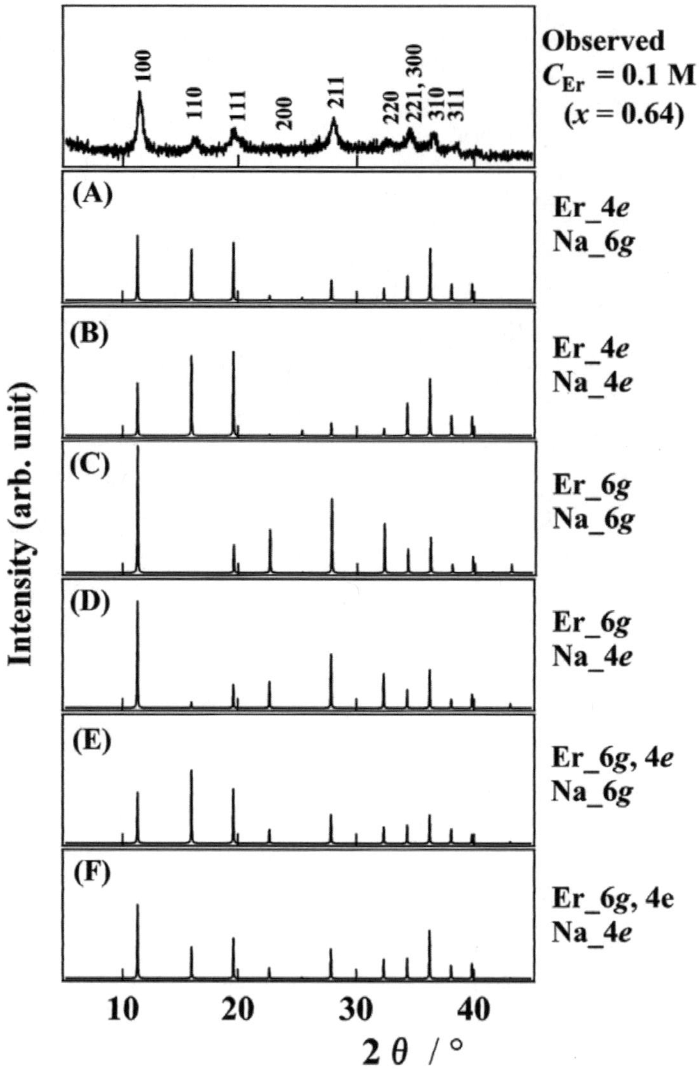

Figure 7. Comparison of the observed XRD pattern of the present Er^{3+}-exchanged Na-GTS with the calculated ones from the cation-distribution models (A)–(F) in Table 1. The diffraction indices are labeled as those in pseudo-cubic system.

4. Conclusion

The Er^{3+}-exchanged Na-GTS samples [$Na_{4(1-x)}Er_{(4/3)x}Ti_4Si_3O_{16}\cdot nH_2O$] with the compositions in the range of $0.28 \leq x \leq 0.64$ have successfully been prepared and characterized by TG-DTA and XRD techniques. The exchange of Na^+ for Er^{3+} up to $x = 0.64$ brings about a slight decrease in the water contents and the dehydration temperatures. The increase in the Er^{3+} composition (x) decreases the lattice parameter a and increases the lattice parameter α to decrease the unit-cell volume V. Moreover, with increasing x, the intensity ratios $I(hkl)/I(100)$ of pseudo-cubic 110, 111, 221+300 (overlapped peak of 221 and 300), 310 and 400 peaks increase, but those of pseudo-cubic 200, 211, 220 and 311 peaks decrease. The comparison of observed and calculated XRD patterns leads to a conclusion that the preferential occupied sites of Er^{3+} and Na^+ ions are the 4*e* and 6*g* sites, respectively.

Acknowledgments
This work was partly supported by a Grant-in-Aid for Scientific Research (C) (No. 24561005) from the Japan Society for the Promotion of Science.

References

[1] Dadachov M S and Harrison W T A 1997 *J. Solid. State Chem.* **134** 409–415
[2] Behrens E A, Poojary D M and Clearfield A 1996 *Chem. Mater.* **8** 1236–1244
[3] Behrens E A and Clearfield A 1997 *Micropor. Mater.* **11** 65–75
[4] Fujiwara K, Tamaki T, Kishimori C, Titorenkova R, Nakatsuka A and Nakayama N 2013 *Trans. Mat. Res. Soc. Japan* **38** 455–458
[5] Kostov-Kytin V, Ferdov S, Kalvachev Yu, Mihailova B and Petrov O 2007 *Micropor. Mesopor. Mater.* **105** 232–238
[6] Fujiwara K, Shioda T, Sugimoto H, Nakatsuka A, Nakayama N, Nikolova R P and Kostov-Kytin V 2010 *Topics Chem. Mater. Sci.* **4** 184–190
[7] Spiridonova D V, Krivovichev S V, Yakovenchuk V N and Pakhomovsky Ya A 2011 *Geol. Ore Dep.* **53** 670–677

IOP Conf. Series: Journal of Physics: Conf. Series **950** (2017) 042049 doi:10.1088/1742-6596/950/4/042049

Exploring silicon allotropy and chemistry by high pressure – high temperature conditions.

O O Kurakevych[1], Y Le Godec[1], T A Strobel[2], D Y Kim[2], W A Crichton[3], and J Guignard[3,4]
[1]IMPMC, UPMC Sorbonne Universités, UMR CNRS 7590, Muséum National d'Histoire Naturelle, IRD UMR 206, 75005 Paris, France
[2]Geophysical Laboratory, Carnegie Institution of Washington, Washington, DC 20015, United States.
[3]European Synchrotron Radiation Facility, 71, Avenue des Martyrs, Grenoble 38043, France.
[4]OMP, Laboratoire Geosciences Environnement Toulouse, Toulouse 31400, France.

E-mail: oleksandr.kurakevych@impmc.jussieu.fr

Abstract. Silicon is the second abundant element, after oxygen, in the earth crust. It is essential for today's electronics because of its ability to show various electronic behaviors that allow covering the numerous fields of cutting-edge applications. Moreover, silicon is not a pollutant and, therefore, is an ideal candidate to replace the actual materials in photovoltaics, like compounds based on the arsenic and heavy metals. It has not replaced them so far because Si is an indirect gap semiconductor and cannot absorb directly the solar photons without thermal agitations of crystal lattice (phonons). This puts it apart from the next-generation applications (light diode, high-performance transistor). That justifies the attempts to create silicon materials with direct gap that can absorb and emit light. Our recent high-pressure studies of the chemical interaction and phase transformations in the Na-Si system, revealed a number of interesting routes to new and known silicon compounds and allotropes. The pressure-temperature range of their formation is suitable for large-volume synthesis and future industrial scaling. The variety of properties observed (e.g. quasi-direct bandgap of open-framework allotrope Si_{24}) allows us to suggest future industrial applications.

1. Introduction

During last years high pressure techniques have been widely used for creation of new solids with advanced mechanical and physical properties [1]. High pressure completely change the thermodynamics and chemistry of elements, known at ambient pressure [2], and allowed obtaining novel (i) *crystal structures,* (ii) *compositions* and even (iii) *nanostructures*. Detailed characterization of these materials have been performed by physical measurements and thermodynamic analysis [3]. For some solids, the whole of properties allows considering them as advanced superhard, semiconductive and even photovoltaic materials. The pressures required for synthesis of many of them belong to the domain easily accessible for industrial scaling (up to 2500 K and 10 GPa).

High pressure often induces new allotropy in the elements [4]. However, most of the crystal structures that form under extreme pressure-temperature conditions cannot be recovered at ambient. Those that remain metastable, can be potentially considered as new materials, similar to synthetic diamond. Recent discoveries of two novel boron allotropes, orthorhombic γ-B_{28} [5,6] and tetragonal (pseudo-cubic) pc-B_{52} [7], allowed us to constuct the boron phase diagram, which has been questioned for more than one century [8], but also considering boron as superhard, second only after carbon in its diamond form(s) [9].

High pressure also completely changes element chemistry. For example, magnesium and carbon, both of which form numerous compounds with other elements, have remarkably low affinity to each other. At ambient pressure, only the reaction of Mg or MgO with hydrocarbons leads to the formation of metastable carbides with reasonable yields, and pure Mg and carbon do not react to form stable compounds at any temperature. It has been demonstrated that under high-pressure conditions (above 5 GPa and 1300 K) thermodynamically stable Mg-C compounds are possible. The new compound synthesized, β-Mg_2C_3 [10], crystallizes in the monoclinic *C2/m* space group, and contains rare allylene-derived C_3^{4-} anions that are isoelectronic with CO_2. At higher pressures, above 15 GPa, antifluorite Mg_2C forms [11]. Thus, after exploring pressure as additional dimension for chemistry of the Mg–C system, four magnesium carbides are now known: (1) tetragonal MgC_2 [12], (2) orthorhombic α-Mg_2C_3 [12], (3) monoclinic β-Mg_2C_3 [10], and (4) cubic Mg_2C [11]. Taking into account that at pressures below 5 GPa the elements do not interact at any temperature, it is quite astonishing to observe such rich chemistry. These results indicate that the Mg-C phase diagram should be completely revised under high pressure [13]. In fact, this system is important for the production of structurally-perfect diamonds with high growth rates as compared to other known metal-carbon systems, and also for the synthesis of unique nitrogen-free and boron-doped semiconductive diamonds.

The application of high pressure for materials synthesis, both in industry and science, have been traditionally limited to superhard materials [14,15], e.g. diamond [16,17], cubic boron nitride cBN [18,19], carbon-metal systems [20], including Mg-C described above [13], nano-cBN [21], nano-diamond [22], cBC_5 [23,24], γ-B_{28} [5,6,25], $B_{13}N_2$ [26-31], B_6O [32-35], high-pressure B_2O_3 [36-38], etc. Some interest has been paid to design [39,40] and understanding [41-43] of high-pressure superconductors. Our recent results allowed us to overcome this "restriction" and to give a new dimension for design of high-pressure materials. Silicon is essential for today's electronics because of its ability to show various electronic behaviors that allow covering the numerous fields of cutting-edge applications, in particular in optoelectronics. For this reason many attempts have been made to design a silicon with direct gap that could efficiently absorb and emit light. Ideal material should have optically allowed direct bandgap [44], close direct & indirect gaps of absolute values from 1 to 2 eV in order to have highest efficiency of cells, and high light absorption in thin films [45,46]. Three principal methodologies may be employed for this, i.e. (1) nanostructuring, (2) low-dimensional materials (silicene, thin films) and (3) new crystal structures, both dense and open-framework [47-50]. The last approach has been explored mainly by *ab initio* metadynamical studies. Although from the general point of view the open-framework structures can hardly be stable at positive pressures as compared to Si-I, they can be stabilized by intercalation at high pressures, similar to silicon clathrates [51,52].

In this paper the recent achievements in synthesis of silicon allotropes will be described. They imply new chemical routes that combine both soft chemistry and high-pressure techniques.

2. Synthesis of dense tetrahedral silicon allotropes

Seven dense (i.e. without large voids) crystal structures of Si are known to be stable at pressures from 1 MPa up to 100 GPa at room temperatures [53-58]; and three more phases are believed to be metastable below 10 GPa [59,60]. At ambient conditions one can recover only Si-I (common and most stable diamond structure) [53], and two metastable Si-III and Si-IV [59]. The crystal forms that exist below 16 GPa have tetrahedral coordination (from ideal in diamond Si-I and lonsdaleite Si-IV) [53,59,61] to strongly distorted (white-tin Si-II and orthorhombic Si-XI [54,58]). Above 16 GPa the stable crystal structures are produced mainly by traditional hexagonal or cubic packing (hp Si-V, hcp Si-VII and ccp Si-X [55,56]).

Figure 1 shows the relative transformations at pressures below 12 GPa. Only three phases can be obtained/recovered at ambient conditions. In the case of diamond silicon (Si-I) the bandgap engineering by nanostructuring [62] is possible. However, the most interesting is exotic Si-III (BC8) phase. In a pure polycrystalline state (with strained lattice) it can be obtained by decompression from above 15 GPa without heating [63] or from pressures above 10 GPa with heating [64]. It is a semimetallic allotrope, whose gap was predicted to open by **(1)** the change of structural order parameter, the x of Wickoff position 16c with (x,x,x) unit cell coordinates in *Ia*-3 space group (No 206) [65], or **(2)** the grainsize of Si-III clusters [66]. Moreover, very recently the nanoparticles of BC8 cubic silicon obtained by a mean of colloidal chemistry were claimed in ref. [67]. The bandgap measured by absorption in visible region was found to be of ~2.8 eV [67], slightly higher than predicted by *ab initio* calculations for the same grainsize [66].

Our recent results on Na-Si interactions at HPHT conditions using large-volume multianvil press and synchrotron radiation at the ID06 beamline of the ESRF, have indicated the possibility to obtain pure polycrystalline strain-free Si-III grains up to 500 μm at pressures as low as 9.5 GPa in the Na-Si system. At lower pressure this phase was also often observed (at least down to 7 GPa in Na-Si system), but pure grains cannot be easily extracted. In a number of samples obtained in our experiments by variation of chemical conditions, Rietveld refinement give x from 0.101 to 0.105. The lowest x corresponds to the sample obtained at room temperature (~13 GPa in our experiments using Paris-Edinburgh cell at IMPMC, 40% yield of Si-III). At the same time the nanoparticles [67], the PXRD pattern may be adjusted to original cubic *Ia*-3 space group using structureless full-profile LeBail analysis, however the pattern shows quite different intensities from theoretical ones. Our attempts to adjust the intensities to experimental one using Rietveld analysis, led to the value of $\langle x \rangle$ as low as of ~0.088. However, not all the intensities are in reasonable agreements, moreover, very short Si-Si bonds appear (17% shorter as compared to expected ~2.4 Å). To avoid these discrepancies, we tried to refine the structure using cubic subgroups of *Ia*-3 space group. The obtained mean values of x's are between 0.070 and 0.088, representing the analogue of x in the original structure, now for 2 or 4 sublattices of Si with more reasonable interatomic distances. More detailed study of this crystallographic point is required. Anyway, one can easily see the possibility of vary x by the methods of synthesis. Here, the grainsize may be considered a crucial point for impact on order parameter x (or $\langle x \rangle$) and even stability of crystal structure, different from microcrystalline bulk counterpart, like it often occurs, e.g. for nano polymorphism of TiO_2 [68] or stability of nanostructured high-pressure form of ZnO [69,70].

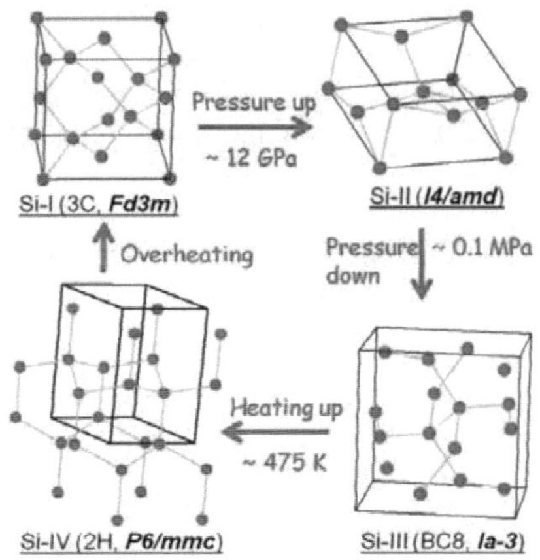

Figure 1. Direct phase transformations between dense Si forms.

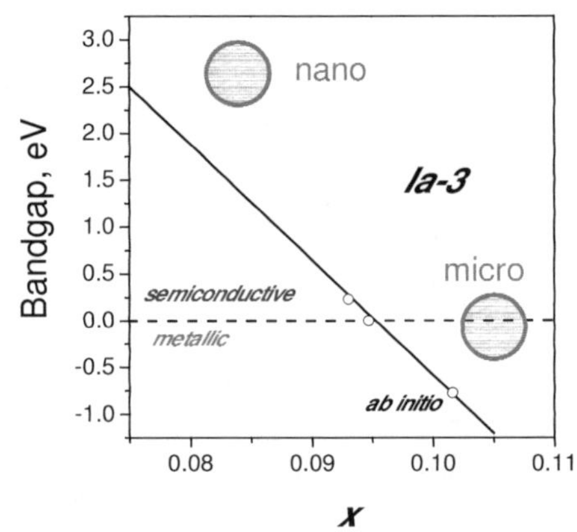

Figure 2. Bandgap of Si-III (BC8) as a function of order parameter, atomic coordinate (x,x,x).

IOP Conf. Series: Journal of Physics: Conf. Series **950** (2017) 042049 doi:10.1088/1742-6596/950/4/042049

Figure 2 shows the relative position in the order parameter x - bandgap E_g coordinates of Si-III nanoparticles [67] vs polycrystalline materials synthesized at high-pressure by direct transformation & in chemical systems (Na-Si in our case). Solid line connect the *ab initio* points [65] (open circles); although the "extrapolated values" may be quite different from true *ab initio* values, the general tendency of gap opening with decreasing x, is expected with quite high probability. The experimentally known materials are represented by large red circles (denoted micro and nano, the size of the red circles approximatively represent the spread/error of experimental values). One can see that, although displaced, the experimental and *ab initio* dependencies of bandgap on x (or $\langle x \rangle$) have very similar slopes (here we can note that usually DFT calculations often lead to underestimated band gap values, which give even more relevance to the tendency shown on Figure 2). So, now semimetallic polycrystalline and semiconductive wide-bandgap semiconductive nanoparticles are known for Si-III. The use of high-pressure chemical reactions seem to be quite promising to fulfill the "gap" of materials with intermediate (direct) bandgaps (especially close to 1-2 eV) by playing the grainsize through the kinetics dependent on p-T-time regime and chemical environments.

3. Synthesis of open-framework silicon allotropes

As it has been already mentioned above, high-pressure is a newly proposed route to the new and known compounds in the Na-Si system [51,71] (Zintl phase Na_4Si_4 and intermetallic clathrates Na_8Si_{46}, $Na_{24+x}Si_{136}$ and Na_4Si_{24}); while the pressure-temperature range of their formation (3-8 GPa) is suitable for large-volume synthesis and future industrial scaling. Moreover, sodium removal at T > 700 K from the common $Na_{24-x}Si_{136}$ [72] or its high-pressure form $Na_{24+x}Si_{136}$ leads to the "clathrate" allotrope Si_{136} with quasidirect bandgap [73] of ~2 eV [74] (Figure 3). Unfortunately, the direct electron transitions at Γ point are not allowed by the cubic symmetry and this silicon form is not a good candidate for photovoltaic applications.

In a similar way, sodium removal from Na_4Si_{24} by heating in vacuum leads to a new allotrope of silicon with zeolite structure, orthorhombic o-Si_{24} [75] (Figure 3). In this case the required temperature is remarkably lower, ~400 K. The zeolite silicon framework deform during this process (+6% along a-axis and -3% along b- and c-axes), but in general silicon structure remain the same (zeolite CAS). The quasi-direct bandgap of ~1.3 eV with allowed direct electronic transitions due to orthorhombic symmetry (*Cmcm* space group) allow considering this material for future photovoltaic applications.

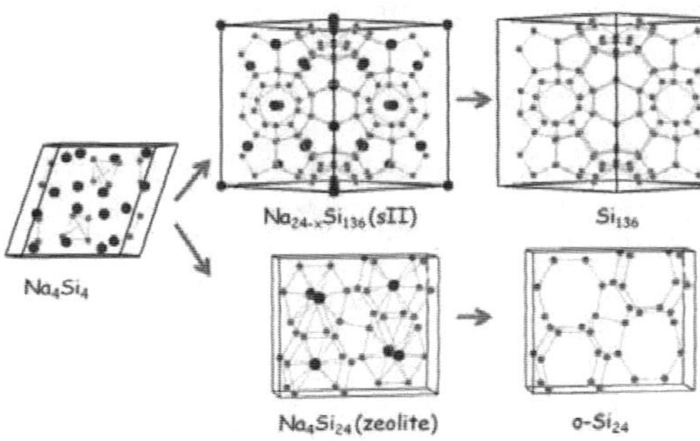

Figure 3. Chemical routes to clathrate silicon allotropes. Top: Synthesis below 0.1 MPa (in Ar or vacuum) [71], bottom: combined synthesis at ~8 GPa (Na_4Si_4+Si) and in vacuum [51,75].

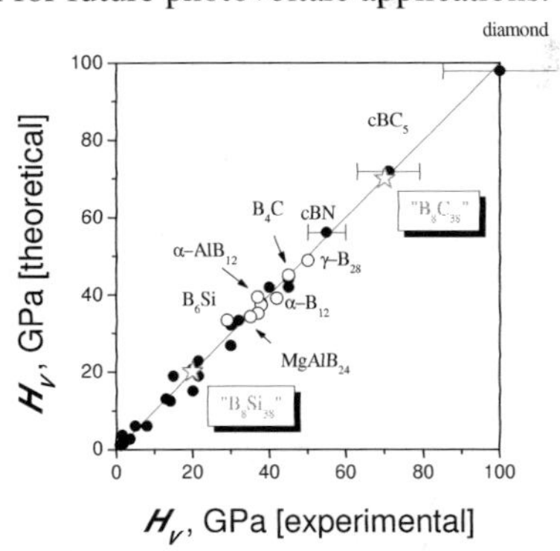

Figure 4. Estimation of clathrate hardness (B-Si & B-C frameworks) using "thermodynamic model of hardness" [76-78].

4. Hardening of silicon clathrates

Silicon clathrates seem to have the most rigid covalent framework as compared to other clathrates (with frameworks produced by other elements of group 14, silicon oxide or water), since carbon clathrates are not known at present time [79]. So they are the hardest known clathrates with highest elastic moduli, including bulk modulus. In particular, the hardness, just like the bulk modulus and other elastic constants, is expected to be close to that of conventional diamond silicon Si-I [73], even in the case of empty-case Si_{136}. The rigidity of silicon framework is defined by Si-Si bondings, and both intercalated metals and voids does not influence it much.

Only one compound known experimentally can be a candidate to be harder than silicon clathrates, i.e. $K_7B_7Si_{39}$ [80]. Here the B atoms replace Si the clathrate framework and one may expect that hardness should become higher than in the case of Si frameworks, expected value of H_V ~10 GPa [76], although still lower than in the case of boron (H_V > 40 GPa) [9,78]. The hardness estimations (in this chapter we talk about the limit Vickers hardness value in the high-load limit [76]) by modelling is not a trivial, but is still possible task. To do that, one may suggest - similarly to the case of known clathrates - the mechanical properties are close to the original diamond phase of the framework element(s). This was confirmed for Si clathrates (e.g. in the experiments on compressibility measurements [81,82]) and predicted for hypothetical carbon clathrates (both carbon [73,82] and carbon-boron frameworks [83]), expected to have the highest hardness among the clathrate family.

Our previous estimations of a number of B-Si phases indicated that most compounds are at the lower boundary of superhardness (conventionally, 40 GPa of Vickers hardness *H*v for cubic boron nitride, second hardest material after diamond on industrial scale). The good correlation of previous estimations using "thermodynamic model of hardness" [77] that typically gives 10% accuracy for a large number of covalent and partially ionic compounds. Our estimates give a value of *H*v ~20 GPa for $K_7B_7Si_{39}$ [80] (strictly speaking, for "B_8Si_{38}" clathrate framework, Figure 4), which should exceed that of silicon nitride Si_3N_4 and lays between diamond silicon Si-I and industrial abrasive silicon carbide SiC. In the case of recently predicted Li-B-C clathrates [83], the hardness should be close to that of diamond-like cBC_x phases [24] (Figure 4). The detailed study of the interstitial-atom impact on hardness of clathrates (both of intercalated and Zintl nature) outstand the scope of present work.

5. Conclusions

Finally, chemical interactions in the Na-Si system are very promising for elaboration of known (Si-III or BC8) or new (zeolite $o\text{-}Si_{24}$) allotropes of silicon in the forms of well crystallized polycrystalline grains of high purity. High pressure chemical routes seem also quite promising, by varying *p-T*-time-concentration protocols for nanostructuring & bandgap engineering, however, unexplored so far. The improvement of mechanical properties of Si framework (and of clathrate in total) is also possible by silicon replacement for boron.

Acknowledgements

The high-pressure syntheses using *in situ* X-ray diffraction were performed at ID06 beamline at the European Synchrotron Radiation Facility (Grenoble, France). We thank B. Baptiste for help in high-temperature studies of clathrate and allotrope stability in vacuum.

References

[1] McMillan P F 2004 *High Press. Res.* **24** 67-86.
[2] McMillan P F 2003 *Chem. Comm.* 919-23.
[3] McMillan P F 2002 *Nat. Mater.* **1** 19-25.
[4] Kurakevych O O, Le Godec Y, Hammouda T and Goujon C 2012 *High Pressure Res.* **32** 30-8.
[5] Oganov A R, Chen J, Gatti C, Ma Y, Ma Y, Glass C W, Liu Z, Yu T, Kurakevych O O and Solozhenko V L 2009 *Nature* **457** 863-7.
[6] Oganov A R, Chen J, Gatti C, Ma Y, Ma Y, Glass C W, Liu Z, Yu T, Kurakevych O O and Solozhenko V L 2009 *Nature* **460** 292.
[7] Kurakevych O O and Solozhenko V L 2013 *J. Superhard Mater.* **35** 60-3.
[8] Solozhenko V L and Kurakevych O O 2013 *Sci. Rep.* **3** art. 2351, doi:10.1038/srep02351.
[9] Solozhenko V L, Kurakevych O O and Oganov A R 2008 *J. Superhard Mater.* **30** 428-9.

[10] Strobel T A, Kurakevych O O, Kim D Y, Le Godec Y, Crichton W, Guignard J, Guignot N, Cody G D and Oganov A R 2014 *Inorg. Chem.* **53** 7020-7.
[11] Kurakevych O O, Strobel T A, Kim D Y and Cody G D 2013 *Angew. Chem. Int. Ed.* **52** 8930-3.
[12] Ruschewitz U 2003 *Coord. Chem. Rev.* **244** 115-36.
[13] Kurakevych O O, Le Godec Y, Strobel T A, Kim D Y, Crichton W A and Guignard J 2014 *J. Phys. Chem. C* **118** 8128-33.
[14] Kurakevych O O 2009 *J. Superhard Mater.* **31** 139-57.
[15] Kurakevych O O and Solozhenko V L 2011 *High Press. Res.* **31** 48-52.
[16] Bundy F P, Hall H T, Strong H M and Wentorf Jr. R H 1955 *Nature* **176** 51-5.
[17] Liander H 1955 *ASEA Journal* **28** 97-8.
[18] Wentorf Jr. H R 1957 *Chem. Phys.* **26** 956-60.
[19] Solozhenko V L, Turkevich V Z and Holzapfel W B 1999 *J. Phys. Chem. B* **103** 2903-5.
[20] Novikov N V 1999 *Diamond Relat. Mater.* **8** 1427–32.
[21] Solozhenko V L, Kurakevych O O and Le Godec Y 2012 *Adv. Mater.* **24** 1540-4.
[22] Irifune T, Kurio A, Sakamoto S, Inoue T and Sumiya H 2003 *Nature* **421** 599-600.
[23] Solozhenko V L, Kurakevych O O, Andrault D, Le Godec Y and Mezouar M 2009 *Phys. Rev. Lett.* **102** 179901.
[24] Solozhenko V L, Kurakevych O O, Andrault D, Le Godec Y and Mezouar M 2009 *Phys. Rev. Lett.* **102** 015506.
[25] Le Godec Y, Kurakevych O O, Munsch P, Garbarino G and Solozhenko V L 2009 *Solid State Comm.* **149** 1356-8
[26] Kurakevych O O and Solozhenko V L 2007 *Acta Crystallogr. Sect. C* **63** i80-2.
[27] Solozhenko V L and Kurakevych O O 2008 *J. Phys. Conf. Ser.* **121** 062001.
[28] Kurakevych O O and Solozhenko V L 2009 *Solid State Comm.* **149** 2169-71.
[29] Solozhenko V L and Kurakevych O O 2009 *J. Solid State Chem.* **182** 1359-64.
[30] Solozhenko V L, Kurakevych O O, Turkevich V Z and Turkevich D V 2009 *J. Superhard Mater.* **31** 1-6.
[31] Solozhenko V L, Kurakevych O O, Turkevich V Z and Turkevich D V 2010 *J. Phys.Chem. B* **114** 5819-22.
[32] Hubert H, Garvie L A J, Devouard B, Buseck P R, Petuskey W T and McMillan P F 1998 *Chem. Mater.* **10** 1530-7.
[33] Solozhenko V L, Kurakevych O O and Bouvier P 2009 *J. Raman Spectr.* **40** 1078-81.
[34] Kurakevych O O and Solozhenko V L 2011 *J. Superhard Mater.* **33** 421-8.
[35] Kurakevych O O and Solozhenko V L 2014 *J. Superhard Mater.* **36** 270-8.
[36] Prewitt C T and Shannon R D 1968 *Acta Crystallogr. B* **24** 869-74.
[37] Mukhanov V A, Kurakevich O O and Solozhenko V L 2008 *J. Superhard Mater.* **32** 71-2.
[38] Solozhenko V L, Kurakevych O O, Le Godec Y and Brazhkin V V 2015 *J. Phys. Chem. C* **119** 20600-5.
[39] Jin C Q, Puzniak R, Zhao Z X, Wu X J, Tatsuski T, Tamura T, Adachi S, Tanabe K, Yamauchi H and Tanaka S 2000 *Phys. Rev. B* **61** 778-83.
[40] Yamanaka S, Enishi E, Fukuoka H and Yasukawa M 2000 *Inorg. Chem.* **39** 56-8.
[41] Mignot J M, Chouteau G and Martinez G 1985 *Physica B+C* **135** 235-8.
[42] Mignot J M, Chouteau G and Martinez G 1986 *Phys. Rev. B* **34** 3150-5.
[43] Connétable D, Timoshevskii V, Masenelli B, Beille J, Marcus J, Barbara B, Saitta A M, Rignanese G M, Mélinon P, Yamanaka S and Blase X 2003 *Phys. Rev. Lett.* **91** 247001.
[44] Zhang P, Crespi V H, Chang E, Louie S G and Cohen M L 2001 *Nature* **409** 69-71.
[45] Archer M D and Bolton J R 1990 *J. Phys. Chem. A* **94** 8028-36.
[46] Botti S, Flores-Livas J A, Amsler M, Goedecker S and Marques M A L 2012 *Phys. Rev. B* **86** 121204.
[47] Xiang H J, Huang B, Kan E, Wei S-H and Gong X G 2013 *Phys. Rev. Lett.* **110** 118702.
[48] Lee I-H, Lee J, Oh Y J, Kim S and Chang K J 2014 *Phys. Rev. B* **90** 115209.

[49] Wang Q, Xu B, Sun J, Liu H, Zhao Z, Yu D, Fan C and He J 2014 *J. Amer. Chem. Soc.* **136** 9826-9.
[50] Mujica A, Pickard C J and Needs R J 2015 *Phys. Rev. B* **91** 214104.
[51] Kurakevych O O, Strobel T A, Kim D Y, Muramatsu T and Struzhkin V V 2013 *Cryst. Grow. Des.* **13** 303-7.
[52] Yamanaka S, Komatsu M, Tanaka M, Sawa H and Inumaru K 2014 *J. Amer. Chem. Soc.* **136** 7717-25.
[53] Debye P and Scherrer P 1916 *Phys. Zeits.* **17** 277-83.
[54] Jamieson J C 1963 *Science* **139** 762-4.
[55] Olijnyk H, Sikka S K and Holzapfel W B 1984 *Phys. Lett. A* **103** 137-40.
[56] Duclos S J, Vohra Y K and Ruoff A L 1987 *Phys. Rev. Lett.* **58** 775-7.
[57] Hanfland M, Beister H and Syassen K 1989 *Phys. Rev. B* **39** 12598-603.
[58] McMahon M I and Nelmes R J 1993 *Phys. Rev. B* **47** 8337-40
[59] Wentorf R H and Kasper J S 1963 *Science* **139** 338-9.
[60] Karki B B, Ackland G J and Crain J 1997 *J. Phys.: Condens. Matter* **9** 8579-89.
[61] Jennings H M and Richman M H 1976 *Science* **193** 1242-3.
[62] Delerue C, Allan G and Lannoo M 1998 *J. Luminescence* **80** 65-73.
[63] Hu J Z, Merkle L D, Menoni C S and Spain I L 1986 *Phys. Rev. B* **34** 4679-84.
[64] Demishev S V, Lunts D G, Nekhaev D V, Sluchanko N E, Samarin N A, Brazhkin V V, Lyapin A G, Popova S V and Mel'nik N N 1996 *JETP* **82** 1159-67.
[65] Malone B D, Sau J D and Cohen M L 2008 *Phys. Rev. B* **78** 035210.
[66] Wippermann S, Vörös M, Rocca D, Gali A, Zimanyi G and Galli G 2013 *Phys. Rev. Lett.* **110** 046804.
[67] Ganguly S, Kazem N, Carter D and Kauzlarich S M 2014 *J. Amer. Chem. Soc.* **136** 1296-9.
[68] Gałyńska M and Persson P 2013 *Int. J. Quant. Chem.* **113** 2611-20.
[69] Baranov A N, Kurakevych O O, Tafeenko V A, Sokolov P S, Panin G N and Solozhenko V L 2010 *J. Appl. Phys.* **107** 073519.
[70] Solozhenko V L, Kurakevych O O, Sokolov P S and Baranov A N 2011 *J. Phys. Chem. A* **115** 4354-8.
[71] Jouini Z, Kurakevych O O, Moutaabbid H, Le Godec Y, Mezouar M and Guignot N 2016 *J. Superhard Mater.* **38** 66-70.
[72] Kasper J S, Hagenmul.P, Pouchard M and Cros C 1965 *Science* **150** 1713-4.
[73] Connétable D 2010 *Phys. Rev. B* **82** 075209.
[74] Gryko J, McMillan P F, Marzke R F, Ramachandran G K, Patton D, Deb S K and Sankey O F 2000 *Phys. Rev. B* **62** R7707-10.
[75] Kim D Y, Stefanoski S, Kurakevych O O and Strobel T A 2015 *Nat. Mater.* **14** 169–73.
[76] Mukhanov V A, Kurakevych O O and Solozhenko V L 2008 *J. Superhard Mater.* **30** 368-78.
[77] Mukhanov V A, Kurakevych O O and Solozhenko V L 2008 *High Press. Res.* **28** 531-7.
[78] Mukhanov V A, Kurakevych O O and Solozhenko V L 2010 *J. Superhard Mater.* **32** 167-76.
[79] Yamanaka S 2010 *Dalton Trans.* **39** 1901-15.
[80] Jung W, Lörincz J, Ramlau R, Borrmann H, Prots Y, Haarmann F, Schnelle W, Burkhardt U, Baitinger M and Grin Y 2007 *Angew. Chem. Int. Ed.* **46** 6725-8.
[81] San-Miguel A, Keghelian P, Blase X, Melinon P, Perez A, Itie J P, Polian A, Reny E, Cros C and Pouchard M 1999 *Phys. Rev. Lett.* **83** 5290-3.
[82] Rey N, Muñoz A, Rodríguez-Hernández P and San Miguel A 2008 *J. Phys. Cond. Matter.* **20** 215218.
[83] Zeng T, Hoffmann R, Nesper R, Ashcroft N W, Strobel T A and Proserpio D M 2015 *J. Amer. Chem. Soc.* **137** 12639-52.

Bonding, elastic and vibrational properties in low and high pressure synthesized diamond-like BCx phases

P Zinin[1,2], X R Liu[1], R Jia[1], S K Sharma[1], L C Ming[1], I Kutuza[2], I Troyan[3]

[1]Hawaii Institute of Geophysics and Planetology, University of Hawaii, Honolulu, HI 96822, U.S.A.
[2]Scientific-Technological Center of Unique Instrumentation, Russian Academy of Sciences, Moscow, Russia
[3]Shubnikov Institute of Crystallography, Russian Academy of Sciences, Moscow Russia

E-mail : zosimpvz@mail.ru

Abstract. Recent studies demonstrate that low pressure chemical vapor deposition at 950 K leads to the synthesis of diamond-like boron carbides with high concentrations of boron (0.66 < x < 4) in which the sp^2 fraction depends on the boron concentration [1]. This indicates that the graphitic BC_3 (g-BC_3) phases obtained by chemical vapor deposition materials are mixtures of diamond-like and graphitic BCx phases. This finding allows us to revise the interpretation of the x-ray diffraction (XRD) patterns of the g-BC_3 phases discussed previously [2, 3]. To support the new interpretation, we conducted a laser heating experiment of the g-BC_3 phase. We found that after laser heating at 1100 K and 25 GPa in a diamond anvil cell (DAC) almost all graphitic layers of the g-BC_3 transform into a cubic structure. The XRD pattern of the cubic BC_3 phase (c-BC_3) can be indexed with a cubic unit cell a = 3.619 (0.165) Å. Measurements of the equation of state of the g-BC_3 phase demonstrated that boron atoms were incorporated into the graphitic B-C network. The linear compressibility along the c axis can be characterized by the value of the linear modulus Bc = 29.2 ± 1.8 GPa. Linear fitting of the experimental data for the a/a_o parameter as a function of pressure gives us the value of the linear elastic modulus along the a axes: Ba = 800 ± 75 GPa.

1. Introduction

Dense BC_x phases with high boron concentration are predicted to be metastable, superhard, and conductors or superconductors depending on boron concentration. However, up to this point, diamond-like boron rich carbide BC_x (dl-BC_x) phases have been obtainable only through high pressure and high temperature treatment, necessitating small specimen volume [3-6]. Recently, it was shown that low pressure synthesis (chemical vapor deposition) of BC_x phases may also lead to the creation of dl-BC_x [1]. The elastic properties of the dl-BC_x phases depend on the carbon sp^2 versus sp^3 content, which decreases with increasing boron concentration, while the boron bonds determine the shape of the Raman spectra of the dl-BC_x after high pressure-high temperature treatment. Does this mean that all the graphitic BC_x phases obtained by chemical vapor deposition should contain carbon atoms in the state of sp^2 hybridization as well as in sp^3? Are boron atoms incorporated into graphitic layers in the dl-BC_x? To answer to these questions we measured the equation of state of the bulk graphitic BC_3 (g-BC_3) in a diamond anvil cell (DAC) up to 12 GPa using energy-dispersive X-ray powder diffraction with synchrotron radiation and a water-alcohol mixture as a pressure medium. We also conducted

IOP Conf. Series: Journal of Physics: Conf. Series **950** (2017) 042050 doi:10.1088/1742-6596/950/4/042050

laser heating of the *g*-BC_3 phases in a DAC to demonstrate that the graphitic structure transforms into the cubic one *c*-BC_3.

2. Method

The precursor, graphitic (*g*-BC_3), was prepared by a high-pressure thermal chemical vapor deposition (HPCVD) method using C_2H_2, BCl_3 (as carbon and boron sources) and H_2 (30 mbar, 1225° C)[7]. The pressure- and temperature-induced phase transformation of *g*-BC_3 was studied at a pressure up to 25 GPa using laser-heated DACs and angle-dispersive powder X-ray diffraction (XRD) at the Advanced Photon Source (APS), Argonne National Laboratory. X-ray diffraction measurements were performed using a symmetrical DAC and a monochromatic synchrotron radiation source (λ = 0.368138 Å) at the 16-ID-B beamline of the HPCAT facility. Each cell had a pair of type I brilliant-cut diamond anvils of 0.3 carat and a culet of 300 μm. A stainless steel, fully hardened, 250-μm-thick gasket was first indented to 40 μm and then a 90 μm diameter hole was drilled, which served as the sample chamber. The sample, which included a tiny ruby chip, was gradually loaded to a desired pressure and then laser-heated to high temperature. To ensure efficient laser-heating in the sample between the diamond-anvils, a thin layer of NaCl (a few microns in thickness) was placed on the upper side as well as the lower side of the sample to provide a good thermal insulation. The sample temperature was monitored using the spectra-radiometric method. The uncertainty of the temperature measurement was reported as ± 100 K. The laser moved stepwise vertically and horizontally to heat the whole sample area and the duration of the laser heating at each point was about 10 seconds. After quenching, each sample was decompressed gradually stepwise to atmospheric pressure. The sample in the gasket was then removed from the cell and rinsed with water to remove the NaCl layers on both sides of the BCx sample for x-ray diffraction and electron microscope analysis. In order to characterize the elasticity of the micron-sized specimens, we studied the equation of state of the bulk graphitic BC_3 (*g*-BC_3) in the DAC up to 12 GPa using energy-dispersive X-ray powder diffraction with synchrotron radiation and a water-alcohol mixture as a pressure medium.

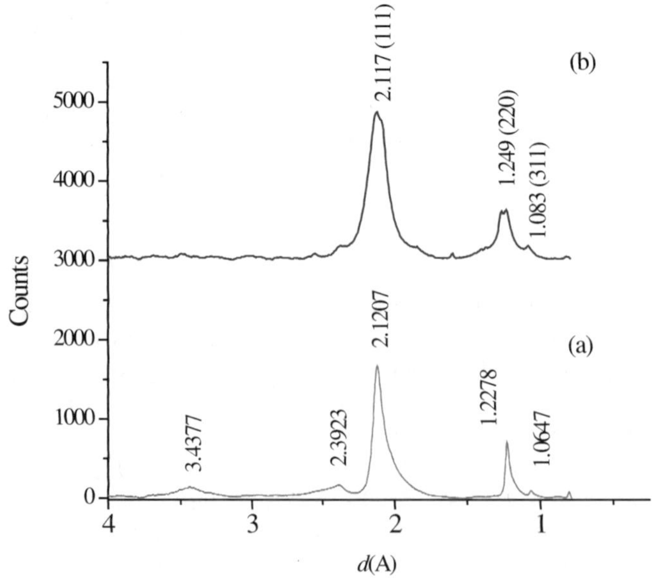

Figure 1. XRD characterization of (a) the starting material, g-BC_3, and (b) *c*-BC_3 after laser heating.

The elemental compositions of the B-C phases were taken from the Ref. [3] determined for an FIB-prepared TEM thin section and examined with an FEI Titan^3 G2 60-300 dual-C_s-aberration-corrected, monochromated (scanning) transmission electron microscope (STEM) at the Lawrence Livermore National Laboratory.

3. Results

We found that the relative intensities of the XRD peaks in the g-BC_3 phase are different from those of graphite (Fig. 1a). Typically, the *(001)* peak of turbostratic graphitic phases is the most intense [8, 9], whereas the most intensive peak in g-BC_3 is the (10) peak [3]. A weak peak at 2.384 Å is believed to be from the B_4C left in the starting material during the process of synthesis [10]. Peak positions of the g-BC_3 were found at d = 3.4377 Å (100), 2.1207 Å (10), 1.2278 Å (11), 1.0647 Å (20). The positions of the peaks on the x-ray spectrum can be indexed using a hexagonal unit cell with a=2.4537±0.0031 Å and c= 13.45±0.37 Å), where indexing was done according to Warren [9]. Parameters of the unit cell of g-BC_3 are close to those of graphite (a=2.463 Å and c= 6.714 Å) (JCPDS # 23-64) if the c-axis is doubled. It was suggested that the weak intensity of the (100) peak was because of very different stacking layers of B-C hexagons along the c-axis [2, 3]. Results of a recent finding [1] allow us to assume that the g-BC_3 is a mixture of diamond-like and graphitic BC_x phases and that the strongest peak at 2.1202 Å on the XRD pattern of the g-BC_3 is an overlap of the (10) graphitic and (111) diamond-like peak of the BC_x phases (d = 3.6725 Å). After laser (1100 K) heating the g-BC_3 at 25 GPa, the spectrum of the BC_3 phase looks similar (Fig. 1b), however the strongest graphitic (100) peak practically disappears from the XRD pattern. Then the interpretation can be different from that in Ref. [2]. The three broad peaks can be indexed with a cubic unit cell a = 3.619 (0.165) Å, indicating that almost all graphitic layers transform into diamond-like structures.

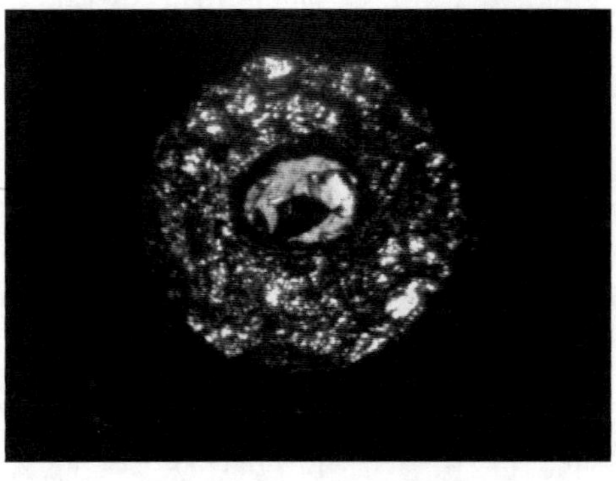

Figure 2. Photo of the g-BC_3 in DAC at 19 GPa.

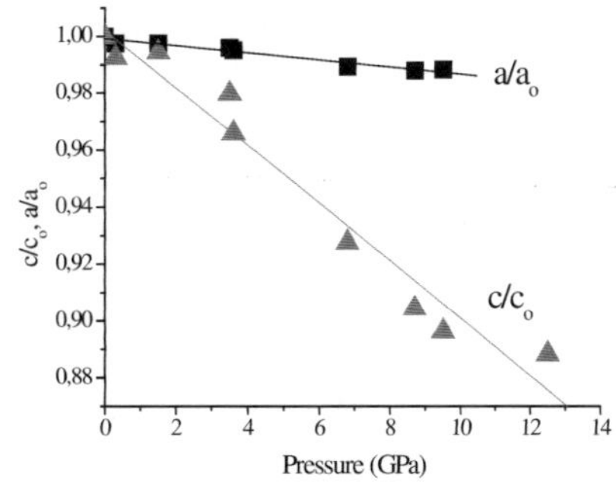

Figure 3. Experimental data for the a- and c-axis lattice parameters of g-BC_3 phase as a function of pressure.

Figure 2 shows the photo of the g-BC_3 in the DAC at 25 GPa. Figure 3 illustrates the behavior of the parameters a/a_0 and c/c_0 of the g-BC_3 under high pressure. Similar to all graphite-like structures, the g-BC_3 phase exhibits much higher compressibility along the c-axis than along the a-axis [11, 12]. The behavior of the a/a_0 and c/c_0 as function of pressure (Fig. 2b) has a linear character within experimental uncertainties (see [11, 13, 14]). The linear compressibility along the c axis of the g-BC_3 can be characterized by the value of the linear modulus B_c = 29.2 ± 1.8 GPa. It is close to that of turbostratic boron-substituted graphite (t-BC), B_c = 25 GPa [12], and is lower than that of graphite, B_c = 37 GPa [11]. The low linear compressibility along c axis of the g-BC_3 compared with that of graphite indicates that it is unlikely that boron atoms intercalate between graphitic planes. Linear fitting of the experimental data for the a/a_0 parameter as a function of pressure (Fig. 3) gives us the value of the linear elastic modulus along the a axes: B_a = 800 ± 75 GPa. This is 20 % softer than that of graphite, where B_a = 1040 GPa [11]. It is reasonable that the modulus B_a of the g-BC_x phases is smaller than that of graphite due to the larger hexagonal lattice parameters a in the layer of g-BC_3 (2.563 Å) as compared to that of graphene (2.562 Å) [15], and it indicates that boron atoms are incorporated into the graphitic BC rings in the g-BC_3 phase. Surprisingly, this value is much smaller than the linear *a*-axis modulus measured for of the turbostratic boron-substituted graphite (t-BC) (B_a = 1400 GPa) [12]. The high value of the B_a modulus of turbostratic boron-substituted graphite (t-BC) [12], which is higher than that of graphite, is probably related to high experimental error.

4. Conclusion

Results of our recent finding allow us to assume that the *g*-BC_3 is a mixture of diamond-like and graphitic BC_x after laser (1100 K) heating at 25 GPa. The spectrum of the BC_3 phase can be indexed with a cubic unit cell a = 3.619 (0.165) Å, indicating that almost all graphitic layers transform into diamond-like structures. Measurements of the equation of state of the graphitic BC phase demonstrate that boron atoms are incorporated into the graphitic B-C network. The linear compressibility along the *c* axis can be characterized by the value of the linear modulus $B_c = 29.2 \pm 1.8$ GPa. The low linear compressibility along the *c* axis of the *g*-BC_3, compared with that of graphite, indicates that it is unlikely that boron atoms intercalate between graphitic planes. Linear fitting of the experimental data for the a/a_0 parameter as a function of pressure gives the value of the linear elastic modulus along the *a* axes as $B_a = 800 \pm 75$ GPa. This is 20 % softer than that of graphite.

Acknowledgement
This work was supported by U.S. DOE grant NO. DE-FG02-07ER46408. The use of the HPCAT facility was supported by DOE-BES, DOE-NNSA (CDAC), NSF, DOD-TACOM, and the W.M. Keck Foundation. The use of Advanced Photon Source was supported by the U.S. Department of Energy, Office of Science, Office of Basic Energy Sciences, under Contract No. W-31-109-Eng-38. I. Kutuza is supported by the Program of the Presidium of RAS on Strongly Compressed Matter.

References

[1] Zinin P V, Liu Y S, Burgess K, Ciston J, Jia R, Hong S M, Sharma S, and Ming L C 2014 *J. Appl. Phys.* **116** 9
[2] Ming L C, Zinin P V, Liu X R, Nakamoto Y, and Jia R 2010 *J. Phys.: Conf. Ser.* **215** 012135
[3] Zinin P V, Ming L C, Ishii H A, Jia R, Acosta T, and Hellebrand E 2012 *J. Appl. Phys.* **111** 114905
[4] Zhuang C Q, Li X Q, Zhao J J, Liu L, and Jiang X 2010 *J. Phys.: Condens. Matter* **22** 215401
[5] Zinin P V, Ming L C, Kudryashov I, Konishi N, Manghnani M H, and Sharma S K 2006 *J. Appl. Phys.* **100** 013516
[6] Zinin P V, Ming L C, Kudryashov I, Konishi N, and Sharma S K 2007 *J. Raman Spectrosc.* **38** 1362
[7] Shirasaki T, Derre A, Menetrier M, Tressaud A, and Flandrois S 2000 *Carbon* **38** 1461
[8] Franklin R E 1951 *Proc. R. Soc. London, Ser. A* **209** 196
[9] Warren B E 1941 *Phys. Rev.* **59** 693
[10] Zinin P V, Ming L C, Sharma S K, Liu Y, and Hong S M 2009 *Diam. Relat. Mat.* **18** 1123
[11] Zhao Y X and Spain I L 1989 *Phys. Rev. B* **40** 993
[12] Solozhenko V L, Kurakevych O O, Solozhenko E G, Chen J H, and Parise J B 2006 *Solid State Commun.* **137** 268
[13] Yagi T, Utsumi W, Yamakata M-a, Kikegawa T, and Shimomura O 1992 *Phys. Rev. B* **46** 6031
[14] Hanfland M, Beister H, and Syassen K 1989 *Phys. Rev. B* **39** 12598
[15] Lowther J E, Zinin P V, and Ming L C 2009 *Phys. Rev. B* **79** 033401

IOP Conf. Series: Journal of Physics: Conf. Series **950** (2017) 042051 doi:10.1088/1742-6596/950/4/042051

Impact sensitivity test of liquid energetic materials

A Tiutiaev, A Dolzhikov, I Zvereva

Samara state technical university, Samara, Russia

E-mail: tyutyaev@mail.ru

Abstract. This paper presents new experimental method for sensitivity evaluation at the impact. A large number of researches shown that the probability of explosion initiating of liquid explosives by impact depends on the chemical nature and the various external characteristics. But the sensitivity of liquid explosive in the presence of gas bubbles increases many times as compared with the liquid without gas bubbles. In this case local chemical reaction focus are formed as a result of compression and heating of the gas inside the bubbles. In the liquid as a result of convection, wave motion, shock, etc. gas bubbles are easily generated, it is necessary to develop methods for determining sensitivity of liquid explosives to impact and to research the explosives ignition with bubbles. For the experimental investigation, the well-known impact machine and the so-called appliance 1 were used. Instead of the metal cup in the standard method in this paper polyurethane foam cylindrical container with liquid explosive was used. Polyurethane foam cylindrical container is easily deforms by impact. A large number of tests with different liquid explosives were made. It was found that the test liquid explosive to impact in appliance 1 with polyurethane foam to a large extent reflect the real mechanical sensitivity due to the small loss of impact energy on the deformation of the metal cup, as well as the best differentiation liquid explosive sensitivity due to the higher resolution method.

1. Introduction

Well known, none of the methods for determining sensitivity liquid explosive to impact can not give an absolute evaluation of the safety. For liquid explosives systems there are features associated with the aggregate state. One problem is the reliable storage of the test sample liquid explosive in the area of mechanical action in the preparation and testing. Metal cup, filled with liquid explosive is used in the standard method [1]. Significant influence on the accuracy and interpretation of the results has the fact that a significant part of the impact energy is expended on the deformation of the cup. Mechanism of initiation is also important. If the gas bubbles exist or arise in the liquid explosive , the initiation occurs at lower energies strike. Focuses of heating and ignition in this case are the result of the compression and heating of the gas in the bubbles. The most correct way out is to improve the complex test sensitivity methods and a detailed theoretical and experimental study explosive initiating by impact. In [2] is developed a mathematical model of thermal and hydrodynamic processes that occur during compression of gas bubbles in liquid explosive.

2.Experimental

Calculations showed that the presence of bubbles can significantly reduce the energy of the external impact for initiation. For the simulation of bubble structure in this paper to determine liquid explosive sensitivity to impact as a container (instead of the cup) to accommodate liquid explosive is used cylinder of polyurethane foam. Cylinder of polyurethane foam has a highly developed porosity, foam is inert and insolubility in most liquid explosives and foam is easily deformed by mechanical actions. Impact machine (figure 1) and roller device (figure 2) were used for determining sensitivity liquid explosive.

Figure 1. Impact machine.

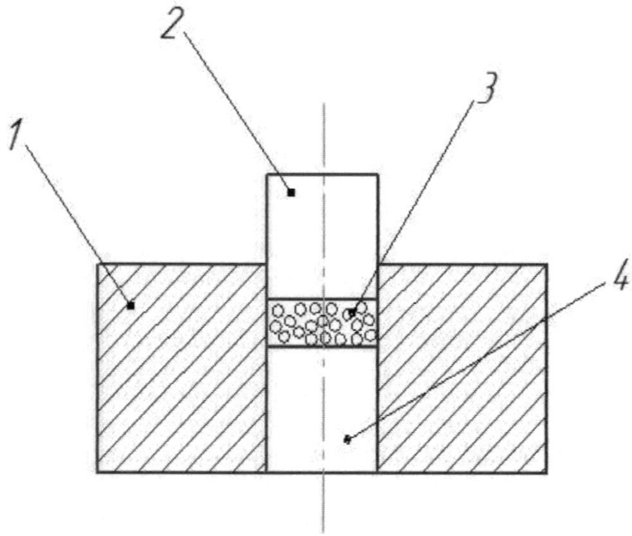

Figure 2. Roller device.
1-coupling; 2,4-upper and bottom rollers; 3- polyurethane foam with liquid explosive.

3.Results and discussion

Tests were made with elastic impact (1), polyurethane foam (1-p) and 1-c (with cup) to estimate the total energy loss. Beat with a load of 10kg was performed with heights ranging 100-600mm. Quantification was performed according to the following:

$$Ead = Eim - Ere,$$

where Ead - the loss or deformation of the foam cap and impact machine; Eim - the impact energy; Ere - the rebound energy. E is calculated from the relationship:

$$E = mgh,$$

where m the weight of the load; h - the height of a reset or rebound.
Obtained experimental data given in table 1 and figure 2.

Table 1. The total energy loss in the device impact machine and roller devices 1, 1-p and 1-c.

Drop height, h ,mm	Impact energy, Eim, J	Energy losses		
		1	1-p	1-c
100	9,81	2,45	3,43	8,83
200	19,62	4,42	5,40	12,26
400	39,24	9,81	10,79	19,52
600	58,86	18,64	20,60	27,96

In device 1-p used foam cylinder with a diameter of 10 mm and a height of 5 mm.

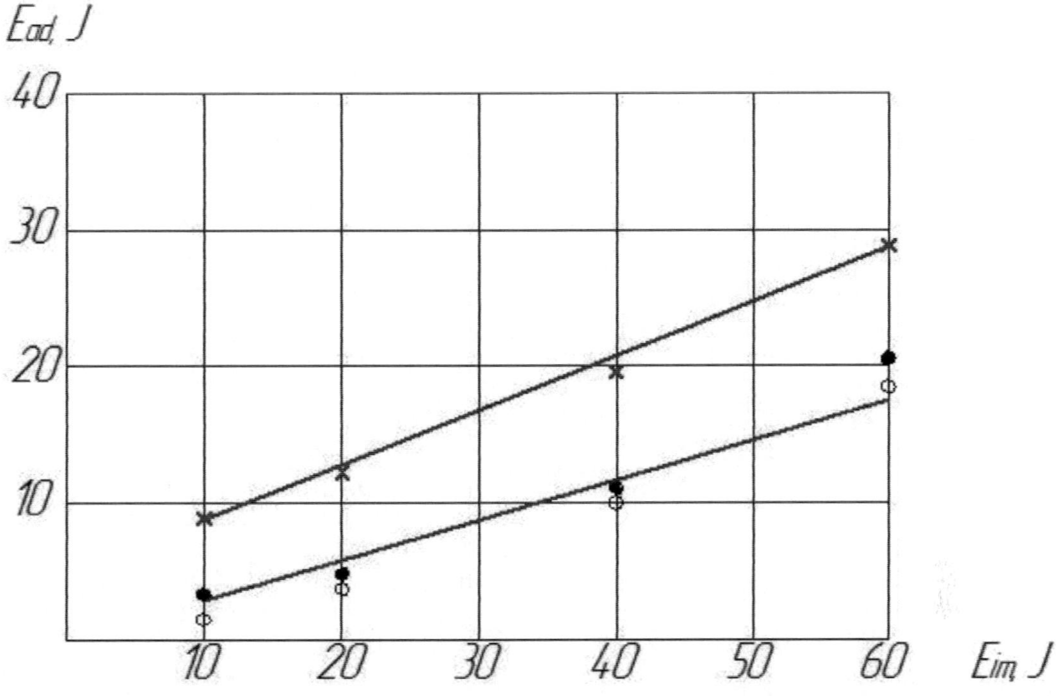

Figure 3. The dependence of the energy absorbed by the impact energy.
○ - device 1, ● - device 1-p, x - device 1-c.

The data in table 1 and figure 3 indicate that at low drop height (100mm), the energy loss in devices 1, 1-p, and 1-c are 25%, 35%, and 90% of the impact energy, respectively. At high drop height, energy loss in 1-p is almost equal to elastic impact energy (the difference is 2.5-3.3%), while the deformation of the cap as 25-40% of the energy expended impact.

Thus, we can conclude that the foam contributes in 4-8 times less distortion during impact as compared with the cup.

Experimental studies on the absorption capacity of the foam were performed on cylinders with a diameter of 10 mm and a height of 10mm 2,4,5,8, 10mm. These results are summarized in table 2.

Table 2. The absorption capacity of the foam cylinder.

Height of the cylinder of foam, mm	2	4	5	8	10
Volume of a cylinder of foam, $m^3 10^{-6}$	0,157	0,314	0,393	0,628	0,785
Weight foam cylinder, mg	6	13	16	25	32
Quantity of a liquid, nitroglycerine, mg	245	490	612	980	1225

The data in table 2 show that the cylinder has a height of 2mm foam has full capacity to absorb about 250mg. However, soft polyurethane has two types of pores: closed and reported. The reported pores can lead to a gradual filtering of the test sample from liquid explosive foam cylinder. In this regard, studies have been made with foam cylinders with benzene, toluene, and glycerol. In the experiments we used samples of foam height from 2 to 10mm. With the help of filter paper and glass, on which were placed foam cylinders, visually with permeability was determined. The results are shown in table 3.

Table 3. Filtration time.

Quantity of a liquid, mg	Filtration time (h, m, s)				
	Height of the foam cylinder (mm)				
	2	4	5	8	10
20	11m	>12h	>12h	>12h	>12h
30	10m	>12h	>12h	>12h	>12h
40	55s	>12h	>12h	>12h	>12h
50	50s	>12h	>12h	>12h	>12h
75	35s	28min	>12h	>12h	>12h
100	25s	4min	10min	14min	>12h

As shown by the test results, there are two kinds of through filtering (time expiration) in dependence on the height and the weight of the liquid foam. In the first case, for large sample (40-100mg) and low height foam cylinder (2mm) filtering is done in a very short time, almost commensurate with the time of making the liquid sample (25-60s). It can be assumed that the height of the cylinder of 2 mm, there is a significant number of through pores communicating height, which due to the gravitational force and the wettability is so rapid filtration.
In the second case was observed penetrating filtration over a longer second time (4-28min.) Thus, the data in Table 3 indicate that for tests carried liquid explosive sample weighing between 50 to 75 mg is optimal foam cylinder height of about 5mm.
Distribution pattern of the liquid sample weighing 75 mg in a cylinder of foam was obtained. It is evident that with the known assumption, this distribution of liquid in a foam cylinder can be considered cylindrical. In this case, the spot diameter is about 6 mm and the depth of penetration of the 3.5-4 mm.
Test using a mass 10 kg can not determine the lower limits of most studied liquid explosive Therefore, to determine the relative frequency of the lower explosion limit of the experiments , we performed experiments with a 2-kg mass, whose results are shown in table 4.

4.Conclusion

Determination of explosion frequency liquid explosive with the construction of complete curves performed on the impact machine and the use of 10kg weight. Experiments were made for different liquid explosives. Experimental data show that the results of tests in 1-p provide an opportunity to get the full frequency curve explosions. The curves have a particular characteristic S-shaped appearance. Also, as in the tests of solid explosives, the difference in sensitivity at the upper limit liquid explosive explosion frequency significantly higher than in the lower limit. As in the case of solid explosives is crossing curves explosion frequency different liquid explosives . Analysis of the results of determining the lower limit of explosions in cups and device 1-p leads to several conclusions.
Firstly, there is the sharp increase in the sensitivity in device 1-p compared with trials in cups.

Table 4. The lower limit of explosion frequency LHE in device 1-c and 1-p.

LHE	1-c		1-p	
	Drop height H_0, mm	Impact energy ,J	Drop height H_0, mm	Impact energy ,J
nitroglycerine	120	2,35	20	0,39

For example, the impact energy corresponding to the lower limit for the particular explosions for nitroglycerine has decreased in 6 times. Absolute values decreased about 2-3 J. This can be attributed to loss of impact energy to deformation of the cap, which is almost identical to the data, where the experimental data loss (40-90%) depending on the drop height.
The method of determining the sensitivity of liquid explosive in device 1-p has a significantly higher resolution than standard. For example, explosion parameters for nitroglycerine when tested with a cap differ in differ from the results of tests in 1-p in 6.3 times. Thus, we can conclude that the test liquid explosive to strike in 1-p to a large extent reflect the real sensitivity to mechanical stress due to the exclusion of loss of impact energy on the deformation of the metal cap, as well as the best differentiation liquid explosive sensitivity due to the higher resolution of the method.

References

[1] "Standard Test Method for Drop-Weight Sensitivity of Liquid Monopropellants", ASTM Designation D2540-93 (2001)
[2] Tyutyaev A, Dolzhikov A, Zvereva I 2014 Phenomenological models of explosive systems initiation at mechanical influences *Proceedings of 17 Seminar on "New Trends in Research of Energetic Materials"* (Pardubice, Czech Republic) pp 428-433

Pressure as driving force in the formation of Fossil Resins: Pressure Induced Changes in *trans*-Communic Acid studied by Raman Spectroscopy

O R Montoro, M Taravillo, I Povedano, J Tortajada and V G Baonza

MALTA-Consolider Team, Departamento de Química-Física I, Facultad de Ciencias Químicas, Universidad Complutense de Madrid, 28040. Madrid, Spain.

E-mail: ormontoro@quim.ucm.es

Abstract. The main purpose of this study is to provide spectroscopic evidence of possible chemical pathways in the formation of fossil resins under high pressure conditions. In this work, we have studied *trans*-communic acid (majority precursor compound in resinites) under high pressure conditions by using a sapphire anvil cell. Raman spectra were measured at atmospheric pressure up to 8 GPa (at room temperature). Slight changes were observed in the spectra. Furthermore, the recovered sample after pressure release was analyzed by FTIR. Our results are not conclusive about a possible chemical pathway of polymerization induced by pressure.

1. Introduction

Fossil resins are organic gemstones which have survived up to now. In general, they derive from natural terpene-based polymers, and are classified in five classes. *Class I* is the most important; its members are composed by polymerization of labdanic acid monomers, primary communic acids (labdatriene diterpenes) [1-3]. From them, Baltic Amber is the most well-known *Class* I resinite.

In Figure 1, we schematize the principal labdatriene isomer in his acid form, i.e. *trans*-communic acid [4], which is the main precursor for a large majority of *Class I* resinites.

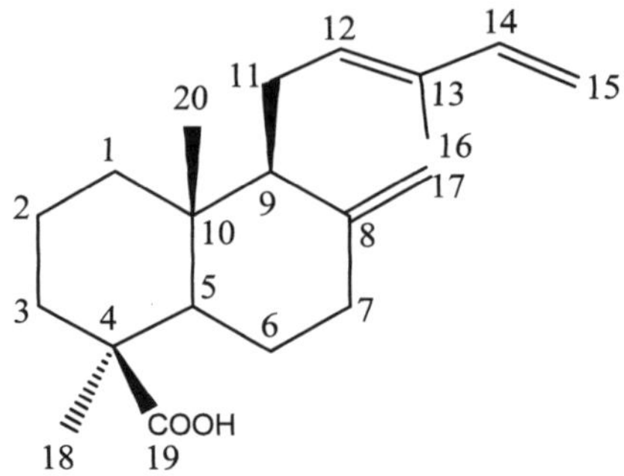

Figure 1. *trans*-Communic acids.

The reactive processes that took place in the formation of fossils resins are barely studied due to their complexity. Current studies conclude that maturing reactions or reactivity schemes of the different *Class I* resinites, seem to occur in two polymerization stages [5,6].

The first suggested step consists of an initial polymerization between the conjugated double bonds of the side-chain olefin (1-2 addition to conjugated double-bonds in *trans*-labdatrienes isomers). Later in the ages, polylabdanoid resinites continue with a more specific reaction resulting in depletion of the C8-C17 exomethylene (exocyclic bond), due to intramolecular cyclation reactions, intramolecular polymerization, among others.

These processes took place throughout ages until formation of fossil resins. Several studies are focused on how the geological environment conditions such as temperature [7], radiation or atmosphere compositions [8], act on trees exudation. However, there is no study considering the pressure as geological condition, even though Baltic Amber samples have been found in the Baltic Sea depths, as well as other resinites buried in under the Earth's surface [9].

Baltic Sea is 57 m depth, on average, and its maximum depth is 459 m. Those are not significant depths in order to generate a substantial high pressure in geological terms, as it can go into the depths of the Pacific Ocean. Nevertheless, fossil resins have been buried under the sea for millions of years. From this point of view, pressure should be taken into account as a formation variable.

In this work, we have studied *trans*-communic acid under pressure conditions by using a sapphire anvil cell (SAC), in order to identify if pressure is an important geological agent in the formation of the fossil resins. Firstly, Raman spectra of the pressurized samples were collected, from room pressure up to 8 GPa. Then FTIR spectra of the recovered sample were recorded at ambient conditions.

2. Experimental

The high pressure experiments of this work were carried out at room temperature in a 350 μm culet SAC. The pressure was determined by Raman spectroscopy using the linear variation of the diamond fundamental phonon [10]. The sample was loaded together with a diamond chip, into a 100-μm-diameter hole drilled on a 200-μm-thick golden gasket pre-indented to 40 μm. In addition to the chip diamond, a ruby chip was loaded as possible alternative to measure the pressure [11]. The *trans-c*ommunic acid itself was used as pressure-transmitting medium.

Raman spectra of the *trans*-communic acid were recorded on a confocal micro-spectrometer (B&W Tek Voyage™, model BWS435-532SY). The spectrometer is coupled to an Olympus BX51 microscope, which allows us to focus the beam spot (~5 μm) onto the sample to register spectra. Excitation is provided by a solid-state laser operating at 532 nm. The spectral range available is 100-3750 cm^{-1} with spectral resolution of about 4 cm^{-1}. But we shall focus on the spectral ranges between 600-1800 and 2800-3200 cm^{-1}. The scattered light is collected on backscattering configuration using a 20x long working-distance objective. The detector consists in a backthinned CCD (Hamamatsu S10141-1107S) thermoelectrically cooled at -20 °C with 2048 × 122 pixels. A Chameleon™ CMLN-13S2C digital camera is used for optical observation. Previously to the temperature study, the laser power influence on the sample was carefully analyzed. All the spectra were registered using laser excitation power of *ca.* 2 mW to avoid unwanted heating.

FTIR spectrum of the *trans*-communic acid was recorded in a Spectra-Tech IR-Plan microscope for reflected light microspectrometry. The microscope uses a 15x IR reflecting lens with Cassegrain optics to focus the infrared beam onto the sample and it is equipped with a liquid-N_2 cooled MCT-A detector.

The molecule used in this study is a pure isomer of communic acid: *trans*-communic acid. Since it is not commercially available, their extraction was carried out from *Cupressus* and *Juniperus* (*Cupressus Sempervirens L.*, *Juniperus Communis L.* and *Juniperus Oxycedrus*) [12,13]. In order to validate the purity of the communic acids ^{1}H-NMR spectroscopy was employed to characterize the *trans*-communic acid [7].

Spectral assignment of the vibrational modes of *trans*-communic acid is based on both, the existing spectral assignments for this molecule [7, 8], and the assignments of different fossil resins found in the literature [14-17].

3. Results and discussion

We have studied *trans*-communic acid under high pressure conditions by using a sapphire anvil cell. Raman spectra were measured at atmospheric pressure up to 8 GPa (at room temperature). Furthermore, when the pressure was released, the recovered sample was analyzed by FTIR.

The study of these spectra is focused on the next spectral regions: a) C-H stretches, and b) modes involving double-bond stretches, with special attention those involving C=C and C=O moieties, and C^{17}-H exomethylene wagging (exocyclic double bond).

3.1. Pressure Induced Changes in trans-Communic Acid studied by Raman Spectroscopy.

Raman spectra at different pressures of *trans*-communic acid show the following results:

3.1.1. C-H stretches región. The vibrational modes present in C-H stretches region (Figure 2) have only been measured until 0.4 GPa, due to the fluorescence background and the weak intensity of these modes above this pressure value. The studied Raman modes are: $\nu(R_3C^9H)$ = 2847 cm^{-1}; $\nu(R_3C^5H$ bridge C) = 2870 cm^{-1}; $\nu_s(CH_2)$ = 2908 cm^{-1}; $\nu_s(CH_3)$ = 2936 cm^{-1}; $\nu_{as}(CH_2)$ = 2952 cm^{-1}; $\nu_{as}(CH_3)$ = 2963 cm^{-1}; $\nu_s(C^8{=}C^{17}H_2$ exocyclic) = 2986 cm^{-1}; $\nu_s(C^{14}H{=}C^{15}H_2$ conjugated) = 3002 cm^{-1}; $\nu(-C^{12}H{=}C^{13}R-)$ = 3036 cm^{-1}; $\nu_{as}(C^8{=}C^{17}H_2$ exocyclic) = 3079 cm^{-1}; $\nu_{as}(C^{14}H{=}C^{15}H_2$ conjugated) = 3094 cm^{-1}.

Only slight changes have been observed around 2850-3100 cm^{-1} at 0.4 GPa (Figure 2). This region involves the $-CH_2-$ and $-CH_3$ stretching modes. The detected changes consist of a broadening and shifting of these peaks to higher wavenumbers with pressure. That could be related to some slight conformational changes in the molecule. No evidence of chemical reaction is observed.

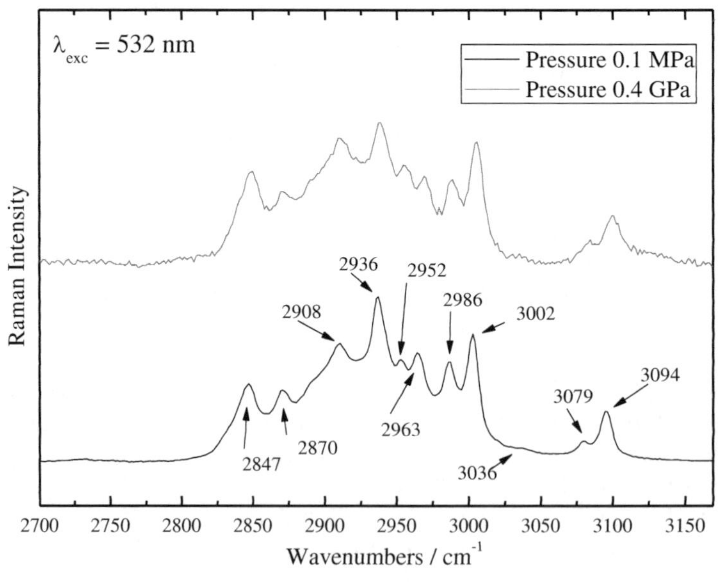

Figure 2. Enlarged Raman spectra of *trans*-communic acid at atmospheric pressure and 0.4 GPa.

3.1.2. Double-bond stretches (C=C), C-H bendings and C-O stretch region. Raman spectra in this region (Figure 3) were measured from atmospheric pressure up to 8 GPa. We have followed the gradual change of the most characteristic Raman bands of trans-communic acid that involve carbon-carbon double bonds C=C (conjugated and exocyclic), with pressure. In general, these modes decrease in intensity as the fluorescence background increases, becoming weak bands.

In this case as well, the previous mentioned factors, band shifting and band broadening, make arduous to follow their evolution pressure.

Despite all these difficulties, we can see how only above 3.3 GPa, the two more characteristic bands ν_s(C=C conjugated) and ν_{as}(C=C conjugated) at 1643 cm^{-1} and 1606 cm^{-1}, respectively, overlap yielding a single broadband.

Other bands, as the ν(C-O) at 1294 cm^{-1} or the C-H bendings, do not show relevant changes. Therefore, we can conclude that there is no evidence of chemical reaction.

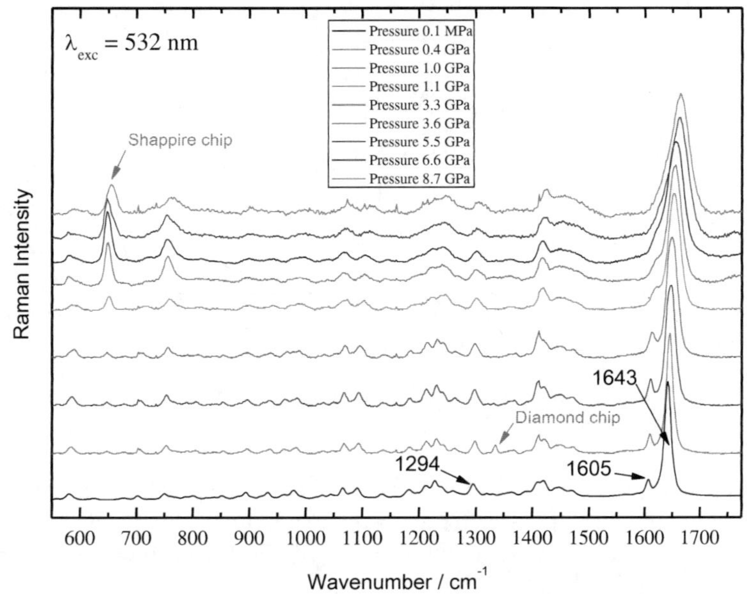

Figure 3. Raman spectra of *trans*-communic acid at selected pressures.

3.2. Pressure Induced Changes in trans-Communic Acid recovered by FTIR.
Before the cell opening, we tried to collect the FTIR spectra of *trans*-communic acid inside the SAC at 8.0 GPa (the sample was stabilized during 109 days at this pressure), but due to the small size of the gasket hole and the spectral sapphire window, the results were unsuccessful.

Consequently, we decided to proceed with the opening of the SAC, in order to collect the FTIR spectrum by reflection of the sample recovered placed directly over the anvil surface. The spectrum was recorded on the SAC to prevent sample losing while handling due to small dimensions.

An important detail about the recovered sample is to highlight that no colour changes were observed.

3.2.1. C-H stretches region. Both FTIR spectra of trans-communic acid, shown in Figure 4, were measured at atmospheric pressure, but recovered sample (plotted in red), was pressurized at 8.0 GPa during 109 days.

The vibrational modes present in C-H stretching region (Figure 4), do not seem to undergo significant changes except for its displacement towards higher frequencies. The IR studied modes are: $\nu(R_3C^9H)$ = 2845 cm^{-1}; $\nu(R_3C^5H$ bridge C) = 2870 cm^{-1}; $\nu_s(CH_2)$ = 2905 cm^{-1}; $\nu_s(CH_3)$ = 2935 cm^{-1}; $\nu_{as}(CH_2)$ = 2952 cm^{-1}; $\nu_{as}(CH_3)$ = 2964 cm^{-1}; $\nu_s(C^8=C^{17}H_2$ exocyclic) = 2986 cm^{-1}; ν_s $(C^{14}H=C^{15}H_2$ conjugated) = 3002 cm^{-1}; $\nu(-C^{12}H=C^{13}R-)$ = 3036 cm^{-1}; $\nu_{as}(C^8=C^{17}H_2$ exocyclic) = 3078 cm^{-1}; $\nu_{as}(C^{14}H=C^{15}H_2$ conjugated) = 3093 cm^{-1}.

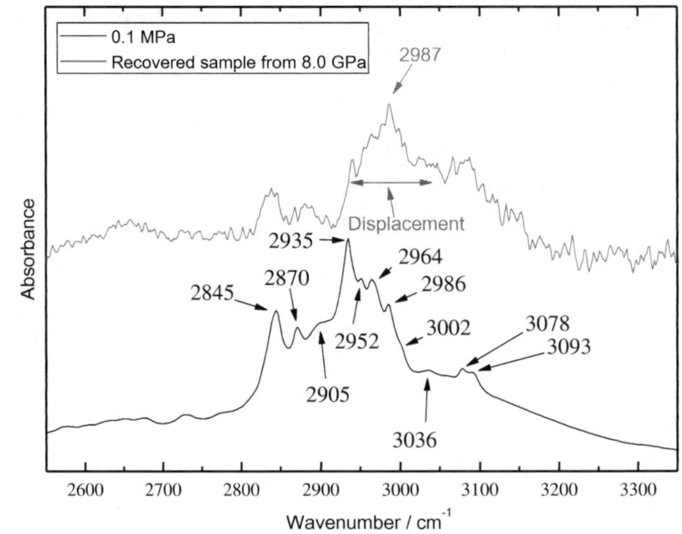

Figure 4. Enlarged FTIR spectra of *trans*-communic acid both at atmospheric pressure, but drawn in red the "Recovered Sample" after pressurizing at 8.0 GPa for 109 days.

3.2.2. Double-bond stretches (C=O and C=C) and C-H bendings region. As the previous section, in this new spectral region, both FTIR spectra of trans-communic acid (Figure 5) were measured at atmospheric pressure, but one of them (displayed in red) is over a recovered sample after having undergone at 8.0 GPa during 109 days.

We have followed the evolution of the more characteristic IR bands of *trans*-communic acid involving the next modes: $\gamma_{o.o.p.}(R_2C{=}CH_2)$ = 892 cm^{-1}; ν_{as}(C=C) = 1606 cm^{-1}; ν_s(C=C) = 1643 cm^{-1}; $\nu(C^8{=}C^{17})$ = 1650 cm^{-1}; ν(C=O) = 1693 cm^{-1}. As well as in the previous region, a band broadening and a displacement of the spectral features until higher wavenumbers with the pressure have occurred.

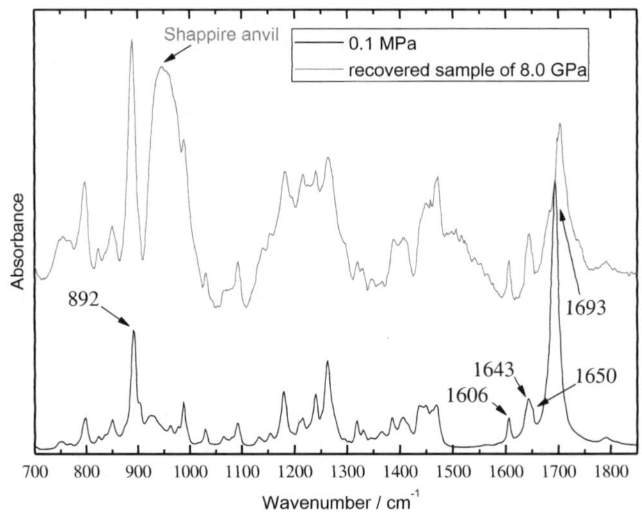

Figure 5. Enlarged FTIR spectra of *trans*-communic acid both at atmospheric pressure, the "recovered sample" (red) was pressurized at 8.0 GPa for 109 days.

4. Conclusions

We have used high pressure techniques (sapphire anvil cells, SAC) and spectroscopic methods (Raman spectroscopy and FTIR) to demonstrate whether or not the pressure can be considered as an important geological agent in the formation of the resinites.

The obtained results show a general broadening in the spectral bands, in addition to a shift to higher wavenumbers with pressure, indicating that sample does not experience significant changes as confirmed by the observation of no colour change in the recovered sample.

According to the spectroscopic results, the *trans*-communic acid experiences slight conformational changes in the explored pressure range but no evidence of pressure-induced polymerization was observed, in contrast high-temperature treatments [7].

We can conclude that within the experimental conditions, pressure cannot be considered as an important geological agent in the formation process of the fossil resins.

Acknowledgments

This work has been funded by the Spanish Ministry of Science and Innovation under Projects CTQ2012-38599-C02-02, CTQ2015-67755-C2, Grupos UCM 910481, MAT2015-71070-REDC, and MALTA-Consolider Ingenio 2010 (CSD2007-00045). Authors wish to thank Dr. José E. Fernández from the Centre of Infrared and Raman Spectroscopy of the Complutense University for kindly allowing use the IR equipment.

References

[1] Anderson K B, Winans R E and Botto R E, 1992 *Org. Geochem.* **18** 829
[2] Lambert J B, Santiago-Blay J A and Anderson K B, 2008 *Angew. Chem. Int. Ed.* **47** 9608
[3] Sargent Bray P and Anderson K B, 2009 Science 326 132
[4] Anderson K B, 1994 *Org. Geochem.* **21** 209
[5] Clifford D J and Hatcher P G, 1995 *Amber, Resinite, and Fossil Resins*, ed K B Anderson and J C Chelling (Washington DC, ACS Symposium Series 617 American Chemical Society) pp 92-104 Chapter 5. ISBN: 978-0841233362
[6] Clifford D J, Hatcher P G, Botto R E, Muntean J V and Anderson K B, 1999 *Org. Geochem.* **30** 635
[7] Montoro O R, Taravillo M, San Andres M, de la Roja J M, Barrero A F, Arteaga P and Baonza V G, 2014 *J. Raman Spectrosc.* **45** 1230
[8] Montoro O R, 2013 *Estudio Espectroscópico de la Formación de Análogos de Resinas Fósiles*, PhD Thesis, Universidad Complutense de Madrid
[9] Slotta R and Ganzelewski M G, 1997 *Bernstein - Tränen der Götter*, ed Verlag Glückauf (Wankendorf, D, Germany) pp 249-268. ISBN: 978-3773906656
[10] Boppart H, Vanstraaten J and Silvera I F, 1985 *Phys. Rev. B* **32** 1423
[11] Piermarini G J, Block S, Barnett J D and Forman R A, 1975 *J. Appl. Phys.* **46** 2774
[12] Pascual-Teresa J, San Feliciano A and Barrero A F, 1973 *An. Quim.* **69** 1065
[13] Pascual-Teresa J, San Feliciano A and Miguel del Corral J M, 1974 *An. Quim.* **70** 1015
[14] Beck C W, 1986 *Appl. Spec. Rev.* **22** 57
[15] Smith G D and Clark R J H, 2004 *J. Archaeol. Sci.* **31** 1137
[16] Vandenabeele P, Edwards H G M and Moens L, 2007 *Chem. Rev.* **107** 675
[17] Brody R H, Edwards H G M and Pollard A M, 2001 *Spectrochim. Acta A* **57** 1325

IOP Conf. Series: Journal of Physics: Conf. Series **950** (2017) 042053 doi:10.1088/1742-6596/950/4/042053

Influence of hydrogen bonding on the elastic properties of molecular glassforming liquids under high pressure

I V Danilov [1, 2], E L Gromnitskaya [1], A G Lyapin[1, 2] and V V Brazhkin[1, 2]

[1]Institute for High Pressure Physics, RAS, Troitsk, Moscow, 142190, Russia
[2]Moscow Institute of Physics and Technology, Dolgoprudny, Moscow Region, 141700, Russia

E-mail: idanilov@hppi.troitsk.ru

Abstract. We present ultrasonic study of bulk modulus of liquid glycerol and propylene carbonate (PC) at room temperature under high pressure up to 0.6 and 1 GPa, respectively. Pressure dependence of the isothermal bulk modulus, recalculated from the ultrasonic adiabatic data, is almost linear for the both liquids with relatively similar slopes, in spite of the initial value of bulk modulus being twice higher for glycerol due to the presence of the hydrogen bonds. The values of pressure derivative (7.8 for glycerol and 8.6 for PC) indicate that the intermolecular interaction in the both liquids is close to the Lennard-Jones potential.

1. Introduction

Intermolecular interaction in molecular liquids with uncharged molecules is caused by the van der Waals forces. However, in many cases, hydrogen bonds, being a complex ion-covalent interaction between the strongly electronegative atoms by means of hydrogen, can produce a significant contribution to the interaction. The energy of the hydrogen bonds is much lower than energies of ion and covalent interactions, but higher than typical energy of the van der Waals intermolecular forces, that can significantly influence on thermodynamic properties of substances, e.g., on the high boiling point of water. From structural point of view, hydrogen bonds may affect orientational correlations of neighbor molecules in liquid and an orientational short range order of molecules in the glass. The different intermolecular potentials (with and without contribution of the hydrogen bonds) should also be directly observed as an expected difference of elastic properties.

Effect of hydrogen bonding on the intermolecular interactions can be illustrated by an example of glycerol $C_3H_8O_3$ and propylene carbonate (PC) $C_4H_6O_3$. Both substances have quite similar size and mass of molecules, densities and temperature dependence of density $\rho(T)$, as well as glass transition temperatures (157 K for PC and 187 K for glycerol) [1,2]. At the same time, each molecule of glycerol has three hydroxyl OH groups, which can be involved into hydrogen bonding, while 3 oxygen atoms in the PC molecule are connected only with carbon in the molecular skeleton and cannot virtually contribute to hydrogen bonds. As a result of this, the velocity of longitudinal ultrasonic waves in liquid glycerol is by a factor >1.5 higher than that in PC [3-5]. Correspondingly, bulk modulus of liquid glycerol is by two times higher than that of PC.

Whereas the difference in elasticity of liquid glycerol and PC is remarkable at normal conditions, it is not clear how this difference should change with pressure. Pressure dependence of bulk modulus is direct test of change of intermolecular potential parameters. In this respect, ultrasonic technique is very adequate method for accurate measurement of elastic characteristics (bulk modulus B for liquids), including pressure derivative of B and its nonlinearity with pressure (see examples for solid molecular systems under pressure in [6-8]).

Here, we present ultrasonic study of bulk modulus of liquid PC and glycerol under high pressure up to 0.6 and 1 GPa, respectively.

2. Experimental setup

Experiments were carried out using a high-pressure ultrasonic piezometer based on piston-cylinder device [9]. During the experiment simultaneous measurements of sample length and transit time of ultrasonic wave were made in pressure range up to ≈1 GPa (here, the maximum pressure is defined by mechanical stability of capsule with liquid) at room temperature 295 K, allowing to calculate density, ultrasonic velocity and bulk modulus of a sample. Liquid glycerol and PC were placed in a teflon capsule (with internal diameter of 16 mm and height of 8 mm) with copper caps (with thickness of ≈1 mm). Glycerol was purchased from Aldrich chemical company and PC from Acros Organics. The purity of both materials was guaranteed to be higher than 99.5%.The estimated pressure uncertainty was ±0.05 GPa. The measurements of the transit time (with uncertainty within 0.001 μs) of longitudinal ultrasonic pulses were performed by the pulsed ultrasonic method using LiNbO3 plates as piezoelectric sensors with carrier frequency 10 MHz. The transit time of ultrasonic pulses was directly measured by the system on the basis of the National Instruments PXI platform. Changes in the ultrasonic signal paths were also measured with uncertainty within 5 μm using dial-type micrometer indicators. The density of samples under pressure was determined from the change in their length. Bulk modulus was calculated from the ultrasonic data using the approximation of homogeneous isotropic medium.

3. Experimental results and discussion

Ultrasonic measurements at megahertz frequencies provide experimental data for adiabatic elastic properties. The relationship between the isothermal bulk modulus B_T and adiabatic modulus B_S is determined by equation

$$\frac{B_S}{B_T} = 1 + \frac{\alpha^2 T B_S}{\rho c_P} = 1 + \alpha\gamma T\,, \tag{1}$$

where α is the coefficient of volume thermal expansion, c_P is the specific heat capacity at constant pressure, ρ is the density of a substance, and $\gamma$is the Gruneisen parameter close to unity for condensed state. The difference between B_T and $B_S$at room temperature is equal to ≈22% for PC and to ≈14% for glycerol, and is relatively large due to the large values of volume thermal expansion coefficients [10] (for glassy states of these substances as well as for majority of solids the difference between B_T and $B_S$is of the order of 1-3%). Figure 1 presents the pressure dependence of the isothermal bulk modulus B_T at room temperature for both studied liquids. Both were recalculated from directly obtained ultrasonic data for adiabatic bulk moduli under pressure, taking into account the data of [10] on the volume thermal expansion of glycerol and PC under pressure and the value of Gruneisen parameterγ ≈1, that was checked at room pressure. The maximum pressure (0.6 GPa) achieved during the measurement of glycerol (figure 1) was restricted by strong absorption of ultrasonic wave in the sample due to approach to the glass transition region. In PC, there is no glass transition in this region and pressure range (up to 1 GPa) is limited only by possibilities of the experimental setup and stability of capsule.

First, although the initial value of B_T is twice higher for glycerol (figure 1), pressure dependence of bulk modulus is quite similar for both substances, and both bulk moduli grow linearly at the relatively similar rate under pressure, $(\partial B_T/\partial P)_T \approx 7.8 \pm 0.2$ for glycerol and $(\partial B_T/\partial P)_T \approx 8.6 \pm 0.2$ for PC. The other important fact is that both pressure derivatives of bulk modulus $B'(P)$ are close to 8. It is known that for central interparticle n-m potential $U = A/V^n - B/V^m$, pressure derivative of bulk modulus is $B'(P) = m + n + 2$ [11]. For Lennard-Jones potential, n=4, m=2, and, so, $B'(P) = 8$. This means that the interaction in liquid glycerol and PC can be approximately described by the Lennard-Jones potential, at least in the considered pressure range, and hydrogen bonds in glycerol, diminishing compressibility with respect to that in PC, do not significantly influence on the character of volume dependence of the van der Waals intermolecular forces. The small difference between the derivatives of bulk moduli for the both liquids results in the fact that the elastic moduli values approach each other with increasing pressure.

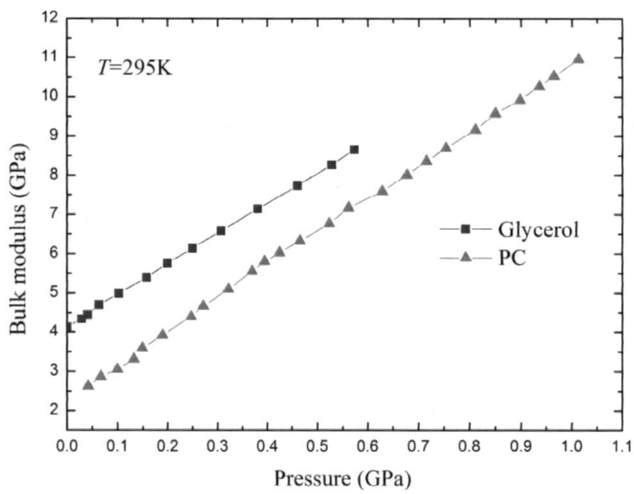

Figure 1. Pressure dependence of the isothermal bulk modulus B_T at room temperature 295 K

The measurements of the relationship between pressure and volume using a high-pressure ultrasonic piezometer are not quite accurate, despite the corrections with regard to the friction, hysteresis, and deformation of the components of the piston-cylinder device [9]. A more accurate equation of state *V(P)* can be obtained by the pressure integration of isothermal compressibility (the inverse bulk modulus).

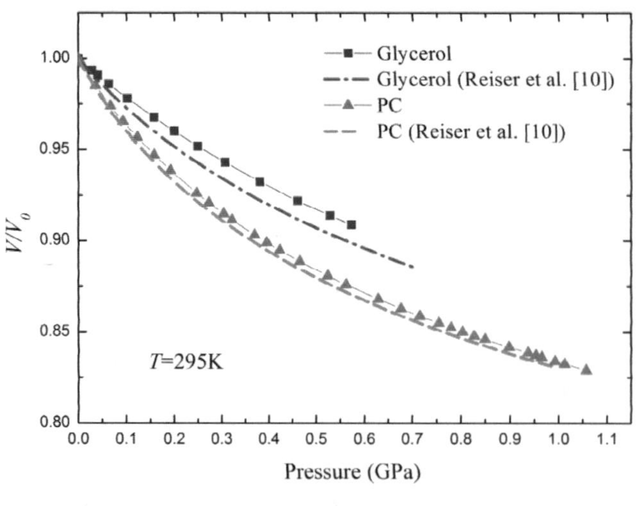

Figure 2. Relative volume change under pressure at room temperature 295 K

Pressure dependence of relative volume V/V_0 is presented in figure 2 for both substances. They show the typical behaviour for liquids at elevated pressures. At 0.6 GPa, volume of glycerol decreases by ≈9.5% and volume of PC - by ≈13%.We compared our data with equation of state for glycerol and PC given by Reiser et al. [10] from direct volumetric measurements. While our results on propylene carbonate are in a perfect coincidence with data from [10], in the case of glycerol we observe some differences.Compressibility of glycerol defined fromthe ultrasonic study is slightly less than that from the volumetric measurements.

Thus, using glycerol and propylene carbonate as model molecular liquids with and without hydrogen bonds, we have shown that, whereas hydrogen bonds make intermolecular interaction stronger, they do not significantly influence on pressure derivatives of bulk modulus under pressure. This implies that glycerol and PC have quite similar functional forms of potential with respect to change of volume and intermolecular distances.

Acknowledgements
We are grateful to Russian Science Foundation (14-22-00093) for the financial support.

References

[1] Schneider U, Lunkenheimer P, Brand R, and Loidl A 1999 *Phys.Rev.* E **59** 6924
[2] Lunkenheimer P, Loidl A 2002 *Chem. Phys.***284** 205
[3] Elmroth M, Borjesson L, Torell L 1992 *Phys. Rev. Lett.* **68** 79
[4] Jeong Y, Nagel S, Bhattacharya S 1986 *Phys. Rev.* A **34** 602
[5] Kondrin M, Gromnitskaya E, Pronin A, Lyapin A, Brazhkin V, Volkov A 2012 *J. Chem. Phys.***137** 084502
[6] Yagafarov O, Gromnitskaya E, Lyapin A, Brazhkin V, Kondrin M, Bogdan A 2010 *Fullerenes, Nanotubes, and Carbon Nanostructures***18** 406
[7] Yagafarov O, Gromnitskaya E, Lyapin A, Brazhkin V 2010 *J. Phys.: Conf. Ser* **215** 012054
[8] Brazhkin V, Gromnitskaya E, Danilov I, Katayama Y, Lyapin A, Popova 2015 *Mater. Res. Express***2**, 025201
[9] Stal'gorova O, Gromnitskaya E, Dmitriev D and Voronov F 1996 *Inst. Exp. Tech.* **39** 880
[10] Reiser A, Kasper G, Hunklinger S 2005 *Phys. Rev.* B**72** 094204
[11] Alexandrov I, Goncharov A, Zisman A, Stishov S 1987 *Sov. Phys. JETP***66** 384

IOP Conf. Series: Journal of Physics: Conf. Series **950** (2017) 042054 doi:10.1088/1742-6596/950/4/042054

Investigation of magnetic and transport properties of $La_{0.8}Sm_{0.2}O_{0.5}F_{0.5}BiS_2$ superconductor under external hydrostatic pressure

G Kalai Selvan[a*] and S Arumugam[b]

[a,b] Centre for High Pressure Research, School of Physics, Bharathidasan University, Tiruchirapalli 620024, India

Email: kalai71309@gmail.com

Abstract. The dc electrical resistivity and magnetization measurements as a function of temperature 1.8-300 K were formed at external hydrostatic pressures on normal and superconducting state of new polycrystalline BiS_2 based layered $La_{0.2}Sm_{0.8}BiS_2O0_{0.5}F_{0.5}$ superconductor. In situ, electrical resistivity measurements, At ambient pressure the superconducting transition temperature (T_c) at ~3 K which are enhanced substantially under external pressure to 10.3 K at 1.74 GPa with positive pressure coefficient, dT_c/dP= 4.19 K/GPa. Further increasing pressure T_c has been decreased upto 6.2 K at 1.96 GPa with negative pressure coefficient dT_c/dP= -2.09 K/GPa. From the normal state resistivity exhibits the strong suppression of the semiconducting behavior in $La/SmO_{0.5}F_{0.5}BiS_2$ compounds suggests that the Fermi surface is in the vicinity of some instability. We report here the enhancement of T_c from 2.83 to 5.86 K with small increase in pressure upto 0.8 GPa with similar pressure coefficient of dT_c/dP= 3.78 K/GPa. However, it should be noted that either chemical substitution or physical pressure on these systems show a dome shaped T_c variation which means that the enhancement may not be achieved for higher doping concentration and (or) pressure s beyond an optimum level. In our results, exhibits which higher T_c may be achieved lies around 1.74 GPa. Above this pressure the T_c may decrease and exhibit a dome shaped behavior similar to other rare –earth systems of BiS_2 and Fe-based superconductors. Substantial enhancement of superconductivity under moderate pressure in studied new BiS_2 based superconductors call for the attention of condensed matter physics community.

1. Introduction

In the past couple of decades, the discovery of a serious of new superconducting compounds in Fe based pnictides $LaFeAsO_{1-x}F_x$ shows the T_c at 26 K and has renewed interest in superconductivity on correlated electron systems in condensed matter physics [1]. The chemical doping and external pressure are two key parameters, which can alter the T_c and innovate the new superconductors. Subsequently, Superconductivity with T_c = 8.6 K has recently been reported in the layered compound $Bi_4O_4S_3$ [2,3]. Following this report, similar 1111 family pnicitde structured other BiS_2-based superconductors including rare earth doped $LnO_{1-x}F_xBiS_2$ (Ln = La, Ce, Pr, Nd, Yb), $La_{1-x}M_xOBiS_2$ (M = Ti, Zr, Hf, Th) and $Sr_{1-x}La_xFBiS_2$, were found with a T_c as high as 10.6 K have been synthesized and studied [4-16]. It may be noted that Sm/Gd-based BiS_2 (which shows high T_c in oxypnictides) has doesn't exhibits superconductivity, while, both x=0 and 0.5 in $SmO_{1-x}F_xBiS_2$ compounds are paramagnetic, exhibiting semiconducting behavior down to 2 K [17], even though the smaller rare-earth ion Yb- based compound is superconducting. Evidently, these compounds share a common BiS_2 layer, which serves as the basic building blocks for this new superconducting family consists of a stacking of alternating BiS_2 bilayers and spacer layers. Quickly followed is the theoretical work based on the first principles band structure calculations [18], which predicts that the dominating bands for the electron conduction as well as for the superconductivity are derived from the Bi *6px* and *6py* orbits. Electrical resistivity and magnetic measurements under pressure revealed that the T_c of BiS_2 based family was sensitive to application of pressure and can be significantly enhance as observed in the cuprate and Fe-based family. Also the same could effectively alter the lattice constants through bond lengths and angles that would certainly affect the electronic and magnetic correlations of a superconductor. In addition to the chemical pressure (La/Sm substitution in pnictide compounds), application of external pressures also played an important role and enhanced the T_c of LaFeAsO/F from 26 K to 43 K (dT_c/dP = 4.25 K/GPa) [19]. Further, a maximum record of pressure dependence of T_c (+9.1 K/GPa) has been reported in FeSe superconductors [20].

The structural changes caused by the application of external hydrostatic pressure results in a suppression of the semiconducting behavior and enhance T_c of $LaO_{0.5}F_{0.5}BiS_2$ to 10.1 K (6.1 K/GPa), $PrO_{0.5}F_{0.5}BiS_2$ to 7.6 K (2.5 K/GPa), $NdO_{0.5}F_{0.5}BiS_2$ to 6.4 K (1.56 K/GPa) and 6.7 K for $CeO_{0.5}F_{0.5}BiS_2$ (2.2 K/GPa) and further increasing pressure T_c decreases slowly with a slop of -0.30 K/GPa in the same way as in layered pnictides and chalcogenides. Here, we report the variation of T_c of the $La_{0.8}Sm_{0.2}O_{0.5}F_{0.5}BiS_2$ samples at ambient and under external hydrostatic pressure studied with dc magnetic susceptibility (0.8 GPa) and electrical resistivity (300-3.5 K) measurements up to ~ 2 GPa. From our results show that optimal doped sample enhanced T_c ~ 10.3 K at an applied pressure 1.74 GPa was found, in comparison with of T_c ~ 3.2 K at ambient pressure. The main goal of this paper is to investigate how the T_c is affected by the external pressure in transport and magnetic measurements in the titled compounds.

2. Experimental

The bulk polycrystalline samples of series $La_{1-x}Sm_xO_{0.5}F_{0.5}BiS_2$ (x=0, 0.2, 0.4, 0.6 and 0.8) were synthesized by solid state reaction route via vacuum encapsulation and characterized as described by Thakur *et al* [17]. The dc magnetization studies of $La_{0.8}Sm_{0.2}O_{0.5}F_{0.5}BiS_2$ samples at ambient pressure and applied pressures up to ~1 GPa by a piston-cylinder pressure cell (Easy Lab Technologies Mcell 10) were performed down to 2 K in an applied magnetic field of 10 Oe zero-field cooling with a PPMS-VSM option installed (Quantum Design, USA). The applied pressure in the cell was calculated by the change of superconducting transition temperature of pure Sn (99.99%). A mixture of Fluorinert FC 70 and FC 77 (1:1) was used as pressure transmitting medium. The pressure dependent resistivity measurements were performed four probe techniques in a CCR-VTI system under pressure were done in the temperature range 4–300 K. The pressure up to ~3 GPa was generated using a hybrid double-cylinder pressure cell. The Daphne 7474 was used as a pressure transmitting medium and the actual pressure inside the cell was calculated using a calibration curve that was previously obtained from the fixed-pressure points of bismuth [21].

3. Results and Discussion

Enhancement in superconductivity has been achieved by means of chemical pressure by Sm doping at La site in $La_{1-x}Sm_xO_{0.5}F_{0.5}BiS_2$ (x=0,0.2,0.4,0.6,0.8 and 1) [17]. The high pressure studies of $LaO_{0.5}F_{0.5}BiS_2$ have been already reported by Kotegawa et al. As follows, we have chosen lower and higher samarium doped samples such as $La_{1-x}Sm_xO_{0.5}F_{0.5}BiS_2$ (x=0.2 and 0.8) for our high pressure studies. The temperature dependent electrical resistivity with applied pressure up to ~2 GPa for $La_{0.8}Sm_{0.2}O_{0.5}F_{0.5}BiS_2$ superconductor is shown in figure.1 with data is recorded from 300 to 2 K. Without applied pressure the normal state resistivity shows the semiconducting behavior down to the onset superconducting transition temperature T_c^{onset} in insert figure.1. The onset T_c is determined as an intersection of two extrapolated lines drawn through 90% and 10% transition curve just above the T_c. With an increase in applied pressure the normal state resistivity is though suppressed but yet exhibits semiconducting behavior. figure. 1(b) shows the expanded part in the temperature range of 2 to 20 K, to clearly visualize the T_c of $La_{0.8}Sm_{0.2}O_{0.5}F_{0.5}BiS_2$ compound for the various applied pressure. At zero pressure $T_c(\rho=0)$ is 3.2 K and which is through sudden increases to 6.8 K at 0.35 GPa pressure with broadening of the superconducting transition width. Unfortunately, the CCR-VTI low temperature system goes up to 3.5 K, we could not able to get the clear T_c^{zero} at this pressure range up to 1.62 GPa. Interestingly for the applied pressure of 1.74 GPa the sudden increases of T_c^{onset} to 10.3 K, the T_c^{zero} is increased sharply to 9.2 K, with relatively sharper superconducting transition in figure.2(left). The highest pressure coefficient dT_c^{onset}/dP for $La_{0.8}Sm_{0.2}O_{0.5}F_{0.5}BiS_2$ is estimated to be around ~4.1 K/GPa at 1.74 GPa. Further, increasing pressure T_c^{onset} get sudden decreasing up to 6.2 K at 1.96 GPa with pressure coefficient of -2.09 K/GPa. On the other hand, the normal state resistivity keep on decreasing with applied pressures up to 1.74 GPa. The sharp increase in T_c under hydrostatic pressure of just around 1 GPa may happen due to structural phase transformation being evident recently in these systems [22, 23]. Also, there is a possibility of the strong electron correlation in BiS_2 superconducting layers being enhanced under pressure. And, the clear T_c trend compared with observed from M vs T measurements under pressure.

A clear onset of T_c was observed at optimal doped $La_{0.8}Sm_{0.2}O_{0.5}F_{0.5}BiS_2$ (0 GPa in figure.1) at ~ 2.8 K by ZFC method (H=10 Oe) corresponding to the superconductivity in this samples. Temperature dependence of resistivity at ambient pressure for $La_{0.8}Sm_{0.2}O_{0.5}F_{0.5}BiS_2$ is shown in the figure. 2. This significant enhancement in T_c could be attributed to the increase in chemical pressure via substitution of smaller Sm ions at the La site. The T_c observed by magnetization is slightly lower than the resistivity superconducting transition temperatures, due to the in homogeneity of the polycrystalline sample. Next we discuss the temperature dependent magnetization for $La_{0.8}Sm_{0.2}O_{0.5}F_{0.5}BiS_2$ compounds under different applied pressure in figure. 2(right). We found that the initially applied hydrostatic pressure, the value of ambient T_c monotonically increases from ~2.8 K (0 GPa) to 5.6 K at 0.8 GPa for $La_{0.8}Sm_{0.2}O_{0.5}F_{0.5}BiS_2$. The insets display the pressure effect on T_c of the magnetization curves for enlarged temperature range from 2 K to 10 K. Surprisingly, it is the first report of magnetization under pressure measurements. The variation of T_c with respect to pressure, the rate of pressure coefficient of (Sm=0.2) optimal doped sample ~3.78 K/GPa under pressure up to 0.8 GPa.

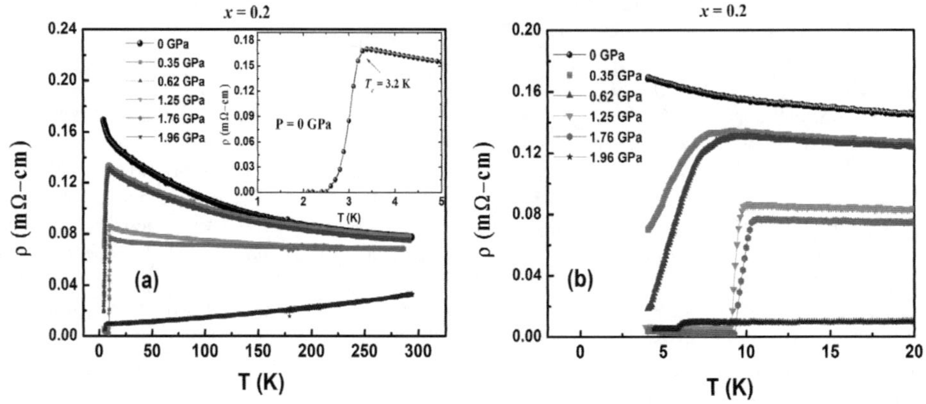

Figure.1. Temperature dependence of the resistivity of $La_{0.8}Sm_{0.2}O_{0.5}F_{0.5}BiS_2$ at various pressures. (a) ρ vs T at ambient pressure, inset shows marking the onset of T_c. (b) An enlarged view of the T_c variation in the low temperature region.

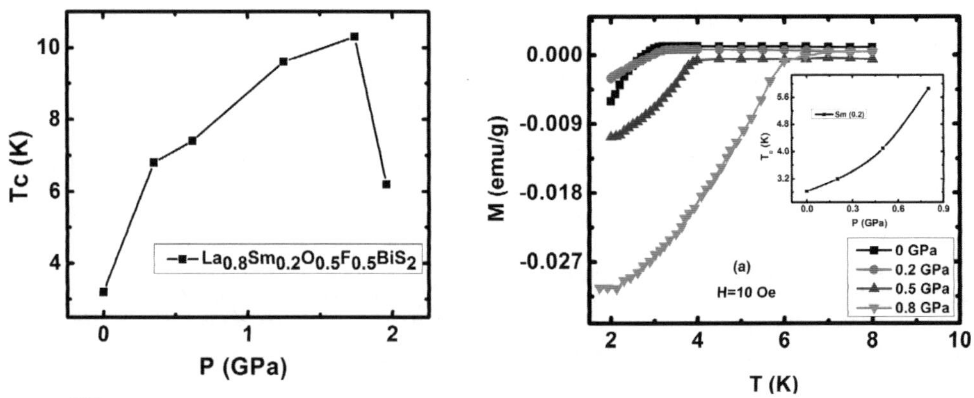

Figure.2. T_c^{onset} vs pressure plots for $La_{0.8}Sm_{0.2}O_{0.5}F_{0.5}BiS_2$ compound (left). The dc magnetization recorded at 10 Oe as a function of temperature for different applied pressures of $La_{0.8}Sm_{0.2}O_{0.5}F_{0.5}BiS_2$ superconducting transition in the low temperature range. Inset of the right side figure show the change in T_c as a function of pressure in the magnetization measurements.

Interestingly, the first principal calculations have suggested that there could be either insulator/semiconductor to metal transition accompanied with superconductivity at low temperature under pressure in the systems [24,25]. The onset T_c initially shows steep increases; however, it gradually decreases above ~ 1.7 GPa. The slope of above 1.7 GPa is ~ - 2.09 K/GPa. The Highest T_c obtained at 1.74 GPa is achieved just after a large suppression of the semiconducting behavior, and T_c monotonically decreases upon approaching the metallic region when the both energy gaps are overlapped. It is possible that under pressure S-Bi-S bond angle along with inter-atomic distances do change, which in turn affect the charge density at Fermi surface and hence a clear change in normal state electrical transport.

4. Conclusions

In conclusion, we reported the impact of hydrostatic pressure on the superconductivity of new samarium doped BiS_2 based new superconductors $La_{0.8}Sm_{0.2}O_{0.5}F_{0.5}BiS_2$. It is found that the T_c of 3.2 K is increased fourfold to above 10.3 K under just above ~1.5 GPa for optimal doped compound with normal state semiconductor to metallic transformation as well. However, at present, the precise mechanism driving the enhancement of T_c with pressure is unknown. Theoretical and high pressure X-ray diffraction experiments under on the BiS_2 compound strictly need to determine whether the pressure-induced enhancement of T_c and the suppression of semiconducting behavior are related to a structural transition.

Acknowledgements

The author S. Arumugam wishes to thank DST, (SERB, TSDP, FIST) BRNS, DRDO, CEFIPRA and UGC for the financial support. G. Kalai Selvan would like to thank UGC- BSR- RFSMS- SRF for the meritorious fellowship(Govt. of India).

References

[1] Kamihara Y, Watanabe T, Hirano M and Hosono H 2008 *J. Am. Chem. Soc.* **130** 3296
[2] Mizuguchi Y, Fujihisa H, Gotoh Y, Suzuki K, Usui H, Kuroki K, Demura S, Takano Y, Izawa H and Miura O 2012 *Phys. Rev. B.* **86** 220510
[3] Singh S K, Kumar A, Gahtori B, Kirtan S, Sharma G, Patnaik S and Awana V P S 2012 *J. Am .Chem. Soc.* **134** 16504
[4] Li B, Xing Z W and Huang G Q 2013 *Europhys. Lett.* **101**, 47002
[5] Jha R, Kumar A, Singh S K and Awana V P S 2013 J. Sup.Novel Mag. **26** 499
[6] Jha R, Kumar A, Singh S K and Awana V P S 2013 *J. Appl. Phys.* **113** 056102
[7] Deguchi K, Mizuguchi Y, Demura S, Hara H, Watanabe T, Denholme S J, Fujioka M, Okazaki H, Ozaki T, Takeya H, Yamaguchi T, Miura O and Takano Y 2013 *Europhys. Lett.* **101** 17004
[8] Kotegawa H, Tomita Y, Tou H, Izawa H, Mizuguchi Y, Miura O, Demura S, Deguchi K and Takano Y 2012 *J. Phys. Soc. Jpn.* **81** 103702
[9] Awana V P S, Kumar A, Jha R, Kumar S, Pal A, Shruti, Saha J and Patnaik S 2013 *Solid State Commun.* **157** 21
[10] Demura S, Mizuguchi Y, Deguchi K, Okazaki H, Hara H, Watanabe T, Denholme S J, Fujioka M, Ozaki T, Fujihisa H, Gotoh Y, Miura O, Yamaguchi T, Takeya H and Takano Y 2013 *J. Phys. Soc. Jpn.* **82** 033708
[11] Mizuguchi Y, Demura S, Deguchi K, Takano Y, Fujihisa H,. Gotoh Y, Izawa H and Miura 2012 O *J. Phys. Soc. Jpn.* **81** 114725
[12] Xing J, Li S, Ding X, Yang H and Wen H H 2012 *Phys. Rev. B.* **86** 214518
[13] Yazici D, Huang K, White B D, Chang A H, Friedman A J and Maple M B 2012 *Philos. Mag.* **93** 673
[14] Lin X, Ni X, Chen B, Xu X, Yang X, Dai J, Li Y, Yang X, Luo Y, Tao Q, Cao G and Xu Z 2013 *Phys. Rev. B* **87** 020504
[15] Yazici D, Huang K, White B D, Jeon I, Burnett V W, Friedman A J, Lum I K, Nallaiyan M, Spagna S and Maple M B 2013 *Phys. Rev. B.* **87** 174512
[16] Maziopa A K, Guguchia Z, Pomjakushina E, Pomjakushin V, Khasanov R, Luetkens H, Biswas P, Amato A, Keller H and Conder K 2014 *J. Phys.: Condens. Matter* **26** 215702
[17] Thakur G, Kalai Selvan G, Zeba H, Gupta L, Samal S, Arumugam S and Ganguli A 2015 *Journal of Inorg chem.* **54** 1076
[18] Usui H, Suzuki K, Kuroki K 2012 *Phys Rev B.* **86** 220501
[19] Takahashi H 2008 *Nature.* **453** 376.
[20] Medvedev S 2009 *Nature Mater.* **8** 630
[21] Kalai Selvan G, Bhoi D, Arumugam S, Midya A and Mandal P 2015 *Supercond Sci Technol.* **28** 015009
[22]Tomita, Ebata M , Soeda H, Takahashi H, Fujihisa H, Gotoh Y, Mizuguchi Y, Izawa H, Miura O, Demura S, Deguchi K and Takano Y 2014 *J.Phys.Soc.Jpn.***83** 63704
[23] Kajitani J, Hiroi T, Omachi A, Miura O and Mizuguchi Y *arXiv*:1404.6361
[24] Jha R, Tiwari B and Awana V P S 2015 *J. Appl. Phys.* **117** 013901
[25] Kalai Selvan G, Kanagaraj M, Esakki Muthu S, Jha R, Awana V P S and Arumugam S 2013 *Phys. Status Solidi RRL* **7** 510.

IOP Conf. Series: Journal of Physics: Conf. Series **950** (2017) 042055 doi:10.1088/1742-6596/950/4/042055

Forcing Cesium into Higher Oxidation States Using Useful hard x-ray Induced Chemistry under High Pressure

D Sneed[1], M Pravica[1], E Kim[1], N Chen[3], C Park[2], M White[1]

[1] Department of Physics and Astronomy, and High Pressure Science and Engineering Center, University of Nevada Las Vegas, 4505 Maryland Parkway, Las Vegas, NV

[2] HPCAT, Geophysical Laboratory, Carnegie Institution of Washington, 9700 South Cass Ave., 434E, Argonne, IL, 60439, USA

[3] Canadian Light Source Inc., 44 Innovation Boulevard, Saskatoon, SK S7N 2V3, Canada

Daniel Sneed: sneedd3@unlv.nevada.edu, Michael Pravica: pravica@physics.unlv.edu

Abstract. This paper discusses our attempt to synthesize higher oxidation forms of cesium fluoride by pressurizing cesium fluoride in a fluorine-rich environment created via the x-ray decomposition of potassium tetrafluoroborate. This was done in order to confirm recent theoretical predictions of higher oxidation forms of CsF_n. We discuss the development of a technique to produce molecular fluorine *in situ* via *useful hard x-ray photochemistry*, and the attempt to utilize this technique to form higher oxidation states of cesium fluoride. In order to verify the formation of the novel stoichiometric species of CsF_n. X-ray Absorption Near Edge Spectroscopy (XANES) centered on the cesium K-edge was performed to probe the oxidation state of cesium as well as the local molecular coordination around Cs.

1. Introduction

Recent theoretical work has predicted that cesium may undergo inner-shell (p-shell) bonding with fluorine under high pressure, forming higher oxidation species CsF_n (n=2-6) [1]. Though inner shell electrons are normally inaccessible for bonding under ambient conditions, pressure can broaden and increase the energy of the 5p shell of cesium, allowing the electrons to interact with a highly oxidizing atom such as fluorine. Thus, though cesium has only one valence electron in the 6s shell to contribute to bonding (giving it a formal oxidation state of 1+), it has been predicted that the highly electronegative fluorine will begin to pull electrons out of the 5p shell causing cesium and fluorine to undergo xenon like bonding under pressures greater than 10 GPa[1].

The primary difficulty in synthesizing new CsF_n compounds is that fluorine is difficult to work due to its extremely reactive nature caused by its high electronegativity. Recent developments in *useful hard x-ray induced chemistry* [2-7] offer a method to produce simple molecules for reaction *in situ* within a diamond anvil cell by decomposing relatively safe, inert, easy-to-handle solid crystalline powders using the highly ionizing and highly penetrating characteristics of hard x-rays (>7keV). Using this technique, multiple species of simple molecules have been formed, including H_2, N_2, O_2, Cl_2, and most recently F_2, all *in situ* within a diamond anvil cell [2-7]. It has also been shown that highly controlled chemistry can be achieved at high pressures via the reaction to produce molecular hydrogen (formed by the x-ray induced decomposition of ammonia borane) and molecular oxygen (formed by the x-ray induced decomposition of potassium perchlorate), which then react to form water; all *in situ* within a diamond anvil cell at high pressure [5]. OF_2 has also been formed by similar methods [7]. This experiment, combined with the successful formation of molecular fluorine via *useful hard x-ray induced chemistry* has given us a novel means to release F_2 and perform fluorine chemistry under extreme conditions all *in situ*.

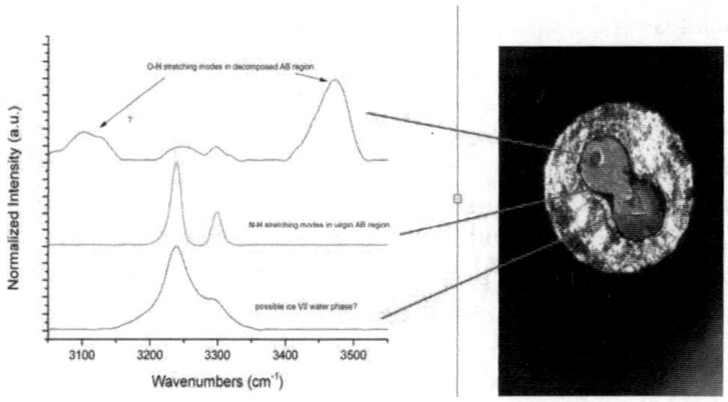

Figure 1: Raman spectra of segregated mixture of ammonia borane and potassium perchlorate showing the H_2O vibrational modes post x-ray irradiation at 5 GPa [fig. 2 of reference 5]

Here we report on experiments that attempt to combine the recent developments of useful x-ray induced chemistry with the theoretical prediction presented by Miao [1] in order to synthesize these higher oxidation states of CsF_n under high pressures. Through the use of Raman spectroscopy to verify the production of molecular fluorine, and cesium K-edge XANES to probe the valence state of cesium as well as local atomic coordination around the cesium atom, we demonstrate the synthesis of these novel CsF_n stoichiometric species via *useful hard x-ray induced chemistry* confined within a diamond anvil cell.

2. Experiment

It has been previously demonstrated that the formation of F_2 can be achieved via the x-ray decomposition of C_6F_{14} [4] and potassium tetrafluoroborate [7], verifying that both compounds could be used as sources for molecular fluorine using *useful x-ray induced chemistry*. The production of molecular fluorine was verified via Raman spectroscopy [4, 7]. We chose to use powdered KBF_4 for ease of mixing with CsF.

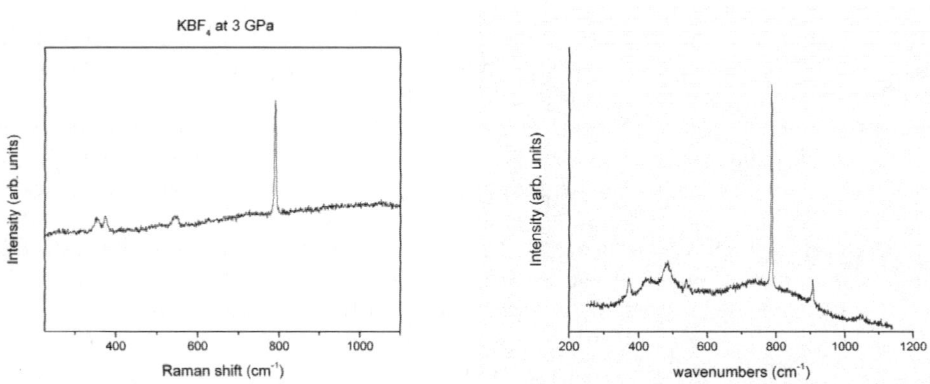

Figure 2: (left) Raman spectra of unirradiated KBF_4, (right) Raman spectra of irradiated KBF_4 showing the fluorine vibron at ~903 cm^{-1} [fig. 2 of reference 7]

The first stage of our experiment consisted of a high pressure X-ray Absorption Near Edge Spectroscopy (XANES) experiment. A panoramic-style diamond anvil cell with a Be gasket pre-indented to 50 μm thickness and subsequently laser-drilled to make a 120 μm hole forming the sample chamber. The hole was loaded with a 1:1 ratio by mass mixture of KBF_4 (Sigma-Aldrich >99%) and CsF (Sigma-Aldrich 99%) under argon. A small 20 μm, thermally-relieved ruby sphere was placed in with the sample for pressure measurements. The cell was sealed and pressurized to ~3 GPa in order to seal the mixture and then irradiated at the 16 BM-B beamline of the Advanced Photon Source of Argonne National Laboratory. "White" x-ray radiation in the 10 – 70 KeV energy range was used to decompose the KBF_4 to produce molecular fluorine, which would be available to react with CsF to potentially synthesize the novel CsF_n stoichiometric species. The sample mixture was placed into the path of an unfocused 80 μm by 80 μm x-ray beam for approximately 8 hours.

The DAC was then pressurized to 16 GPa and placed in the 16 BM-D beamline at the Advanced Photon Source of Argonne National Laboratory. The cell was placed in radial transmission geometry to the beam path so that absorption measurements through the beryllium gasket containing the sample could be made. A focused 8 μm by 8 μm monochromatic x-ray beam centered at the 35.985 keV Cs K-edge was used to probe the sample. All measurements were made in transmission mode. The monochrometer was scanned from 400 eV below the edge to 800 eV above the edge. The energy resolution of the monochromator was 8-10 eV. Photon counting was performed pre- and post- sample using a mixture of argon and xenon gas-filled ion chambers for incident and transmitted x-ray intensity measurements. Measurements were conducted at 16, 53, and 79 GPa. Seven scans were performed at the three different pressures for signal averaging the spectra.

The second stage of our experiments consisted of performing a high pressure XANES experiment of pure CsF for comparison purposes. A panoramic-style DAC with a Be gasket was pre-indented to 50 μm thickness and a 120 μm hole was laser drilled and then loaded with CsF (Sigma-Aldrich 99%) under argon. A small 20 μm, thermally relieved ruby sphere was placed in with the sample for pressure measurements. The cell was pressurized to 2.7 GPa in order to ensure that the sample was in its low pressure face-centered cubic (fcc) NaCl structure [8]. The experimental geometry and instrumental setup were identical to the first experiment. Measurements were made at 2.7, 12.6, 25, 43, 55, and 78 GPa respectively. Three scans each were performed for each different pressures and signal averaged.

3. Results and discussion

The XANES spectrum of the irradiated mixture displays a demonstrable change in the near edge spectrum of the 79 GPa measurement, which can be seen in the left pane of figure 3; this is indicative of a change to the local molecular structure around the Cs atom. The control measurement of the pure unirradiated CsF did not show the same change to the near edge structure, as shown in the right pane of figure 3, which is to be expected as the body-centered cubic (bcc) CsCl structure of CsF is predicted to be stable up to 200 GPa [1]. CsF undergoes one known structural transition from the low pressure face-centered cubic (fcc) NaCl structure to the high pressure bcc CsCl structure at approximately 4.8 GPa [8], which can be seen to occur in Figure 3 between 2.7 and 12.6 GPa by the down shift in energy of the near edge features at 36025 eV and 36075 eV, which is indicative an increase of the Cs-F bond length, and is to be expected in the transition from the fcc to bcc structure in CsF.

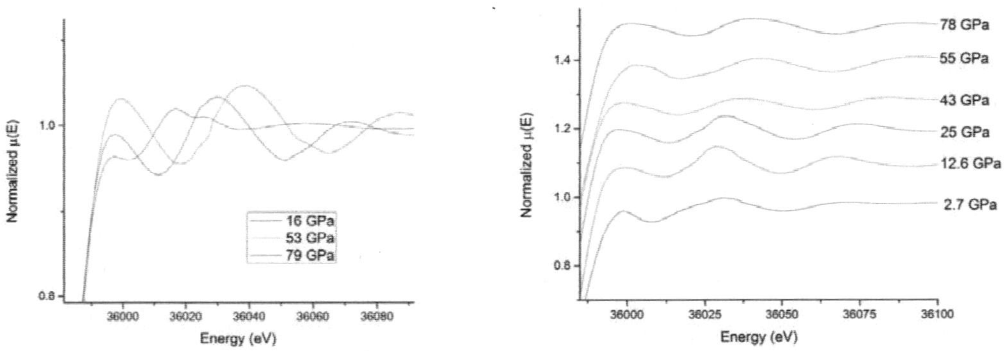

Figure 3: (Left) XANES spectra of the irradiated mixture of $CsF+KBF_4$ up to 79 GPa . (Right) XANES spectra of pure CsF up to 78 GPa

A similar argument can be made for the near edge spectral changes of the irradiated mixture at 79 GPa (left panel of Fig. 3), though the different features of the near edge spectrum change in a much more complicated manner. When comparing the spectrum at 53 GPa to the spectrum at 79 GPa in the left panel of Figure 3, it can be seen that the features at 36030 eV and 36090 eV can be seen to decrease in energy to 36015 eV and 36080 eV respectively, while a new shoulder feature is seen to emerge at 36025 eV. This, along with the changes in relative intensities, suggests that there is far more occurring within the sample that requires further experimentation to verify and properly classify. It should also be noted that visible changes also occurred within the sample chamber of the x-ray photochemistry experiment that did not occur in the pure CsF control experiment. The sample went from being a white semi-translucent color, to dark brown, providing further evidence for a chemical change in the sample.

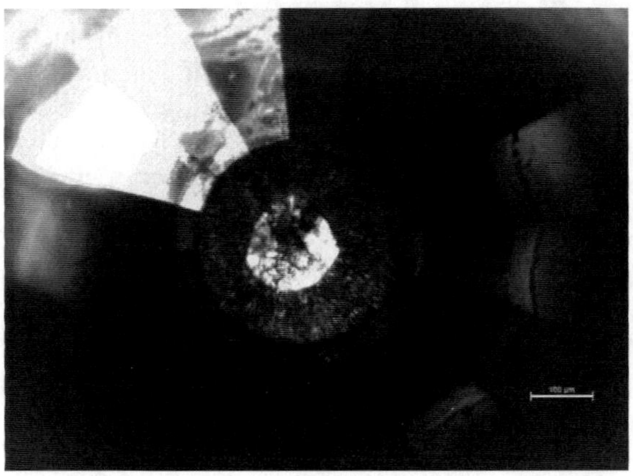

Figure 4: Image of post irradiated mixture of CsF+KBF4 at 79 GPa

4. Conclusions

An attempt to synthesize novel stoichiometric species of CsF_n (n=2-6) was made by pressurizing a mixture of cesium fluoride and KBF_4 that was irradiated in order to create a fluorine-rich atmosphere via *useful hard x-ray photochemistry*. XANES was utilized to probe potential oxidation state changes of the cesium atom as well as changes in the local molecular structure in order to verify the production of new stoichiometric species of cesium fluoride in higher oxidation states. Our experiments show the potential formation of a new molecular structure, though further x-ray diffraction and XANES experiments along with theoretical study using the density-functional theory are neccessary in order to verify and properly classify these changes. Though we cannot definitively state that we have formed a new species of CsF_n, these experiments provide tantalizing evidence of the possibility that we have formed a new species of CsF_n and give us strong impetus to investigate this further.

5. Acknowledgments

We thank Yonggang Wang with HPCAT for all of his help with the experiments. We acknowledge support from the Department of Energy National Nuclear Security Administration under Award Number(s) DE-NA0000979. and from the DOE Cooperative Agreement No. DE-FC08-01NV14049 with the University of Nevada, Las Vegas.. HPCAT operation is supported by DOE-NNSA under Award No. DE-NA0001974 and DOE-BES under Award No. DE-FG02-99ER45775. HPCAT is supported by DOE-BES, DOE-NNSA, NSF, and the W.M. Keck Foundation. Use of the Advanced Photon Source is supported by the DOE Office of Science, Office of Basic Energy Sciences, under Contract No. DE-AC02-06CH11357

References

[1.] Miao M Caesiumin high oxidation states and as a p-block elemen *Nature Chem.* Vol 5 pp 846-852 (2013) doi: http://dx.doi.org/10.1038/nchem.1754

[2.] Pravica M Popov D Sinogeikin S Sneed D Guardala G Smith Q X-ray induced mobility of molecular oxygen at extreme conditions *Applied Phys. Letters (2013)* http://dx.doi.org/10.1063/1.4836475

[3.] Pravica M Sneed D Smith Q Bai L High pressure X-ray photochemical studies of carbon tetrachloride: Cl2 production and segregation *Chem. Phys. Letters* Volume 590 18 December (2013) pp 74-76 doi: http://dx.doi.org/10.1016/j.cplett.2013.10.056

[4.] Pravica M Sneed D White M Wang Y Note: Loading method of molecular fluorine using x-ray induced chemistry *Review of Scientific Instruments*, 85, 086110 (2014), doi: http://dx.doi.org/10.1063/1.4893384

[5.] Pravica M Sneed D White M Wang Y Communication: A novel method for generating molecular mixtures at extreme conditions: The case of hydrogen and oxygen *J. Chem. Phys.* 141 091101 (2014) doi: http://dx.doi.org/10.1063/1.4894402

[6.] Pravica M Liu Y Bai L Hydrazine at High Pressures *Chem. Phys. Letters*, 555 pp 113-115 (2013) doi: http://dx.doi.org/10.1016/j.cplett.2012.10.079

[7.] Pravica M White M Wang Y A novel method for generating molecular mixtures at extreme conditions: The case of fluorine and oxygen Accepted for publication, *APS SCCM conference precedings* (2015).

[8.] Merrill Leo Behavior of the AB-type compounds at high pressures and high temperatures *Journal of Physical and Chemical Reference Data* 6 1205-1252 (1977) doi: http://dx.doi.org/10.1063/1.555565

IOP Conf. Series: Journal of Physics: Conf. Series **950** (2017) 042056 doi:10.1088/1742-6596/950/4/042056

The photochemical reaction of hydrocarbons under extreme thermobaric conditions

Aleksandr Serovaiskii [1,2,3], Anton Kolesnikov [1], Elena Mukhina [1,2], Vladimir Kutcherov [1,2]

[1] Department of Physics, Gubkin Russian State University of Oil and Gas, Moscow, Russia

[2] Department of Energy Technology, Royal Institute of Technology, Stockholm, Sweden

E-mail: alexandrserov@gmail.com

Abstract. The photochemical reaction of hydrocarbons was found to play an important role in the experiments with the synthetic petroleum conducted in Diamond Anvil Cell (DAC). Raman spectroscopy with a green laser (514.5 nm) was used for *in situ* sample analysis. This photochemical effect was investigated in the pressure range of 0.7-5 GPa, in the temperature interval from the ambient conditions to 450°C. The power of laser used in these experiment series was from 0.05 W to 0.6 W. The chemical transformation was observed when the necessary threshold pressure (~2.8 GPa) was reached. This transformation correlated with the luminescence appearance on the Raman spectra and a black opaque spot in the sample was observed in the place where the laser focus was forwarded. The exposure time and laser power (at least in the 0.1-0.5 W range) did not play a role in the 0.1-0.5 GPa range.

1. Introduction

Hydrocarbon photochemistry has received serious attention in the last few years due to its importance in the chemistry of cosmic water ices in space [1] and the higher layers of Jovian planets [2], [3]. High pressure studies have also touched on this problem. Interesting results were found for unsaturated hydrocarbons, especially butadiene. It was found that using laser radiation at 488 nm or 458 nm and 20 mW for 2-3 hours switched the dimerization pathway of the butadiene transformation to a selective polymerization [4]. While special experiments are still rare in which significant efforts are made to change the reaction pathways, the use of laser radiation as an analytical tool is a standard procedure.
Raman laser spectroscopy became a routine analytical technique for DAC experiments very long ago due to its high sensitivity to most components and relatively low diamond quality requirements. Heating from Raman laser is typically almost negligible, even for rubies, which have a high absorption rate. A couple degrees of temperature rise during ruby irradiation with a 150 mW (~75 mW) green laser (532.8 nm) was found to occur during the investigation of ethane (which will be published elsewhere).
At the same time, if a chemical reaction takes place, laser radiation should accounted for, especially if thermobaric stability limits are being investigated.

Acetylene was studied by [5] in which a few milliwatts of laser light was used above 3 GPa. Exceeding this low threshold led to polymerization. Below this pressure and the corresponding ambient

IOP Conf. Series: Journal of Physics: Conf. Series **950** (2017) 042056 doi:10.1088/1742-6596/950/4/042056

temperature orthorhombic phase, the cubic phase was stable with several tens of milliwatts of 530.9 and 647.1 nm Kr laser irradiation.

A uniform acetylene color change [6] due to the polymerization was observed at 3.5 GPa and ambient temperature inside an Inconel X-750 gasket between diamonds. Laser radiation with a 0.3 mm focus diameter radiated the entire sample hole. The effects of radiation on the polymerization products were investigated at -196°C using laser strengths of a few milliwatts at different wavelengths. No spectral change was observed with 676.4 nm and 647.1 nm wavelengths, while the background intensity increased with 476.2 nm.

[7] paid great attention to give away the photochemical and heating potential of the laser radiation of the benzene phase and chemical transformations. The Raman laser point was defocused, and the laser power was 20 mW (it was optimal to have Raman signals). To confirm the hypothesis concerning the benzene transformation line in the phase diagram found by laser testing, separate experiments were carried out without Raman laser radiation. The laser effect was found to shift the reaction pressure threshold from 23 to 16 GPa when the sample was radiated for 23 hours with 70 mW using a 458 nm laser [8].

[9] emphasized that it was the Raman laser that caused methane to change into graphite-like soot at 2.5-5.0 GPa, 639°C and a hard, clear solid film at 10-13 GPa, 675°C. Blue 488 nm and 458 nm lasers were used. Soot formation was observed using a 458 nm laser with a 40 mW power output. This formation occurred more quickly using a 488 nm laser at 200 mW. The paper did not describe whether the wavelength, power, or the combination of the two accelerated the transformation. The formation of a brown spot of soot correlated with the incident laser spot. It was observed that with color change it was not possible to find the soot signals. Therefore, there is a possibility that soot was not the first substance formed, though soot signals were observed after cooling the samples to the ambient temperature.

[10] did not mention the laser properties, but Raman spectroscopy was the only analytical technique used for the methane resistive heating experiments. The methane dissociation temperatures in their experiments were even lower than in [9]. The reactions were observed at 3.17 GPa and 520°C, 7.36 GPa and 450°C, as well as at 560°C and 0.86 GPa.

[11] did not observe a n-heptane reaction using a 532 nm laser at 120 mW power. The authors measured the laser power near the sample, and it appeared to be 30 mW. This difference between the output laser power value and the power that reached the sample is a typical issue in experiments where the optical pathway diminishes the laser power output. The question in this case is how much it diminishes. [11] reported no chemical changes in n-heptanes with pressures up to 16 GPa at ambient temperature. The same setup was used for two further studies [12], [13], and no photochemistry effects were noted for n-hexane and n-pentane. For pentane, 15 mW of laser power reached the sample at the highest pressure tested (17 GPa) and ambient temperature.

It is difficult to estimate the influence of a Raman laser on the transformation of samples in experiments in which infrared lasers were used to heat the sample. Raman laser irradiation of the same incident point of on the sample as the heating laser may have some influence, but this question has not been investigated thus far. In some papers, the type of Raman laser used was not mentioned [14], and in others[15], [16], no laser power information was presented.

In experiments in which the synthetic petroleum stability was investigated under the Earth's crust at thermobaric conditions (to be published elsewhere), the photochemical reaction was found to play an important role.

2. Experimental details

Synthetic petroleum synthesized in Fischer-Tropsch process (figure 1) and (table 1) was received from ZAO "Zeosit".

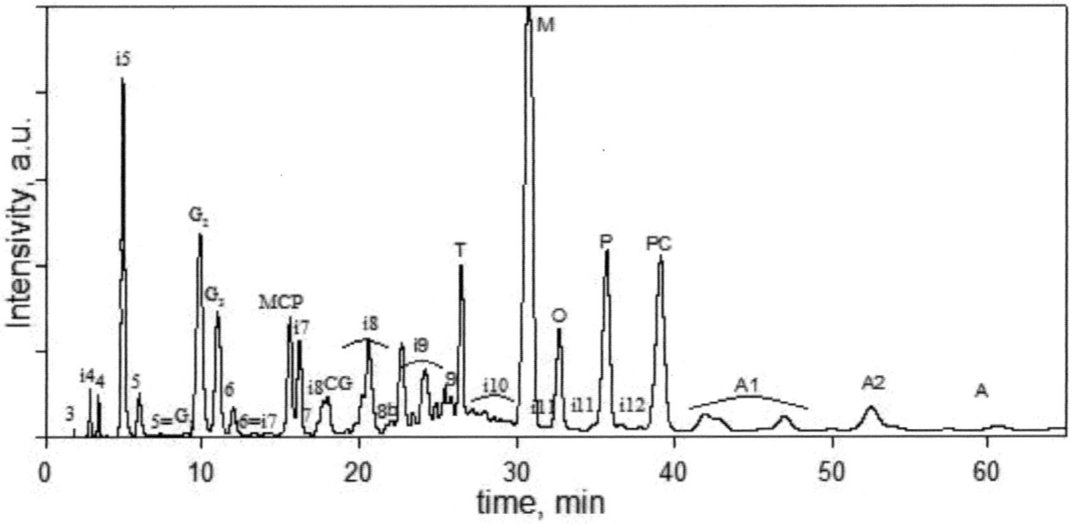

Figure 1. Chromatogram of synthetic petroleum using in Fischer-Tropsch process

Table 1. Components of synthetic petroleum using in Fischer-Tropsch process

Symbol	Compounds	Symbol	Compounds
3	propane C_3H_8	8	n-C_8, n-octane, C_8H_{18}
i4	i-C_4, i-butane, $CH(CH_3)_3$	MCG	methylcyclohexane
4	n-C_4, n-butane, C_4H_{10}	B	benzene, C_6H_6
4=	butylenes, C_4H_8	i9	i-C_9, i-nonanes
i5	i-C_5, i-pentane (2-methylbutane)	9	n-C_9, n-nonane, C_9H_{20}
5	n-C_5, n-pentane, C_5H_{12}	T	toluene, C_6H_5-CH_3
5=	pentenes, C5H10	i10	i-C_{10}, i-decanes
G	2,2-dimethylbutane	M	ethylbenzene+meta-xylene+para-xylene, C_6H_5-C_2H_5+ meta-C_6H_5-$(CH_3)_2$ + para-C_6H_5-$(CH_3)_2$
G1	2,3-dimethylbutane	i11	i-C_{11}, i-undecanes
G2	2-methylpentane	O	ortho-xylene, ortho-C_6H_5-$(CH_3)_2$
6	n-C_6, n-hexane, C_6H_{14}	P	ethyltoluens
6=	hexenes, C_6H_{12}	i12	i-C_{12}, i-dodecanes
i7	i-C_7, i-heptanes	PC	pseudocumene, 1,2,4-trimethylbenzene
MCP	methylcyclopentane	A1	ethylxylenes
7	n-C_7, n-heptane, C_7H_{16}	A2	durol, 1,2,4,5-thetramethylbenzene
i8	i-C_8, i-octanes	A	aromatic hydrocarbons C_{11+}
CG	cyclogexane, C_6H_{12}		

This synthetic petroleum was loaded in the Diamond Anvil Cells (DACs) with internal resistive heaters made of platinum wires. The temperature was measured by Pt/Pt-Rh thermocouples. Ruby and Sm:YAG were used as internal pressure and temperature sensors. Raman spectroscopy with a green Ar^+-laser (514.5 nm) was used for *in situ* sample analysis. The temperature increased slowly during the experiments (60-70°C/h). Periodically, the pressure was checked and corrected if necessary. The temperature range was 25-450°C, and the pressure range was 0.5-1.5 GPa. The details of the experiments are described in [17].

3. Results and discussion

A series of experiments with the synthetic petroleum was performed with starting pressures of 1, 3 and 5 GPa. Without laser radiation, the synthetic petroleum remained stable over the entire thermobaric range investigated in this study, from 25 to 450°C and 1-5 GPa. During the experiments, it was discovered that the hydrocarbons only reacted to the laser radiation at specific thermobaric conditions. This observation was noticed because the appearance of luminescence in the Raman spectra showed that the hydrocarbon peaks were no longer visible (figure 2 b,c).

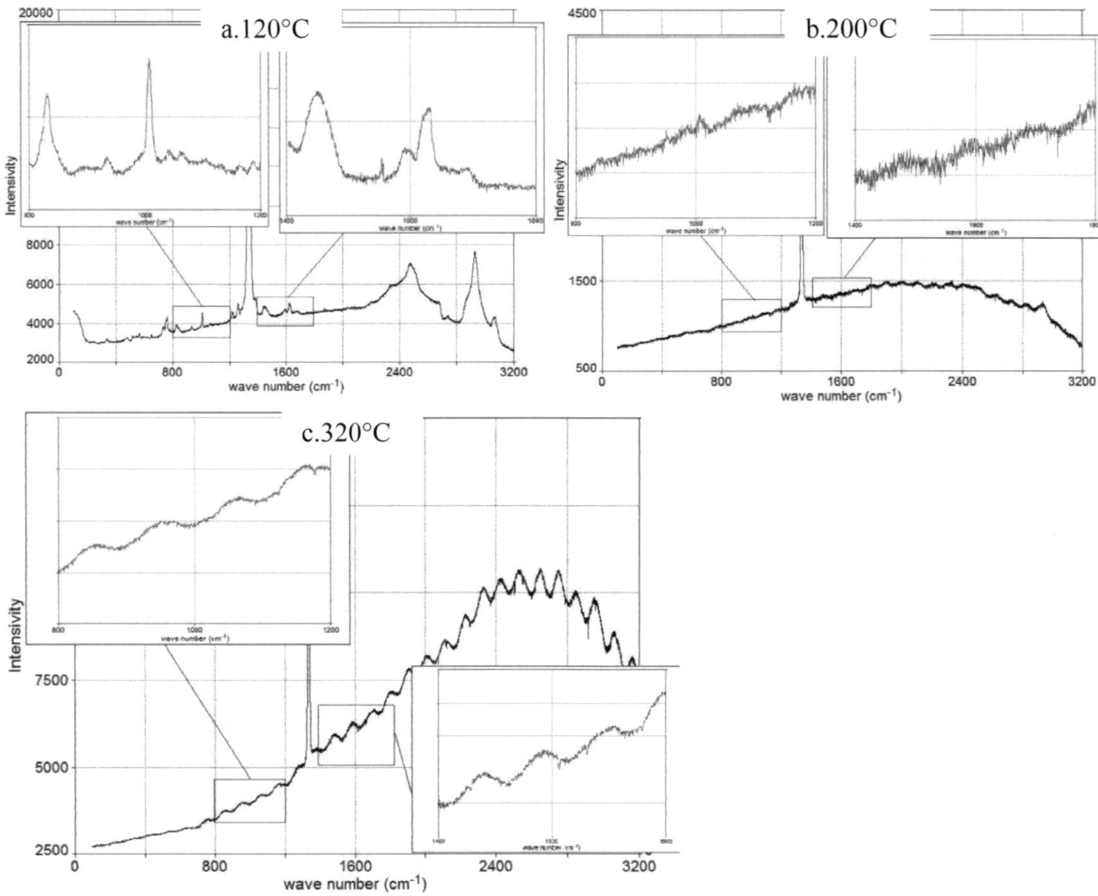

Figure 2. Fluorescence in Raman spectra during the heating of the synthetic petroleum at 2.8 GPa.

This fluorescence phenomenon was also observed by [16] when methane was heated above 2700°C without a coupler at 23.1 GPa. This was interpreted as the formation of ultra-dispersive diamonds.

A black opaque spot in the hydrocarbon sample was found by optical observations of the focal point of the laser on the sample. If the sample was a liquid at certain thermobaric conditions, the black spot began to dissolve in the sample after a certain amount of time, and in this case, the sample color did not change (figure 3).

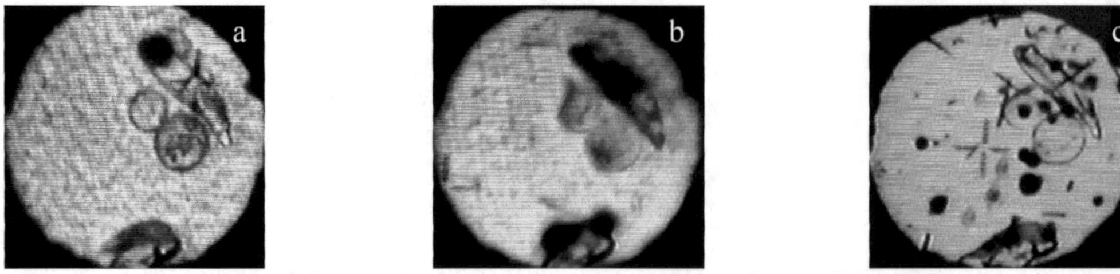

Figure 3. Reaction of the synthetic petroleum during heating at 170°C, 2.8 GPa in response to 0.5 W Raman Ar^+-laser radiation. a) Solid sample and black spot after one laser shot; b) brownish

IOP Conf. Series: Journal of Physics: Conf. Series **950** (2017) 042056 doi:10.1088/1742-6596/950/4/042056

area around Sm:YAG and ruby main spots during the dissolving process; c) fully liquefied sample and multiple black dots in radiated spots.

Laser induced transformations took place during loading runs at 100°C and 4.5 GPa (0.2 W laser power), 160°C and 5.0 GPa (0.15 W), and 170°C and 2.8 GPa (0.5 W). At the same time, with lower pressures, much higher temperatures were been reached without synthetic petroleum transformations: 220°C at 1.2 GPa (0.5 W), 320°C at 0.7 GPa (0.4 W), and 420°C at 1.15 GPa (0.5 W).

The initiation of a photochemical reaction in the sample depended on the pressure, temperature and laser power. The key factor was the pressure inside the cell. The photochemical reaction took place at very high pressures (10 GPa and more) even at an ambient temperature and with a low laser power (0.15 W). However, no reactions occurred at low pressures (less than 1.5 GPa) even at 450°C and with a laser power of 0.3-0.4 W.

The hydrocarbon photochemical reactions were more intense when the Raman laser was focused on ruby and Sm:YAG that was placed in the DAC to control the pressure and temperature. Both the ruby and Sm:YAG were more opaque and better absorbers of laser radiation. The background intensity of the spectra increased with each successive accumulation.

The time of exposure of the laser radiation was not found to play a role. If the threshold pressure was overcome, than even a second of radiation with 0.1 W power was enough to cause the reaction. On the contrary, at a pressure of 1.15 GPa, the sample was radiated with a 0.5 W laser for 2 hours without any sign of transformation, and the Raman spectra did not change.

4. Conclusions

During the investigation of the synthetic petroleum received from Fischer-Tropsch, laser radiation was observed to induce chemical transformations if the necessary threshold pressure (~2.8 GPa) was reached. The exposure time and laser power did not play a role in the 0.1-0.5 GPa range. The temperature appeared to be important, and its critical value changed with the pressure. More experiments are necessary to find a clear correlation between the pressure and critical temperature as well as to identify fluorescent reaction products.

5. Acknowledgment

The authors gratefully acknowledge the support of the Deep Carbon Observatory. A.S. would like to thank DAAD for the Ivan Gubkin Scholarship.

References

[1] Cook A M *et al* Photochemistry of Polycyclic Aromatic Hydrocarbons in Cosmic Water Ice: The Role of PAH Ionization and Concentration *The Astrophysical Journal* 2015 **799**(1) p. 14.

[2] Dobrijevic M *et al* Key reactions in the photochemistry of hydrocarbons in Neptune's stratosphere *Planetary and Space Science* 2010 **58**(12) p. 1555-1566.

[3] Gladstone G, Allen R M and Yung Y L Hydrocarbon photochemistry in the upper atmosphere of Jupiter *Icarus* 1996 **119**(1) p. 1-52.

[4] Citroni M *et al* Laser-induced selectivity for dimerization versus polymerization of butadiene under pressure *Science* 2002 **295** p. 2058-2058.

[5] Aoki K *et al* High-pressure Raman study of liquid and crystalline C_2H_2 *The Journal of Chemical Physics* 1988 **88**(8) p. 4565-4568.

[6] Aoki K *et al* Raman study of the solid-state polymerization of acetylene at high pressure *J. Chem. Phys.* 1988 **89** p. 529-529.

[7] Cansell F Fabre D and Petitet J P Phase transitions and chemical transformations of benzene up to 550 °C and 30 GPa *Journal of Chemical Physics* 1993 **99** p. 7300-7300.

[8] Ciabini L *et al* High pressure photoinduced ring opening of benzene *Phys Rev Lett* 2002 **88**(8) p. 085505.

[9] Culler T S and Schiferl D New chemical reactions in methane at high thermobaric conditions *Journal of Physical Chemistry* 1993 **97** p. 703-706.
[10] Chen J Y *et al* In Situ Raman Spectroscopy Study on Dissociation of Methane at High Temperatures and at High Pressures *Chinese Physics Letters* 2008 **25**(2) p. 780-782.
[11] Kavitha G and Narayana C Raman Scattering Studies on n-Heptane under High Pressure *The Journal of Physical Chemistry B* 2006 **110**(17) p. 8777-8781.
[12] Kavitha G and Narayana C Raman Spectroscopic Investigations of Pressure-Induced Phase Transitions in n-Hexane *The Journal of Physical Chemistry B* 2007 **111**(51) p. 14130-14135.
[13] Kavitha G and Narayana C Pressure-Induced Structural Transition in n-Pentane: A Raman Study *The Journal of Physical Chemistry B* 2007 **111**(25) p. 7003-7008.
[14] Hirai H *et al* Polymerization and diamond formation from melting methane and their implications in ice layer of giant planets *Physics of the Earth and Planetary Interiors* 2008 **174**(1-4) p. 242-246.
[15] Benedetti L R *et al* Dissociation of CH_4 at High Pressures and Temperatures : Diamond Formation in Giant Planet Interiors? *Science* 1999 **286** p. 100-102.
[16] Zerr A *et al* Decomposition of alkanes at high pressures and temperatures *High Pressure Research* 2006 **26**(1) p. 23-32.
[17] Cernik R J *et al* An in situ high pressure-high temperature powder diffraction study of the formation of a precursor phase of bismuth manganite *Ceramics International* 2010 **36**(8) p. 2315-2321.

Formation of amorphous hydrogenated carbon from compression of benzene

Ayako Shinozaki[1,2], Hiroyuki Kagi[2], Koichi Mimura[1]

[1]Department of Earth and Planetary Sciences, Graduate School of Environmental Studies, Nagoya University, Nagoya 464-8601, Japan

[2]Geochemical Research Center, Graduate School of Science, The University of Tokyo, Hongo, Tokyo 113-0033, Japan

E-mail: shinozaki.ayako@f.mbox.nagoya-u.ac.jp

Abstract. Infrared spectra of benzene were measured at pressures up to 25.3 GPa at room temperature using a diamond anvil cell. With decompression from pressures above 18.7 GPa, a new peak of C-H vibration mode appeared at a lower frequency than that of benzene. This indicates that an irreversible chemical change occurred at a pressure that was lower than that reported in previous studies. The IR spectra of the recovered sample after compression up to 20.3 GPa are comparable with those of the amorphous hydrogenated carbon reported in the previous study that was formed by compression of benzene up to approximately 30 GPa.

1. Introduction

Pressure-induced chemical reactions of organic compounds have been intensely researched [1,2], with pressure-induced irreversible chemical reactions found for many aromatic hydrocarbons such as phenanthrene [3], picene [4], pyrene [5], and coronene [6] as well as benzene. At room temperature, benzene, the simplest aromatic hydrocarbon, crystallized to form phase I at 0.07 GPa and then transformed to phase II at approximately 4 GPa [7, 8]. Several phase changes of benzene were reported with further compression, although these remain controversial [9-11]. In addition to the structural changes, benzene was theoretically predicted to undergo a pressure-induced chemical reaction [12-17]. Precise chemical analysis of the recovered samples using gas chromatography/mass spectrometry (GC/MS) after compression revealed that oligomerization of benzene occurred above 13 GPa at room temperature; naphthalene, biphenyl, benzene dimers, and terphenyl formed while most of the benzene remained [18]. Above approximately 20 GPa, an irreversible chemical reaction was observed using infrared (IR) spectra under high pressure; it was considered to cause the formation of polymers such as amorphous hydrogenated carbon (a:C-H) [19, 20]. Recently, a one-dimensional sp^3 carbon nanomaterial was found in a recovered sample after compression at approximately 20 GPa at room temperature [21]. Such pressure-induced oligomerization and amorphization are likely to occur upon the decrease of the intermolecular distance between the benzene molecules with compression that leads to the intermolecular interaction that exceeds the reaction threshold [15, 18, 22]. In this study, IR spectra of benzene were measured upon compression and decompression at room temperature to investigate the formation pressure of a:C-H as well as the influence of the maximum pressure on IR spectra of the amorphous hydrogenated carbon.

2. Experiments

High-pressure experiments were performed at room temperature using a diamond anvil cell (DAC) with a pair of type Ia diamond of 450 μm and 300 μm culets. A stainless steel and Re were used for the gasket after pre-compression to approximately 40–150 μm thickness. A sample hole with a diameter of 150–200 μm was drilled into the gasket. Distilled liquid benzene (99.5% purity, Kishida Chemical Co., Ltd.) was loaded in the sample hole without pressure medium. Pressure was measured using the ruby fluorescence method. IR absorption spectra in the mid-IR region (750 cm^{-1} to 7800 cm^{-1}) were measured using a Fourier transform IR spectrometer (FT-IR: Spectrum 2000; PerkinElmer Inc.) equipped with a globar source, a KBr beam splitter, and a liquid nitrogen cooled mercury cadmium telluride (MCT) detector at a resolution of 4 cm^{-1} at the University of Tokyo. In total, six runs were conducted with the maximum pressure changing between 6.3 GPa and 25.3 GPa.

3. Results and Discussions

Figures 1a and 1b show the representative IR spectra in the C-H stretching region with increasing pressure and decreasing pressure, respectively. The benzene C-H stretching peaks broadened and shifted to higher frequencies with increasing pressure up to 21.9 GPa. During decompression, a new broad peak gradually appeared at approximately 2900 cm^{-1}, at a lower wavenumber than the benzene C-H stretching peaks even though, the new peak was not observed at the maximum pressure. The absorbance of the new broad peak relative to that of the benzene C-H peaks increased with decreasing pressure. Based on its frequency and peak shape, the new C-H peak can be assigned as arising from a:C-H that was reported in previous studies [19,20]. No new peaks were detected between 1350 cm^{-1} and 1800 cm^{-1} both with increasing and decreasing pressure.

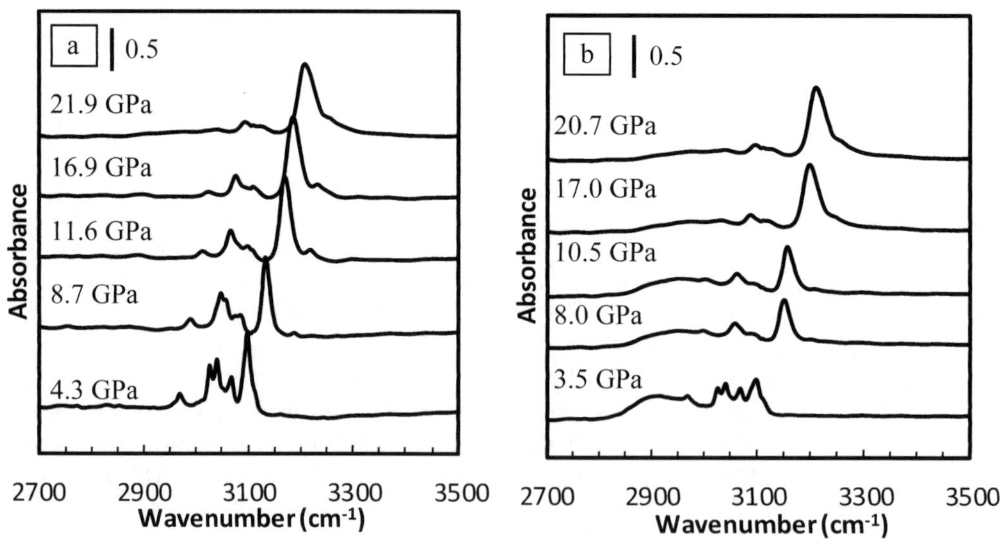

Figure 1. Representative IR spectra of C-H stretching mode region upon (a) compression and (b) decompression.

To investigate the minimum pressure of the a:C-H formation, IR spectra of several samples were measured after decompression from each maximum pressure conditions, with the maximum pressure ranging from 6.7 GPa to 25.3 GPa. Figure 2 shows the representative obtained IR spectra in the 2700–3200 cm^{-1} range collected after decompression to approximately 1.2–2.2 GPa. The new broad peak appears from the samples compressed up to >18.7 GPa but was not observed in the IR spectra of the samples compressed up to 16.2 GPa. This result suggests that the minimum pressure of a:C-H formation is between 16.2 and 18.7 GPa, which is lower than the approximately 21 GPa value reported in the previous study. [19]. The absorbance of the C-H peak due to a:C-H increases with increasing the maximum pressure relative to the absorbance of the CH stretching peaks derived from benzene, observed between approximately 3000 cm^{-1} and 3120 cm^{-1}. This indicates that the yield of a:C-H increased with increasing pressure. The peaks from benzene still remained after the compression up to 25.3 GPa, indicating that polymerization of benzene took place partially.

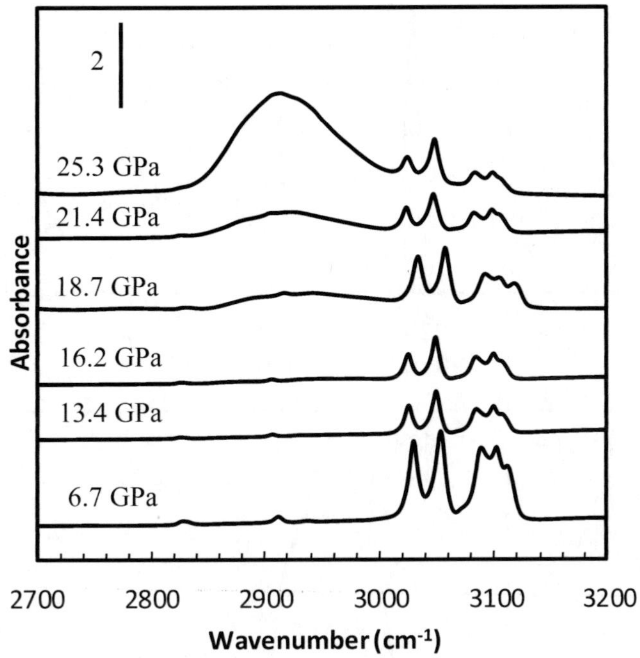

Figure 2. Representative IR stretching mode of C-H stretching mode region in six samples with the maximum pressure changing between 6.7 GPa and 25.3 GPa. The spectra were obtained around 1.2–2.2 GPa during decompression.

The solid material was observed when the samples were recovered at ambient condition from >18.7 GPa. The samples were picked from DAC and vacuumed to remove the starting material and other volatile products. Figure 3 shows the representative IR spectrum of the recovered sample compressed up to 20.3 GPa. Intense C-H stretching mode peaks were observed at approximately 2900 cm−1 and a weaker C-H peak was observed at 3030 cm−1, with several peaks being observed between 850 cm−1 and 1800 cm−1. The spectrum is comparable with that reported by Pruzan et al. in 1990, where the benzene was compressed up to approximately 30 GPa [19]. It is suggested that for maximum pressures in the 20.3–30 GPa range, the structure of a:C-H was not significantly changed, regardless of the maximum pressure. In contrast, the spectrum is partly different from that reported by Chiabini et al. in 2002 where the peaks arising from the C-O stretching mode and the C≡O stretching mode were not observed [20]. Such differences may be induced by the difference in the surrounding conditions of sample recovery from DAC: DAC was opened in the atmosphere in this study, while it was opened in an inert atmosphere in the study by Chiabini et al. [20]. In addition, in the study of Chiabini et al, benzene was compressed up to 50 GPa and was therefore likely to transform to the amorphous hydrogenated carbon completely [20], while the unreacted benzene remained during the compression and decompression in this study.

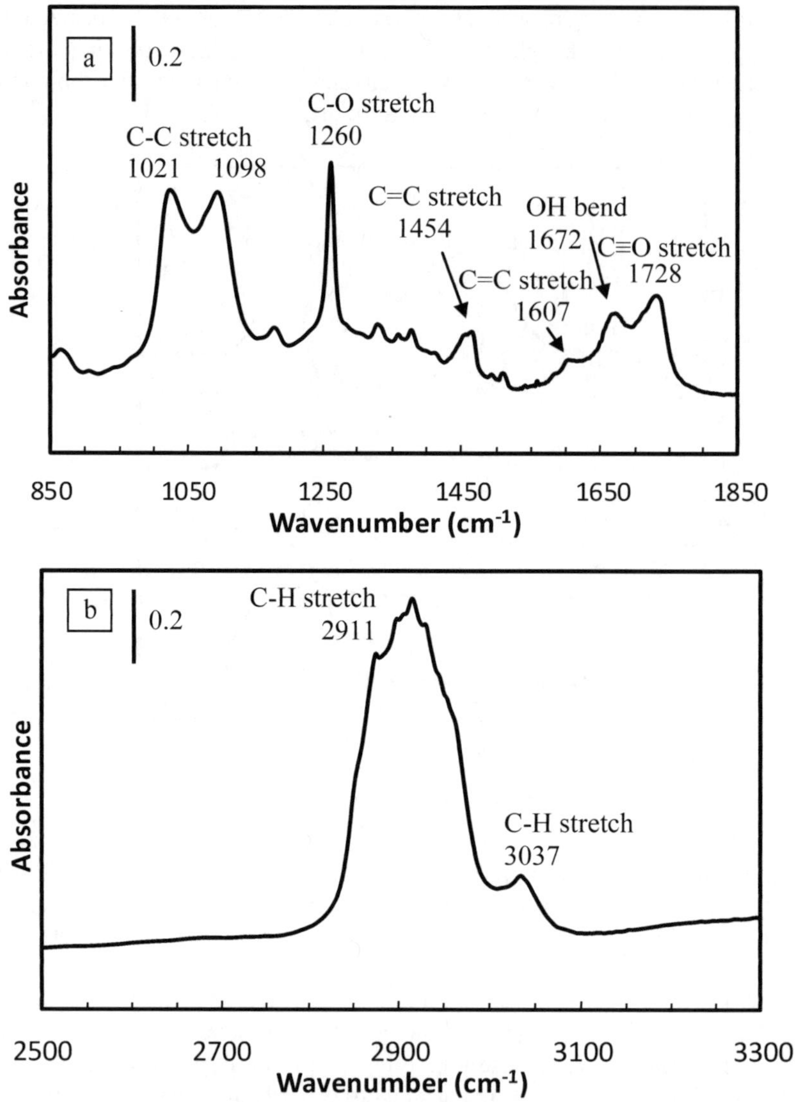

Figure 3. Representative IR spectrum of the recovered sample after compression up to 20.3 GPa. (a) between 850 cm^{-1} and 1850 cm^{-1}, (b) between 2500 cm^{-1} and 3300 cm^{-1}. Peaks are tentatively assigned to the amorphous phase of benzene of a previous study [19]. A slight peak at 1672 cm^{-1} can be assigned as an O-H bending mode of absorbed water.

In the present study, an obvious change in IR spectra was not observed for the sample compressed up to 16.2 GPa. Namely, the irreversible chemical reaction was not detected at pressures lower than 18.7 GPa. In contrast, GC/MS measurements of the recovered samples reported that dimerization and trimerization of benzene were observed at the pressure higher than 13 GPa [18]. This difference is likely due to the difference of detection limits between GC/MS and IR spectra. The yield of the dimer and trimer reaction products is much smaller than the detection limit of the IR spectra. Therefore, pressure-induced dimers and trimers are likely to have been formed even in the present experiments, despite the lack of their detection in the IR spectra. The C-H vibration peaks of the amorphous phase were observed at approximately 2900 cm^{-1} and 3030 cm^{-1}, at a lower frequency than that of the benzene C-H vibrational peaks. These peaks can be considered as vibrations arising from aliphatic hydrocarbons such as the one-dimensional sp^3 carbon nanomaterial found from the recovered sample of the benzene compressed up to approximately 20 GPa [21] rather than arising from aromatic compounds [19].

Acknowledgements
This study was financially supported by JSPS KAKENHI Grant number 25287147. A. Shinozaki is supported by a JSPS Research Fellowship for Young Scientists.

References

[1] Aoki K, Usuba S, Yoshida M, Kakudate Y, Tanaka K and Fujiwara S 1988 *J. Chem. Phys.* **89** 529
[2] Citroni M, Ceppatelli M, Bini R and Schettino V 2002 *Science* **295** 2058
[3] Huang QW, Zhang J, Berlie A, Qin ZX, Zhao XM, Zhang JB, Tang LY, Liu J, Zhang C, Zhong GH, Lin HQ and Chen XJ 2013 *J. Chem. Phys.* **139** 104302
[4] Fanetti S, Citroni M, Malavasi L, Artioli GA, Postorino P and Bini R, 2013 *J. Phys. Chem. C* **117** 5343
[5] Sun B, Dreger ZA and Gupta YM, 2008 *J. Phys. Chem. A* **112** 10546.
[6] Zhao XM, Zhang J, Berlie A, Qin ZX, Huang QW, Jiang S, Zhang JB, Tang LY, Liu J, Zhang C, Zhong GH, Lin HQ and Chen XJ 2013 *J. Chem. Phys.* **139** 144308
[7] Piermari GJ, Mighell AD, Weir CE and Block S 1969 *Science* **165** 1250
[8] Akella J and Kennedy GC 1971 *J. Chem. Phys.* **55** 793
[9] Thiery MM and Leger JM, 1988 *J. Chem. Phys.* **89** 4255
[10] Cansell F, Fabre D and Petitet JP 1993 *J. Chem. Phys.* **99** 7300
[11] Ciabini L, Gorelli FA, Santoro M, Bini R, Schettino V and Mezouar M, 2005 *Phys. Rev. B* **72** 094108
[12] R. Engelke, Hay PJ, Kleier DA and Wadt WR, 1983 *J. Chem. Phys.* **79** 4367
[13] Engelke R, Hay PJ, D.A. Kleier, W.R. Wadt, 1984 *J. Ame. Chem. Soc.* **106** 5439
[14] Engelke R 1986 *J. Ame. Chem. Soc.* **108** 5799.
[15] Root S and Gupta YM 2009 *J. Phys. Chem. A* **113** 1268
[16] Rogachev AY, Wen XD and Hoffmann R 2012 *J. Ame. Chem. Soc.* **134** 8062
[17] Qin Y, Huo RP and Zhang X 2014 *Chem. Phys. Lett.* **610** 192.
[18] Shinozaki A, Mimura K, Kagi H, Komatu K, Noguchi N and Gotou H 2014 *J. Chem. Phys.* **141** 084306
[19] Pruzan P, Chervin JC, Thiery MM, Itie JP, Besson JM, Forgerit JP and Revault M 1990 *J. Chem. Phys.* **92** 6910
[20] Ciabini L, Santoro M, Bini R, and Schettino V 2002 *J. Chem. Phys* **116** 2928.
[21] Fitzgibbons TC, Guthrie M, Xu E, Crespi VH, Davidowski SK, Cody GD, Alem N and Badding JV 2015 *Nat. Mat.* **14** 43.
[22] Ciabini L, Santoro M, Gorelli FA, Bini R, Schettino V and Raugei S, 2007 *Nat. Mat.* **6** 39

IOP Conf. Series: Journal of Physics: Conf. Series **950** (2017) 042058 doi:10.1088/1742-6596/950/4/042058

Charge density wave in hydrogen at high pressure

Ioan B. Magdău and Graeme J. Ackland

CSEC, SUPA, School of Physics and Astronomy, University of Edinburgh, UK

E-mail: i.b.magdau@sms.ed.ac.uk, gjackland@ed.ac.uk

Abstract. We present extensive molecular dynamics (MD) simulations investigating numerous candidate crystal structures for hydrogen in conditions around the present experimental frontier (400 GPa). Spontaneous phase transitions in the simulations reveal a new structure candidate comprising twofold coordinated chains of hydrogen atoms. We explain the electronic structure of this phase in terms of a charge density wave and calculate its experimental signature. In detailed tests of the accuracy of our calculation, we find that k-point sampling is far more important in MD than in static calculations, because of the freedom it give the atoms to rearrange themselves optimally for the given sampling.

1. Introduction

Hydrogen is typically located atop group 1A in the periodic table, and while the H_2 molecule has no analogue among the alkali metals, one might expect similarities to appear if a non-molecular phase is obtained at extreme conditions.

At pressures where the binding energy is close to zero, the structure is held together by the external pressure rather than chemical bonding. For the alkali metals in this pressure range, complex phases are observed in which the electronic structure can be assigned to two classes. The relatively simple electride structures involve the electron being located in the interstices between atoms playing the role of pseudo-anion [1–4]. Alternately, the atoms arrange themselves so as to perturb a free electron gas into a charge-density wave, opening a gap (or pseudogap) at the Fermi surface - the signature here being a clustering of diffraction peaks at $2k_F$ [4–10].

Static experiments using diamond anvil cells are now reaching equivalent pressures in hydrogen, where the work done in compression is greater than the chemical binding energy. Measurements at these pressures are restricted to spectroscopy and conductivity, so density functional calculations are used to determine the structure. Despite the wide variety of methods used, including different treatment of the quantised nuclear motion and different approaches to the exchange-correlation functional, the same set of candidate structures invariably appears, albeit with considerable variation in the stability range. Furthermore, more accurate theoretical treatments invariably come at high computational cost, so at elevated temperature enhanced finite sampling errors can eliminate theoretical improvements.

At present there is no good consensus about the best trade-off: typical calculations have errors of up to 100 GPa and a few hundred Kelvin. Thus the practical role of calculations is to propose candidate structures for test against experiment.

2. Calculations

For analysis we use the radial distribution function (RDF), angular distribution function (ADF), projected mean square displacement (MSD), phonon and vibrational mode projection; details of which are given in the SM. In addition we calculate average electron density of states (DoS) and average XRD patterns, all explained in SM.

3. Results and discussions

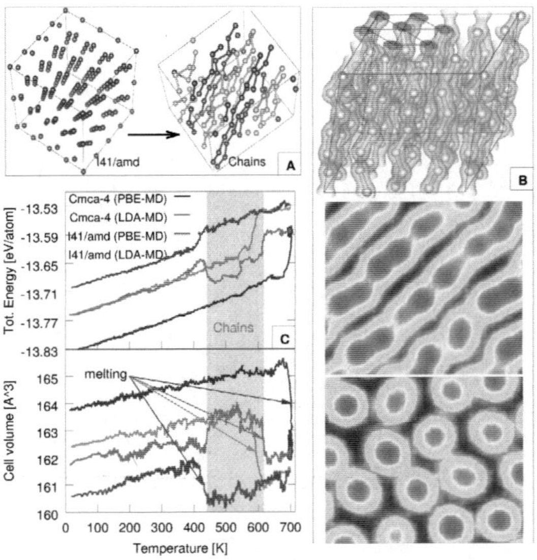

Figure 1. (A) Transformation of $I41/amd$ into chains wrapping through periodic boundaries. (B) MD snapshots showing the charge density iso-surface (top) and the ELF [11] (bottom) associated with *Chains*. (C) Cell volume and total energy upon heating at 400 GPa (SM). Shaded in red is the region where *Chains* appears.

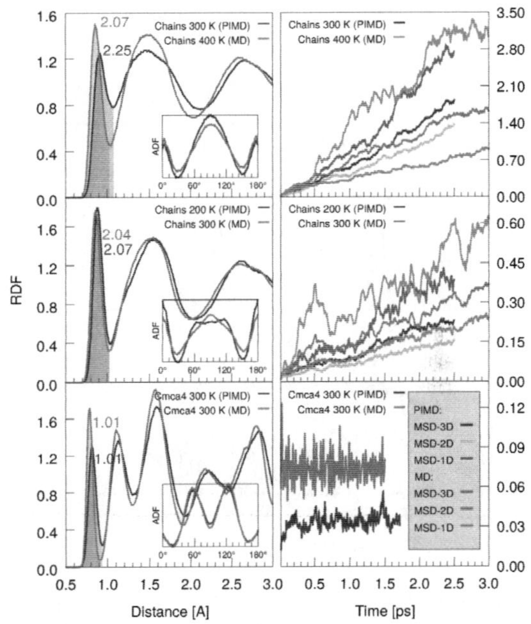

Figure 2. The left panels show RDFs and ADFs (insets) at 400 GPa and temperatures as labeled. Shaded is the integrated area under the first peak of RDF. The right panels depict the corresponding MSD calculated as total displacement (3D), displacement along chains (1D) and between chains (2D); normalised according to dimensionality.

We performed 17 short MD simulations starting with different structures found using static structure searches [12–14]. Simulations starting in molecular phases retained molecular character, but all high-symmetry structures are unstable and immediately undergo total reconstruction at finite temperature. Surprisingly, they adopt rather similar structures with twofold coordinated chains of hydrogen threading through the simulation box, subject to the constraints of periodic boundary conditions (Figure 1). This *Chains* structure, has on average 2D hexagonal symmetry in the plane perpendicular to the chains, while being essentially a collection of charge tubes containing diffusive hydrogen nuclei in the chain direction. Similar structures have been identified before in structure searches as transition states between the molecular and atomic phases [14] and probably the same structure was reported in previous PIMD simulations [15, 16]. Its ubiquity marks it out as a candidate high temperature structure worthy of further study.

IOP Conf. Series: Journal of Physics: Conf. Series **950** (2017) 042058 doi:10.1088/1742-6596/950/4/042058

Figure 1 also shows that the liquid is denser than any candidate solid structure at 400 GPa, confirming the negative slope of the melting curve [17]. $Cmca-4$ would give a much steeper slope than *Chains*, which could be measured experimentally. Using LDA exchange-correlation, both phases melt at lower temperatures, indicating that choice of exchange-correlation functional can change melting temperatures by around 100 K (corresponding to about 10 meV).

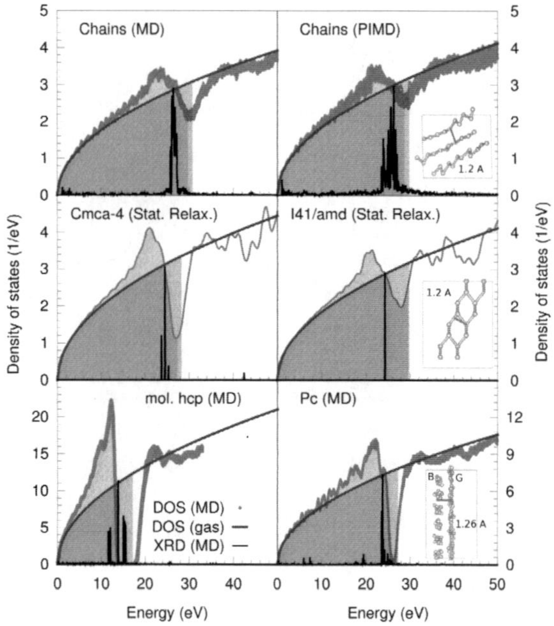

Figure 3. Density of electronic states (DoS in red) from multiple snapshots in the MD/PIMD at 300 K and from static relaxations (middle panels). The pressures are: *mol. hcp* (50 GPa), *Pc* (275 GPa), $Cmca-4$, $I41/amd$ and *Chains* (400 GPa). Blue lines are the analytic DoS for a free electron gas at the same density. Shaded regions depict the occupied states. In solid black we show the simulated XRD powder pattern (calculated with GDIS [18]) averaged over the MD/PIMD snapshots. Only the relative intensities are meaningful, and the peak positions are plotted in units of energy of a free electron with equivalent wavelength. The insets show the length scale corresponding to the most intense peak.

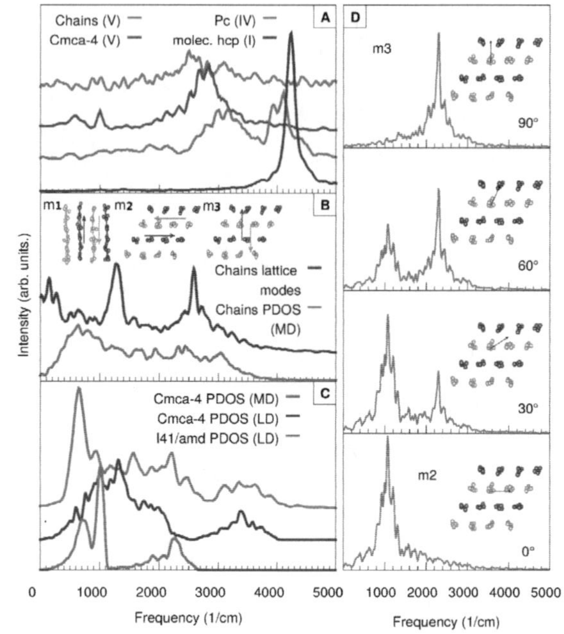

Figure 4. (A) Fourier transformed projection of the MD velocities onto molecular stretches [19–21] for the candidates of phases I, IV and beyond. (B) Phonon density of states (PDoS) extracted from VACF together with projections onto the most symmetric modes of the *Chains* (also depicted in the insets), all extracted from MD trajectories. (C) PDoS calculated from MD (300 K) and LD (0 K) for $Cmca-4$ and $I41/amd$. (D) illustrates how the the most symmetric modes emerge as the eigenvector is rotated from 0° to 90° .

We find that diffusion in the *Chains* structure is significantly higher than in the competing molecular metallic phases (Figure 2). Studying the atomic trajectories in detail we identify 3 main mechanisms; in order of importance: a) chains sliding "freely" along their length (included in MSD-1D), b) atoms migrating to neighbour chains (MSD-2D) and c) atoms swapping position within the chain (included in MSD-1D), sometimes by forming a short-lived rotating molecule. The diffusive nature of *Chains* makes it a higher entropy structure which might stabilize it over $Cmca-4$ at higher temperature. For both *Chains* and $Cmca-4$ the effect of quantum protons

(PIMD) on MSD, RDF and ADF is similar to increasing the temperature by about 100 K in classical MD. The RDF confirms that *Chains* has coordination number 2 with a peak at 0.9 A, meaning the molecules are no longer well defined. We notice in the ADF a well defined peak at 90° which means that the pseudo-molecules are pointing towards one-another. $Cmca-4$ can be viewed as a similar structure, but with the chains broken into well defined molecules, oriented at angles with respect to the chain direction. We can say *Chains* is a high entropy version of $Cmca-4$.

Figure 3 shows the electronic density of states (DoS) taken from individual snapshots of MD. DoS is extremely well fitted by a free-electron gas model, with a pronounced dip at the Fermi energy. To open a pseudogap at the Fermi surface of a free electron gas it is necessary to have an arrangement of atoms which will strongly scatter electrons at k_F. Such an arrangement will also strongly scatter X-rays at $q = 2k_F$. In figure 3 we relate the X-ray scattering vector to an equivalent energy using $E_q = \hbar^2 q^2/8m$. We have used GDIS software [18] to plot the simulated X-ray diffraction pattern (XRD) implied by our simulations. It can be seen that the *Chains* structure self-organizes in such a way that a large number of symmetry-unrelated peaks are co-located at $E_q \lesssim E_f$.

This is identical to the situation in the alkali metals, in particular for the incommensurate phases [10, 22] which typically have chains of atoms in a "host" framework. It is thus clear that our *Chains* structure can be interpreted as a Fermi-surface perturbation from a free electron gas. This nearly-free electron picture, stabilization by a charge density wave, appears to be valid for other candidate hydrogen structures. In the case of Phase-IV candidate, Pc, the charge density wave creates a layered structure (insets to figure 3) which can accommodate molecules and hexamers. This could explain the preferences of most structure candidates in the pressure range close to metalization to arrange in layers or polymers. It is peculiar that even at 50 GPa, the molecular structure *mol.hcp.* appears to have a nearly free electron DoS. This might be a side effect of the plane wave approximation employed in DFT.

In figure 4 we show our full phonon spectrum (PDOS), calculated from the fourier transform of the velocity autocorrelation function for $Cmca-4$ and *Chains* and from lattice dynamics (LD) for $Cmca-4$ and $I41/amd$. The spectra obtained for $Cmca-4$ at 0 K and 300 K are consistent suggesting the phonons are harmonic. While $I41/amd$ presents three distinct bands in the PDOS at 0 K, the *Chains* structure has only one broad band of frequencies with no gap.

Because *Chains* is metalic and only stable in MD, there is no way to identify the Raman modes, we can only impose restrictions based on symmetry. We therefore calculate the frequencies for the most symmetric phonon modes, involving oscillations of chains, by projecting the MD velocities onto those motions, and taking the FT of the resulting trajectory (see SM for details). There are three main modes involving chain oscillations in the 3 Cartesian directions as depicted in figure 4. We also demonstrate that our method correctly picks up the eigen-frequencies by investigating the spectrum with respect to different projection vectors.

Since the *Chains* have no molecular modes, so there are no vibronic modes above 3000 cm^{-1} in the PDOS. The highest frequency modes in the calculation are around 2500 cm^{-1}, which comprise atoms vibrating within a chain. Occasionally, we do see interatomic distances comparable with the free molecule, but these are short lived. *Chains* is a natural continuation of the molecular phases on the path to dissociation, as illustrated by the vibrational modes (Figure 4) given by the different structure candidates in order of increasing pressure.

4. Simulation shortcomings

The solid phases of hydrogen discovered so far at room temperature are stabilised by entropy, therefore MD simulations are crucial for understanding the properties of possible candidates. These calculations are expensive, however, and we need to consider trade-offs.

A large size of the cell is important to reduce finite size effects, a lengthy run is needed for

IOP Conf. Series: Journal of Physics: Conf. Series **950** (2017) 042058 doi:10.1088/1742-6596/950/4/042058

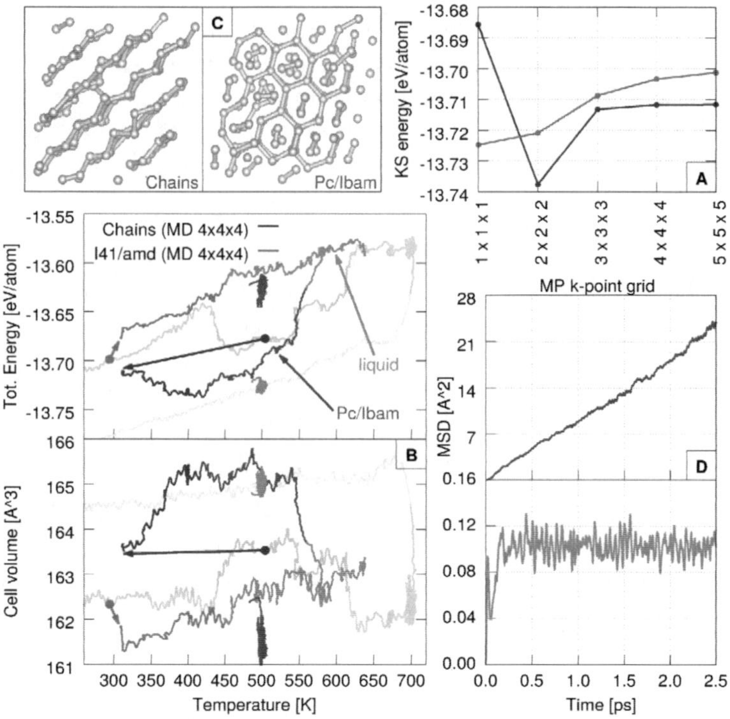

Figure 5. (A) Single point calculations of energy convergence with k-point sampling for *Chains* and *I*41/*amd*. (B) Heating at 400 GPa with increased k-points (4x4x4), starting from snapshots of (2x2x2) MD runs (Figure 1). The colors code the different structures: *Cmca*−4 (green), *Chains* (blue), *Pc*/*Ibam* (brown), *liquid* (purple). (C) Transitions from *Chains* to *Pc*/*Ibam* through snapshots from previous MD runs. (D) MSD of MD simulations started in *I*41/*amd* and *Cmca* − 4, at 400 GPa and 500 K with 2x2x2 k-points.

equilibration and good sampling of the energy surface. At the same time, a reasonably small time step is required to capture the hydrogen molecular vibrations. In addition, for PIMD runs, a large number of beads is necessary to sample the quantum wave function of the nuclei. These are all supplementary to the actual DFT settings that mainly include the choice of functional, electronic energy cut-off and k-point sampling. The accuracy of the PBE functional for hydrogen has vastly been disputed in the literature, however, we emphasize that in the case of MD using PBE is a reasonable trade-off to minimize the other more important shortcomings.

We draw attention here to yet another convergence issue that has not been fully investigated before. It is common practice to run large MD calculations with small k-point sets, as appropriate for the large cell size. In MD runs with a single k-point, we noticed that nuclei can rearrange in real space such that the energy of the electrons is minimized at Γ, but not at unsampled points in the zone. We demonstrate the effect using calculations (shown in Figure 5A) on snapshots taken from our previously discussed MD heating runs (Figure 1) comparing the pre-transition *I*41/*amd* structure with the post-transition *Chains*. We see that using Γ-point only sampling, *I*41/*amd* has the lowest energy, whereas in the actual simulation, using a 2x2x2 grid, *Chains* is lower. For higher K-point sampling, the *Chains* snapshots are still lower in energy than *I*41/*amd*, but by a smaller amount. This suggests that although the *Chains* configuration may be optimised to match the *k*-point grid, it is still a valid candidate structure. Previous MD calculations have, at best, investigated *k*-point convergence with respect to static structures and often details of *k*-points are not even quoted. The 20 meV uncertainty suggests that *k*-point sampling, rather than functional or quantum-proton effects, may be the dominant source of error in previous MD calculations.

Figure 5 also compares the previous MD runs (Figure 1) with a subsequent MD heating simulation with a denser MP grid, in which *I*41/*amd* melts at 450 K and *Chains* transforms to a *Pc*/*Ibam* type structure. In constant pressure and temperature simulations at 400 GPa and

500 K, $I41/amd$ melts whereas $Cmca-4$ is stable. This new results suggest that the stability ranking at these conditions is: $Cmca-4$, $Pc/Ibam$, $Chains$, $I41/amd$. Of course, we must study all competitive candidates, leaving the experiment to decide.

5. Conclusions

We have found and investigated a candidate structure for high-temperature solid hydrogen under pressure. The polymeric $Chains$ structure has twofold coordinated hydrogen atom, and is stabilized by a charge density wave similar to the incommensurate structures in the simple metals. In experiment, it would be characterized by the disappearance of high frequency vibronic peak. Up to 400 GPa, $Chains$ is clearly more stable than any network atomic structure, but its stability relative to molecular structures such as $Cmca-4$ cannot be reliably determined using our methods.

Acknowledgments

We thank E.Gregoryanz, A.Hermann and M.Martinez-Canales for many useful discussions. IBM thanks the Cray XC30 for computing time (Archer at EPCC) and EPSRC for a studentship.

References

[1] Ma Y, Eremets M, Oganov A R, Xie Y, Trojan I, Medvedev S, Lyakhov A O, Valle M and Prakapenka V 2009 *Nature* **458** 182–185
[2] Gatti M, Tokatly I V and Rubio A 2010 *Phys. Rev. Lett.* **104** 216404
[3] Marqués M, Ackland G J, Lundegaard L F, Stinton G, Nelmes R J, McMahon M I and Contreras-Garcia J 2009 *Phys. Rev. Lett.* **103** 115501
[4] Marqués M, McMahon M I, Gregoryanz E, Hanfland M, Guillaume C L, Pickard C J, Ackland G J and Nelmes R J 2011 *Phys. Rev. Lett.* **106** 095502
[5] Heine V 2000 *Nature* **403** 836–837
[6] Arapan S, Skorodumova N V and Ahuja R 2009 *Phys. Rev. Lett.* **102** 085701
[7] Arapan S, Mao H and Ahuja R 2008 *Proc. Natl. Acad. Sci.* **105** 20627
[8] Reed S K and Ackland G J 2000 *Phys. Rev. Lett.* **84** 5580
[9] Degtyareva V F 2006 *Phys. Uspek.* **49** 369–388
[10] Ackland G J and Macleod I R 2004 *New J. Phys.* **6** 138
[11] Savin A, Jepsen O, Flad J, Andersen O K, Preuss H and von Schnering H G 1992 *Angew. Chem. Int. Ed. Engl.* **31** 187–188
[12] Pickard C J and Needs R J 2007 *Nat. Phys.* **3** 473–476
[13] Pickard C J, Martinez-Canales M and Needs R J 2012 *Phys. Rev. B* **85** 214114
[14] Geng H Y, Song H X, Li J and Wu Q 2012 *J. Appl. Phys.* **111** 063510
[15] Biermann S, Hohl D and Marx D 1998 *Solid State Commun.* **108** 337–341
[16] Biermann S, Hohl D and Marx D 1998 *J. Low Temp. Phys.* **110** 97–102
[17] Howie R T, Dalladay-Simpson P and Gregoryanz E 2015 *Nat. Mater.* **14** 495–499
[18] Fleming S and Rohl A 2005 *Z. Kristallogr.* **220** 580–584
[19] Magdău I B and Ackland G J 2013 *Phys. Rev. B* **87** 174110
[20] Ackland G J and Magdau I B 2013 *High Pressure Res.* 1–7
[21] Magdău I B and Ackland G J 2014 *J. Phys.: Conf. Ser.* **500** 032012
[22] McMahon M I and Nelmes R J 2004 *Phys. Rev. Lett.* **93** 055501

IOP Conf. Series: Journal of Physics: Conf. Series **950** (2017) 042059 doi:10.1088/1742-6596/950/4/042059

Theory of high pressure hydrogen, made simple

Ioan B Magdău, Floris Balm and Graeme J Ackland

CSEC, SUPA, School of Physics and Astronomy, The University of Edinburgh, Edinburgh EH9 3JZ, United Kingdom

E-mail: i.b.magdau@sms.ed.ac.uk, gjackland@ed.ac.uk

Abstract. Phase I of hydrogen has several peculiarities. Despite having a close-packed crystal structure, it is less dense than either the low temperature Phase II or the liquid phase. At high pressure, it transforms into either phase III or IV, depending on the temperature. Moreover, spectroscopy suggest that the quantum rotor behaviour disappears with pressurisation, without any apparent phase transition. Here we report molecular dynamics (MD) calculations which show that all these phenomena can be understood in a classical model with simple treatment of exchange-correlation. We also discuss the thermodynamics of the phase boundaries of Phase I.

1. Introduction

In 1935 Wigner and Huntington proposed a metallic modification of hydrogen as an atomic phase at high pressure. They calculated that metallisation would occur at about tenfold compression, which is remarkably close to modern estimates. They also quoted a transition pressure of 25GPa, which has proved over an order of magnitude too low, and has been widely ridiculed. This should stand as testament that calculations of volumes are more reliable than pressures.

The experimental search for crystalline metallic hydrogen continues, and with numerous non-metallic phases reported, it has become clear that a different physical picture from Wigner and Huntingdon's "atomisation begets metallisation" is required.

The structure of solid hydrogen up to 250GPa is now generally agreed upon. Phase I is a hexagonal close-packed molecular liquid. Here, the high-school picture of H_2 as a dumbbell molecule is misleading, because at low pressure H_2 behaves as a free molecular rotor, and the $J = 0$ quantum ground state is spherical. Hence, phase I can be thought of as simply close packing of spherical molecules. As pressure is increased the molecules interact with each other and J ceases to be a good quantum number. At low temperature this leads to a "broken symmetry" Phase II, where the rotation has stopped. There have been numerous predictions for the crystal structure of this phase, which is normally regarded as a structure near to hcp in which molecular orientation is fixed. At high temperature, the melt line shows a maximum around 900K/70GPa[1, 2, 3]; if pressure is increased further the melting temperature drops, meaning the liquid is denser than the close-packed Phase I.

At high pressures, where the mechanical work of compression (PV) approaches the molecular binding energy, complex phases are observed. Theory predicts that these are based around a new motif - weakly bound molecules arranged into hexagonal trimers[4, 5, 6]. Distinct molecules are still observed, albeit with short-lived, weaker and longer bonds. In the low temperature "Phase III" all molecules are in such trimers, however, at high temperature "Phase IV" appears to comprise alternating layers of trimers and relatively freely rotating molecules. [5, 7].

These phases can all be reproduced by considering classical protons interacting via forces calculated in the density functional theory. The calculated pressure and temperature phase boundaries depend on the details of the calculation which typically involve trade offs between system size, exchange correlation treatment, pseudopotential fidelity, quantisation of nuclear motion and k-point sampling.

It is widely assumed that quantum protons and advanced treatment of exchange-correlation for van der Waals bonding are essential to describe bonding in solid hydrogen [8]. So it is particularly curious that the "quantum rotor" phase I is well explained by MD using classical rotors[1]. Given that the melting point maximum was predicted by classical MD simulation with remarkable accuracy, one has to believe that if those quantum effects are essential, they are exactly compensated by some incorrect piece of classical physics.

In this paper we examine what is the minimal physics required for an understanding of the hydrogen phase diagram. We do not aspire to obtain accurate pressures or temperatures, rather the questions to address are:

(i) Is there a simple explanation for the mixed atomic-molecular phase IV?

(ii) Does Phase I comprise free rotors at all pressures?

(iii) Why is the melt denser than the close packed solid phase I?

2. A simple packing model

Despite the wide range of candidate structures reported from ab initio structure search techniques[4, 9, 10], three motifs seem to be sufficient to capture the physics of the phase diagram: rotating molecules, non-rotating molecules, and layers comprising short-lived molecules where the time-averaged indistinguishable atomic positions form hexagonal layers. Structure search reveals numerous possible small distortion which MD or path integral methods show are eliminated by elevated temperatures and/or quantum behaviour of the protons[5, 6, 11]. Our simple model considers three objects:

- S: Spherical molecules, which correspond to a $J = 0$ quantum rotor ground state, or a time-averaged classical free rotor. S objects have high volume and high internal entropy. The zero of energy is taken as the covalent bond strength.
- R: Rod-like molecules, corresponding to the non-rotating type, with length similar to the diameter of S. R molecules have smaller volume than S, and zero internal entropy or energy.
- A: Spherical "atoms", about half that of S. They have no internal entropy, and high internal energy due to the broken bond.

In our model, phase I comprises S-type objects, phase II R-like, phase III R-type (better packed, but elongated), phase IV is a binary mixture of A and S. The liquid contains S, R and A objects according to their Boltzmann weights. A packing fraction and an internal energy is assigned to each crystal structure. The total free energy of each phase is then the sum of the energies of the crystal and its component objects. We have considered other crystalline arrangements within the model, and the only competitive structure is close-packing involving a single H atom (labeled "Atomic" in figure 1).

We can model the hydrogen molecule as two spherical atoms with centers separated by about 0.74Å. In the absence of external pressure, the density is 23 cm^3/mol, (about 39 $\AA^3$ per molecule) indicating that the molecules are far apart compared to their bond length. At low pressure, hydrogen is extremely compressible, which implies that in this regime close-packing of a rigid spheres is not a good approximation. The close-packed density for such S-objects, i.e. where the intermolecular distance becomes close to double the bohr radius, is reached around 200GPa[7, 12, 13] where phase transitions start to happen.

It may seem curious that there is no explicit treatment of compressibility within a phase. This is because to calculate phase boundaries using the geometric model presented here requires only that the compressibility of competing phases is similar at the phase boundary (c/f figure 2), and the additional terms serve only to complicate the model.

The hcp structure has a packing fraction of 0.7405, which is maintained under a affine deformation to nonideal c/a by transforming spheres into oblate spheroids. Naively, one might expect a transformation of all molecules directly from the S to A at a pressure which favours the smaller objects. In fact, the model predicts that a mixed S-A structure would have lower free energy for a range of pressure provided it adopts the AB_2, MgB_2 structure, one of only two structure which has a denser packing than hcp for binary hard spheres. The time-averaged structure of Phase IV is precisely this close-packed AB_2 structure, space group P6/mmm[5, 6]. Therefore the stability of this apparently complex structure can be attributed to an exceptionally simple packing model.

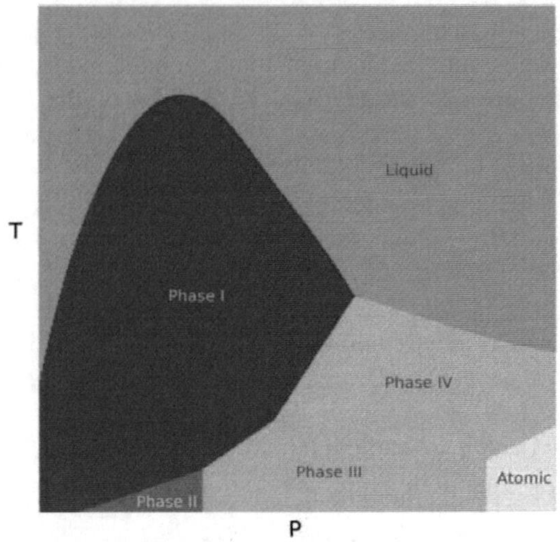

Figure 1. Example phase diagram from the simple packing model. Code is available online[14], where parameters can be adjusted for fit or sensitivity analysis.

Thus our model for Phase IV regards it as close packing of binary hard spheres: S and A objects. This picture of Phase IV as an entropically stabilised phase is consistent with classical and quantum molecular dynamics the "G" site atoms temporarily pair up to form weak molecules, but over time sample the fully symmetric phase space. If the proton motions are treated quantum mechanically, either by path integral sampling or through consideration of tunneling probabilities, the many-proton wavefunction attains $P6/mmm$ symmetry directly and the weak molecules manifest as a correlation in the proton wavefunction.

The transition between phases I and II occurs at much lower densities, indicating that it is driven by the interplay between quadrupole interactions and entropy, as well as packing.

A phase diagram corresponding to this model is shown in figure 1, and the thermodynamic explanation for the phase transitions in this picture is:

- I $\rightarrow$ II. Energy is gained by the quadrupole-quadrupole interaction. Volume decreases due to the rods being smaller than spheres, despite inefficient packing. Entropy favours free rotor phase I.
- I $\rightarrow$ IV. Volume decreases due to denser packing of two sphere types. Entropy favours free rotor phase I.
- I $\rightarrow$ liquid. Conventional hard sphere melting transition at low pressure, with an increased population of type R and A objects at high pressure causing the liquid to become denser than the solid.
- III$\rightarrow$IV. Entropy gain from rotation and volume decrease due to packing atoms.

3. A classical-proton AIMD model for the free and arrested rotor phases

We have carried out extensive ab initio MD simulations on the phase boundaries around Phase I using the CASTEP code with PBE exchange and classical protons[14]. We carry out NPT simulations at 100GPa, starting at 0K in $P63m$. NPT enables us to determine the c/a ratio of hexagonal phases, and the volume changes on transition. The rapid heating rates and finite system size mean that the transition temperatures will be an overestimate of their true thermodynamic values. Phase transitions are observed at this level of theory in qualitative accordance with experimental data.

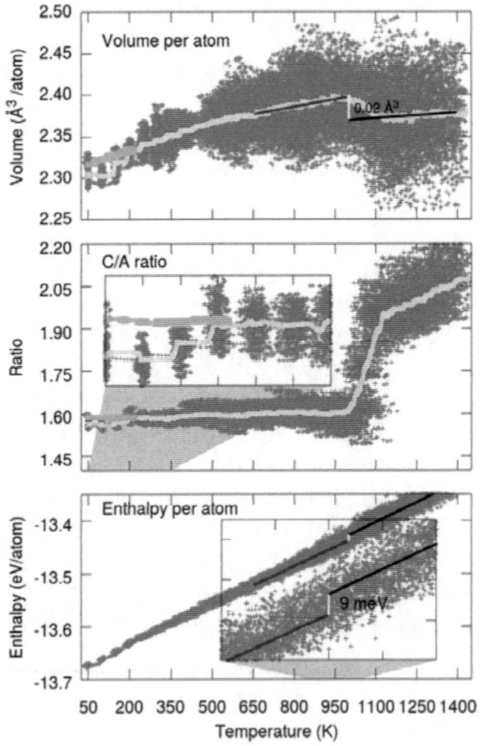

Figure 2. Data from 288-atom heating NPT simulations. Red crosses: instantaneous data from 13.5ps, 0.5fs/step, calculation with Γ k-point sampling at 100GPa heated at 50K each 0.5ps. Green lines: running average over a 100fs window. Orange lines: running average with (4x4x4) k-point sampling heated at 25K/150fs. Yellow lines: estimated changes at melting, from linear regression fits.

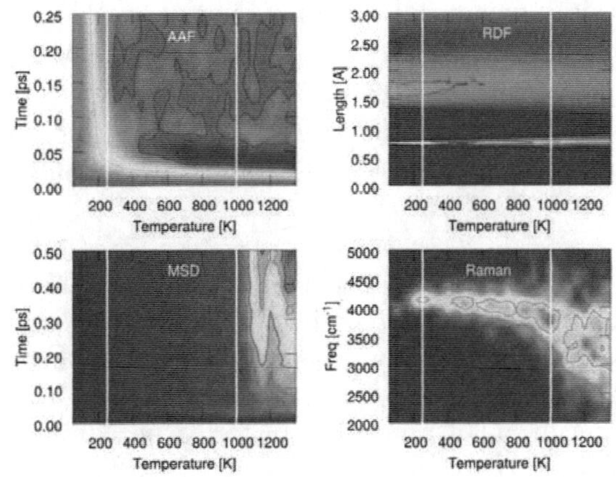

Figure 3. Analysis of heating MD runs at 100GPa (a) AAF vs temperature. Below 150K, in phase II, there is no rotation. Above 150K, in phase I, AAF vanishes within 10 fs, but only at at higher temperature, is rotation through 90 degrees free enough to see anticorrelation at 40fs. (b) RDF: Phase II has two molecular neighbours peaks at 1.7 and 2.1 Å, which merge in phase I. all peaks broaden with T. (c) MSD is low in Phase II, indicating molecular libration. In Phase I, it is higher due to rotations. Above 1100K, a linear increase in MSD indicates melting. (d) Raman vibron frequency obtained via projection[15], dropping with increasing T, and further at the melting transition. This is consistent with experimental data[3].

On heating at 100GPa, (figure 2) we find that the symmetry-broken phase II transforms into the molecular rotor phase I at 100K, ultimately melting at 1100K. The close-packed phase has the lowest density of the three. The phase transitions are evident to movie visualisation. $II \rightarrow I$ most clearly shown in the disappearance of AAF (figure 3a), while melting becomes evident in the mean squared displacement (MSD figure 3c).

Direct measurement of quadrupole interaction energy from the electronic structure is swamped by fluctuations in the MD, but we can measure it via orientational correlations. The quantity EQQ (figure 4) is the interaction energy of two linear quadrupoles pointing in directions, k, k', separated by a vector r. The molecular quadrupole moment for hydrogen is around $Q = 0.26D\mathring{A}$ [16], and the ensemble average of EQQ is the quadrupole-quadrupole energy.

$$< EQQ >=< (3Q^2/4r^5)[35(k.r)^2(k'.r)^2 - 5(k.r)^2$$
$$-5(k'.r)^2 + 2(k.k')^2 - 20(k'.r)(k.r)(k.k') + 1] >$$

IOP Conf. Series: Journal of Physics: Conf. Series **950** (2017) 042059 doi:10.1088/1742-6596/950/4/042059

A series of simulations of pressurisation at 600K (Fig 5) showed phase I persisting until about 200GPa at which point it transformed into a disordered structure. This is not the expected melting transition, as evidenced by the *enhanced* AAF, indicating the arrest of the rotors, and the disappearance of the strong vibron. The structure may be similar to the chain-like structure reported elsewhere in this volume.

Angular autocorrelation function (AAF), radial distribution function (RDF), and mean square displacement (MSD) graphs confirm the broken symmetry - free rotor - liquid sequence with temperature and the weakening of the vibron with increased pressure, followed by extreme broadening and further softening, plus suppression of molecular rotation in Phase III.

3.1. k-point convergence

Previous work has reported both $Pca21$ and $P63m$ as the most stable structure for phase II in this model. Preliminary calculations with a single k-point found a transition to $P63m$, however at full k-point convergence $Pca21$ is more stable. This conclusion was tested in static relaxations of MD snapshots, and is the likely resolution to the historical discrepancy. The description of phase I appears to be converged with Γ-point sampling, but k-point sampling issues also appear in the melting. Sampling with a single k-point we observed melting at 1100K; this calculation is essentially a repeat of previous work[1]. However, using a denser k-point mesh systematically raised the melting point by several hundred degrees. It is unusual to require dense k-point sampling in molecular crystal, but careful convergence is needed to properly resolve the quadrupole moments in phase II, which are crucial to the ordering. Meanwhile, in the liquid phase, free diffusion allows the atoms to arrange themselves in a way most favourable for minimising the Γ-point electron energy at the expense of unsampled contributions from other parts of the Brilloiun Zone.

3.2. Lattice parameters, local and long ranged order

Our simulation spontaneously transformed from $Pca21$ to hcp for phase I. We find the c/a ratio to be essentially independent of temperature, but to decrease with increased pressure from ideal to about 1.57 at 150GPa. This, and the PV equation of state, are in good agreement with the experiment[17].

Phase I has no long range order, but the negative ⟨EQQ⟩ (Fig. 4) indicates short ranged order and increasing coupling between the rotors with pressure of a few meV, increasing approximately as $V^{-\frac{5}{3}}$ with pressure. As the correlation increases, there will be a gradual breakdown in J as a good quantum number and consequent disappearance of the distinctive roton bands. Although there will be a distinctive signature in spectroscopy, this correlation between rotors does not become long-ranged and so does not constitute a thermodynamic phase transition.

At low temperature, EQQ is sufficient to cause long-range ordering in the broken-symmetry phase II: The binding due to EQQ is up to 30meV, enough to stabilise $Pca21$ below 100K.

4. Conclusions

We have presented a model which distills the essential physics determining the phase stability in hydrogen. Phases II and III are energetically favored, through quadrupole interaction and electron delocalisation respectively. Phases I and IV are entropically favoured, primarily due to the high entropy of rotating molecules, which can be understood classically or quantum mechanically. The J=0 quantum rotor is an intrinsically large object: the higher compressibility of the liquid compared to phase I can be understood by the increased population of "smaller" molecules in higher energy states.

The success of the classical MD models can be attributed to this near equivalence of quantum uncertainty described by the wavefunction, and classical uncertainty described by the entropy.

IOP Conf. Series: Journal of Physics: Conf. Series **950** (2017) 042059 doi:10.1088/1742-6596/950/4/042059

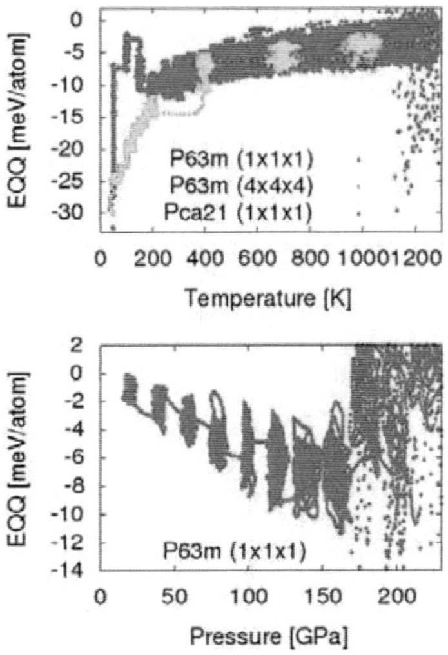

Figure 4. EQQ for heating at 100 GPa (top) and pressurising at 600 K (bottom). Phase II is unstable with Γ-point sampling (red), but stabilised by slower heating and more k-point sampling (green). The blue dot shows EQQ for $Pca21$.

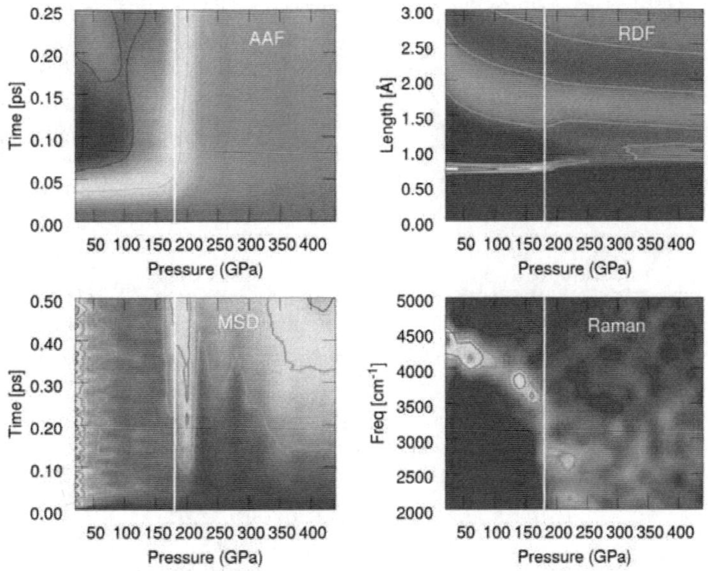

Figure 5. Analysis of pressure dependence at 600K: (a) AACF (b) RDF (c)MSD (d) Raman vs P. Two simulations were started from the equilibrated cell at 100GPa, 600K (phase I) and pressurised/depressurised adiabatically by 20GPa every 0.5ps. The simulations were run with Γ k-point and 288 atoms.

The free rotors in phase I become increasingly correlated as pressure increases, due to the strengthening quadrupole moment and steric effect. The many-body wavefunction for this behaviour is not presented here, but clearly the excited states observable in spectroscopy should not have the characteristic overtones of a free rotor. This change in behaviour does not indicate a phase transition.

Acknowledgements
We thank EPSRC for support, under UKCP grant K01465X and a studentship for IBM.

References
[1] Bonev S A, Schwegler E, Ogitsu T and Galli G 2004 *Nature* **431** 669–672
[2] Eremets M I and Trojan I 2009 *JETP letters* **89** 174–179
[3] Howie R T, Dalladay-Simpson P and Gregoryanz E 2015 *Nat. Mater.* **14** 495–499
[4] Pickard C J, Martinez-Canales M and Needs R J 2012 *Phys. Rev. B* **85** 214114
[5] Magdău I B and Ackland G J 2013 *Phys. Rev. B* **87** 174110
[6] Liu H and Ma Y 2013 *Phys. Rev. Lett.* **110** 025903
[7] Eremets M and Troyan I 2011 *Nature materials* **10** 927–931
[8] Li X Z, Walker B, Probert M I, Pickard C J, Needs R J and Michaelides A 2013 *J.Phys: CM* **25** 085402
[9] Pickard C J and Needs R J 2007 *Nat. Phys.* **3** 473–476
[10] Geng H Y, Song H X, Li J and Wu Q 2012 *J. Appl. Phys.* **111** 063510
[11] Ackland G J and Magdau I B 2015 *Cogent Physics*
[12] Caillabet L, Mazevet S and Loubeyre P 2011 *Phys. Rev. B* **83** 094101
[13] Mao H k and Hemley R J 1994 *Rev.Mod.Phys.* **66** 671
[14] *http://dx.doi.org/10.7488/ds/364*
[15] Ackland G J and Magdau I B 2013 *High Pressure Res.* 1–7
[16] Orcutt R H 1963 *J.Chem.Phys.* **39** 605–608
[17] Loubeyre P, LeToullec R, Hausermann D, Hanfland M, Hemley R, Mao H and Finger L 1996 *Nature* **383** 702–704

IOP Conf. Series: Journal of Physics: Conf. Series **950** (2017) 042060 doi:10.1088/1742-6596/950/4/042060

Raman scattering and X-ray diffraction studies on phase III of solid hydrogen

Y Akahama[1], Y Mizuki [1], S Nakano[2], N Hirao[3] and Y Ohishi[3]
[1]Graduate School of Material Science, University of Hyogo, 3-2-1 Kouto, Kamigohri, Hyogo, 678-1297 Japan
[2]NIMS, 1-1 Namiki, Tsukuba, Ibaraki 305-0044, Japan
[3]Japan Synchrotron Radiation Research Institute, 1-1-1 Kouto, Mikazuki, Sayo, Hyogo, 679-5198 Japan

E-mail: akahama@sci.u-hyogo.ac.jp

Abstract. The structure of the phase III of solid hydrogen has been investigated from Raman scattering and X-ray powder diffraction experiments at low temperature based on the diamond anvil Raman gauge at pressures up to 300 GPa and 190 GPa, respectively. Observed splitting of low frequency Raman bands at around 230 GPa has suggested distinct lattice deformation of the hcp-like structure in phase III while X-ray diffraction data proposed that the molecular center was located in the vicinity of lattice points of hcp up to 190 GPa.

1. Introduction

Recently, the phase diagram of solid hydrogen around room temperature at multimegabar pressures has been updated by the observation of the phase IV [1-5]. The phase IV is considered to be semiconductor with two kind of molecular sites. On the other hand, H_2 molecules in the phase III are considered to be an orientationally ordered state [6,7]. Spectroscopic studies reported the stability of phase III up to 340 GPa at low temperature below 100 K [4,8-10]. An X-ray diffraction study revealed that the molecular center in phase III was located in the vicinity of lattice points of hcp up to 183 GPa [11]. For this phase, theoretical studies predicted the transformation to metallic state without structural transitions around 400 GPa before breakdown to a monatomic solid [12,13]. Up to now, many structures are proposed for the ordered phase, such as *Cmca* [13,14] and *C2/c* [15,16]. The proposed structures basically differ in the orientation of molecules in a unit cell each other. However, the structure of phase III still remains unidentified experimentally.

Purpose of this study is to get the structural information of phase III of solid hydrogen. In this paper, we present the results of Raman scattering and X-ray powder diffraction experiments of solid hydrogen at pressures up to 300 GPa and 190 GPa, respectively and discuss the structure of phase III.

2. Experimental

Merrill-Bassett diamond anvil cells were used for high-pressure generation. Diamond anvils with typical top surface of 100 μm and 20 μm were used for diffraction experiments and Raman spectroscopy, respectively. Raman scattering experiments were carried out at pressure up to 300 GPa at low temperature of 90 K. An He-Ne laser (632.8 nm, 35 mW) or an Ar ion laser (514.5 nm, 50 mW) was used as a excitation source for Raman measurements. X-ray experiments at 100 K were carried out using monochromatic SR source on the BL10XU station at SPring-8 with monitoring the transition into phase III by *in situ* measurement of Raman vibron frequencies of hydrogen. Details of X-ray experiments have been described in elsewhere[11]. The diamond anvil Raman gauge calibrated by using the equation of state of Pt at room temperature [17-19] was used to estimate pressures at low temperatures.

3. Results and Discussion

3.1. Raman spectroscopy

In order to collect Raman spectra of phase III at pressure up to 300 GPa, three experimental runs were performed. Figure 1 shows the pressure change of the vibron frequency together with our previous data [10]. Present data are in good agreement with our previous results. Corresponding to the transition to phase III, the frequency drops by ~100 cm^{-1} at 165 GPa. With increasing pressure the vibron is softened with a coefficient of -5.07 cm^{-1}/GPa. However, the tendency becomes moderate above 230 GPa. The behavior, which was observed in our previous study, suggests certain change in the crystal lattice of phase III. However, the previous study at 100 K by Loubeyre *et al.* [9] has reported a linear pressure dependence up to 316 GPa. The difference in the behavior of the vibron frequency comes from the pressure determination method. Though they also used the pressure shift of the high frequency edge of the optical phonon band of the diamond anvil for pressure estimation, the pressure shift was calibrated to 200 GPa by using the pressure shift of the vibron and pressure more than 200 GPa was extrapolated from the calibration. Actually, in the present study, the vibron frequency decreases to 3355 cm^{-1} at pressure of 296 GPa but the value has been observed at 316 GPa by them. If the behaviour of the vibron above 230 GPa is true, certain change may be expected to the pressure dependence of low frequency Raman active modes.

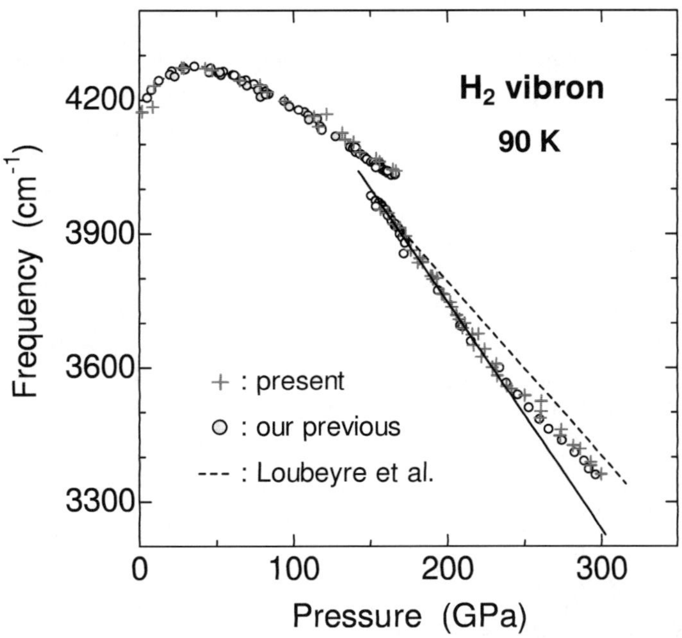

Figure 1. Pressure dependence of the vibron frequency together with our previous data [10]. Broken line corresponds to previous data by Loubeyre *et al.* [9] at 100 K. The vibron in phase III was softened with a coefficient of -5.07 cm^{-1}/GPa.

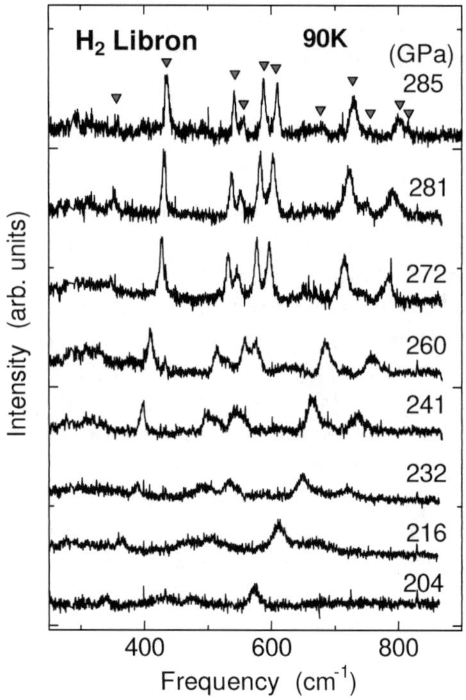

Figure 2. Pressure evolution of low frequency Raman bands: librons in the phase III. Clear splitting of Raman bands was observed above 230 GPa.

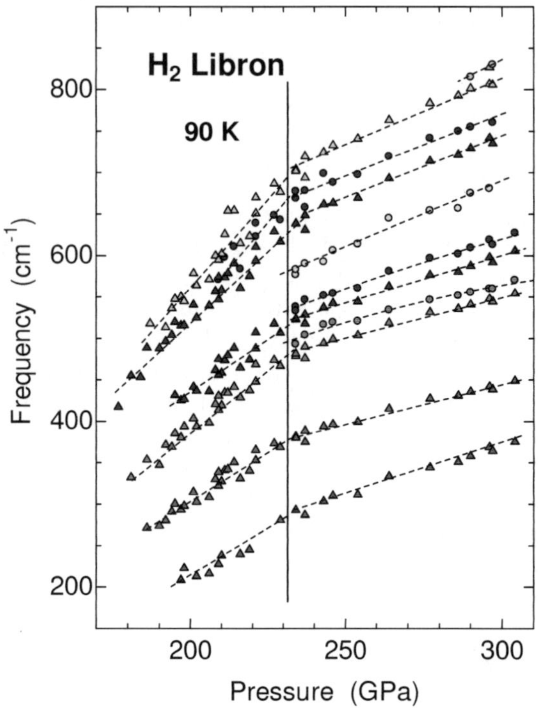

Figure 3. Pressure dependence of each libron frequency. Broken lines are guides to the eye. Pressure dependence of frequency changed around 230 GPa.

Pressure evolution of low frequency Raman spectrum of phase III is illustrated in figure 2. Low frequency modes, which mainly correspond to librons, are considerably broad and weak in intensity at pressure below 200 GPa. With increasing pressure, most of the bands split into two or three sharp bands at around 230 GPa. At 300 GPa eleven libron bands were identified below the frequency of 900 cm^{-1}. The number of libron bands is consistent with previous studies [9,10] and may suggests a large unit cell. Since one Raman active vibron, one lattice phonon, one IR active vibron and two IR active librons have been observed up to now, the primitive cell of phase III would involved at least 4 molecules. Figure 3 shows the pressure change of frequency of observed bands. Though the frequency of each mode increases with increasing pressure, the tendency gradually becomes moderate, corresponding to the splitting around 230 GPa. The behavior is coincident with the anomaly of the vibron frequency. Clear splitting of Raman bands means asymmetrization of the hcp structure. Therefore, distinct change from the hcp-like structure occurs in phase III at around 230 GPa.

3.2. X-ray diffraction measurements

In order to collect X-ray powder diffraction patterns of phase III, we performed three different experimental runs in this study. The maximum pressure was extended to 190 GPa. In each run, only two diffraction peaks were observed in this phase. The observed *d*-values at each pressure are in good agreement with our previous report [11] shown in figure 4. The peaks correspond to the 100 and 101 diffraction lines of the hcp structure. The results meant that the molecular center is located in the vicinity of lattice points of hcp at least to 190 GPa. Therefore, assuming the hcp lattice, lattice constants *a* and *c* were estimated from the *d*-values. Obtained lattice constants are showed in the figure 5 as a function of pressure, together with our previous data [11]. After the II-III phase transition, the lattice constant *c* is more compressible compared with *a*. Present results reproduced our previous data. Considering the hydrogen molecule has an ellipsoidal shape and each molecular axis is ordered to tilt against the *c* axis of the hcp structure, the lattice parameter, *c* should contract. As a result, the *c*/*a* ratio decreased to 1.55-1.56 at the transition and reached 1.54 at 190 GPa.

The decrease of the *c*/*a* ratio in the phase III was theoretically predicted by Ashcroft [12] and Kitamura *et al.* [14]. According to the calculation by Kitamura *et al.*, the *c*/*a* ratio decreases from 1.583 in phase II (ordering of angular-moment states) at 130 GPa to 1.575 in phase III (classical orientational ordering of rotating hydrogen molecules) at 180 GPa because of an increase in the angle between the *c*-axis and the molecular axis. Though the present value at 190 GPa is larger than the theoretical one, the predicted behavior of the *c*/*a* ratio is consistent with the present result. The decrease in the *c*/*a* ratio with pressure in phase III would be related to the progress of orientational ordering, which was also inferred from the IR absorption study by Hanfland *et al.* [7]. The absorption intensity of IR vibron increased with increasing pressure up to 175 GPa. They pointed out that the absorption may continued to rise even further with increasing pressure above 175 GPa. Therefore, the structural behavior supports a classical feature of the orientational order of rotating hydrogen molecules for phase III [12,14,20].

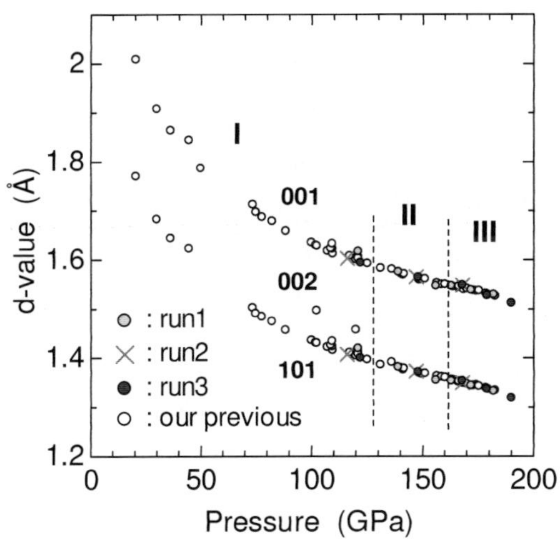

Figure 4. The *d*-values of observed two diffraction peaks from solid hydrogen at 100 K as a function of pressure together with our previous data [11].

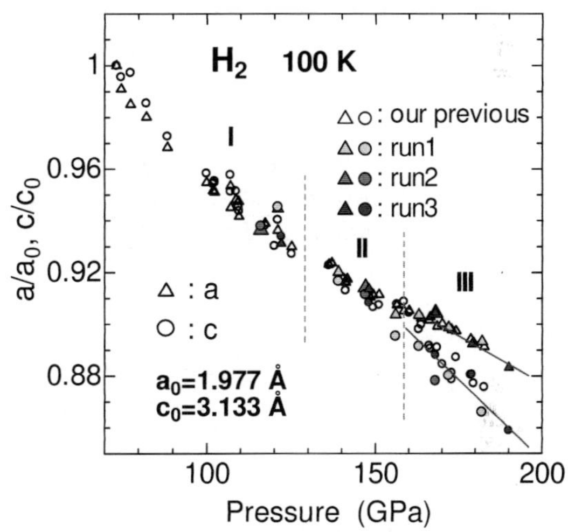

Figure 5. Pressure dependence of lattice constants *a* and *c*, which were estimated from *d*-values assuming the hcp structure. Solid lines are guides to the eye.

Figure 6 shows molar volume of solid hydrogen as a function of pressure, together with our previous data [11]. The dependence shows an anomaly at the II-III phase transition. From the result, the volume reduction, ΔV, of the transition was estimated to be 1% or less. The volume of the phase III is 2.33 cm^3/mol just after the transition and tends to show a larger decrease with increasing pressure. The tendency may result from an advance of the orientational ordering with pressure. The solid line shows the equation of state of phase I, V_0= 23.0 cm^3/mol, K_0= 0.275(6) GPa, and K'_0= 6.39(3), which was estimated from the pressure-volume data by a least-squares fit of Vinet fomula.

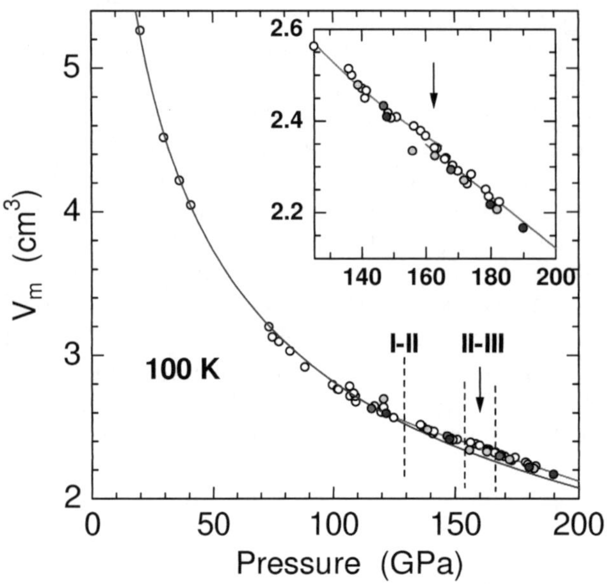

Figure 6. Molar volume of solid hydrogen as a function of pressure, together with our previous data [11]. The blue solid line shows the equation of state of phase I, V_0= 23.0 cm^3/mol, K_0= 0.275(6) GPa, and K'_0= 6.39(3).

In conclusion, from Raman scattering studies we obtained a structural information that a distinct lattice deformation of the hcp-like structure occurred at around 230 GPa in phase III. This strongly suggests necessity of X-ray diffraction experiments above the pressure for determining the structure of phase III.

Acknowledgments
This work is supported by Grant-in-Aid for Scientific Research (C) (No. 25400381) from the Japan Society for the Promotion of Science. X-ray diffraction measurements were done under Proposal No. 2013A1086, 2013B1172, 2014A1245, 2014B1124 & 2015A1297 of SPring-8.

References
[1] Eremets M I and Troyan I A 2011, *Nature Materials* **10** 927
[2] Howie R T, Guillaume C L, Scheler T, Goncharov A and Gregoryanz E 2012 *Phys. Rev. Lett.* **108** 125501
[3] Zha C S, Liu Z and Hemley R 2012 *Phys. Rev. Lett.*, **108** 146402
[4] Zha C S, Liu Z, Ahart M, Boehler R and Hemley R 2013 *Phys. Rev. Lett.* **110** 217402
[5] Loubeyre P, Occelli F and Dumas P 2013 *Phys. Rev. B.* **82** 134101
[6] Mao H K and Hemley R J 1994 *Rev. Mod. Phys.* **66** 671
[7] Hanfland M, Hemley R J and Mao H.-K 1993 *Phys. Rev. Lett.* **70** 3760
[8] Goncharov A F, Gregoryanz E, Hemley R J and Mao H.-K 2001 *Proc. Natl. Acad. Sci.* **98** 14234
[9] Loubeyre P, Occelli F and Letoullec R 2002 *Nature* **416** 613
[10] Akahama Y, Kawamura H, Hirao N, Ohishi Y and Takemura K 2010 *J. Phys.: Conf. Series* **215** 012056
[11] Akahama Y, Nishimura M, Kawamura H, Hirao N, Ohishi Y and Takemura K 2010 *Phys. Rev. B* **82** 060101(R)
[12] Ashcroft N W 1991 *Molecular Systems under High Pressure*, ed R Pucci and G Piccitto (Amsterdam: Elsevier p 201

[13] Johnson K and Ashcroft N W 2000 *Nature* **403** 632
[14] Kitamura K, Tsuneyuki S, Ogitsu T and Miyake T, 2000 *Nature* **404** 259
[15] Pickard C J and Needs R J, 2007 *Nature Phys.* **3** 473
[16] Tse J S, Klug D D, Yao Y, Page Y L and Rodgers J R 2008 *Solid State Comm.* **145** 5
[17] Akahama Y and Kawamura H 2007 *High Pressure Research* **7** 473
[18] Akahama Y and Kawamura H 2006 *J. Appl. Phys.* **100** 043516
[19] Akahama Y and Kawamura H 2010 *J. Phys.: Conf. Series* **215** 012195
[20] Mazin I I, Hemley R J, Goncharov A F, Hanfland M and Mao H K 1997 *Phys. Rev. Lett.* **78** 1066